T0192414

Mathematical Engineering

Today, the development of high-tech systems is unthinkable without mathematical modeling and analysis of system behavior. As such, many fields in the modern engineering sciences (e.g. control engineering, communications engineering, mechanical engineering, and robotics) call for sophisticated mathematical methods in order to solve the tasks at hand.

The series Mathematical Engineering presents new or heretofore little-known methods to support engineers in finding suitable answers to their questions, presenting those methods in such manner as to make them ideally comprehensible and applicable in practice.

Therefore, the primary focus is—without neglecting mathematical accuracy—on comprehensibility and real-world applicability.

To submit a proposal or request further information, please use the PDF Proposal Form or contact directly: Dr. Thomas Ditzinger (thomas.ditzinger@springer.com)

Indexed by SCOPUS, zbMATH, SCImago.

Antonio Palacios

Mathematical Modeling

A Dynamical Systems Approach to Analyze Practical Problems in STEM Disciplines

Antonio Palacios
Nonlinear Dynamical Systems Group
Department of Mathematics
San Diego State University
San Diego, CA, USA

ISSN 2192-4732 ISSN 2192-4740 (electronic)
Mathematical Engineering
ISBN 978-3-031-04731-2 ISBN 978-3-031-04729-9 (eBook)
https://doi.org/10.1007/978-3-031-04729-9

This Springer imprint is published by the registered company Springer Nature Switzerland AG
The registered company address is: Gewerbestrasse 11, 6330 Cham, Switzerland

To Daniel, for letting me learn how to be a father and a son,
To Lydia, for showing me the meaning of unconditional,
To Grace and Will for their friendship and support.

Preface

This text is designed to introduce upper division undergraduates and first-year graduate students to mathematical modeling. The models represent an idealization of real-life phenomena that arise in STEM (Science, Technology, Engineering, and Mathematics) fields. The science in STEM typically includes the natural sciences: biology, physics, and chemistry, while mathematics appears as a standalone subject. The phenomena encountered in STEM, can change in space and time, and are frequently complex and often not totally understood. Mathematical modeling provides a means to better understand the processes and unravel some of the complexities. This gives a natural synergistic relationship between the various STEM fields as research expands in the future. Examples include population dynamics in biological and economical systems; the global positioning system; predator-prey ecosystems; crystal oscillators; fluxgate magnetometers; compartmental models to study the spread of infectious diseases, e.g., COVID-19; laser systems; hybrid systems, such as networks of gyroscopic systems; cascade arrays for generating multifrequency patterns in signal processing; precision timing devices; pattern formation in spatial-temporal models, e.g., Turing patterns; flame dynamics; agent-based models of bubble formation and evolution in fluidization processes.

Qualitative and quantitative methods are developed throughout the book to derive and solve mathematical models. The actual methods involve ideas and techniques from dynamical systems theory. More contemporary methods, i.e., equivariant bifurcation theory, are also employed to study models that posses symmetry. It is well known that symmetry alone can restrict the type of solutions of systems of ordinary and partial differential equations, which often serve as models of complex systems. So, it is reasonable to expect that certain aspects of the collective behavior of a complex system can be inferred from the presence of symmetry alone. Thus, ideas and methods from equivariant bifurcation theory can be used to model, analyze, and predict the behavior of mathematical models without having closed-form analytical solutions and without the need of computer simulations. Inverse problems, and fitting a model to experimental data, are also covered. Mathematical software such as MATLAB is used to eliminate tedious calculations and for developing a better understanding of solution sets. Specialized topics in model reduction and simplification, which has always been of great interest to the STEM community, are also included.

There are several distinguishing features of the textbook. The incorporation of real-life data in the derivation of certain models, e.g., estimating the radius of explosion of an atomic bomb based on sequences of snapshots published after the explosion, is one example. Another example includes the derivation of a discrete model for reproducing the life cycle of sockeye salmon based on data that describes the growth of the population of sockeye salmon over a long period of time. Many of the mathematical models include results from experiments that serve to validate the models. For instance, the presentation of the compartmental model for COVID-19 includes experiments in support of the model. An entire chapter is dedicated to hybrid models or complex networks, which have some parts that are modeled as discrete event systems, while other parts are modeled with continuous (differential or differential-algebraic) equations. Many phenomena in STEM fields are subject to random or stochastic effects. These phenomena rarely are completely deterministic, so they cannot be accurately modeled with difference or differential equations. One method of introducing these variations into the corresponding mathematical models is by adding some type of noise. Thus, the book also includes analysis and simulations of stochastic models, including the introduction of noise into a differential equation, which leads to the field of stochastic differential equations, Ornstein-Ullenbeck processes for studying the effects of colored noise, and analysis of the Fokker-Planck equation for describing the probability that a system is in some state at a given time.

Another distinguishing feature is the fact that many of the models presented in the book are innovative models that were derived as a result of multiple projects from an ongoing collaboration with scientists and engineers from the Naval Warfare Information Center (NIWC), San Diego. For instance, the running model of a fluxgate magnetometer, which appears in many chapters as it includes features that involve, continuous models, bifurcation theory, delay, and hybrid models, was derived and analyzed, as part of a 7-year project that led to its fabrication and a new prototype for highly sensitive magnetic and electric field sensors. I would like to thank, in particular, Dr. Visarath In and Dr. Patrick Longhini from NIWC. Furthermore, none of the collaborative projects would have been possible without the active participation of students from San Diego State University. I also wish to acknowledge the financial support provided by several agencies to conduct the necessary research work that serves as the foundation of some of the models and technologies discussed in this book, including Army Research Office, Department of Defense, Department of Energy, the National Science Foundation, the National Security Agency, the Office of Naval Research, and the Naval Information Warfare Center, San Diego.

The book is intended for a broad audience. For interdisciplinary scientists in biology, physics, and chemistry who might be interested in learning the skills to derive a mathematical representation for explaining the evolution of a real system. For engineers who might be interested in the bridge between the derivation of mathematical models and the methods to study and explain the behavior of engineering systems, e.g., magnetic sensor systems. For applied mathematicians and physicists who want to develop a broader idea of applications from the field of Dynamical Systems with Symmetry and Equivariant Bifurcation theory to model, analyze, and predict the behavior of systems.

Indeed, the book employs rigorous theorems from the field of dynamical systems to study the existence and stability of invariant sets, including, equilibrium or steady states, periodic orbits, and chaotic sets. It does not include, however, rigorous proofs of those theorems, as those proofs are beyond the scope of a textbook in mathematical modeling. Instead, theorems are employed to carry out in-depth analysis of the behavior of a system while leveraging rigorous proofs and referring readers to the appropriate sources for those proofs.

The book could be adapted for either an undergraduate-level or graduate-level course. In the former case, material can be drawn from Chap. 2 Algebraic Models; Chap. 3 Discrete Models; Chap. 4 Continuous Models, and Chap. 5 Bifurcation Theory. In the latter case, additional material can be used from Chap. 7 Delay Models; Chap. 8 Spatio-Temporal Models; and Chap. 9 Stochastic Models. If time permits it, material from Chap. 6 Network-Based Modeling; and from Chap. 10 Model Reduction could also be incorporated into a graduate-level course.

Mathematics is the alphabet with which God has written the universe.
Galileo Galilei.

San Diego, CA, USA
2021

Antonio Palacios

Contents

Chapter 1
Introduction

STEM (science, technology, engineering, and mathematics) fields are one of the most rapidly expanding and diverse areas in the world. The science in STEM typically includes the natural sciences: biology, physics, and chemistry, while mathematics appears as a standalone subject. The problems encountered in STEM are frequently complex and often not totally understood. Mathematical models provide a means to better understand the processes and unravel some of the complexities. This gives a natural synergistic relationship between the various STEM fields as research expands in the future. The mathematical tools provide ways for developing a better qualitative and quantitative understanding of some biological problems, while the STEM-related problems often stretch the techniques that mathematicians must use to find solutions.

In this book we develop qualitative and quantitative methods to derive and solve mathematical models. The models, in turn, can be used to analyze and describe a wide range of STEM fields phenomena that change in space or time, or in both. Examples include: population dynamics in biological and economical systems; the global positioning system; predator-prey ecosystems; crystal oscillators; fluxgate magnetometers; pharmokinetic models; laser systems; hybrid systems, such as networks of gyroscopic systems; cascade arrays for generating multifrequency patterns in signal processing; precision timing devices; pattern formation in spatial-temporal models, e.g., Turing patterns; flame dynamics; agent-based models of bubble formation and evolution in fluidization processes. We also include topics in model reduction and simplification, which has always been of great interest to the engineering community. We employ ideas and methods from dynamical systems theory to study the behavior of models. But we also include more contemporary methods, i.e., equivariant bifurcation theory, to study systems that posses symmetry.

Let's start first with an overview of what constitutes a mathematical model.

A. Palacios, *Mathematical Modeling*, Mathematical Engineering,
https://doi.org/10.1007/978-3-031-04729-9_1

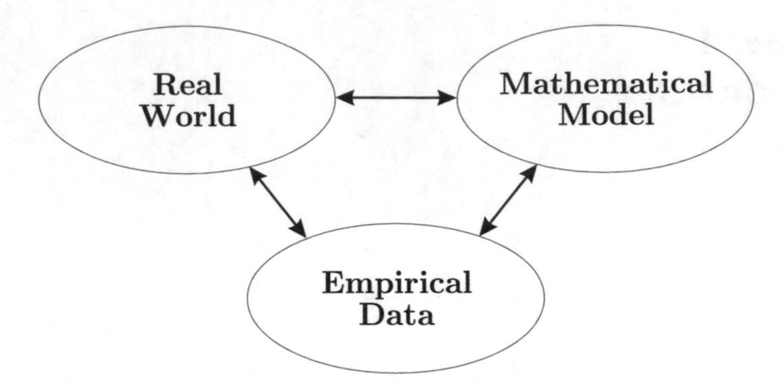

Fig. 1.1 Conceptualization of mathematical modeling

1.1 So What Is a Mathematical Model?

A *mathematical model* is a representation of a real system. An idealization of natural or artificial phenomena that allows us to investigate and understand the underlying principles that govern those phenomena. Mathematical modeling is often an iterative process to help one obtain a better understanding of some observation from the "Real World." Figure 1.1 illustrates the critical steps to creating a mathematical model. The problem from the "Real World" is abstracted into some symbolic idea, which is often expressed as mathematical equations or "Mathematical Model." "Empirical Data" are collected about the system of interest, and these data are compared to the output from the mathematical model. Through an iterative process, one obtains better approximations and obtains greater insight into the underlying principles from the original problem from the "Real World."

The essence of a good mathematical model is that it is simple in design and exhibits the basic properties of the real system that one is attempting to understand. The model should be testable against empirical data. The comparisons of the model to the real system should ideally lead to improved mathematical models. The model may suggest improved experiments to highlight a particular aspect of the problem, which in turn may improve the collection of data. More importantly, a good model should be able to lead to predictions of unexpected behavior. Thus, modeling itself is an evolutionary process, which continues toward learning more about certain processes rather than finding an absolute reality.

Consider, for instance, the *real world* problem of describing the behavior of a neuron cell. In the year 1939, Alan Hodgkin and Andrew Huxley started a series of experiments to get insight into the fundamentals of nerve cell excitability. At the time, it was only known that a cell body was connected to a long "cable" known as an *axon*. Hodgkin and Huxley conjectured that the axon served as a conductor of electrical signals generated by the nerve cell. To prove their conjecture, they collected *empirical data* by inserting a fine capillary electrode intro the nerve fiber of the squid giant axon. The experiments led to the first intracellular recording of an *action potential* [1–4].

The joint work between Alan Hodgkin and Andrew Huxley culminated between the years 1949–1952 with the publication [5] of a *mathematical model* that describes how action potentials in nerve cells are generated and propagated. The model consists of a system of nonlinear Partial Differential Equations (PDEs) that approximates the electrical characteristics of excitable cells such as neurons and cardiac myocytes. The model has the form:

$$
\begin{aligned}
C_m \frac{dV_m}{dt} &= \frac{a}{2R}\frac{\partial^2 V}{\partial x^2} + g_{N_a} m^3 h (V_{N_a} - V_m) + g_K n^4 (V_K - V_m) + g_L (V_L - V_m)\\
\frac{dm}{dt} &= \varphi(T)[\alpha_m - m(\alpha_m + \beta_m)]\\
\frac{dh}{dt} &= \varphi(T)[\alpha_h - h(\alpha_h + \beta_h)]\\
\frac{dn}{dt} &= \varphi(T)[\alpha_n - n(\alpha_n + \beta_n)],
\end{aligned}
\tag{1.1}
$$

where V_m is the action potential, C_m is the membrane capacitance that quantifies the ability of a neuron to store ions or charges, g_{N_a} and g_K are the sodium and potassium conductances per unit area, g_L is the leak channel for any other type of ions to flow through the membrane. V_{N_a}, V_K and V_L are the threshold values at which the corresponding ions (sodium, potassium and leaked) start to flow. The variables m and h and n are dimensionless quantities that model sodium channel activation, while n models potassium channel activation. The parameters α and β are related to the steady-state values for activation and inactivation of the channels. φ is a time coefficient that controls how quickly the channels respond. Finally, the first term on the right, $\frac{a}{2R}\frac{\partial^2 V}{\partial x^2}$, is the external current, I_{applied} that is applied to a neuron to excite it into initiating an action potential. a is the radius of the axon and R is its resistance as a "cable" that conducts electricity.

The Hodgkin-Huxley model (1.1) is a space-time model because it accounts for the voltage propagation of a nerve impulse through the spatial location, x, along the axon of the cell. The voltage can be measured at any given time, t, so it is deemed a spatial-time model.

Hodgkin and Huxley solved their mathematical model for both stationary and propagating action potentials using what might best be described as a "brute force" method. The iterative solution for the propagating action potential, whose results are shown in Fig. 1.2a, took a few weeks and many thousands of rotations of the mechanical calculator crank, see Fig. 1.2b.

Alan Lloyd Hodgkin and Andrew Fielding Huxley received the 1963 Nobel Prize in Physiology or Medicine for their contributions.

Nowadays, we know a lot more details about the fundamental structure of a neuron. In addition to the nerve cell and axons, there are also dendrites, which serve to collect signals traveling from other neurons, see Fig. 1.3. More importantly, the Hodgkin-Huxley model has led to a much deeper understanding of the structure and dynamics of nerve cells.

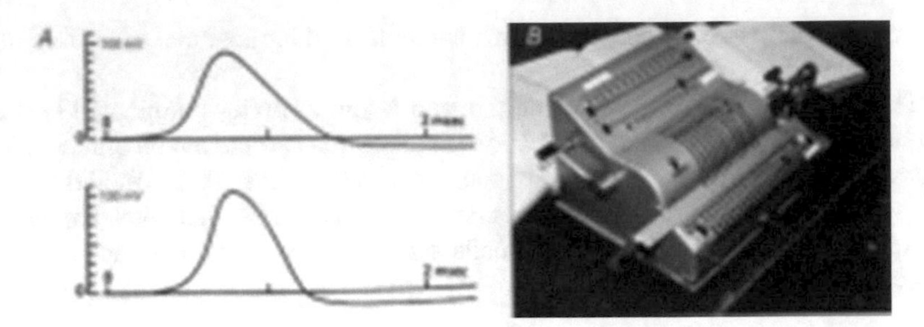

Fig. 1.2 Validation of the mathematical model proposed by Hodgkin and Huxley to describe the generation and propagation of an action potential in a neuron or nerve cell. (Top a) is the computer simulation from the mathematical model. (Bottom a) is the measured action potential in squid giant axons. **b** Is a picture of the Brunsviga-20 mechanical computer that was used for the simulations. *Source* [5]

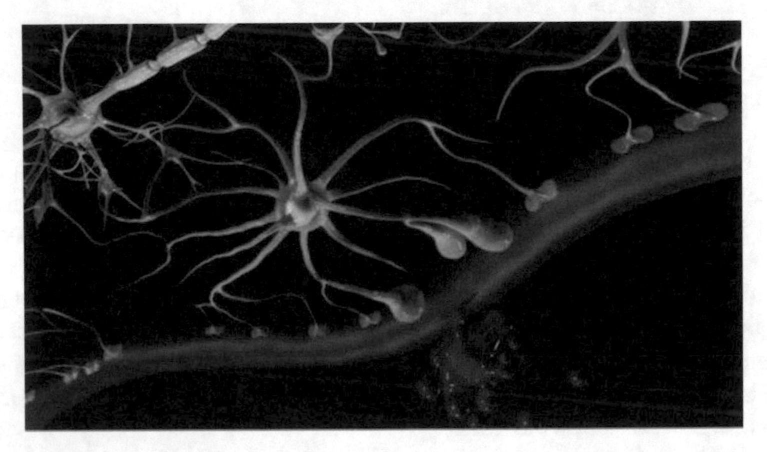

Fig. 1.3 The fundamental structure of a neuron or nerve cell consists of an excitable membrane cell, dendrites to collect electrical signals traveling from other neurons and axons to transmit electrical signals onto dendrites of another cell. *Source* Astrocyte Pharmaceuticals

1.2 State Variables and Parameters

Mathematical models, in general, are made up of state variables and parameters. State variables are the quantities that we measure to characterize the evolution, in space and time, of a system. Parameters are similar to state variables except that they are used to "tune up" a model to operate within a certain range of values. That is, they can be treated as "variables" which are held fixed during the course of an experiment. For instance, in the Hodgkin-Huxley model (1.1), the action potential $V_m(t)$ is the state variable, which measures the response of the neuron at any given time. All other "variables" that appear in the model are considered parameters.

The distinction between stave variables and parameters can be subtle since there are also instances where parameters are allowed to vary during the run of an experiment. For instance, John Rinzel at NIH [6] developed a model for bursting behavior by suggesting that certain parameters (previously treated as fixed or constant) could also vary. The model is of the form

$$\frac{d\mathbf{x}}{dt} = \mathbf{f}(\mathbf{x},\varepsilon)$$
$$\frac{d\varepsilon}{dt} = g(\mathbf{x},\varepsilon).$$

In this formulation, the state variable is $\mathbf{x} = (x, y)$. Since the parameter ε is assumed to evolve very slowly, compared to the true state variables, then ε can be treated as a variable.

1.3 Methods and Challenges

The seminal work by Lorenz in 1963 [7], and later by May in 1976 [8–10], have lead scientists and engineers to recognize that nonlinear systems can exhibit a rich variety of dynamic behavior. From simple systems, such as the evolution of single species [11], an electronic or biological oscillator [12, 13], to more complex systems such as chemical reactions [14], climate patterns [15], bursting behavior by a single neuron cell [16], and flucking of birds [17, 18], *Dynamical Systems* theory provides quantitative and qualitative (geometrical) methods to study these and many other complex systems that evolve in space and/or time. Regardless of the origins of a system, i.e, Biology, Chemistry, Engineering, Physics, or even the Social Sciences, dynamical systems theory seeks to explain the most intriguing and fundamental features of spatio-temporal phenomena.

In this book, we employ methods from Dynamical Systems theory to study the *long-term* behavior of the mathematical models that serve to analyze spatio-temporal phenomena in STEM fields. By long-term behavior we mean solution sets such as equilibrium points, periodic points, and their stability properties, as well as more complicated solutions, including collective behavior and chaos. The standard methods are, in general, suitable to unravel the complexity of most mathematical models. But there is also a wide range of contemporary problems, such as network-based modeling of *hybrid* systems, that are not amenable to analysis by traditional techniques. Hybrid systems consist of a combination of discrete and continuous systems. For instance, a network of a discrete number of units in which each unit is governed by a continuous model. A representative example is that of networks of coupled oscillators or a network of interconnected gyroscopes.

One of the major challenges in the analysis of network systems is that the corresponding models are high-dimensional. That is, they contain a large number of equations. A general approach to get insight into the behavior of complex network systems has been to derive a detailed model of its individual parts, connect the parts and note that the system contains some sort of symmetry, then attempt to exploit this symmetry in order to simplify numerical computations. This approach can result in very complicated models that are difficult to analyze even numerically. In the example of a network of gyroscopes, each individual unit is governed by a four-dimensional nonautonomous system of Ordinary Differential Equations (ODEs). A network of n gyroscopes would be governed by a coupled nonautonomous ODE system of dimension $4n$, which can be a daunting task to study when n is large. Furthermore, prediction of complicated dynamics is practically impossible.

In this book we include ideas and methods to study systems with symmetry, i.e., Equivariant Bifurcation Theory, which allow us to formulate, directly, the appropriate models and analyze their long-term behavior. This approach, while nonstandard, is not entirely new among the mathematics community. However, there is much less familiarity with the techniques of *symmetry-breaking bifurcation*, developed by Golubitsky and Stewart [19–21], as they apply to mathematical modeling in STEM fields.

1.4 Model Reduction

One of the topics that has continuously attracted strong interest in mathematical modeling is that of *Reduced Order Models*. Typically, the methods for deriving a model, either through first principles or through data fitting, or through any other valid technique, lead to many more equations, variables or parameters, that may be needed to describe the behavior of a given phenomenon. In general, the larger the number of equations and variables in a model the greater the complexity of its analysis. Sometimes, computer simulations of the original model may suggest the presence of long-term behavior that can be captured by a low-dimensional system. Thus, reduced order modeling refers to the process of reducing a mathematical model to a simpler form that can be more amenable to analysis, yet still capture that same low-dimensional behavior without compromising the overall results and interpretation of the solutions.

The Morris-Lecar model, for instance, is a two-dimensional "reduced" version of the Hodgkin-Huxley model through the following system of Ordinary Differential Equations (ODEs):

$$\begin{aligned} C\frac{dV}{dt} &= -g_{Ca}m_\infty(V)h(V - V_{Ca}) - g_K w(V - V_K) - g_L(V - V_L) + I_{\text{applied}} \\ \frac{dw}{dt} &= \frac{1}{\tau_w(V)}(w_\infty - w), \end{aligned} \tag{1.2}$$

where V is the action potential, w is the recovery variable, which is equivalent to the normalized potassium-ion conductance, g_{Ca} and g_K are the calcium and potassium conductances per unit area, g_L is the leak channel for any other type of ions to flow through the membrane. I_{applied} is the external current that is applied to a neuron to excite it into initiating an action potential. m_∞ and w_∞ are conductance functions that model the open-close probabilities of the channels.

The Morris-Lecar model (1.2) describes the same three ionic currents as those of the Hodgkin-Huxley model except that in this formulation the sodium channel is omitted, but it can be included with a little manipulation of the recovery process.

The process of reduction and simplification may include different techniques depending on the type of model that may be involved, i.e., PDE vs ODE. In this book we explore various techniques for model reduction and simplification. The first technique for model reduction is known as the *Galerkin Projection*. This technique is about reducing an evolution equation or PDE model to an ODE model by "projecting" the PDE equations onto a basis of eigenfunctions. The eigenfunctions might be known a priori but, in many other cases, they may need to be "extracted" directly from a data set of experimental or numerical measurements, i.e., empirically. A popular technique for extracting those eigenfunctions is known as the *Proper Orthogonal Decomposition*. Another approach is the *Center Manifold Reduction*. This technique is applicable to reduce both PDEs or ODEs onto an *invariant manifold*, assumed to be lower-dimensional than the original model. The technique is a local reduction since it works around the location of an equilibrium point. It is commonly used to reduce models that posses symmetry. The last topic on modal reduction is that of *normal forms*. Typically, after a model has been reduced it may still contain terms that are not essential to the long-term dynamics. Eliminating those terms should still lead to the same qualitative behavior. This process of simplification of the model equations is not trivial, we cover the basic ideas and procedure for deriving normal forms.

References

1. A.L. Hodgkin and A.F. Huxley. Propagation of electrical signals along giant nerve fibres. Proc. Royal Soc. Lond. B Biol. Sci. **140**, 177–183 (1952)
2. A.L. Hodgkin, A.F. Huxley, Currents carried by sodium and potassium ions through the membrane of the giant axon of Loligo. J. Physiol. **116**, 449–472 (1952)
3. A.L. Hodgkin, A.F. Huxley, The components of membrane conductance in the giant axon of Loligo. J. Physiol. **116**, 473–496 (1952)
4. A.L. Hodgkin, A.F. Huxley, The dual effect of membrane potential on sodium conductance in the giant axon of Loligo. J. Physiol. **116**, 497–506 (1952)
5. A.L. Hodgkin, A.F. Huxley, A quantitative description of membrane current and its application to conduction and excitation in nerve. J. Physiol. Lond. **117**, 500–544 (1952)
6. J. Rinzel, A formal classification of bursting mechanisms in excitable systems, in *Proceedings of the International Congress of Mathematics* (1986)
7. E.N. Lorenz, Deterministic nonperiodic flow. J. Atmos. Sci. **20**, 130 (1963)
8. R.M. May, Biological populations with no overlapping generations: stable points, stable cycles, and chaos. Science **186**, 645–647 (1974)
9. R.M. May, Biological population obeying difference equations: stable points, stable cycles, and chaos. J. Theor. Biol. **51**, 511–524 (1975)
10. R.M. May, Simple mathematical models with very complicated dynamics. Nature **261**, 459–467 (1975)
11. R. Pearl, The growth of populations. Quart. Rev. Biol. **2**, 532 (1927)
12. A.T. Winfree. *Geometry of Biological Time* (Springer, 2001)
13. A.T. Winfree, *When Time Breaks Down: The Three-Dimensional Dynamics of Electrochemical Waves and Cardiac Arrhythmias* (Princeton University Press, 1987)

14. B.P. Belousov, Oscillation reaction and its mechanisms (in Russian). *Sbornik Referatov po Radiacioni Medicine* (1958), p. 145
15. P.M. Green, D.M. Legler, C.J. Miranda V, J.J. O'Brien. The north American climate patterns associated with the el niño-southern oscillation. Technical report, Center for Ocean-Atmospheric Prediction Studies Report 97-1 (1997)
16. E. Izhikevich, Neural excitability, spiking and bursting. Int. J. Bif. Chaos **10**(6), 1171–1266 (2000)
17. C. Reynoldsr=, Flocks, herds, and schools: a distributed behavioral model. Comput. Graph. **21**25 (1987)
18. J. Toner, T. Yuhai, Flocks herds and schools: a quantitative theory of flocking. Phys. Rev. E **58**, 4828–4858 (1998)
19. M. Golubitsky, I.N. Stewart, Patterns of oscillations in coupled cell systems, in *Geometry, Mechanics, and Dynamics* ed. by P. Holmes, A. Weinstein. (Springer, New York, 2002), p. 243
20. M. Golubitsky, I. Stewart, *The Symmetry Perspective* (Birkháuser Verlag, Basel, Switzerland, 2000)
21. M. Golubitsky, I.N. Stewart, D.G. Schaeffer, *Singularities and Groups in Bifurcation Theory Vol. II*, vol. 69 (Springer, New York, 1988)

Chapter 2
Algebraic Models

In this chapter we study the simplest form of a mathematical model, a *linear model*. Then we study some popular methods to fit a linear model through a data set, also known as the linear least square approximation. This method is very popular in the analysis of statistical data. As a representative example, we consider a data set where the relation between ambient temperature and the chirping rate of crickets is examined. The least square fitting method is then extended to nonlinear functions. As a case study, we examine a mathematical model for a Global Positioning System or GPS. The chapter ends with an introduction to allometric models, which serve to study the relationship between body properties such as weight, size, and metabolism.

2.1 Temperature and the Chirping of a Cricket

Consider the snapshot of a snowy tree cricket shown in Fig. 2.1.

We ask the fundamental question:

Is it possible to tell the temperature of the environment by listening to the chirping of a cricket?

According to Dolbear's law, the answer is yes. This law, formulated by Amos Dolbear in a 1897 publication [1], states that temperature and the rate at which crickets chirp form a linear relationship. The relationship can be described, in words, as:

Count the number of chirps in a minute and divide by four, then add forty.

Dolbear wrote this relationship, mathematically, through the following formula:

A. Palacios, *Mathematical Modeling*, Mathematical Engineering,
https://doi.org/10.1007/978-3-031-04729-9_2

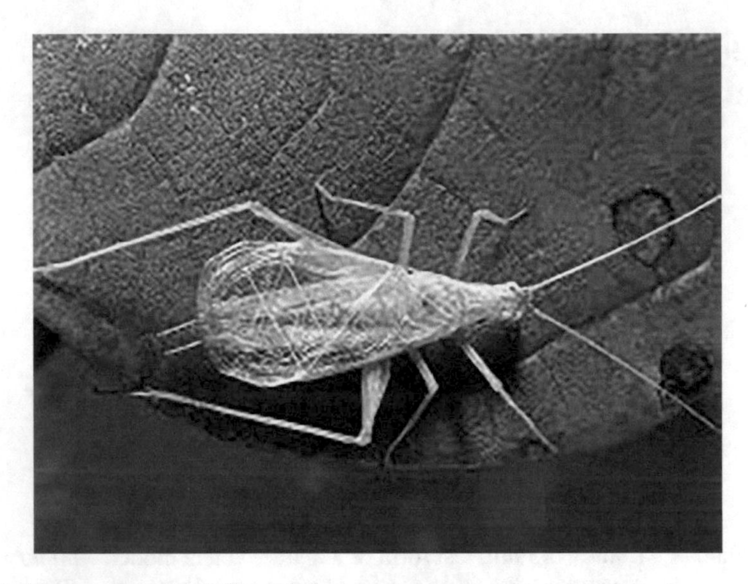

Fig. 2.1 A snowy tree cricket, *Oecanthulus niveus*

$$T = 50 + \frac{N - 40}{4},$$

where N is the *independent variable*, representing the number of chirps per minute and T is the *dependent variable*, describing temperature. Observe that this formula reduces to the word description, divide by four and add forty : $T = N/4 + 40$.

In a follow-up study, "Further Notes on Thermometer Crickets", Carl Bessey and Edward Bessey [2], derived a slightly different linear relationship between the number of chirps and temperature. Through careful observations and measurements, which included wind speed, tree heights, etc., they arrived at the following formulation:

$$T = 60 + \frac{N - 92}{4.7},$$

which is slightly off from Dolbear's model. In fact, if we simplify this latest model we get $T = N/4.7 + 40.426$, which confirms a subtle deviation from Dolbear's model. Figure 2.2 shows a comparison of the temperature predicted by each of the linear models.

At first glance, the values predicted by the Bessey's model seem to hover closer to the data measurements, while those of the Dolber model seem to overshoot the actual data values. This result might be due to the careful observations and measurements that were used by Bessey to derive their model. Consequently, the observations seem to suggest that the Bessey's model is more accurate than the Dolber's model. An obvious question that arises immediately, is how can we quantify the accuracy of

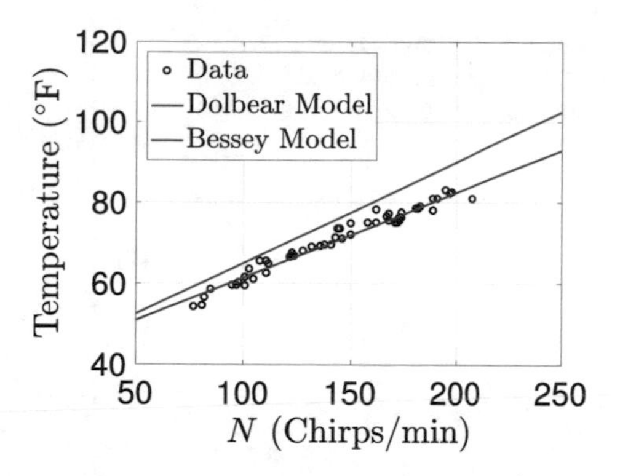

Fig. 2.2 Comparison of two linear models, Dolbear's and Bessey's, which describe the relationship between temperature and number of chirps of a snowy tree cricket

the two linear fits? That is, how can we, mathematically, describe how much better the Bessey's model is? The answer to this question is the subject of the following section.

2.2 Least Squares Fitting of Data

The most common approach to measure, mathematically, the accuracy of the fit of the output of a mathematical model to a data set is through the sum of square errors. The basic idea is to find the optimal choice of parameter values of the model that can minimize the errors. This approach is also known as *least sum of square errors*. Next, we discuss the details of the approach, first for the case of linear models, and, then, for the case of a quadratic polynomial. The latter case will lead us to a generalization of the approach to nonlinear models.

2.2.1 *Linear Least Squares Fit*

Let us consider the cricket thermometer problem as a starting reference. While looking at Fig. 2.2, let's assume the data points (black circles) to be described by a collection of $n + 1$ data points of the form:

$$\{(x_i, y_i)\}_{i=0}^{n} = \{(x_0, y_0), (x_1, y_1), \ldots, (x_n, y_n)\}.$$

We wish to quantify the accuracy of the approximation of both, Dolber's and Bessey's models. These models, or any other linear model, can be written in the general form

$$y(x) = a_1 x + a_0, \tag{2.1}$$

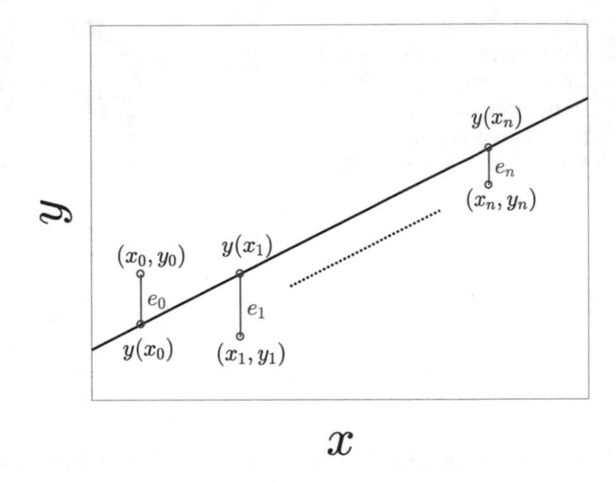

Fig. 2.3 Individual errors, e_i, between a linear model and the data points

where a_1 is one parameter that represents the slope of the linear model, and a_0 is a second parameter that describes the y-intercept of the linear fit. Both parameters are, so far, unknown, but their actual values will determine the accuracy of the linear fit.

The fundamental idea of the *least squares best fit* is to find the optimal values of the unknown parameters, a_0 and a_1, that will minimize the cumulative error between the y_i values of the data points and the y values of the line, $y(x_i)$. The errors, at each individual data point, are depicted in Fig. 2.3.

These individual errors can be described, more formally, as

$$e_i = y_i - y(x_i), \quad i = 0, \ldots, n.$$

Evaluating Eq. (2.1) at each data point x_i, we can rewrite the individual errors as

$$e_i = y_i - (a_1 x_i + a_0), \quad i = 0, \ldots, n.$$

One can visualize the linear fit moving up and down through the data points as the parameters a_0 and a_1 are varied. As the line moves, the error changes dynamically. The trick is then to find the optimal parameter values that minimizes the cumulative error:

$$E(a_0, a_1) = e_0^2 + e_1^2 + \cdots + e_n^2 = \sum_{i=0}^{n} e_i^2. \tag{2.2}$$

In this formulation, the square of the individual errors is used in order to prevent errors of opposite sign canceling out with each other and yielding a misleading zero-sum error. Equation (2.2) is a function of two variables, a_0 and a_1, whose minimum can be computed using standard multi-variable Calculus. We first write

$$E(a_0, a_1) = \sum_{i=0}^{n} \left[(a_0 + a_1 x_i) - y_i\right]^2.$$

Then we minimize $E(a_0, a_1)$ by computing its critical values, which correspond to the first partial derivatives of the function E with respect to a_0 and a_1 set to zero:

$$\frac{\partial}{\partial a_0} E(a_0, a_1) = 2 \sum_{i=0}^{n} [(a_0 + a_1 x_i) - y_i] \quad = 0$$
$$\frac{\partial}{\partial a_1} E(a_0, a_1) = 2 \sum_{i=0}^{n} x_i [(a_0 + a_1 x_i) - y_i] = 0.$$

Rearranging these expressions we get

$$\sum_{i=0}^{n} a_0 + \sum_{i=0}^{n} a_1 x_i \quad = \sum_{i=0}^{n} y_i$$
$$\sum_{i=0}^{n} x_i a_0 + \sum_{i=0}^{n} x_i a_1 x_i = \sum_{i=0}^{n} x_i y_i,$$

which leads to the *normal equations* for the unknowns a_0 and a_1

$$\begin{pmatrix} (n+1) & \sum_{i=0}^{n} x_i \\ \sum_{i=0}^{n} x_i & \sum_{i=0}^{n} x_i^2 \end{pmatrix} \begin{pmatrix} a_0 \\ a_1 \end{pmatrix} = \begin{pmatrix} \sum_{i=0}^{n} y_i \\ \sum_{i=0}^{n} x_i y_i \end{pmatrix}.$$

These normal equations represent a linear system of equations which can easily be solved to yield the desired optimal parameters:

$$a_0 = \bar{y} - a_1 \bar{x}, \quad a_1 = \frac{\sum_{i=0}^{n} (x_i - \bar{x})(y_i - \bar{y})}{\sum_{i=0}^{n} (x_i - \bar{x})^2},$$

where

$$\bar{x} = \frac{1}{n+1} \sum_{i=0}^{n} x_i, \quad \bar{y} = \frac{1}{n+1} \sum_{i=0}^{n} y_i.$$

2.2.2 *Quadratic Least Squares Fit*

The linear least square approximation can be readily extended to the case of a quadratic polynomial of the form $p_2(x) = a_0 + a_1 x + a_2 x^2$. As before, the cumulative error is defined as:

$$E(a_0, a_1, a_2) = \sum_{i=0}^{n} \left[a_0 + a_1 x_i + a_2 x_i^2 - y_i\right]^2.$$

Again, from standard Calculus, we minimize the error function $E(a_0, a_1, a_2)$ by computing the critical values at the minimum, which correspond to the first partial derivatives with respect to a_0, a_1, and a_2, set to zero:

$$\begin{aligned}
\frac{\partial}{\partial a_0} E(a_0, a_1, a_2) &= 2\sum_{i=0}^{n} \left[(a_0 + a_1 x_i + a_2 x_i^2) - y_i\right] &= 0 \\
\frac{\partial}{\partial a_1} E(a_0, a_1, a_2) &= 2\sum_{i=0}^{n} x_i \left[(a_0 + a_1 x_i + a_2 x_i^2) - y_i\right] &= 0 \\
\frac{\partial}{\partial a_2} E(a_0, a_1, a_2) &= 2\sum_{i=0}^{n} x_i^2 \left[(a_0 + a_1 x_i + a_2 x_i^2) - y_i\right] &= 0.
\end{aligned}$$

Collecting coefficients of the unknown parameters, we get

$$\begin{aligned}
a_0(n+1) + a_1 \sum_{i=0}^{n} x_i + a_2 \sum_{i=0}^{n} x_i^2 &= \sum_{i=0}^{n} y_i \\
a_0 \sum_{i=0}^{n} x_i + a_1 \sum_{i=0}^{n} x_i^2 + a_2 \sum_{i=0}^{n} x_i^3 &= \sum_{i=0}^{n} x_i y_i. \\
a_0 \sum_{i=0}^{n} x_i^2 + a_1 \sum_{i=0}^{n} x_i^3 + a_2 \sum_{i=0}^{n} x_i^4 &= \sum_{i=0}^{n} x_i^2 y_i.
\end{aligned}$$

Rewriting these equations in vector-matrix form, we arrive at the *normal equations* through a linear system of equations for the unknowns a_0, a_1, and a_2:

$$\begin{pmatrix}
(n+1) & \sum_{i=0}^{n} x_i & \sum_{i=0}^{n} x_i^2 \\
\sum_{i=0}^{n} x_i & \sum_{i=0}^{n} x_i^2 & \sum_{i=0}^{n} x_i^3 \\
\sum_{i=0}^{n} x_i^2 & \sum_{i=0}^{n} x_i^3 & \sum_{i=0}^{n} x_i^4
\end{pmatrix}
\begin{pmatrix} a_0 \\ a_1 \\ a_2 \end{pmatrix}
=
\begin{pmatrix}
\sum_{i=0}^{n} y_i \\
\sum_{i=0}^{n} x_i y_i \\
\sum_{i=0}^{n} x_i^2 y_i.
\end{pmatrix}.$$

A closed-form solution to this system of equations can be found by use of a matrix notation in terms of the monomials x_i. This little trick is discussed next.

2.2.3 General Discrete Least Squares

We now consider the general case of a mth **degree polynomial** of the form

$$p_m(x) = a_0 + a_1 x_i + a_2 x_i^2 + \cdots + a_m x_i^m, \quad i = 0, \ldots, n,$$

If we let $p_m(x_i) = y_i$, then we can write the values of y_i, $i = 0, \ldots, n$ in matrix notation as

$$\underbrace{\begin{pmatrix} 1 & x_0 & x_0^2 & \cdots & x_0^m \\ 1 & x_1 & x_1^2 & \cdots & x_1^m \\ 1 & x_2 & x_2^2 & \cdots & x_2^m \\ 1 & x_3 & x_3^2 & \cdots & x_3^m \\ \vdots & \vdots & \vdots & \vdots & \vdots \\ 1 & x_n & x_n^2 & \cdots & x_n^m \end{pmatrix}}_{A} \underbrace{\begin{pmatrix} a_0 \\ a_1 \\ \vdots \\ a_m \end{pmatrix}}_{\mathbf{a}} = \underbrace{\begin{pmatrix} y_0 \\ y_1 \\ y_2 \\ y_3 \\ \vdots \\ y_n \end{pmatrix}}_{\mathbf{y}},$$

where A is an $(n+1) \times (m+1)$ *Vandermonde matrix.*, $\mathbf{a}$ is an $(m+1) \times 1$ vector, and $\mathbf{y}$ is an $(n+1) \times 1$ vector. In short, we can rewrite the above equations as a linear system of the form

$$A\mathbf{a} = \mathbf{y}. \tag{2.3}$$

The trick to find a solvable system of equation is to multiply both sides of Eq. (2.3) by the transpose of A, i.e., by A^T, to get

$$A^T A\mathbf{a} = A^T\mathbf{y}. \tag{2.4}$$

Since A^T is an $(m+1) \times (n+1)$ matrix, it follows that $A^T A$ is a square matrix of dimensions $(m+1) \times (m+1)$. In fact, direct computations show

$$A^T A = \begin{bmatrix} n+1 & \sum_{i=0}^{n} x_i^1 & \cdots & \sum_{i=0}^{n} x_i^m \\ \sum_{i=0}^{n} x_i^1 & \sum_{i=0}^{n} x_i^2 & \cdots & \sum_{i=0}^{n} x_i^{m+1} \\ \vdots & \vdots & \ddots & \vdots \\ \sum_{i=0}^{n} x_i^m & \sum_{i=0}^{n} x_i^{m+1} & \cdots & \sum_{i=0}^{n} x_i^{2m} \end{bmatrix}.$$

while $A^T\mathbf{y}$ yields,

$$A^T\mathbf{y} = \begin{bmatrix} \sum_{i=0}^{n} y_i \\ \sum_{i=0}^{n} x_i y_i \\ \sum_{i=0}^{n} x_i^2 y_i \\ \vdots \\ \sum_{i=0}^{n} x_i^m y_i. \end{bmatrix}$$

Comparing the right-hand side of the expressions for $A^T A$ and $A^T\mathbf{y}$, we can then observe that Eq. (2.4) is the generalization of the *normal equations*. We could then, in principle, solve Eq. (2.4) directly for $\mathbf{a}$ to get the optimal parameter values

$$\mathbf{a} = (A^T A)^{-1}(A^T\mathbf{y}).$$

In practice, there are, however, more efficient algorithms, e.g., QR-algorithm, that can be used to solve for $\mathbf{a}$. Those algorithms are beyond the scope of this book.

2.2.4 Cricket Model Revisited

We will now examine different levels of approximation to the data set of the cricket thermometer problem shown in Fig. 2.2. Let that data set be $(\mathbf{x}, \mathbf{y})$, where $\mathbf{x} = \{x_0, x_1, \ldots, x_n\}^T$ and $\mathbf{y} = \{y_0, y_1, \ldots, y_n\}^T$. We consider various polynomial approximations of degree m, i.e.,

$$p_m(x) = a_0 + a_1 x_i + a_2 x_i^2 + \cdots + a_m x_i^m, \quad i = 0, \ldots, n.$$

The *Vandermonde matrix* associated with each individual polynomial can be written as

$$A_m = \begin{bmatrix} | & | & | & & | \\ | & | & | & & | \\ \vec{1} & \mathbf{x} & \mathbf{x}^2 & \cdots & \mathbf{x}^m \\ | & | & | & & | \\ | & | & | & & | \end{bmatrix},$$

where $\mathbf{x}^j = [x_0^j, x_1^j, \ldots, x_n^j]^T$. In the crickett problem, the $\mathbf{x}$ vector corresponds to the number of chirps per minute, $N = [N_1, N_2, ..., N_n]^T$, while the $\mathbf{y}$ vector represents temperature measurements, $T = [T_1, T_2, ..., T_n]^T$. In the Appendix we show MATLAB code that can be used to readily create the Vandermonde matrices A_m.

Linear Fit

This is the case where $m = 1$, which yields the following *Vandermonde matrix*:

$$A_1 = \begin{pmatrix} 1 & N_1 \\ 1 & N_2 \\ \vdots & \vdots \end{pmatrix}.$$

The left-hand side of the normal equation (2.4) becomes:

$$A_1^T A_1 = \begin{pmatrix} 52 & 7447 \\ 7447 & 1133259 \end{pmatrix}.$$

Solving Eq. (2.4) for **a** in MATLAB through the command:

$$A_1 \backslash T,$$

we get $a_0 = 39.7441$ and $a_1 = 0.2155$. We can then write the least square linear approximation in the form

$$T(N) = 0.2155\, N + 39.7441.$$

Quadratic Fit

In the quadratic fit, $m = 2$, and the corresponding Vandermonde matrix becomes

$$A_2 = \begin{pmatrix} 1 & N_1 & N_1^2 \\ 1 & N_2 & N_2^2 \\ \vdots & \vdots & \vdots \end{pmatrix},$$

which leads to

$$A_2^T A_2 = \begin{pmatrix} 52 & 7447 & 1133259 \\ 7447 & 1133259 & 1.8113 \times 10^8 \\ 1133259 & 1.8113 \times 10^8 & 3.0084 \times 10^{10} \end{pmatrix}.$$

Solving Eq. (2.4) for **a** in MATLAB through the command:

$$A_2 \backslash T,$$

we get $a_0 = 27.8489$, $a_1 = 0.39625$, and $a_2 = -0.00064076$. The desired quadratic approximation becomes:

$$T(N) = -0.00064076\, N^2 + 0.39625\, N + 27.8489.$$

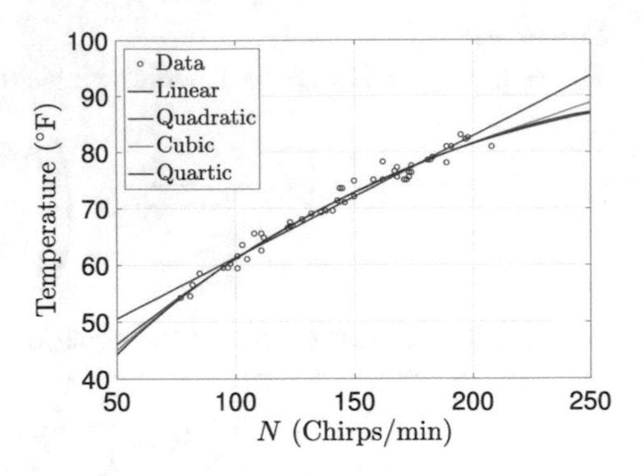

Fig. 2.4 Least square fitting of polynomials of order $m = 1$ up to $m = 4$ for the cricket thermometer data

Cubic and Quartic Fits
A similar set of calculations yields the best cubic and quartic fits:

$$
\begin{aligned}
T_{cubic}(N) &= 0.0000018977\,N^3 - 0.001445\,N^2 + 0.50540\,N + 23.138 \\
T_{quartic}(N) &= -0.00000001765\,N^4 + 0.00001190\,N^3 - 0.003504\,N^2 \\
&\quad +0.6876\,N + 17.314.
\end{aligned}
$$

Results and comparison of each of the four polynomial fits are shown in Fig. 2.4. In the next section we discuss different quantitative measurements that will help compare the accuracy of the various models.

2.2.5 *Model Selection*

Looking back at the data set and all the polynomial fits shown in Fig. 2.4, we ask the obvious question:

Which model is the best fit to the data?
One possible way to answer this question is by comparing the sum of square errors (SSE) for each polynomial fit. The SSE for the different polynomial fits are shown in Table 2.1. Accordingly, the quartic approximation shows the smallest SSE.

The SSE values shown in Table 2.1 suggest that a higher degree polynomial might better approximate the cricket thermometer data. However, a higher degree polyno-

Table 2.1 Three different model selection criteria: Sum of Square Errors, Bayesian Information Criteria, and Akaike, are applied to the cricket thermometer problem

	Linear	Quadratic	Cubic	Quartic
SSE	108.8	79.08	78.74	**78.70**
BIC	46.3	33.65	**33.43**	37.35
AIC	189.97	**175.37**	177.14	179.12

mial involves more parameters, and, consequently, higher complexity. In addition to the SSE values, there are two other criteria for model selection which are quite popular, especially among the statistics community. One is the Bayesian Information Criterion (BIC) and the other one is the Akaike Information Criterion (AIC).

Definition 2.1 (*Bayesian Information Criterion*) Let n be the number of data points, SSE be the sum of square errors, and let k be the number of parameters in the model.

$$BIC = n\ln(SSE/n) + k\ln(n).$$

Definition 2.2 (*Akaike Information Criterion*)

$$AIC = 2k + n(\ln(2\pi SSE/n) + 1).$$

The results of applying these two criteria to the cricket thermometer data are also shown in Table 2.1. According to the BIC criterion, the best model is the cubic fitting, while according to the AIC criterion, the best fit is the quadratic polynomial.

2.2.6 *Nonlinear Discrete Least Squares*

Consider again a set of data points

$$\{(x_i, y_i)\}_{i=0}^{n} = \{(x_0, y_0), (x_1, y_1), \dots, (x_n, y_n)\}.$$

In the previous section we derived a formula for finding the Least Squares best fit to an mth **degree polynomial**, $p_m(x)$, evaluated at the points x_i. In this section, we consider a more general (nonlinear) model function $y = f(x, \mathbf{a})$, which depends on the variable x and on n parameters $\mathbf{a} = (a_1, a_2, \dots, a_n)$. The error function, $E(\mathbf{a})$, which needs to be minimized, satisfies:

$$E(\mathbf{a}) = \sum_{i=0}^{n} e_i^2 = \sum_{i=0}^{n} [y_i - f(x_i, \mathbf{a})]^2, \quad i = 1, 2, \dots, n.$$

A necessary condition for the error E to achieve its minimum value is for its gradient to be zero, i.e., $\nabla E = 0$. Since the model contains n parameters there are n gradient equations:

$$\frac{\partial E}{\partial a_j} = 2\sum_i e_i \frac{\partial e_i}{\partial a_j} = 0, \quad j = 1, 2, \dots, n.$$

So far the derivation is similar to that of the general discrete least squares model. However, this time the system is nonlinear. Thus, the gradient equations cannot be solved, in general, for a closed form solution because the derivatives $\frac{\partial e_i}{\partial a_j}$ are functions of both the independent variable and the parameters. An alternative approach is to solve for the parameters iteratively through successive approximations:

$$a_j \approx a_j^{k+1} = a_j^k + \Delta a_j,$$

where k is the iteration number and $\Delta\mathbf{a}$ is a shift vector. At each iteration, the function f is linearized to a first-order Taylor polynomial expansion about $\mathbf{a}^k$. That is,

$$f(x_i, \mathbf{a}) \approx f(x_i, \mathbf{a}^k) + \sum_j \frac{\partial f(x_i, \mathbf{a}^k)}{\partial a_j}(a_j - a_j^k) = f(x_i, \mathbf{a}^k) + \sum_j J_{ij} \Delta a_j,$$

where J_{ij} is the Jacobian matrix of $f(x, \mathbf{a})$. Observe that the Jacobian satisfies

$$J_{ij} = -\frac{\partial e_i}{\partial a_j}.$$

The individual errors e_i can be approximated as follows:

$$e_i = y_i - f(x_i, \mathbf{a}) = \left(y_i - f(x_i, \mathbf{a}^k)\right) + \left(f(x_i, \mathbf{a}^k) - f(x_i, \mathbf{a})\right) \approx \Delta y_i - \sum_{s=1}^{n} J_{is} \Delta a_s,$$

where $\Delta y_i = y_i - f(x_i, \mathbf{a}^k)$. Substituting these expressions into the gradient equations, they become

$$-2 \sum_{j=1}^{n} J_{ij} \left(\Delta y_i - \sum_{s=1}^{n} J_{is} \Delta a_s\right) = 0,$$

which can be rewritten as

$$\sum_{i=1}^{n} \sum_{s=1}^{n} J_{ij} J_{is} \Delta a_s = \sum_{i=1}^{n} J_{ij} \Delta y_i, \quad j = 1, 2, \ldots, n.$$

Finally, the normal equations can be written in matrix form

$$(\mathbf{J}^T \mathbf{J})\mathbf{a} = \mathbf{J}^\mathbf{T} \mathbf{y}. \tag{2.5}$$

2.2.7 Gauss-Newton's Method

A special case of the normal equations for the nonlinear least squares fitting is the Gauss-Newton's method, which is an iterative method to compute approximate solutions to a system of equations

$$\mathbf{f}(\mathbf{x}) = 0,$$

where $\mathbf{f} : D \subseteq \mathbb{R}^n \to \mathbb{R}^n$ is a vector-valued differentiable function on the domain D of the form

$$\mathbf{f}(\mathbf{x}) = \begin{bmatrix} f_1(\mathbf{x}) \\ f_2(\mathbf{x}) \\ \vdots \\ f_n(\mathbf{x}) \end{bmatrix}, \quad \mathbf{x} \in D,$$

Solving the normal Eq. (2.5) for $\mathbf{a}$ leads to

$$\mathbf{a} = (\mathbf{J}^T\mathbf{J})^{-1}\mathbf{J}^\mathbf{T}\mathbf{y},$$

which simplifies to

$$\mathbf{a} = \mathbf{J}^{-1}\mathbf{y}.$$

For the purpose of data fitting, the goal is to find a set of parameters $\mathbf{a}$ such that the function $f(x, \mathbf{a})$ best fits the data points (x_i, y_i). In this case,

$$\mathbf{y} = \mathbf{y} - \mathbf{f}(\mathbf{x}, \mathbf{a}),$$

and $\mathbf{a^{k+1}} = \mathbf{a^k} + \mathbf{a}$. Substituting and solving for $\mathbf{a^{k+1}}$ yields

$$\mathbf{a^{k+1}} = \mathbf{a^k} + \mathbf{J}^{-1}\mathbf{f}(\mathbf{a^k}).$$

This is an iterative process by which a new set of parameters $\mathbf{a^{k+1}}$ is obtained and updated based on previous values. The process starts with an initial value $\mathbf{a^0}$.

For the purpose of solving $\mathbf{f}(\mathbf{x}) = 0$, the fitting points satisfy $y_i = 0$, so that $\mathbf{y} = -\mathbf{f}(\mathbf{x})$. Furthermore, the unknowns $\mathbf{x}$ play the role of the parameters $\mathbf{a}$. Then substituting $\mathbf{a^{k+1}} = \mathbf{a^k} + \mathbf{a}$ with $\mathbf{x^{k+1}} = \mathbf{x^k} + \mathbf{x}$ and solving for $\mathbf{x^{k+1}}$ leads to

$$\mathbf{x^{k+1}} = \mathbf{x^k} - \mathbf{J}^{-1}\mathbf{f}(\mathbf{x^k}). \tag{2.6}$$

Equation (2.6) is the Gauss-Newton method. It is an iterative process by which a new iterate $\mathbf{x^{k+1}}$ is obtained from the previous one. The process starts with an initial guess $\mathbf{x^0}$ and it continues for $k = 1, 2, \ldots$ until convergence (which is not always guaranteed) can be achieved.

Gauss-Newton method can also be interpreted as a discrete model. Details of the analysis of these type of models, including convergence to long-term solutions, can be found in the next chapter.

2.3 The Global Positioning System

The Global Positioning System (GPS) consists of twenty four satellites orbiting the earth, while transmitting signals that can be used to calculate the position of objects (e.g., cell phones or users) on the earth, see Fig. 2.5.

Precision timing is essential for a GPS to work accurately. In this section we show the algebraic equations that are used in GPS location. We also show how the Implicit Function Theorem is used to determine the accuracy of location.

2.3.1 *Algebraic Equations for GPS Location*

Let $(x_i, y_i, z_i), i = 1, \ldots, 4$, be the coordinates of four of the twenty four satellites S_i. Let (x, y, z) be the coordinates of the object or user. GPS works under the assumption that an object or user is located at the intersection of four spheres, see Fig. 2.5, whose algebraic equations are

$$\begin{array}{l} (x - x_1)^2 + (y - y_1)^2 + (z - z_1)^2 = r_1^2 \\ (x - x_2)^2 + (y - y_2)^2 + (z - z_2)^2 = r_2^2 \\ (x - x_3)^2 + (y - y_3)^2 + (z - z_3)^2 = r_3^2 \\ (x - x_4)^2 + (y - y_4)^2 + (z - z_4)^2 = r_4^2, \end{array} \tag{2.7}$$

where r_i (also known as the “slant range”) is the true distance from satellite S_i to the object or user (Fiig. 2.6).

Fig. 2.5 The Global Positioning System consist of twenty four satellites that orbit the earth, while transmitting signals that can be used to locate objects on earth. *Source* Wikipedia

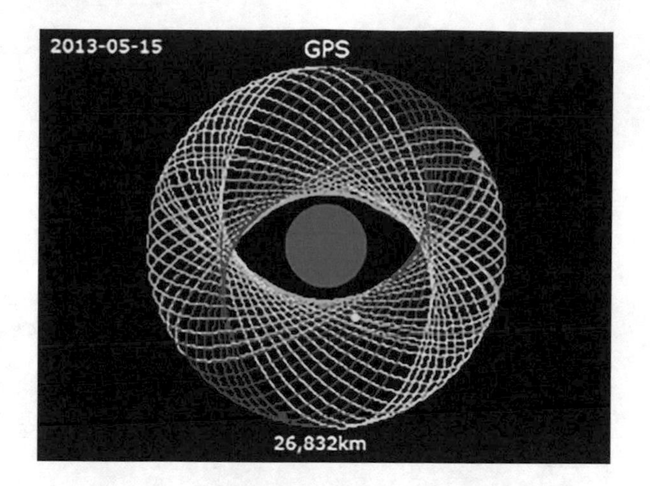

Fig. 2.6 GPS works under the assumption that an object is located at the intersection of four spheres defined among the twenty four satellites that orbit earth. *Source* Wikipedia

It can be shown that the true distance r_i from the user to the satellite S_i can be calculated as

$$r_i = c(\Delta t_i - \Delta t_{i,\text{prop}}),$$

where c is the speed of light in a vacuum, Δt_i is the transit time of the signal, and $\Delta t_{i,\text{prop}}$ is the propagation delay caused by the atmosphere. Since all the precision timing devices (i.e., atomic clocks) in the satellites run independently, then Δt_i is only an approximated quantity. Indeed, cesium or rubidium atomic clocks aboard the satellites experience phase drift due to material imperfections, electronic noise, frequency changes due to radiation, temperature and power supply variations. The quantity Δt_i aboard satellite S_i is approximated as follows

$$\Delta t_i = \Delta t_{i,\text{pseudo}} + \Delta t_{i,\text{drift}} + \Delta t_{\text{rec.clock}},$$

where $\Delta t_{i,\text{pseudo}}$ is the difference between the received satellite broadcast time and the receiver's current time, $\Delta t_{i,\text{drift}}$ is the amount of time that the atomic clock differs from GPS standard time, $\Delta t_{\text{rec.clock}}$ is the amount of time that the receiver clock differs from GPS standard time. Observe that $\Delta t_{\text{rec.clock}}$ is independent of i. Substituting into the equation for r_i we get

$$r_i = c(\Delta t_{i,\text{pseudo}} + \Delta t_{i,\text{drift}} - \Delta t_{i,\text{prop}} + \Delta t_{\text{rec.clock}}).$$

Substituting into Eq. (2.7), we get

$$\begin{aligned}
(x - x_1)^2 + (y - y_1)^2 + (z - z_1)^2 &= c^2(t_1 + d)^2 \\
(x - x_2)^2 + (y - y_2)^2 + (z - z_2)^2 &= c^2(t_2 + d)^2 \\
(x - x_3)^2 + (y - y_3)^2 + (z - z_3)^2 &= c^2(t_3 + d)^2 \\
(x - x_4)^2 + (y - y_4)^2 + (z - z_4)^2 &= c^2(t_4 + d)^2,
\end{aligned} \tag{2.8}$$

where $t_i = \Delta t_{i,\text{pseudo}} + \Delta t_{i,\text{drift}} - \Delta t_{i,\text{prop}}$ and $d = \Delta t_{\text{rec.clock}}$. Observe that the new algebraic system of equations (2.8) is a system of four equations in four unknowns, the user coordinates (x, y, z) and time $\Delta t_{\text{rec.clock}}$.

2.3.2 *Solution via Gauss-Newton's Method*

We now apply the Gauss-Newton's method discussed previously in Sect. 2.2 to find an approximate solution to the model Eq. (2.8). We seek a solution to Eq. (2.8) which can be rewritten as

$$\mathbf{f}(\mathbf{x}) = 0,$$

where

$$\mathbf{f}(\mathbf{x}) = \begin{bmatrix} f_1(\mathbf{x}) = (x - x_1)^2 + (y - y_1)^2 + (z - z_1)^2 - c^2(t_1 + d)^2 \\ f_2(\mathbf{x}) = (x - x_2)^2 + (y - y_2)^2 + (z - z_2)^2 - c^2(t_2 + d)^2 \\ f_3(\mathbf{x}) = (x - x_3)^2 + (y - y_3)^2 + (z - z_3)^2 - c^2(t_3 + d)^2 \\ f_4(\mathbf{x}) = (x - x_4)^2 + (y - y_4)^2 + (z - z_4)^2 - c^2(t_4 + d)^2 \end{bmatrix}, \quad \mathbf{x} \in \mathbb{R}^3.$$

The parameter values for (x_i, y_i, z_i) and t_i employ up to ten digits of precision in order to achieve numerical accuracy. For brevity, we list them on Table 2.2 up to five digits.

The speed of light constant for the calculations is $c = 299792458.0$ m/sec. We can now solve numerically for (x, y, z) and d using Gauss-Newton's iterative scheme:

$$\mathbf{x}^{\mathbf{k+1}} = \mathbf{x}^{\mathbf{k}} - \mathbf{J}^{-1}\mathbf{f}(\mathbf{x}^{\mathbf{k}}),$$

where in this case $\mathbf{x} = (x, y, z, d)$ and the Jacobian is given by

$$J = \begin{bmatrix} 2(x - x_1) & 2(y - y_1) & 2(z - z_1) & -2c^2(t_1 + d) \\ 2(x - x_2) & 2(y - y_2) & 2(z - z_2) & -2c^2(t_2 + d) \\ 2(x - x_3) & 2(y - y_3) & 2(z - z_3) & -2c^2(t_3 + d) \\ 2(x - x_4) & 2(y - y_4) & 2(z - z_4) & -2c^2(t_4 + d) \end{bmatrix}.$$

Table 2.2 Parameters values for solving GPS algebraic equations numerically

i	x_i (m)	y_i (m)	z_i (m)	t_i (s)
1	1.87637×10^6	-1.06414×10^7	2.42697×10^7	0.07234
2	1.09766×10^7	-1.30814×10^7	2.03511×10^7	0.06730
3	2.45851×10^7	-4.33502×10^6	9.08630×10^6	0.06738
4	3.85413×10^6	7.24857×10^6	2.52663×10^7	0.07651

Using initial conditions $\mathbf{x^0} = (x, y, z, d) = (1000, 1000, 50000, 0)$, the iterative method converges to a solution given by

$$(x, y, z, d) = (-39.74783, -134.27414, -9413.62455, -0.18517),$$

where the location coordinates (x, y, z) are measured in km and d in sec.

2.3.3 Accuracy

In the previous section we showed that the true distance r_i from the user to the satellite S_i was calculated by subtracting the current time on the user's end from the satellites time stamp, and then multiplying this difference by the speed of light. But since Δt_i are only approximations, we would like to know what happens when these time stamps are slightly perturbed. Conversely, if we need the location (x, y, z) of a user to be within a specified degree of accuracy, we would like to know how much deviation on the Δt_i can the system tolerate.

To answer these questions, we make use of the implicit function theorem.

Theorem 2.1 *Let $f : \mathbb{R}^{n+m} \to \mathbb{R}^m$ be a continuously differentiable function. Let $\mathbb{R}^{n+m}$ have coordinates $(\mathbf{x}, \mathbf{y})$. Fix a point $(\mathbf{a}, \mathbf{b}) = (a_1, \ldots, a_n, b_1, \ldots, b_m)$ with $F(\mathbf{a}, \mathbf{b}) = \mathbf{0}$, where $\mathbf{0} \in \mathbb{R}^m$. If the Jacobian matrix*

$$J_{f,\mathbf{y}}(\mathbf{a}, \mathbf{b}) = \left[\frac{\partial f_i}{\partial y_j}(\mathbf{a}, \mathbf{b})\right]$$

is invertible, then there exists an open set U of $\mathbb{R}^m$ containing $\mathbf{a}$ such that there exists a unique continuously differentiable function $g : U \to \mathbb{R}^m$ such that

$$g(\mathbf{a}) = \mathbf{b}$$

and

$$f(\mathbf{x}, g(\mathbf{x})) = \mathbf{0}, \quad \forall \in U.$$

Moreover, the partial derivatives of g in U are given by

$$\frac{\partial g}{\partial x_j}(\mathbf{x}) = \left[J_{f,\mathbf{y}}(\mathbf{x}, g(\mathbf{x}))\right]^{-1}\left[\frac{\partial f}{\partial x_j}(\mathbf{x}, g(\mathbf{x}))\right].$$

Applying the implicit function theorem to our GPS equations, we get

$$\begin{bmatrix} \frac{\partial x}{\partial t_1} & \frac{\partial x}{\partial t_2} & \frac{\partial x}{\partial t_3} & \frac{\partial x}{\partial t_4} \\ \frac{\partial y}{\partial t_1} & \frac{\partial y}{\partial t_2} & \frac{\partial y}{\partial t_3} & \frac{\partial y}{\partial t_4} \\ \frac{\partial z}{\partial t_1} & \frac{\partial z}{\partial t_2} & \frac{\partial z}{\partial t_3} & \frac{\partial z}{\partial t_4} \\ \frac{\partial d}{\partial t_1} & \frac{\partial d}{\partial t_2} & \frac{\partial d}{\partial t_3} & \frac{\partial d}{\partial t_4} \end{bmatrix} = \begin{bmatrix} \frac{\partial f_1}{\partial x} & \frac{\partial f_1}{\partial y} & \frac{\partial f_1}{\partial z} & \frac{\partial f_1}{\partial d} \\ \frac{\partial f_2}{\partial x} & \frac{\partial f_2}{\partial y} & \frac{\partial f_2}{\partial z} & \frac{\partial f_2}{\partial d} \\ \frac{\partial f_3}{\partial x} & \frac{\partial f_3}{\partial y} & \frac{\partial f_3}{\partial z} & \frac{\partial f_3}{\partial d} \\ \frac{\partial f_4}{\partial x} & \frac{\partial f_4}{\partial y} & \frac{\partial f_4}{\partial z} & \frac{\partial f_4}{\partial d} \end{bmatrix}^{-1} \begin{bmatrix} \frac{\partial f_1}{\partial t_1} & \frac{\partial f_1}{\partial t_2} & \frac{\partial f_1}{\partial t_3} & \frac{\partial f_1}{\partial t_4} \\ \frac{\partial f_2}{\partial t_1} & \frac{\partial f_2}{\partial t_2} & \frac{\partial f_2}{\partial t_3} & \frac{\partial f_2}{\partial t_4} \\ \frac{\partial f_3}{\partial t_1} & \frac{\partial f_3}{\partial t_2} & \frac{\partial f_3}{\partial t_3} & \frac{\partial f_3}{\partial t_4} \\ \frac{\partial f_4}{\partial t_1} & \frac{\partial f_4}{\partial t_2} & \frac{\partial f_4}{\partial t_3} & \frac{\partial f_4}{\partial t_4} \end{bmatrix}.$$

Direct computations yield

$$\begin{bmatrix} \frac{\partial x}{\partial t_1} & \frac{\partial x}{\partial t_2} & \frac{\partial x}{\partial t_3} & \frac{\partial x}{\partial t_4} \\ \frac{\partial y}{\partial t_1} & \frac{\partial y}{\partial t_2} & \frac{\partial y}{\partial t_3} & \frac{\partial y}{\partial t_4} \\ \frac{\partial z}{\partial t_1} & \frac{\partial z}{\partial t_2} & \frac{\partial z}{\partial t_3} & \frac{\partial z}{\partial t_4} \\ \frac{\partial d}{\partial t_1} & \frac{\partial d}{\partial t_2} & \frac{\partial d}{\partial t_3} & \frac{\partial d}{\partial t_4} \end{bmatrix} = - \begin{bmatrix} 2(x-x_1) & 2(y-y_1) & 2(z-z_1) & -2c^2(t_1+d) \\ 2(x-x_2) & 2(y-y_2) & 2(z-z_2) & -2c^2(t_2+d) \\ 2(x-x_3) & 2(y-y_3) & 2(z-z_3) & -2c^2(t_3+d) \\ 2(x-x_4) & 2(y-y_4) & 2(z-z_4) & -2c^2(t_4+d) \end{bmatrix}^{-1} \begin{bmatrix} -2c^2(t_1+d) & 0 & 0 & 0 \\ 0 & -2c^2(t_2+d) & 0 & 0 \\ 0 & 0 & -2c^2(t_3+d) & 0 \\ 0 & 0 & 0 & -2c^2(t_4+d) \end{bmatrix}.$$

Substituting parameter values we get

$$\begin{bmatrix} \frac{\partial x}{\partial t_1} & \frac{\partial x}{\partial t_2} & \frac{\partial x}{\partial t_3} & \frac{\partial x}{\partial t_4} \\ \frac{\partial y}{\partial t_1} & \frac{\partial y}{\partial t_2} & \frac{\partial y}{\partial t_3} & \frac{\partial y}{\partial t_4} \\ \frac{\partial z}{\partial t_1} & \frac{\partial z}{\partial t_2} & \frac{\partial z}{\partial t_3} & \frac{\partial z}{\partial t_4} \\ \frac{\partial d}{\partial t_1} & \frac{\partial d}{\partial t_2} & \frac{\partial d}{\partial t_3} & \frac{\partial d}{\partial t_4} \end{bmatrix} \approx \begin{bmatrix} 1.27\,10^9 & -1.20\,10^9 & 2.22\,10^8 & -3.00\,10^8 \\ 2.51\,10^8 & 1.23\,10^8 & -1.90\,10^7 & -3.54\,10^8 \\ 1.73\,10^9 & -1.90\,10^9 & 8.06\,10^8 & -6.38\,10^8 \\ -5.48 & 5.49 & -2.41 & 1.39 \end{bmatrix}.$$

To address the first question, we consider first the x-coordinate of the user. Through the chain rule we can find an approximate value of the change in the x-coordinate as a function of changes in times Δt_i. That is

$$\Delta x \approx \frac{\partial x}{\partial t_1}\Delta t_1 + \frac{\partial x}{\partial t_2}\Delta t_2 + \frac{\partial x}{\partial t_3}\Delta t_3 + \frac{\partial x}{\partial t_4}\Delta t_4.$$

If we allow the times to vary within an ε window, i.e., $|\Delta t_i| < \varepsilon$, then

$$|\Delta x| \leq \left(\left|\frac{\partial x}{\partial t_1}\right| + \left|\frac{\partial x}{\partial t_2}\right| + \left|\frac{\partial x}{\partial t_3}\right| + \left|\frac{\partial x}{\partial t_4}\right| \right) \varepsilon.$$

Using the parameter values from our example, we find

$$|\Delta x| \leq 3 \times 10^9\,\varepsilon.$$

This means that if the time stamps Δt_i are slightly perturbed by ε sec, then the location of the x-coordinate will vary by at most $3 \times 10^9\,\varepsilon$. It follows that if the time stamps from the satellites were to be accurate to within 10^{-9} s (i.e., a nanosecond) then the x-coordinate would be accurately computed to within 3 m. A similar set of calculations can be carried out for the other coordinates.

We now address the second question. Using the same parameter values, let's assume we would like the accuracy in the calculation of the x-coordinate to be within 300 m. Then the time stamps would need to be accurate to within 10^{-8} s.

2.4 Allometric Models

Allometry is the study of the relationships between body properties, such as weight, size, metabolism, shape, and, overall anatomy. For instance, the bones in the skeleton of an elephant, see Fig. 2.7, compared to those of a tiger, tend to be ticker since the body size of an elephant is significantly bigger. This type of relationship between bone thickness and body size of animals is an example of an allometric scaling.

Typically, the scaling in those relationships follow a **power law**, which is formulated in the form of an **allometric model** of the form

$$\boxed{y = \kappa x^m,} \tag{2.9}$$

where x and y represent some type of body properties, e.g., weight, size, or shape, m is a scaling exponent of the power law, and κ is just a multiplicative parameter. In logarithmic form, Eq. (2.9) can be rewritten as

$$\ln(y) = \ln(\kappa x^m) = m\ln(x) + \ln(\kappa).$$

If we introduce the change of variable: $X = \ln(x)$ and $Y = \ln(y)$, $b = \ln(\kappa)$ then it is easier to see that a power law scaling, expressed in logarithmic form, becomes a straight line relationship:

$$Y = mX + b. \tag{2.10}$$

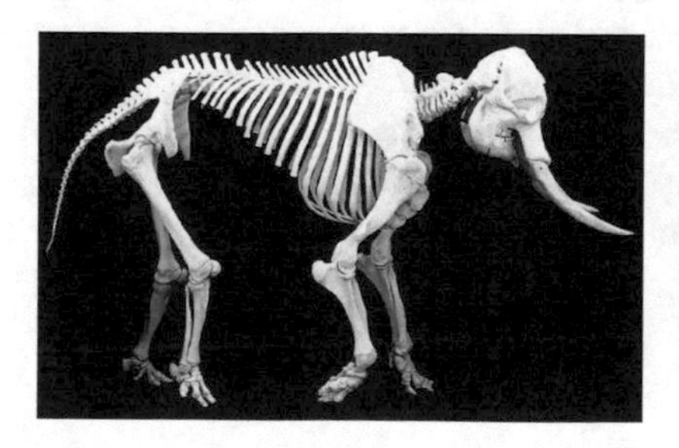

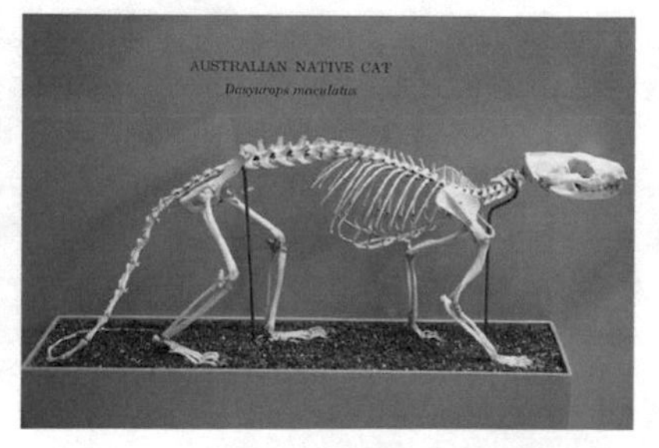

Fig. 2.7 Relationship of thicker bones to size can be observed by comparing the skeletons of (left) an elephant to that of a (right) tiger. This is an example of an allometric scaling. *Source* Wikipedia

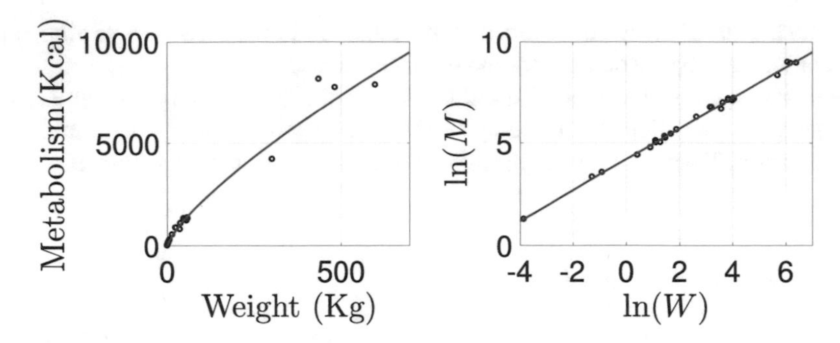

Fig. 2.8 Allometric scaling between metabolism, measured in calories, and body weight, measured in kilograms, for a wide range of animals. (Left) Original data set. (Right) Algorithmic scaling of the data shows a good fit with a straight line

As an example, we consider next the relationship between weight and metabolism between different types of animals.

2.4.1 *Kleiber's Law*

Common sense suggests that the metabolism (measured in calories) of certain types of animals varies with size. Indeed, in "Body Size and Metabolic Rate", Max Kleiber [3] asks whether a horse can produce more heat per day per kilogram of body weight than a rat?".

Again, common sense suggests that the answer is *Yes*. However, as Kleiber observes, the rate of heat produced per unit body weight of big animals is less than that of small animals. In fact, Fig. 2.8(left) shows the relationship between weight and metabolism for a wide range of animals that include: mice, rats, guinea pigs, rabbits, cats, macque, dogs, goats, chimpanzees, sheep, and cows. The data points (black circles) clearly show that metabolism increases as weight increases. The relationship can be seen to be nonlinear since the data points do not seem to be aligned in a straight line. But when the data points are plotted in a logarithmic scale, as is shown in Fig. 2.8(right), they do seem to fit a straight line.

The above observations suggest that the data set, metabolism vs. weight, could be described by an allometric model of the form (2.9). If we let $y = M$ to represent metabolism, and $x = W$ to represent Weight, then we could write an allometric model, in logarithmic form, as:

$$\ln(M) = m \ln(W) + \ln(\kappa). \tag{2.11}$$

The following MATLAB script can be used to compute the parameters m and κ for the best linear fit to the logarithmic data:

```
% Allometric Model for Metabolism vs Weight
clear all;
close all;
clc;

load 'metabolism.data'
xdata  = metabolism(:,1);
ydata  = metabolism(:,2);

% Linear Least Squares Fit to Logarithmic Data
Y = log(ydata);          % Logarithm of y-data
X = log(xdata);          % Logarithm of x-data

% Find Parametmeters k, mu to Model y = k*x^mu
p = polyfit(X,Y,1);     % Linear fit to X and Y
mu  = p(1)                % Scaling exponent
k   = exp(p(2))           % Multiplicative factor
```

Upon running the script we get:

$$\kappa = 66.82, \qquad m = 0.7565,$$

which leads to an allometric model of the form:

$$M = 66.82W^{0.7565}. \tag{2.12}$$

Applying the nonlinear least squares best fit method, also described earlier on) we get the best fit parameters $m = 63.86$ and $m = 0.7685$, which are not that different from those of the linear fit.

So, what is Kleibar's law?

Kleiber's Law is related to the actual value of the scaling exponent m. When $m = 1$ we get $M = \kappa W$, which means that the metabolism rate is directly proportional to weight. When $m < 1$, then the metabolism rate decreases as a function of weight. In fact, the particular case $m = 2/3$ corresponds to heat loss through skin, as it corresponds to surface area over volume ratio. The case $m = 3/4$ is known as **Kleiber's Law**.

2.5 Dimensional Analysis

Mathematical models are made up of state variables and parameters, which are quantities that serve to characterize a physical phenomenon. A *dimension* is a measure of each physical quantity. A *unit* is a way to assign a number to those dimensions. For instance, the dimension *mass* (M) can be measured in units of kilograms (kg) or

Table 2.3 Some commonly used dimensions and their units

Dimension	Length	Mass	Time	Force
Symbol	L	M	T	$F = MLT^{-2}$
SI Unit	m (m)	kg (kg)	s (s)	$N = \text{kg} \cdot \text{m/s}^2$
BG Unit	ft (feet)	lb (pound)	s (s)	bf (pound-force)

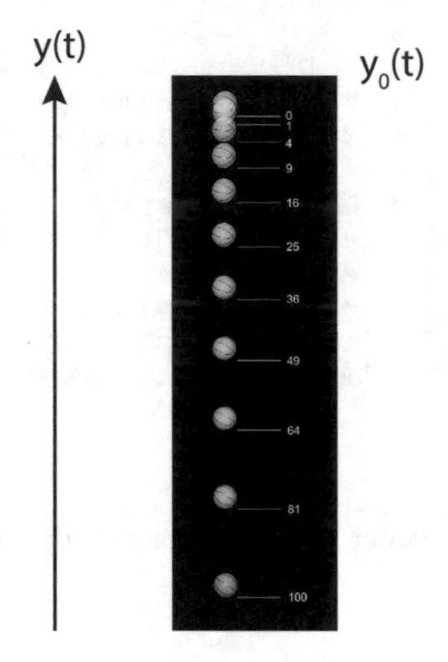

Fig. 2.9 Free fall of a basketball

pounds (lb), while the dimension *length* (L) can be measured in units of meters (m) or feet (ft). Some of the most common dimensions are listed in Table 2.3.

2.5.1 Free Fall of an Object

For instance, consider a basketball falling free, i.e., based on gravity, from an initial vertical position, y_0, and initial velocity, v_0, as is shown in Fig. 2.9.

A mathematical model for the vertical displacement of the basketball, which is measured by the state variable y as a function of the independent variable time, t, is given by

$$y = y_0 + v_0 t - \frac{1}{2} g t^2, \tag{2.13}$$

where g is gravity. We can rewrite Eq. (2.13) in terms of the quantities that appear in each term:

$$[L] = [L] + \left[\frac{L}{T}\right][T] - \left[\frac{L}{T^2}\right][T^2]. \tag{2.14}$$

Observe that each additive term in Eq. (2.14), or Eq. (2.13), has exactly the same dimensions of length, L. The process of *nondimensionalization* is about removing the units from the physical variables or parameters. This results in *nondimensionalized* model equations, also known as *dimensionless* equations. This process can be accomplished through a suitable change of variables. There are few different methods to figure out the correct change of variables. One, is simply by inspection of the underlying variables. We discuss this method first and then a more rigorous approach.

2.5.2 Inspection Method

Consider again the dimensional version of the free falling basketball Eq. (2.13). We want to write this equation in dimensionless form. By direct inspection we can see that the displacement and time variables can be made dimensionless through the following change of variables:

$$z = \frac{y}{y_0}, \quad \tau = \frac{v_0 t}{y_0}.$$

Observe that these quantities are dimensionless since:

$$[z] = \frac{[L]}{[L]}, \quad [\tau] = \frac{[LT^{-1}][T]}{[L]}.$$

Direct substitution of the change of variables, z and τ, into Eq. (2.13) leads to:

$$z = 1 + \tau + \frac{1}{2\mu}\tau^2, \quad \mu = \frac{v_0^2}{g y_0}. \tag{2.15}$$

We can quickly verify that Eq. (2.15) is dimensionless. The variables z and τ were already shown to be dimensionless. Now, the parameter μ yields:

$$[\mu] = \frac{[L^2T^{-2}]}{[LT^{-2}][L]}.$$

Thus, every term in Eq. (2.15) is dimensionless. If we compare directly Eq. (2.13) with Eq. (2.15), we can describe the nondimensionalization process as a transformation of a dimensional model with 5 variables into a dimensionless model with 3 variables. In functional form, we have:

$$f(y, y_0, v_0, g, t) \Longrightarrow F(z, \tau, \mu) = 0,$$

where f is the original dimensional model and F is the dimensionless version. In addition to the obvious advantage of having to work with less variables, a dimensionless model can be very useful in interpreting results, data analysis, and understanding similarity properties, among many other features.

In general, the process of nondimensionalization seeks to transform a dimensional model, f, with n variables, of the form

$$f(x_1, x_2, \ldots, x_n) = 0,$$

into a dimensionless model, F, with $k < n$ variables, of the form

$$F(\Pi_1, \Pi_2, \ldots, \Pi_k) = 0.$$

The formal process, including how to determine the number, k, of non-dimensional variables is the subject of the following theorem.

2.5.3 Buckingham Pi Theorem and Method

The following theorem provides a rigorous approach to determining the number of dimensionless variables.

Theorem 2.2 (Buckingham Pi Theorem) *Let $x_1, x_2, \ldots, x_n$ be n dimensional variables of a physical phenomenon expressed analytically by an implicit functional form:*

$$f(x_1, x_2, \ldots, x_n) = 0.$$

If m is the minimum number of fundamental dimensions required to describe the n variables, then there will be m primary variables and the remaining variables can be expressed as $(n - m)$ dimensionless and independent quantities or Pi groups, $\Pi_1, \Pi_2, \ldots, \Pi_{n-m}$. The functional relationship reduces to:

$$F(\Pi_1, \Pi_2, \ldots, \Pi_k) = 0, \quad k = n - m.$$

Exponent Method
This is a systematic method to determine the dimensionless Pi groups, as follows:

1. List all variables of the problem.
2. Express each variable in terms of fundamental dimensions.
3. Determine the number k of terms.
4. Select as many repeating variables as the number of fundamental dimensions. Exclude the dependent variable since the repeating variables might appear in various Pi groups.
5. Form individual Pi groups by multiplying each non-repeating variable by the product of the repeating variables, each raised to an exponent.

Table 2.4 Dimensions and units of the variables involved in a free fall basketball

Variable	y	y_0	v_0	g	t
Dimension	L	L	LT^{-1}	LT^2	T
Unit	m	m	m/s	m/s^2	s

Let us apply the exponent method to the free fall basketball problem. In Step 1, we identify $n = 5$ variables: y, y_0, v_0, g, and t. In Table 2.4 we perform Step 2 by expressing each variable in terms of fundamental dimensions

We observe a total of $m = 2$ fundamental dimensions, one for L and one for T. This means that in Step 3 we can calculate $k = n - m = 5 - 2 = 3$ Pi groups of dimensionless variables. In Step 4 we must select $m = 2$ repeating variables. We must make sure that all of the fundamental dimensions are included in the group of repeating variables. Since L and T are the fundamental dimensions, we can choose y_0 and t as repeating variables. Next we perform Step 5 to determine each dimensionless group.

Group Π_1
Let

$$\Pi_1 = y_0^a t^b y = L^a T^b L.$$

In order for Π_1 to be dimensionless, we must have

$$a + 1 = 0, \qquad b = 0.$$

Thus, $a = -1$ and $b = 0$, which yields:

$$\Pi_1 = \frac{y}{y_0}.$$

Group Π_2
Let

$$\Pi_2 = y_0^a t^b v_0 = L^a T^b (LT^{-1}).$$

In order for Π_2 to be dimensionless, we must have

$$a + 1 = 0, \qquad b - 1 = 0.$$

Thus, $a = -1$ and $b = 1$, which yields:

$$\Pi_2 = \frac{v_0 t}{y_0}.$$

Group Π_3
Let

$$\Pi_3 = y_0^a t^b g = L^a T^b (LT^{-2}).$$

In order for Π_3 to be dimensionless, we must have

$$a + 1 = 0, \qquad b - 2 = 0.$$

Thus, $a = -1$ and $b = 2$, which yields:

$$\Pi_3 = \frac{g y_0 t}{v_0^2} \tau^2.$$

Observe that $\Pi_1 = z$, $\Pi_2 = \tau$, and $\Pi_3 = \frac{\tau^2}{\mu}$, correspond to the results obtained by the inspection method.

2.5.4 Allometric Model of Atomic Bomb Blast

In a series of papers, Geoffrey Taylor [4, 5] derived an allometric model that could accurately predict the radius, and, more importantly, the power of the Trinity test in White Sands, NM. What is most remarkable about those papers is that Taylor was able to obtain those very accurate results from studying photographs of the ball of fire produced during the explosion, as is shown in Fig. 2.10.

Furthermore, the article was considered a serious violation of national security since, at the time of publication, the amount of energy released by the atomic explosion was a highly guarded top secret by the U.S. government. In this section, we will discuss the process that allowed Taylor to derive the allometric model. But, first, we start with a dimensionless analysis.

There are $n = 4$ variables: radius of the atomic blast, R, energy released by the blast, E, time t, and ambient air density, ρ. Their dimensions and units are listed in Table 2.5.

The table shows a total of $m = 3$ fundamental dimensions for L, M, and T. It follows that there is $k = n - m = 4 - 3 = 1$ Pi group of dimensionless variables. Assuming the radius of the blast, R, to be the dependent variable, and all other $m = 3$ variables to be repeating, we let

$$\Pi = E^a t^b \rho^c R = (ML^2T^{-2})^a T^b (ML^{-3})^c L.$$

In order for Π to be dimensionless, we must have:

$$\begin{array}{lll} (L) & 2a - 3c + 1 & = 0, \\ (M) & a + c & = 0, \\ (T) & -2a + b & = 0. \end{array}$$

Solving this linear system we get:

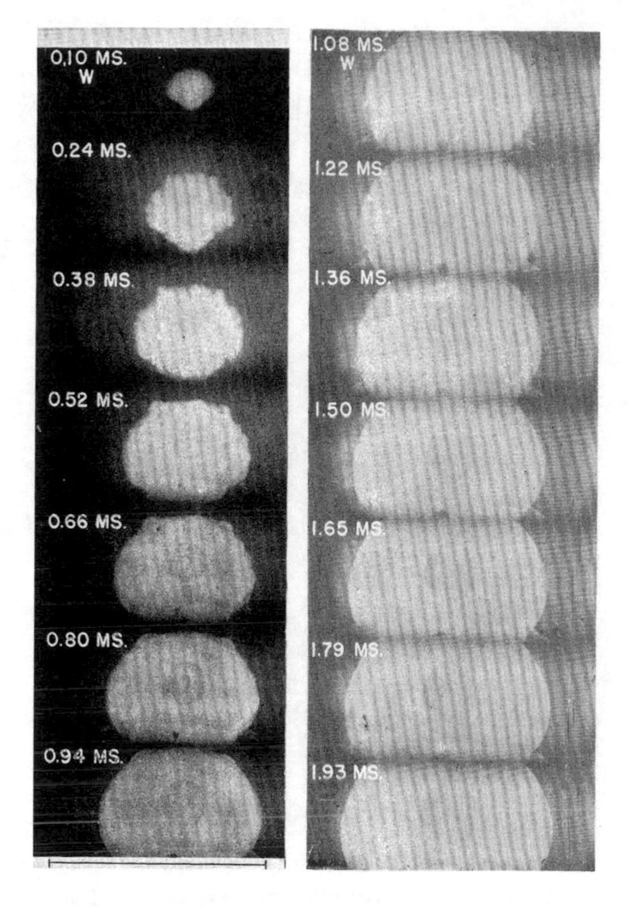

Fig. 2.10 Sequential photographs of the atomic blasts produced by the Trinity test in White Sands, NM

Table 2.5 Dimensions and units of an atomic blast

Variable	R	E	t	ρ
Dimension	L	ML^2T^{-2}	T	ML^{-3}
Unit	m	$\text{kg} \cdot \text{m}^2/\text{s}^2$	s	kg/m^3

$$a = -\frac{1}{5}, \quad b = -\frac{2}{5}, \quad c = \frac{1}{5}.$$

The dimensionless variable becomes

$$\Pi = E^{-1/5} t^{-2/5} \rho^{1/5} R.$$

Solving for R we get an allometric model of the form

$$R = \kappa \left(\frac{Et^2}{\rho} \right)^{1/5}. \tag{2.16}$$

Taylor shows that the scaling factor κ is , approximately one. In addition, the density of air at 1310 m, which is the elevation of White Sands, NM, is also, approximately, one. Thus, setting $\kappa = 1$ and $\rho = 1$, Eq. (2.16) becomes:

$$R = (Et^2)^{1/5}.$$

In logarithmic form we get

$$\ln(R) = \frac{2}{5} \ln(t) + \frac{1}{5} \ln(E).$$

Using the data from the sequence of photographs of the blast, we can get the best linear least square fit to be:

$$\ln(R) = 0.4024 \ln(t) + 6.4038,$$

which is very close to the analytical result of the allometric model. A very critical observation, which is obtained by comparing these last two equations, is the fat that the linear fit to the data yields the y-intercept:

$$\frac{1}{5} \ln(E) = 6.4038.$$

This is critical information because while solving for E, Taylor was able to estimate the energy released by the atomic blast, namely:

$$E = e^{5\times 6.4038} = e^{32.02} = 8.05 \times 10^{13} \text{ J}.$$

The actual amount of energy measured by scientists at Los Alamos was 9×10^{13} J. This serves to validate the information obtained by Taylor from a simple series of snapshots. Figure 2.11 shows the fits of the data to the *allometric model*.

2.6 Exercises

Exercise 2.1 Calculate the least-squares approximation to the data shown in Table 2.6 below by a function of the form

$$f(x) = a_1 + a_2 x + a_3 \sin(123(x - 1))$$

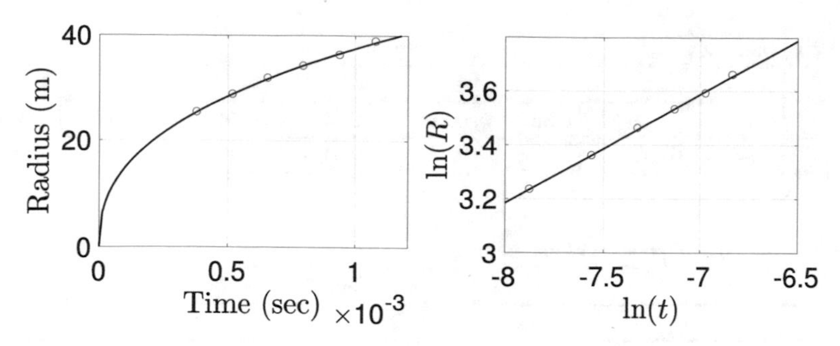

Fig. 2.11 The graph on the left shows the best fitting allometric relationship of the atomic blast, while the graph on the right shows the linear fit to the logarithm of the data

Table 2.6 Data set for least-squares approximation

x_i	1	2	3	4	5	6	7	8	9	10	11
f_i	0.0	0.6	1.77	1.92	3.31	3.52	4.59	5.31	5.79	7.06	7.17

Table 2.7 Sales projections

Months	Sales in USD
0	200,000
24	248,000
48	296,000
72	344,000
96	392,000
120	440,000

Exercise 2.2 Table 2.7 shows sales projections of a company per month. Use the least-squares approximation to fit a linear model the data shown in the table.

Exercise 2.3 Consider the following data set

$$\{(x_i, y_i)\}_{i=0}^{3} = \{(1, 6), (2, 5), (3, 7), (4, 10)\}.$$

Analytically, calculate the parameters a_0 and a_1 that will lead to the best linear fit model of the form

$$y(x) = a_1 x + a_0.$$

Exercise 2.4 Consider the following data set shown in Table 2.8:

(a) Apply the linear least squares algorithm to fit a model of the form

$$y(t) = a_1 t + a_0.$$

Table 2.8 Data set

i	1	2	3	4	5	6	7	8
t_i	0.10	0.23	0.36	0.49	0.61	0.74	0.87	1.00
y_i	0.84	0.30	0.69	0.45	0.31	0.09	−0.17	0.12

Table 2.9 Size of a population of antelopes

Time, t_i, in years	Population size, y_i, in hundreds
1	3
2	4
4	6
5	11
8	20

(b) Apply the nonlinear least squares algorithm to fit a model of the form

$$y(t) = a_1 e^{a_0 t}.$$

(c) Compare the results of the two models.

Exercise 2.5 Male fiddler crabs (*Uca pugnax*) possess an enlarged major claw which can be used for fighting or threatening other male crabs. For a particular species of fiddler crab, the following allometric model has been derived:

$$M_c = A(w)^r,$$

where M_c represents the mass of the major claw, and w describes the body mass of the crab. Parameters are: $A = 0.036$, and $r = 1.356$.

(a) Rewrite the allometric model in logarithmic form.
(b) Determine the slope of the line while plotting $\ln(M)$ and $\ln(Aw^r)$.
(c) What is the approximate predicted major claw mass for a male fiddler crab with a mass of 400 mg?
(d) What is the expected body mass of a male fiddler crab whose major claw mass is 53.0 mg?

Exercise 2.6 The size of a population of antelope at various times is shown in Table 2.9.

Apply the nonlinear least square fitting, i.e., Gauss-Newton method, to fit an exponential model of the form

$$y_i = ae^{bt_i},$$

where a and b are the parameters that need to be found through the nonlinear fitting process.

Table 2.10 Data set

i	1	2	3	4	5	6
x_i	1.2	2.3	3.0	3.8	4.7	5.9
y_i	1.1	2.1	3.1	4.0	4.9	5.9

Exercise 2.7 Consider the following case of a simple logistic function

$$y(t) = \frac{K}{1 + e^{-r(t-t_0)}},$$

where K is a maximum value which is achieved by $y(t)$ as $t \to \infty$, r is the logistic growth rate, and t_0 is the value at which the midpoint $K/2$ is achieved.

Write a MATLAB code to generate a data set of $y(t) +$ noise, where "noise" can be generated by a random variable. Then apply the nonlinear least square fitting to fit the values of K, r and t_0 that best fit the data set.

Exercise 2.8 Repeat the previous exercise with a parametrized circumference function

$$s(t) = \big(c_x + r\cos(t), c_y + r\sin(t)\big),$$

where (c_x, c_y) is the center of the circumference and r its radius.

Exercise 2.9 Consider the following data set shown in Table 2.10:

(a) Apply the linear least squares algorithm to fit a model of the form

$$y(t) = a_1 x + a_0.$$

(b) Plot the results, including the data set and the approximated line, in MATLAB.
(c) Compare the results of the data set and the approximated straight line.

Exercise 2.10 The planarian *Schmidtea mediterranea* is a freshwater triclad that lives in southern Europe and Tunisia [6]. It is a model for regeneration, stem cells and development of tissues such as the brain and germline. Estimates of metabolic rate are based on studying the relationship between dry and wet mass. Tabletab:sch shows this relationship.

Use a linear least square fitting to derive an allometric model for wet mass as a function of dry mass, of the form

$$W_M = A(D_M)^r,$$

where W_M represents wet mass, D_M is dry mass, and A and r are parameters to be found through the least squares fitting of the data (Table 2.11).

Table 2.11 Wet vs dry mass in Schmidtea mediterranea

Wet Mass (mg)	Dry Mass (mg)
0.090	0.051
0.101	0.064
0.139	0.088
1.176	0.324
1.506	0.417
2.486	0.592
5.452	1.446
4.434	1.231
7.726	2.022
14.834	3.976
16.124	4.259
10.379	2.637
12.827	3.388
5.920	1.574
0.815	0.254
0.091	0.030
0.070	0.035
2.957	0.798
0.185	0.084
0.288	0.114
0.536	0.160
2.199	0.604
4.300	1.149
7.970	2.001
11.192	3.082
16.003	4.387
17.250	4.604
13.389	3.469

Exercise 2.11 Assume that last night, at dusk, you heard 31 chirps in 14 s from the nearest cricket in the backyard. Use the chirping crickets and temperature model to estimate the outside temperature.

Exercise 2.12 The following algebraic model allows us to compute the future value FV of an initial or present value, PV, which is assume to increase at the grow rate g, over a period of time T.

$$FV = PV(1 + g)^T.$$

What will the population of India be in year 2030 if the population in 1985 was estimated to be 751 million and the growth rate is expected to remain at 2.5% a year for the entire period?

Exercise 2.13 The pressure drop, $\Delta p = p_1 - p_2$, along a straight pipe of distance D has been experimentally studied, and it is observed that for laminar flow of a given fluid and pipe, the pressure drop varies directly with the distance, l, between pressure taps. Assume that Δp is a function of D and l, the velocity, V, and the fluid viscosity, $\tilde{\mu}$. Use dimensional analysis to deduce how the pressure drop varies with pipe diameter.

Exercise 2.14 A cylinder with a diameter, D, floats upright in a liquid. When the cylinder is displaced slightly along its vertical axis it will oscillate about its equilibrium position with a frequency, ω. Assume that this frequency is a function of the diameter, D, the mass of the cylinder, m, and the specific weight, γ, of the liquid. That is,

$$\omega = f(D, m, \gamma).$$

Determine, with the aid of dimensional analysis, how the frequency is related to these variables. If the mass of the cylinder were increased, would the frequency increase or decrease?

Exercise 2.15 Consider the physical quantities s, v, a, and t, with dimensions $[s] = L$, $[v] = LT^{-1}$, $[a] = LT^{-2}$, and $[t] = T$. Determine whether each of the following equations is dimensionally consistent:

(a) $s = v\,t + 0.5a\,t^2$.
(b) $s = v\,t^2 + 0.5a\,t$.
(c) $s = \sin\,(a\,t^2/s)$.

Exercise 2.16 A student is trying to remember some formulas from geometry. In what follows, assume S is area, V is volume, and all other variables are lengths. Determine which formulas are dimensionally consistent:

(a) $V = \pi r^2 h$.
(b) $A = 2\pi r^2 + 2\pi r h$.
(c) $V = 0.5b\,h$.
(d) $V = \pi d^2$.
(e) $V = \pi d^3/6$.

Exercise 2.17 Consider the physical quantities m, s, v, a, and t, with dimensions $[m] = M$, $[s] = L$, $[v] = LT^{-1}$, $[a] = LT^{-2}$, and $[t] = T$. Assuming each of the following equations is dimensionless consistent, find the dimension of the quantity on the left-hand side of the equation:

(a) $F = m\,a$.
(b) $K = 0.5\,m\,v^2$.
(c) $p = m\,v$.

(d) $W = m\,a\,s$.
(e) $L = m\,v\,r$.

Exercise 2.18 Suppose quantity s is a length and quantity t is a time. Suppose the quantities v and a are defined by

$$v = \frac{ds}{dt}, \qquad a = \frac{dv}{dt}.$$

(a) What is the dimension of v?
(b) What is the dimension of a?
(c) What are the dimensions of $\int v\,dt$ and $\int a\,dt$?

Exercise 2.19 Use the future-value model above to determine what must be put in the bank today if we want to have $20,000 in the bank in 10 years if we expect the interest rate to be 5 percent.

Exercise 2.20 Solve each of the following problems by applying Newton's method.

(a) $e^{-x} - x = 0$.
(b) $x^3 - x - 1 = 0$.
(c) $e^{-x^2} - \cos x = 0$.

Exercise 2.21 Consider the following vector-valued function:

$$\mathbf{F}(x_1, x_2) = \begin{bmatrix} f_1(x_1, x_2) \\ f_2(x_1, x_2) \\ f_3(x_1, x_2) \end{bmatrix} = \begin{bmatrix} x_1 - 0.4 \\ x_2 - 0.8 \\ x_1^2 - x_1^2 - 1 \end{bmatrix}.$$

(a) Apply the Gauss-Newton's method to find the zero contours of f_1, f_2, and f_3. That is, find the approximate values of (x_1, x_2) such that $\mathbf{F}(x_1, x_2) = [0, 0, 0]^T$.
(b) Sketch the zero contours on the (x_1, x_2) plane.

References

1. A.E. Dolbear, The cricket as a thermometer. Am. Nat. **31**, 970–971 (1897)
2. C.A. Bessey, E.A. Bessey, Further notes on thermometer crickets. Am. Nat. **32**, 263–264 (1898)
3. Max Kleiber, Body size and metabolic rate. Physiol. Rev. **24**, 511–541 (1947)
4. Geoffrey Taylor. The formation of a blast wave by a very intense explosion. I. Theoretical discussion. Proc. Roy. Soc. Lond. **201**(1065), 159–174 (1950)
5. Geoffrey Taylor. The formation of a blast wave by a very intense explosion. II. The atomic explosion of 1945. Proc. Roy. Soc. Lond. **201**(1065), 175–186 (1950)
6. A.HZ. Harrath, M. Charni, R. Sluys, F. Zghal, S. Tekaya. Ecology and distribution of the freshwater planarian schmidtea mediterranea in tunisia. Italian J. Zool. **71**(3), 233–236 (2004)

Chapter 3
Discrete Models

This chapter considers mathematical models of phenomena that change in discrete periods of time. Examples include various population models and economic interest models. These representative examples introduce basic concepts for analysis of discrete models, such as equilibria, periodic points, and their stability properties. Similar analytic tools are extended in subsequent chapters to study other types of models. These qualitative methods allow the study of long-term behavior in discrete models, showing that even very simple mathematical models can produce very complicated behavior.

Discrete dynamical models are appropriate when information is available at evenly spaced time intervals or dynamic events occur at discrete times, such as seasonal reproduction and death. Most students first encounter discrete models in high school, while working principle and interest problems; however, discrete models can exhibit complex behavior with very simply described dynamics (May [1]). We present a variety of models to illustrate how discrete dynamical models simulate real world problems and how to analyze them. For more detailed studies of the theory of discrete dynamical systems the reader should consider the texts by Aligood et al. [2] or Devaney [3].

3.1 Malthusian Growth Model

The simplest growth model is the *discrete Malthusian growth model*, which has a constant rate of growth, r, (representing births or deaths) and satisfies the equation:

$$\boxed{P_{n+1} = P_n + rP_n = (1+r)P_n,} \tag{3.1}$$

where P_0 represents the initial population.

A. Palacios, *Mathematical Modeling*, Mathematical Engineering,
https://doi.org/10.1007/978-3-031-04729-9_3

Table 3.1 U.S. Population (early years) in millions from Year 1790 to Year 1870. Source: Census Bureau

Year	Census	Model	% Error	Year	Census	Model	% Error
1790	3.929	3.929	–	1840	17.069	17.553	2.83
1800	5.308	5.300	−0.15	1850	23.192	23.679	2.10
1810	7.240	7.150	−1.24	1860	31.433	31.942	1.62
1820	9.638	9.645	0.08	1870	39.818	43.090	8.22
1830	12.866	13.012	1.13				

This model states that the population at the next time period, P_{n+1}, depends on the current population, P_n, plus the net per capita growth rate, r times the current population. Iterating the model we get:

$$\begin{aligned} P_1 &= (1+r)P_0, \\ P_2 &= (1+r)P_1 = (1+r)^2 P_0, \\ &\vdots \\ P_n &= (1+r)P_{n-1} = (1+r)^n P_0. \end{aligned}$$

In this particular case, we can find a closed-form solution:

$$P_n = (1+r)^n P_0, \tag{3.2}$$

which is valid at any time, n, any initial condition, P_0, and any growth (or decay) rate, r.

Example 3.1 As an example of a Malthusian growth model, we consider the population of the United States from Year 1790 to Year 1870. Table 3.1 shows the actual population growth drawn from the Census Bureau.

Applying the nonlinear least squares best fit, discussed earlier in Chap. 2, to the entire population data set collected from U.S. Census Bureau, we find the best fitting model to be:

$$P_n = (1.1460)^n 16.35,$$

which shows a growth rate of $r = 0.1460$ and initial population $P_0 = 16.35$ (in millions). Figure 3.1 shows a plot of the original data set and the nonlinear Malthusian model fitting.

In this case, the SSE criterion yields $SSE = 2876$, which is a relative large value. Nevertheless, the model is very simple as it contains only two parameters, the growth rate, r, and the initial population, P_0.

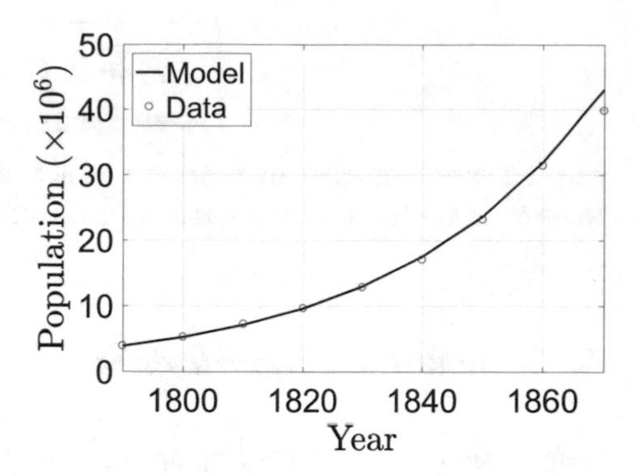

Fig. 3.1 Fitting of a Malthusian growth model to the U.S. population

3.2 Economic Interest Models

Most students first encounter discrete models in high school, while learning the basics of *principal and interest*. These models are closely related to the Malthusian growth population models, where capital in an account or loan balance replaces population. In this section we review the basic compound interest problem and simple amortization of a loan. We show that both problems can be described as discrete systems. The corresponding models become more complex when interest becomes variable or capital value is tied to inflation indices.

3.2.1 Compound Interest

The classic compound interest problem assumes an original deposit of P_0, the *original principal*, and that the bank offers an interest rate, r per year. *Compound interest* is the capital accrued from the interest paid on the original deposit over multiple years. Assuming interest is paid annually, the principal in the account satisfies the discrete model:

$$P_{n+1} = P_n + rP_n = (1+r)P_n,$$

where P_n is the principal in the nth year. This is the same model as the Malthusian growth model, (3.1), so it has the general solution:

$$P_n = (1+r)P_{n-1} = (1+r)^n P_0.$$

This model is easily modified to account for the bank compounding the interest more frequently, k times/year. For example, if the interest is compounded monthly, $k = 12$, then the annual rate r is replaced by $i = r/k = r/12$. In general, the compound interest model is written:

$$P_t = (1+i)^{kt} P_0, \tag{3.3}$$

where t is the number of years for which the principal is being calculated. This solution is known as the *compound interest formula.*

3.2.2 Loans and Amortization

People often need to borrow money to make larger purchases, such as a car or house. Financial institutions loan this money for a fixed interest rate over a fixed period of time. The borrower pays (usually monthly) a fixed amount that is determined by what is called an *amortization table*.

Let P_0 be the amount borrowed at an annual interest rate, r, for a term of N years. We want to compute the monthly payment, d, that the borrower must pay. The discrete model is based on the interest assessed on the *principal balance* during the month, P_n, minus the payment made by the borrower. Let $i = r/12$, then the equation describing the new principal balance is:

$$P_{n+1} = (1+i)P_n - d, \tag{3.4}$$

where the first term represents the previous balance with the monthly interest charged and d is the monthly payment deducted from the remaining loan amount.

Iterating Eq. (3.4) gives:

$$\begin{aligned}
P_1 &= (1+i)P_0 - d,\\
P_2 &= (1+i)P_1 - d \;=\; (1+i)^2 P_0 - (1+i)d - d,\\
P_3 &= (1+i)^3 P_0 - (1+i)^2 d - (1+i)d - d,\\
&\vdots\\
P_n &= (1+i)^n P_0 - (1+i)^{n-1} d - \cdots - (1+i)d - d\\
&= (1+i)^n P_0 - d\big[(1+i)^{n-1} + \cdots + (1+i) + 1\big].
\end{aligned}$$

The term in brackets above relates to the finite partial sum when studying *geometric series*. Using telescoping series, a closed form solution is available for the loan model (3.4):

$$P_n = (1+i)^n P_0 - d\,\frac{(1+i)^n - 1}{i}. \tag{3.5}$$

Since there are $12N$ equal payments and $P_{12N} = 0$, it follows from (3.5) that the monthly payments, d, are:

$$d = P_0 \frac{i(1+i)^{12N}}{(1+i)^{12N} - 1}. \tag{3.6}$$

As these formula are readily calculated and because they are so useful for households trying to design a budget, there are ubiquitous sources available for computing amortization tables. This information is critical for banks deciding how much they can extend in loans based on a client's income and credit information. MatLab has a financial function in its Financial Toolbox that computes relevant values with the function `amortize`. The web has many amortization calculators based on these formulae.

Example 3.2 Suppose a person wants to purchase a car, where the current interest rate is 3.6% for a 5 year loan. Find the monthly payment if this person must borrow $50,000 to purchase a luxury car and determine the total amount paid over the 5 years. If the person can only afford a maximum monthly payment of $600, then what is the maximum amount of capital that the person can borrow.

In the first scenario, $P_0 = \$50,000$, $i = 0.036/12 = 0.003$, and $N = 5$. From (3.6), it follows that d gives:

$$d = 50,000 \times \frac{0.003(1.003)^{60}}{(1.003)^{60} - 1} \approx \$911.83.$$

Thus, the monthly payment must be about $911.83. Since there are 60 payments made over the loan, then the cost of purchasing this car is $54,709.80.

With the limit on the monthly payment capped at $600, Eq. (3.6) is solved for P_0. It follows that

$$P_0 = d\,\frac{(1+i)^{12N} - 1}{i(1+i)^{12N}} \leq 600 \times \frac{(1.003)^{60} - 1}{0.003(1.003)^{60}} \approx 32,900.94.$$

Thus, the maximum amount that can be borrowed for the monthly payments up to $600 is $32,900.94.

The loan model (3.4) also represents a Malthusian growth model with emigration. However, the loan model has fixed interest and payments, while population models rarely have fixed growth rates or emigration rates, so (3.4) is an approximate model usually valid over short periods of time. Banks have used variable interest loans in the past, but they were not popular.

Exercise 3.2 examines a basic model for annuity plans, which is similar to the discrete loan model, except the employee (and possibly the employer) make fixed monthly contributions. Unlike standard loans, annuities generally use indices pegged to markets and/or bonds, so the interest and base capital is variable, which complicates the model. The extension of the annuity model to populations gives a Malthusian growth model with immigration. The inclusion of variability in rates of interest or growth in economic or population models complicates the analysis.

3.3 Time-Dependent Growth Rate

Close examination of the Census data shows that the growth rate of the U.S. population is not exactly a constant. Indeed, in 1800, 1900, and 2000, the growth rates declined, respectively, from 36.4%, 21.0%, down to 9.71%. The lowest growth rate recorded was 7.2% and it happened during the Great Depression. This means that a Malthusian growth model with constant rate should be considered only during short periods of time. One way to adjust for variations in the growth rate is to introduce time as a dependent variable in the model. This issue is discussed next.

3.3.1 General Population Model

In the formulation of the Malthusian model of U.S. growth, the population at time $n + 1$ is assumed to depend only on the current population at time n. This type of model can be generalized as:

$$P_{n+1} = f(P_n),$$

where $f(P)$ is known as the *updating function*. Then, given an initial population, P_0, we could iterate the function f to generate future values of the population. This generalization is a *discrete dynamical model* in the form of a first order difference equation and is *autonomous*, as it only depends on P. Notice that the iteration of the model yields a time-evolving population even though time does not appear explicitly in the model. Autonomous models are the most common models used with animal populations, as their populations most often depend on their existing populations.

In an alternative formulation, the population at time $n + 1$ might depend on both, the population at the current time n and on time itself. This type of model can be generalized as

$$P_{n+1} = f(t_n, P_n),$$

where the updating function f now has time as an explicit variable. This generalization is a *discrete dynamical model* in the form of a first order, *nonautonomous*, difference equation since time appears explicitly. The inclusion of time might complicate analysis but human populations have significant time-varying changes, which render this type of model preferable.

3.3.2 Nonautonomous Malthusian Growth Model

We have seen that over the course of time the growth rate of the U.S. population has actually declined. One way to introduce a declining growth rate is to rewrite the Malthusian growth model as a nonautonomous discrete model of the form:

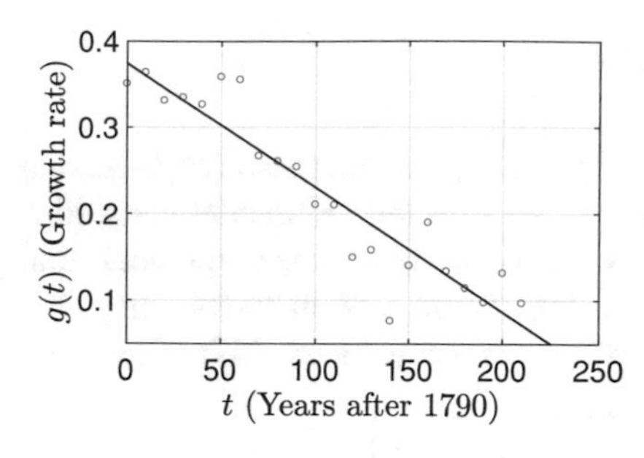

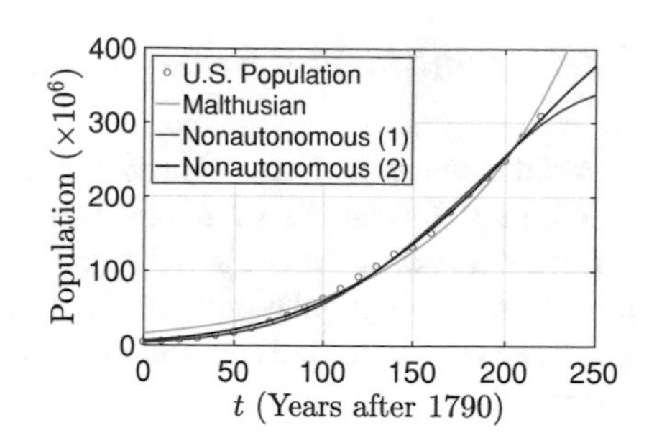

Fig. 3.2 (Left) Least squares best linear growth rate computed directly from the U.S. Census data. (Right) Census data with the best fitting Malthusian growth model and the nonautonomous Malthusian growth models

$$P_{n+1} = P_n + (a + bt_n)P_n, \tag{3.7}$$

in which the constant growth rate, r, is replaced by a time-dependent rate, $r(t) = a + bt_n$. A linear least squares fit through the previously recorded growth rates yields: $a = 0.3744$ and $b = -0.01439$, so that:

$$r(t_n) = 0.3744 - 0.01439n,$$

where n is measured in decades after 1790. Observe that b being negative implies a declining growth rate. Setting P_0 as a parameter, a nonlinear fit of the model yields $P_0 = 3.758$ with a $SSE = 740.1$, which is substantially better than $SSE = 2876$ found earlier in the nonlinear fit to the discrete Malthusian growth model. A graph of this best fitting linear growth rate is shown in Fig. 3.2(left) with a simulation of this model shown on the right.

Alternative, we can set three parameters: a, b, and P_0 and perform again the nonlinear fit of the model, which yields: $a = 0.2961$, $b = -0.009675$, and $P_0 = 6.305$, so that:

$$P_{n+1} = (1.2961 - 0.009675n)P_n.$$

This last model fit has a $SSE = 326.9$, which is substantially lower. See Appendix A.2.1 for MatLab programs to find this nonlinear best fit. Figure 3.2(right) shows that both nonautonomous Malthusian growth models are quite accurate. Furthermore, the SSE indicates that fitting the model with three parameters yields a smaller error, even though the approximation for earlier years is not as good. Future growths are, however, not as accurately fit by neither of the two nonautonomous Malthusian growth models. In fact, the simple Malthusian growth model with constant growth rate performs better for longer periods of time.

3.3.3 Logistic and Beverton-Holt Models

Malthusian growth models assume that populations can increase (or decrease) indefinitely. In real-life situations, other factors, such as environmental constrains can regulate the growth of populations. In this section we introduce two models that incorporate such regulations. The first model, *logistic growth*, includes a negative quadratic term that limits growth due to crowding or resource availability. It has the form:

$$P_{n+1} = P_n + r P_n \left(1 - \frac{P_n}{M}\right), \tag{3.8}$$

where M is a parameter also know as the *carrying capacity*, which represents the sustainability of growth due to crowding or resource availability. Once again, applying a nonlinear least squares fit to the U.S. Census data, we get

$$P_{n+1} = P_n + 0.2245 P_n \left(1 - \frac{P_n}{451.7}\right), \tag{3.9}$$

with initial population $P_0 = 8.575$. In this case, we get $SSE = 557.4$, which is slightly worse fit than the *nonautonomous Malthusian growth model.*

Another popular model that incorporates growth regulation is the *Beverton-Holt model.* This model is commonly found in ecological problems, and it has the form:

$$P_{n+1} = \frac{a P_n}{1 + \frac{P_n}{b}}. \tag{3.10}$$

An advantage of this model, over logistic growth, is that the updating function remains positive for any population, $P_n > 0$. Both models admit a closed-form solution. Applying a nonlinear least squares fit to the U.S. Census data, we get

$$P_{n+1} = \frac{1.2307 P_n}{1 + \frac{P_n}{2110.4}}, \tag{3.11}$$

with initial population $P_0 = 8.261$, which is very similar to that of the logistic model. In this case, we get $SSE = 519.5$, which is also very similar to the logistic model.

Figure 3.3 compares the best three-parameters fits to the U.S. Census data, including: Malthusian growth, logistic growth, and Beverton-Holt models. A closer look

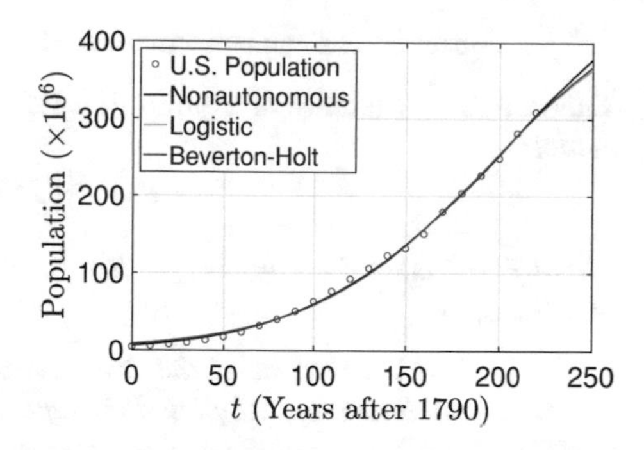

Fig. 3.3 U.S. population data and best fitting models that include: nonautonomous Malthusian, logistic, and Beverton-Holt models

at the graphs show that these fits are not as accurate for long-term predictions of population growth. Appendix A.2.1 shows the MatLab code for finding these best fitting and producing the graph in Fig. 3.3.

3.4 Qualitative Analysis of Discrete Models

In this section we study, *qualitatively*, the *long-term* solutions of an autonomous discrete model. By qualitative, we mean without having an exact solution of the model. Thus, consider the *general autonomous discrete model:*

$$p_{n+1} = f(p_n).$$

Some of the simplest types of long-term behavior of a discrete model are *equilibrium* points, in which the iterative process produces the same solution at each step. That is:

$$p_e = f(p_e).$$

Equilibrium points correspond to the limit as $n \to \infty$ of a closed-form solution for p_n, should such solution exist. In practice, such solution does not exist, yet we can still determine the long-term behavior by solving $p_e = f(p_e)$. Now, in general, $f(p)$ is nonlinear, so this equation must be solved algebraically or numerically. Geometrically, equilibrium solutions, p_e, are found when $f(p)$ crosses the *identity map*. Note: Any *closed population model*, meaning no migration into or out from another population source, must have a trivial equilibrium, $p_e = 0$.

After the equilibrium points have been calculated, the next step is to determine their local stability properties by computing the first derivative of $f(p)$ and evaluating it at each equilibrium point. The stability properties are then inferred by the sign and magnitude of the derivative. If the sign is positive, then nearby solutions monotonically move away while staying on the same side of the equilibrium point. If the sign is negative, then solutions have an *oscillatory behavior*, so they jump back and forth across the equilibrium point.

These observations can be formalized through the following theorem.

Theorem 3.1 (Stability of Equilibrium Points) *Consider the discrete dynamical model:*

$$p_{n+1} = f(p_n), \tag{3.12}$$

where $f(p_n)$ is a differentiable updating function. Suppose that p_e is an equilibrium of this model.

- *If $f'(p_e) > 1$, solutions of the discrete dynamical model grow away from the equilibrium (monotonically), and the equilibrium is unstable.*
- *If $0 < f'(p_e) < 1$, solutions of the discrete dynamical model approach the equilibrium (monotonically), and the equilibrium is stable.*
- *If $-1 < f'(p_e) < 0$, solutions of the discrete dynamical model oscillate about the equilibrium and approach it, and the equilibrium is stable.*
- *If $f'(p_e) < -1$, solutions of the discrete dynamical model oscillate about the equilibrium but move away from it, and the equilibrium is unstable.*

Proof Assume $p = p_e$ to be an equilibrium and let δ be a small, i.e., $|\delta_n| \ll 1$, perturbation around it. We wish to determine if the equilibrium point p_e is stable under the small perturbation δ_n, so let

$$p_n = p_e + \delta_n.$$

Substituting into Eq. (3.12) we get

$$p_e + \delta_{n+1} = f(p_e + \delta_n).$$

Expanding the right hand side of this last equation as a Taylor series about p_e we get

$$p_e + \delta_{n+1} = f(p_e) + f'(p_e)\delta_n + O(|\delta_n|^2).$$

Since $f(p_e) = p_e$, we obtain

$$\delta_{n+1} = f'(p_e)\delta_n.$$

This last equation is in the same form as that of Malthusian growth. An exact solution is given by

$$\delta_n = (f'(p_e))^n \delta_0.$$

It follows that if $|f'(p_e)| < 1$ then $\delta_n \to 0$ as $n \to \infty$. This implies that the perturbation dies out and so the equilibrium p_e is stable. On the other hand, if $|f'(p_e)| > 1$ then $\delta_n \to \infty$ as $n \to \infty$. In this latter case the perturbation grows up and the equilibrium point p_e is unstable. □

Example 3.3 (*U.S. Population Models*) The analysis of the U.S. population models begins with finding the equilibria. The best fitting *discrete logistic population* model for the U.S., Eq. (3.9), and *Beverton-Holt population* model for the U.S., Eq. (3.11), are *closed discrete autonomous population* models, so have one equilibrium at $P_e = 0$. They have another equilibrium, called the *carrying capacity equilibrium*, where the population of these models eventually approach with sufficient iterations.

The general *logistic population model* is:

$$P_{n+1} = P_n + r P_n \left(1 - \frac{P_n}{M}\right).$$

The equilibria are found algebraically by solving:

$$P_e = P_e + r P_e \left(1 - \frac{P_e}{M}\right) \qquad \text{or} \qquad r P_e \left(1 - \frac{P_e}{M}\right) = 0.$$

Solving for P_e yields two equilibrium points: the *extinction equilibrium*, $P_e = 0$ and the *carrying capacity equilibrium*, $P_e = M$.

The general *Beverton-Holt model* satisfies:

$$P_{n+1} = \frac{a P_n}{1 + \frac{P_n}{b}}.$$

The equilibria are found algebraically by solving:

$$P_e = \frac{a P_e}{1 + \frac{P_e}{b}} \qquad \text{or} \qquad P_e \left(\frac{P_e}{b} + 1 - a\right) = 0.$$

Solving for P_e yields two equilibrium points: the *extinction equilibrium*, $P_e = 0$ and the *carrying capacity equilibrium*, $P_e = b(a - 1)$.

From the equilibrium analysis above, Eq. (3.9) gives a *carrying capacity equilibrium*, $P_e = 451.7$. Thus, this model predicts that the U.S. population will level off at 451.7 million. Similarly, Eq. (3.11) gives a *carrying capacity equilibrium*, $P_e = 2110.5(0.23065) = 486.8$. Thus, this model predicts that the U.S. population will level off at 486.8 million, similar to the logistic population model.

Figure 3.4 shows the intersections of the *updating functions* for the logistic and Beverton-Holt population models with the identity map give the equilibria for the models.

For very low populations, the population has plenty of resources and should grow according to the Malthusian growth model, i.e., exponentially. It follows that often the *extinction equilibrium* has populations growing away from this equilibrium, so it is unstable. On the other hand, populations tend to approach the *stable carrying capacity equilibrium*.

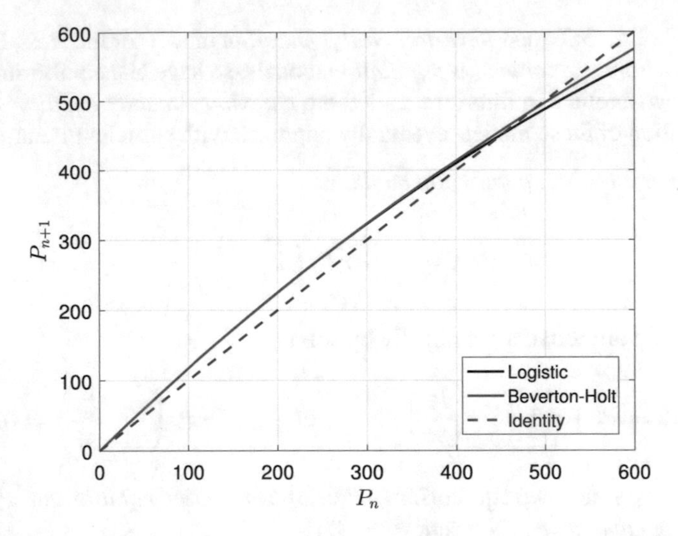

Fig. 3.4 A graph of the updating functions for the logistic and Beverton-Holt models compared to the identity map

Stability

The stability of the equilibria is determined by evaluating the first derivative of the updating function at the equilibria. The *updating function* of the best fitting logistic population model for the U.S. satisfies:

$$P_{n+1} = f(P_n) = P_n + 0.2245P_n \left(1 - \frac{P_n}{451.7}\right),$$

which has a derivative that satisfies:

$$f'(P) = 1.2245 - \frac{0.4490P}{451.7}.$$

At the *extinction equilibrium*, $P_e = 0$, we have

$$f'(0) = 1.2245 > 1,$$

which shows this equilibrium is unstable. At the *carrying capacity equilibrium*, $P_e = 451.7$, we have

$$f'(451.7) = 0.7755 < 1,$$

which shows this equilibrium is stable.

The *updating function* of the best fitting *Beverton-Holt model* for the U.S. satisfies:

$$P_{n+1} = b(P_n) = \frac{1.23065 P_n}{1 + \dfrac{P_n}{2110.5}},$$

which has a derivative that satisfies:

$$b'(P) = \frac{1.23065}{\left(1 + \dfrac{P}{2110.5}\right)^2}.$$

At the *extinction equilibrium*, $P_e = 0$,

$$b'(0) = 1.23065 > 1,$$

which shows this equilibrium is unstable. At the *carrying capacity equilibrium*, $P_e = 486.8$ million, we have

$$b'(486.8) = 0.81257 < 1,$$

which shows this equilibrium is stable.

3.5 Ricker's Model of Salmon Population

A drawback of the logistic growth model is the fact that the updating function yields negative values for large populations. To circumvent this problem, Ricker [4] introduced a new discrete model of the form:

$$\boxed{P_{n+1} = R(P_n) = a P_n e^{-bP_n},} \qquad (3.13)$$

where a and b are positive parameters. The updating function, R, in Eq. (3.13) yields values that are similar to those of the logistic for low population sizes, but it remains positive for large populations.

3.5.1 Salmon Population in the Skeena River

Salmon follow a life cycle with meticulous precision and timing. The cycle begins in freshwater where eggs are hatched and alevins emerge. These are tiny fish with the yolk sac of the egg attached to their bellies. Once they have consumed all of the yolk sac, the alevins emerge as fry. Eventually, fry migrate downstream towards the oceans. Sockeye fry, for instance, tend to migrate to a lake, spending 1–2 years before

Fig. 3.5 Sockeye salmon in their breeding colors

Table 3.2 Four year averages of Skeena River sockeye salmon (population in thousands)

Year	Population	Year	Population	Year	Population
1908	1,098	1924	706	1940	528
1912	740	1928	510	1944	639
1916	714	1932	278	1948	523
1920	615	1936	448		

migrating to sea, where they may spend 1–7 years. Then they return to freshwater and prepare for spawning, see Fig. 3.5. Females build nests, or redds. Eventually, both the males and females die.

The periodic life cycle of breeding and dying of salmon can be described as a discrete process. We will use the *Skeena River* population data from Table 3.2 to find the best logistic growth and Ricker's updating functions.

An alternate method of finding the best fitting population model is to organize the population data into P_n and P_{n+1}, then apply a simple curve fitting algorithm to find the updating function. The following MATLAB script automates this process.

```
1   function salmon
2
3   close all;
4   clear all;
5   clc
6
7   load 'pndata.data';
8   load 'pn1data.data';
9
10  x = linspace(0,1200,50);
11
12  options = optimset;
13  LogGrow = inline('p(1)*x+p(2)*x.^2','p','x');
14  RickGrow = inline('q(1).*x.*exp(q(2)*x)','q','x');
15
16  a = 1.0;
17  b = 0.5;
18  param0 = [a,b];
19  param  = lsqcurvefit(LogGrow,param0,pndata,pn1data,[],[],options);
20
```

```
21  c = 1.0;
22  d = 0.0;
23  qparam0 = [c,d];
24  qparam  = ...
        lsqcurvefit(RickGrow,qparam0,pndata,pn1data,[],[],options)
25
26  figure(1);
27  plot(pndata,pn1data,'bo','MarkerSize',12);
28  hold on;
29  plot(x,LogGrow(param,x),'k-','LineWidth',3);
30  hold on;
31  plot(x,RickGrow(qparam,x),'r-','LineWidth',3);
32  grid on;
33  axis([0 1200 0 1000]);
34
35  % Set up fonts and labels for the Graph
36  legend('Salmon Population','Logistic','Ricker',...
37      'location','northwest');
38  fontlabs = 'Times New Roman';
39  xlabel('$P_n$','FontSize',16,'FontName',fontlabs, ...
40      'interpreter','latex');
41  ylabel('$P_{n+1}$','FontSize',16,'FontName',fontlabs, ...
42      'interpreter','latex');
43  set(gca,'FontSize',36);
```

The nonlinear least squares method readily finds the best updating functions for the logistic growth model to be:

$$P_{n+1} = 1.3277\, P_n - 0.0006146\, P_n^2, \tag{3.14}$$

and Ricker's models is:

$$P_{n+1} = 1.5344\, P_n e^{-0.0007816\, P_n}. \tag{3.15}$$

The sum of the square errors for the Logistic and Ricker's models are: $SSE_{log} = 128,980$ and $SSE_{Ricker} = 124,519$, respectively. The results of the fitting methods are shown in Fig. 3.6. The figure shows that both fitting methods yield very similar results for population sizes of up to a million. In addition, the SSE values show that Ricker's updating function fits the data only 3.6% better than the logistic updating function.

Below is the MatLab function for minimizing the *sum of square errors* for the *Ricker's updating function.* A similar process is followed to best fit the *logistic updating function.*

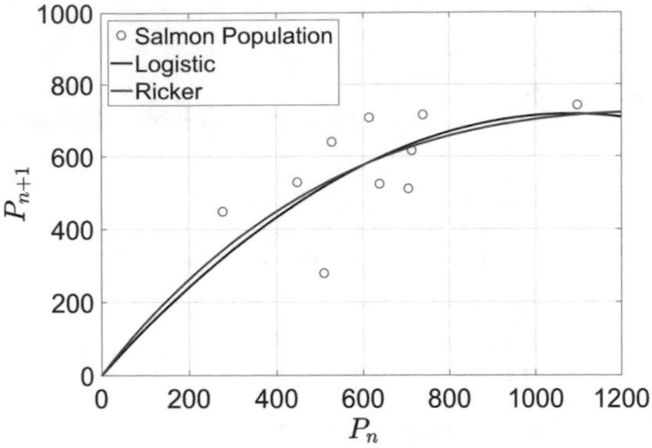

Fig. 3.6 Population data of sockeye salmon from the Skeena river and the best fitting logistic growth and Ricker's updating functions

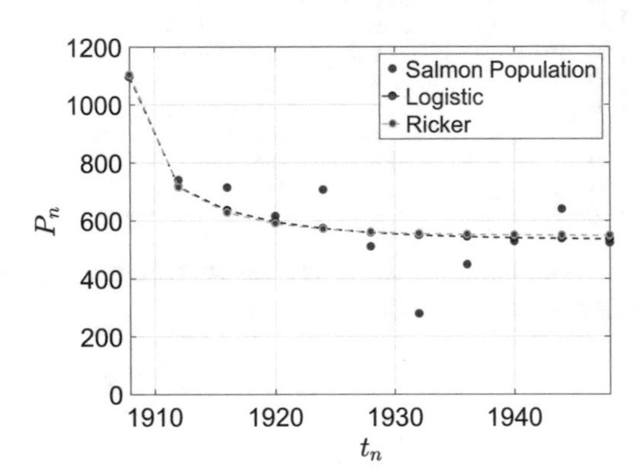

Fig. 3.7 Population data of sockeye salmon from the Skeena river and simulations of the time evolution of the populations using two models: logistic growth and Ricker's model

```
1  function J = sal_ric(p0,pndata,pn1data)
2  % Least Squares fit to Ricker's model
3  N = length(pndata);
4  err = [pn1data - p0(1)*pndata.*exp(-p0(2)*pndata)];
5  J = err*err'; % Sum of square errors
6  end
```

The data for P_{n+1} vs. P_n is entered with an initial parameters `p0 = [a,b]`:
`p1 = fminsearch(@sal_ric,p0,[],pndata,pn1data)`

We can now employ the best fitting updating functions for the *logistic* and *Ricker's models* to simulate the time evolution of the salmon populations. In addition, the initial population is varied to find the least sum of square errors when compared to the time series data. Figure 3.7 shows the best fitting simulations. The best fitting initial values are $P_0 = 1096.8$ for *logistic growth* and $P_0 = 1103.7$ for *Ricker's model*, which are both quite close to the actual population data. The least sum of square errors for these fits are $SSE_{log} = 120,918$ and $SSE_{Ricker} = 126,428$, indicating that the time series simulation with the logistic growth model is about 4.6% better than the Ricker's model.

3.5.2 Analysis of the Ricker's Model

In this section we analyze the general Ricker's Model:

$$P_{n+1} = R(P_n) = aP_n e^{-bP_n}.$$

Existence of Equilibria
The equilibria are found by solving $R(P_e) = P_e$ for P_e. That is, solving

$$P_e = aP_e e^{-bP_e} \quad \text{or} \quad P_e(ae^{-bP_e} - 1) = 0.$$

It follows that the equilibria are

$$P_{e_1} = 0 \quad \text{and} \quad P_{e_2} = \frac{\ln(a)}{b},$$

with $a > 1$ required for a positive equilibrium. The first equilibrium is the *extinction equilibrium*, and the second one is the *carrying capacity*.

The stability of the equilibria are found by evaluating the derivative of the updating function at the equilibria. The derivative of the Ricker updating function is given by:

$$R'(P) = a(P(-be^{-bP}) + e^{-bP}) = ae^{-bP}(1 - bP).$$

At the *extinction equilibrium*, $P_{e_1} = 0$,

$$R'(0) = a.$$

If $0 < a < 1$, then $P_{e_1} = 0$ is stable and the population goes to *extinction* (and there is no positive equilibrium). If $a > 1$, then $P_{e_1} = 0$ is *unstable* and the population grows away from the equilibrium.

At the *carrying capacity equilibrium* (assuming $a > 1$), $P_{e_2} = \frac{\ln(a)}{b}$ and

$$R'(\ln(a)/b) = ae^{-\ln(a)}(1 - \ln(a)) = 1 - \ln(a).$$

Once again the stability of this equilibrium only depends on the value of the parameter a. There are three possible behaviors near this equilibrium. If $1 < a < e \approx 2.7183$, then the solution of Ricker's model is *stable* and *monotonically approaches* the equilibrium $P_{e_2} = \ln(a)/b$. If $e < a < e^2 \approx 7.389$, then the solution of Ricker's model is *stable* and *oscillates as it approaches* the equilibrium $P_{e_2} = \ln(a)/b$. Finally, if $a > e^2 \approx 7.389$, then the solution of Ricker's model is *unstable* and *oscillates as it grows away* from the equilibrium $P_{e_2} = \ln(a)/b$.

This analysis is applied to the best *Ricker's model* for the Skeena sockeye salmon population from 1908 to 1952, Eq. (3.15). From the analysis above, the equilibria are

$$P_{e_1} = 0 \quad \text{and} \quad P_{e_2} = \frac{\ln(1.535)}{0.000783} = 547.3.$$

The derivative is

$$R'(P) = 1.535e^{-0.000783P}(1 - 0.000783P).$$

It follows that at the extinction equilibrium, $P_{e_1} = 0$, $R'(0) = 1.535 > 1$, so this equilibrium is *unstable* (as expected). At the carrying capacity equilibrium, $P_{e_2} = 547.3$, the derivative, $R'(547.3) = 0.571 < 1$, so this equilibrium is *stable* with solutions monotonically approaching the equilibrium, as observed in the simulation.

A similar analysis is performed for the *logistic model*, given by Eq. (3.14). For this model, the equilibria are

$$P_{e_1} = 0 \quad \text{and} \quad P_{e_2} = 533.2.$$

The derivative is

$$F'(P) = 1.3277 - 0.001229\,P.$$

It follows that at the extinction equilibrium, $P_{e_1} = 0$, $F'(0) = 1.3277 > 1$, so this equilibrium is *unstable* (as expected). The carrying capacity from this model is $P_{e_2} = 533.2$, which has $F'(533.2) = 0.6716 < 1$. Hence, this equilibrium is *stable* with solutions monotonically approaching the equilibrium, as observed in the simulation.

Both the *Ricker's* and *logistic models* provide very similar updating functions passing through the Skeena River salmon data. From these updating functions with the best fitting P_0, the discrete dynamical model simulations give very similar solutions. The carrying capacity equilibria are separated by only a few percent with both showing the same monotonic stability. Yet the large P_n behavior of these models from their updating functions is dramatically different.

3.6 Heat Exchange

Consider the following puzzle. A glass of milk is held at 0°C temperature and an identical glass of water is held at 100°C temperature. We would like to heat the glass milk to a temperature >50°C using only the heat from the glass of water. The heat exchange would leave the glass of water at a temperature < 50°C.

Assume there is no heat exchange with the environment or with any other external source. One extra glass is available and heat capacities per unit volume of the water and the milk are assumed to be the same. The heat exchanged can be accomplished through the following steps:

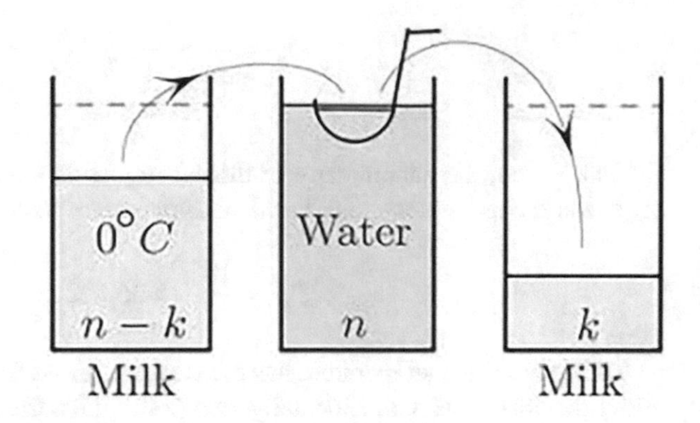

Fig. 3.8 Heat exchange problem. A glass of milk is colder than the water at the beginning of the heat exchange process and hotter than the water at then end. *Source* Mark Levi [5]

Step 1 Scoop $1/nth$ of the milk into a ladle.
Step 2 Dip the ladle in the hot water until the temperatures equalize.
Step 3 Dump the warmed milk into the extra glass.
Step 4 Repeat steps 1–4 n times.

The heat transfer process is illustrated in Fig. 3.8.

Eventually, after n iterations, all of the milk ends up in the extra glass. Dipping the 0 °C milk ladle in the warm water reduces the temperature by a constant factor because the heat of n units of water spreads equally among the $n + 1$ units of liquid. Let T_k be the temperature of the water after k repetitions. Assuming $T_0 = 100\,°\text{C}$ to be the initial temperature of the water, then a discrete model for the cooling process of water can be written as

$$\boxed{T_{k+1} = \frac{n}{n+1} T_k.} \tag{3.16}$$

Equation (3.16) can also be rewritten as

$$T_{k+1} = \frac{1}{1 + \frac{1}{n}} T_k.$$

Observe that this equation is in the form of Malthusian growth, with growth rate

$$r = \frac{1}{1 + \frac{1}{n}}.$$

Thus, a closed form solution is given by

$$T_n = \frac{1}{\left(1 + \frac{1}{n}\right)^n} T_0.$$

As $n \to \infty$, the denominator of this last equation approaches e. Also, all of the milk is in the extra glass, and the water temperature reaches the equilibrium state

$$T_\infty \approx \frac{100}{e} \approx 36.8°\text{C}.$$

Thus, the milk's temperature has increased to about 63°C, which is considerable above the 50°C goal. Coincidentally, this is also the perfect temperature for cooking salmon, while 36.8°C is the temperature of the human body. A mathematical curiosity from this derivation is that we can write e as follows

$$e \approx 1 + \frac{T_{\text{salmon}}}{T_{\text{body}}}.$$

3.7 Newton's Method

Suppose we need to find a solution to the equation $g(x) = 0$. Then Newton's method says we should consider the map

$$x_{n+1} = x_n - \frac{f(x_n)}{f'(x_n)}. \tag{3.17}$$

Observe that Eq. (3.17) is the single-variable version of Gauss-Newton's method discussed in the previous chapter.

To calibrate the method, let's write down the Newton map for solving the following equation $g(x) = x^2 - 4$ for its roots. Direct substitution into Eq. (3.17) leads to

$$x_{n+1} = x_n - \frac{x_n^2 - 4}{2x_n}. \tag{3.18}$$

The equilibrium points of this map are the solutions to

$$x - \frac{x^2 - 4}{2x} = x,$$

which yields $x_{e1} = 2$ and $x_{e1} = -2$. We now proceed to determine the stability of each of these equilibrium points by computing the derivative of

Table 3.3 Newton's method applied to find the roots of $g(x) = x^2 - 4$

n	x_n
0	1.0
1	2.5
2	2.05
3	2.0006
4	2.0

$$f(x) = x - \frac{x^2 - 4}{2x},$$

which satisfies

$$f'(x) = 1 - \frac{x^2 + 4}{2x^2}.$$

At the first equilibrium point $x_{e1} = 2$ we get

$$f'(2) = 0,$$

which shows that $x_{e1} = 2$ is stable. A similar calculation shows that $x_{e2} = -2$ is also stable. This means that Newton's method can converge to both roots of the equation $g(x) = x^2 - 4$. Which one is actually found depends on the initial conditions. For instance, iterating Eq. (3.18) starting from $x_0 = 1$ shows, see Table 3.3 rapid convergence towards $x_{e1} = 2$.

3.8 Periodic Points and Bifurcations

It is also possible for discrete models to exhibit long-term behavior in the form of periodic points. Consider the following example of a population dynamics.

Example 3.4 The population P_n, at time t_n, of a certain species is governed by the following logistic model.

$$P_{n+1} = r P_n \left(1 - \frac{P_n}{C}\right),$$

where C represents the carrying capacity of the environment. When $r = 3.4$ and $C = 5000$, computer iterations with initial conditions $P_0 = 102$ produce the time series graph shown in Fig. 3.9. The first few points represent transient behavior but, eventually, the iterations settle into a period-two cycle at {42108, 2, 2598}.

To determine the stability properties of the cycle, we need to formalize first the concept of periodicity.

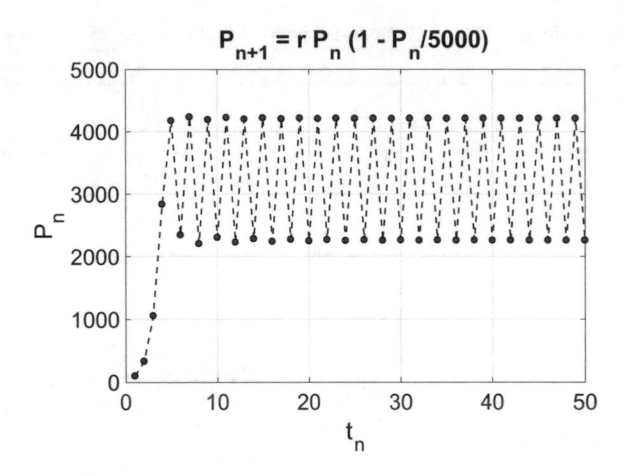

Fig. 3.9 Period-2 cycle in a discrete model of population dynamics

Definition 3.1 Consider a general discrete model of the form

$$x_{n+1} = f(x_n, \lambda), \tag{3.19}$$

where λ is a parameter. The orbit of a point x_1 is

$$O(x_1) = \{x_1, x_2, x_3, \ldots\},$$

where $x_2 = f(x_1)$, $x_3 = f(x_2) = f^2(x_1)$, and so on. A point x_1 is a periodic point, with period k, if there is an integer k such that the orbit satisfies $O(x_1) = \{x_1, x_2, \ldots, x_k\}$, where $x_2 = f(x_1), x_3 = f(x_2) = f^2(x_1), \ldots, x_k = f^{k-1}(x_1), x_{k+1} = f^k(x_1) = x_1$, and k is the smallest such integer.

The above definition implies that k is the minimum number of iterates that is required for the orbit to repeat a point. A related concept is that of *eventually periodic points*.

Definition 3.2 A point x_1 is an eventually periodic point, with period k, if there is an integer N such that

$$f^{n+k}(x_1) = f^n(x_1), \quad \forall n \geq N,$$

and k is the smallest such integer.

Example 3.5 Consider the tent map:

$$x_{n+1} = T(x_n) = \begin{cases} 2x_n, & x_n \leq 1/2 \\ 2(1 - x_n), & 1/2 \leq x_n \end{cases}$$

Let $x_1 = \frac{1}{7}$. The orbit of this initial point under the Tent map is

$$O(\frac{1}{7}) = \left\{\frac{1}{7}, \frac{2}{7}, \frac{4}{7}, \frac{6}{7}, \frac{2}{7}, \frac{4}{7}, \frac{6}{7}, \dots\right\}$$

Consequently, $x_1 = \frac{1}{7}$ is an eventually periodic point.

3.8.1 Chain Rule and Stability

The stability properties of periodic points can also be determined with a theorem analogous to that of equilibrium points, i.e., Theorem 3.1, applied to $f^k(x)$ as follows.

Theorem 3.2 (Stability of Periodic Points) *Consider the discrete dynamical model (3.19). Let x_1 be a period-k point. Then*

- *If $|(f^k)'(x_1)| < 1$, solutions of the discrete dynamical model approach the periodic orbit (monotonically), and the periodic orbit is stable.*
- *If $|(f^k)'(x_1)| > 1$, solutions of the discrete dynamical model grow away from the periodic orbit (monotonically), and the periodic orbit is unstable.*

We could now proceed to compute the stability of the period-2 cycle using Theorem 3.2. This would require, however, that we compute $f^2(x)$ and its derivate explicitly, which can be a little cumbersome due to algebraic difficulties. For higher order cycles, with arbitrary period k, we would need to compute $f^k(x)$ directly, which can be a more daunting task. An alternative approach, which circumvents algebraic difficulties, is to use the chain rule for computing derivatives of composition of functions:

$$(f \circ g)'(x) = f'(g(x))g'(x).$$

Applying this rule to the kth iterate of a map, f^k, we can compute its derivative through a simpler form, as follows

$$\begin{aligned}(f^k)'(x_1) &= (f(f^{k-1}))'(x_1) = f'(f^{k-1}(x_1))(f^{k-1})'(x_1)\\ &= f'(f^{k-1}(x_1))f'(f^{k-2}(x_1)) \cdots f'(x_1).\end{aligned}$$

Substituting $x_k = f^{k-1}(x_1)$, $x_{k-1} = f^{k-2}(x_1)$, and so on, we get a simpler form for the desired derivative of the kth iterate of the map

$$(f^k)'(x_1) = f'(x_k)f'(x_{k-1}) \cdots f'(x_1).$$

We now illustrate the use of this formula.

Example 3.6 Consider again the population model of the previous example

$$P_{n+1} = rP_n\left(1 - \frac{P_n}{C}\right).$$

Recall that a period-2 cycle $\{x_1, x_2\} = \{4, 2108, 2, 2598\}$ was found at $r = 3.4$ and $C = 5000$. To study its stability we compute the derivative of

$$f(x) = rx\left(1 - \frac{x}{C}\right),$$

and get

$$f'(x) = r - 2r\frac{x}{C}.$$

Applying Theorem 3.2, we determine the stability of the period-2 cycle by computing

$$|(f^2)'(x_1)| = |f'(x_1) \cdot f'(x_2)| = |f'(4, 2108) \cdot f'(2, 2598)| = 0.76.$$

Consequently, we conclude that the period-2 cycle $\{4, 2108, 2, 2598\}$ is locally stable.

The MATLAB code used to generate the time-series diagram of Fig. 3.9 is shown below.

```
1  clear all
2  clc
3
4  %Discrete Model: P(n+1)=r*P(n)(1-P(n)/5000)
5  r=3.4;          %parameter r
6  N=200;          %number of iterations
7  P(1)=100;       %initial value
8
9  for k=1:N-1
10     P(k+1)=r*P(k)*(1-P(k)/5000);
11 end
12 t=1:N;
13 figure(1)
14 plot(t(1:50),P(1:50),'k--o','LineWidth',2,...
15     'MarkerSize',8,'markerfacecolor','r')
16 title('{P_{n+1} = r P_n (1 - P_n/5000)}')
17 xlabel('{t_n}')
18 ylabel('{P_n}')
19 set(gca,'FontSize',30);
20 grid on;
```

3.8.2 *Period Doubling*

The birth of the period-2 corresponds to what is known as a *period-doubling* bifurcation. To investigate this concept further, we can simplify first the number of parameters in the logistic model through the substitution $P_n = Cx_n$. This leads to a one-parameter model of the form

$$x_{n+1} = rx_n(1 - x_n). \tag{3.20}$$

We know that equilibrium points correspond to solutions of

$$rx(1 - x) = x,$$

which are $x_{e1} = 0$ and $x_{e2} = 1 - 1/r$. Observe that the branch of nontrivial equilibrium points x_{e2} yields negative values when $0 < r < 1$. From a population dynamics standpoint, these values make no sense. But from a mathematical standpoint, these points can still be considered, for completeness purposes, as valid. Stability properties are determined by Theorem 3.1, i.e., through the derivative of $f(x) = rx(1 - x)$, which satisfies

$$f'(x) = r - 2rx.$$

At the extinction point $x_{e1} = 0$ we get

$$f'(0) = r.$$

At the nontrivial equilibrium $x_{e2} = 1 - 1/r$ we obtain

$$f'(1 - 1/r) = 2 - r.$$

It follows that on the interval $0 \leq r < 1$ the extinction point $x_{e1} = 0$ is stable while the nontrivial equilibrium point $x_{e2} = 1 - 1/r$ is unstable. On the interval $1 < r < 3$, the stability properties of these two equilibrium points changes, however. Thus, $x_{e1} = 0$ becomes unstable and $x_{e2} = 1 - 1/r$ stable. Furthermore, it can be shown that all initial conditions x_0 on the interval $0 < x_0 < 1$ lie in the *basin of attraction* of the nontrivial equilibrium $x_{e2} = 1 - 1/r$, so there can be no periodic cycles of prime period > 1 for $1 < r < 3$.

Let's examine in more detail how the exchange of stability between x_{e1} and x_{e2} occurs. At $r = 1$, the branch of nontrivial equilibrium points $x_{e2} = 1 - 1/r$ meets the extinction equilibrium x_{e1}, while the derivative of $f(x)$ satisfies

$$f'(r = 1, x = 0) = 1.$$

This implies that the stability Theorem 3.1 cannot be applied. It cannot be applied because when the derivative of a general discrete model (3.19) satisfies $|f'(r_c, x_e)| = 1$, a bifurcation occurs. In this case, the type of bifurcation, in which two branches of equilibrium points exchange stability, is known as a transcritical bifurcation. In chapter 5 we study the analytical conditions that allow us to identify and classify a transcritical bifurcation, as well as other types. Returning to the logistic model (3.20), the transcritical bifurcation occurs at the point $(r_c, x_e) = (1, 0)$.

Now, let's consider the period-2 cycle, which corresponds to solutions of $f^2(x) = f(f(x)) = x$, mainly

$$r\,[rx(1-x)(1-rx(1-x))] = x.$$

Expanding and factoring out the equilibrium points we get

$$-x\left[x-\left(1-\frac{1}{r}\right)\right]\left[-r^3x^2+\left(r^3+r^2\right)x-\left(r^2+r\right)\right]=0$$

The period-2 cycle is found by solving the quadratic part of this last equation, which yields

$$x_{1,2}=\frac{1}{2}\left[\left(1+\frac{1}{r}\right)\pm\frac{1}{r}\sqrt{(r-3)(r+1)}\right].$$

Observe that $x_{1,2}$ exist (i.e., are real-valued) only when $r \geq 3$. At $r = 3$ we have

$$f'(1-1/r) = -1.$$

This is the birth point of the period-2 cycle that corresponds to a *period-doubling* bifurcation. Furthermore,

$$(f^2)'(1-\frac{1}{r}) = f'(1-1/r)f'(1-1/r) = +1.$$

Past the critical point $r > r_c$, where $r_c = 3$, the derivative at the nontrivial equilibrium $x_{e2} = 1 - 1/r$ satisfies $|f'(1-1/r)| > 1$, so now both equilibrium points are unstable. The slope of the graph of $f^2(x)$ at the nontrivial equilibrium point becomes greater than one and this graph intersects the line $y = x$ at two new points either side of the nontrivial equilibrium point. Since the new equilibrium points of $f^2(x)$ are not equilibrium points of $f(x)$, they form a new period-2 cycle. This mechanism, i.e., where $(f^2)(x_e) = +1$, corresponds to a pitchfork bifurcation for the updating function $f^2(x)$. Details of analytical conditions for identifying and classifying this type of bifurcation can be found in Chap. 5. It can be shown that the slope of $f^2(x)$ is less than one at each of the new period-2 points. Thus, the period-2 cycle is stable. To find the exact interval of stability of the period-2 cycle we must solve

$$|f'(1-1/r)f'(1-1/r)| < 1.$$

Direct substitution and simplification yields

$$|r^2 - 2r - 4| < 1.$$

There are two cases to consider. We start with

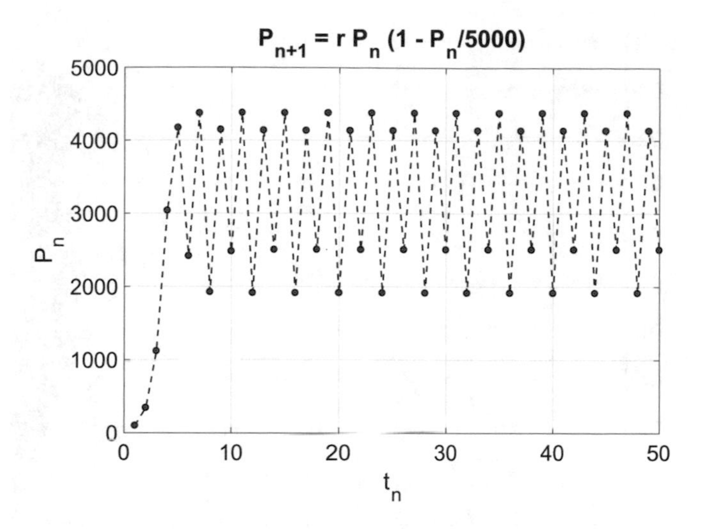

Fig. 3.10 Period-4 cycle in a discrete model of population dynamics

$$-1 < r^2 - 2r - 4, \qquad \text{or} \qquad 0 < (r + 1)(r - 3).$$

Since $r > 0$, the only possible solution is for $3 < r$. The second case is

$$r^2 - 2r - 4 < 1, \qquad \text{or} \qquad r^2 - 2r - 5 < 0,$$

whose solution is $1 - \sqrt{6} < r < 1 + \sqrt{6}$.

Combining these two cases, we find the interval of stability for the period-2 cycle to be

$$3 < r < 1 + \sqrt{6}.$$

As r increases further, beyond the threshold value of $r = 1 + \sqrt{6}$, the period-2 cycle changes from stable to unstable, and a new period-4 cycle is born. For instance, Fig. 3.10 illustrates a period-4 cycle, $\{x_1, x_2, x_3, x_4\} = \{1, 9141, 4, 1347, 2, 5044, 4, 3750\}$, found at $r = 3.5$.

Similar calculations show that the period-4 cycle is stable on the interval $r_1 < r \leq r_2$, where $r_1 = 1 + \sqrt{6}$ and $r_2 = 1 + \sqrt{3 + r_1}$. This process repeats itself into what is known as a *period-doubling cascade*. In fact, it can be shown that a period-k cycle is stable on the interval $r_{k_1} < r \leq r_k$, where $r_k = 1 + \sqrt{3 + r_{k-1}}$. The limit

$$\lim_{k \to \infty} r_k = r_\infty = 3.61547$$

is known as the Feigenbaum number. The intervals of stability of the period-k cycles show that bifurcations occur faster and faster as r increases, so that convergence towards r_∞ is geometric. Furthermore, the distance between successive transitions shrinks by a constant factor

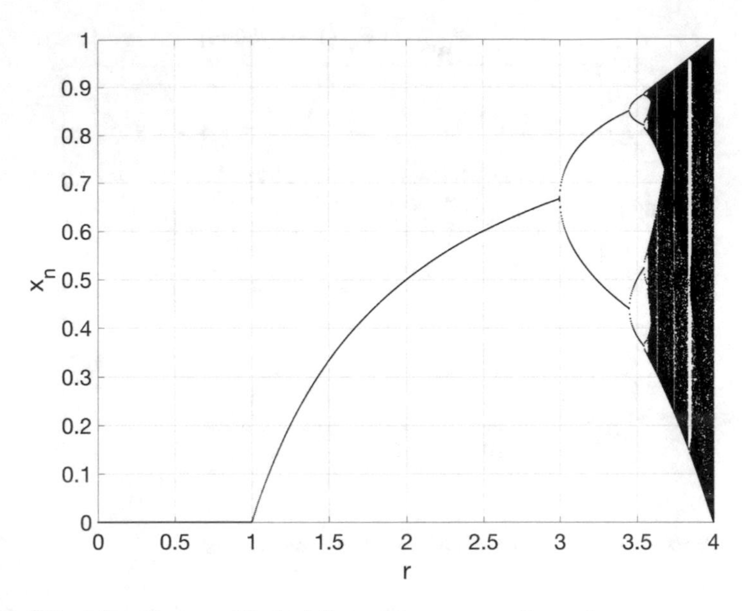

Fig. 3.11 Bifurcation diagram of the logistic model $x_{n+1} = rx_n(1 - x_n)$

$$\delta = \lim_{k\to\infty} \frac{r_k - r_{k-1}}{r_{k+1} - r_k} = 4.669,$$

which is a universal constant known as Feigenbaum constant.

It can also be shown that while the period-2 cycle remains stable, every point of the interval $0 < x < 1$, except for the unstable nontrivial equilibrium point and its pre-images, is in the basin of attraction of the period-2 cycle. Consequently, there are no other periodic cycles except for the period-2 cycle and the two equilibrium points $x_{e1} = 0$ and $x_{e2} = 1 - 1/r$. Similarly, as the period-doubling cascade evolves into period-k cycles, every point of the interval $0 < x < 1$, except for equilibria, cycles of lower order and their pre-images, is in the basin of attraction of the period-k orbit.

Figure 3.11 is a bifurcation diagram for the logistic model and it serves to visualize the sequence of changes in solution types as the bifurcation parameter r varies.

Figure 3.12 shows a zoom-in region of the previous bifurcation diagram, in which we can observe in more detail the period-doubling cascade and the complexity that follows it.

Notice that the period-doubling cascade described above started at an equilibrium point and the cycles included only points of even period. But it is also possible to have cycles of odd period. For instance, direct calculations while solving $f^3(x) = x$ yield a polynomial

$$(r^6 + 5r^5 + 3r^4 + r^3)x^3 + (3r^6 + 4r^5 + r^4)x^4 - (3r^6 - r^5)x^5 + r^6x^6 = 0.$$

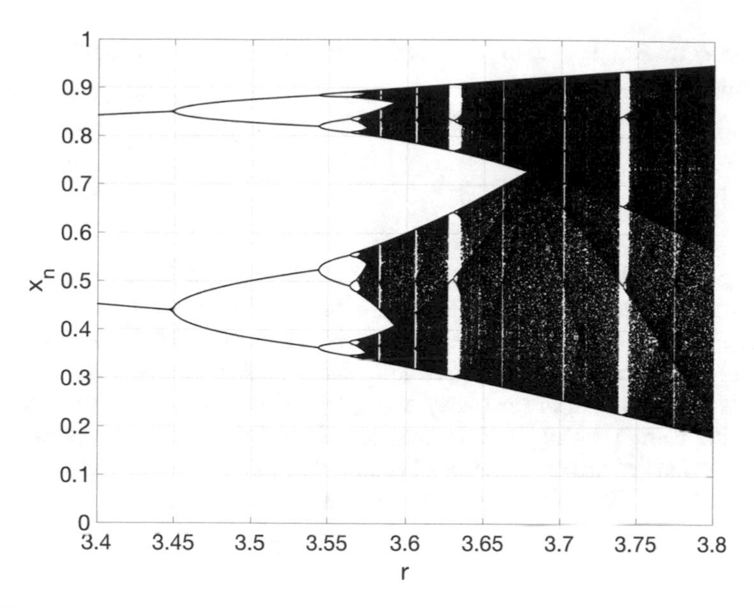

Fig. 3.12 Zoom in regions of the bifurcation diagram of the logistic model $x_{n+1} = rx_n(1 - x_n)$

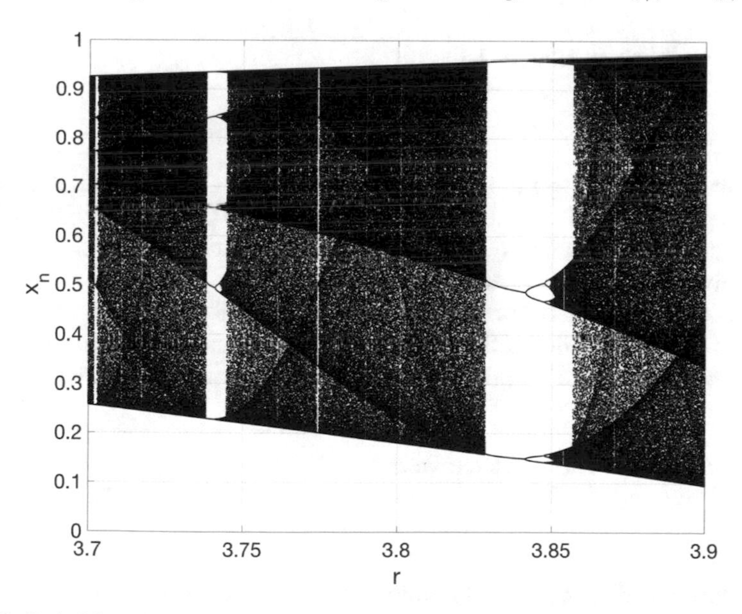

Fig. 3.13 Period-3 cycle in the logistic model $x_{n+1} = rx_n(1 - x_n)$ occurs near $r = 3.8284$

Solving this polynomial one can show that a period-3 cycle starts at $r = 1 + 2\sqrt{2} = 3.828427$. Figure 3.13 shows a small window of parameter space where the period-3 cycle can be observed.

The MATLAB code used to generate the bifurcation diagram of Fig. 3.11 is shown in the Appendix.

3.9 Chaos

The long-term behavior of mathematical models that we have discussed so far includes, mainly, equilibrium points and periodic cycles or orbits. It is also possible for very simple mathematical models to display much more complicated behavior [6] known as *chaos*. Loosely speaking, *chaotic behavior* is aperiodic behavior in which the orbits do not repeat. Instead, they linger around points of the phase space without specific pattern. Nearby points may be able to follow a chaotic orbit but only for brief periods of time or iterations because they exhibit *sensitive dependence* on initial conditions. We elaborate on this point next.

3.9.1 Sensitive Dependence

Consider again the logistic model

$$x_{n+1} = rx_n(1 - x_n).$$

Figure 3.14 depicts two orbits of the logistic model. One with initial population $P_0 = 0.1$ and one with initial condition $Q_0 = 0.10001$. At the beginning of the iterative process, the orbit of Q_0 follows that of P_0 quite closely but, eventually, they start to diverge after a few iterations.

The observed divergence of the two nearby orbits in Fig. 3.14 is not due to numerical error. It is, mainly, due to the intrinsic *nonlinear* behavior of the system, known as *sensitive dependence* on initial conditions. The following definition formalizes this concept.

Definition 3.3 Consider a discrete model

$$x_{n+1} = f(x_n, \lambda).$$

An initial condition x_0 is said to exhibit sensitive dependence on initial conditions if there is a constant $d > 0$, and some integer k, and a neighborhood $N_\varepsilon(x_0) = \{x \in \mathbb{R} : |x - x_0| < \varepsilon\}$ such that

$$|f^k(x) - f^k(x_0)| \geq d.$$

This definition simply says that two nearby orbits show sensitive dependence if they separate some distance d after a certain number of iterations k. In general, the

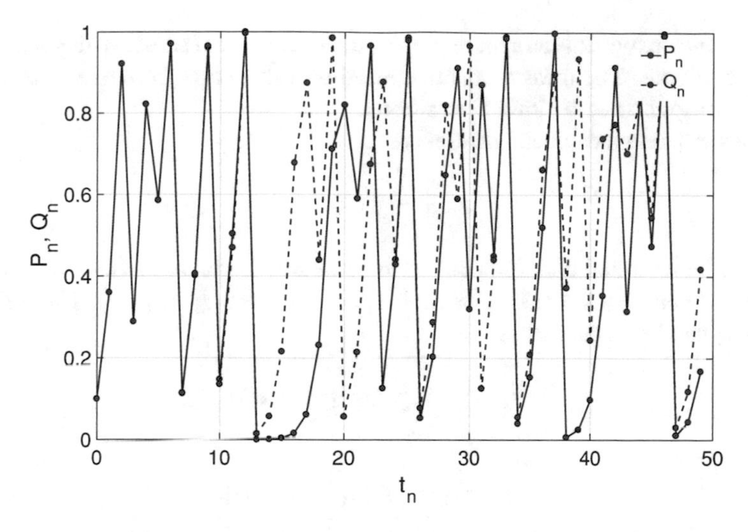

Fig. 3.14 Sensitive dependence in the logistic model. Two orbits, one with initial condition $P_0 = 0.1$ and one with $Q_0 = 0.10001$ evolve, for a short time, close to one another but they, eventually diverge. This behavior is intrinsic to the logistic model and it is due to the nonlinearities in the system. Parameter $r = 4.0$

closer the point x is to x_0 the larger the number of iterations k will need to be in order to observe the divergence in the orbits. The definition does not say anything, however, about the nonlinearities in the model. In fact, consider for instance the following model

$$x_{n+1} = rx_n,$$

whose solution is $x_n = r^n x_0$. Let $r > 1$ and let d be the initial separation of the initial conditions, i.e., $d = |x - x_0|$. Then

$$|f^k(x) - f^k(x_0)| = |r^n x - r^n x_0| = r^n |x - x_0| > |x - x_0| = d.$$

Consequently, linear models that are governed by exponential growth, e.g., Malthusian growth models, can also exhibit sensitive dependence on initial conditions. This statement does not contradict the previous assertion of the nonlinearities being the source of the observed sensitive dependence on the logistic growth model. The difference is in the fact that the orbit of the logistic growth model is bounded.

3.9.2 Lyapunov Exponents

Sensitive dependence and boundedness of orbits are the hallmarks of chaotic behavior. But before we can define chaos in more technical terms, we need to introduce first Lyapunov exponents as a more precise measurement of sensitive dependence on initial conditions. The measurement is based on calculating the average separation rate of orbits with nearby initial conditions.

Consider again the general discrete model

$$x_{n+1} = f(x_n, \lambda).$$

Let x_1 be an initial condition and y_1 be a nearby initial condition, i.e., $y_1 \approx x_1$. After one iteration of the model, we have $x_2 = f(x_1)$ and $y_2 = f(y_1)$. The derivative of f at x_1 can be calculated as

$$y_2 - x_2 \approx f'(x_1)(y_1 - x_1),$$

which implies

$$|y_2 - x_2| \approx |f'(x_1)||y_1 - x_1|.$$

It follows that the separation between the two new points, y_2 and x_2 is, approximately, given by $|f'(x_1)|$. Similarly, after two iterations

$$|y_3 - x_3| \approx f'(x_2)(y_2 - x_2) \approx f'(x_2)f'(x_1)(y_1 - x_1),$$

which leads to

$$|y_3 - x_3| \approx |f'(x_2)f'(x_1)||y_1 - x_1|.$$

Then the average separation rate per iteration is $A = |f'(x_2)f'(x_1)|^{\frac{1}{2}}$, while the separation rate after two iterates is $A^2 = |f'(x_2)f'(x_1)|$.

In general, after n iterations, we get

$$|y_{n+1} - x_{n+1}| \approx |f'(x_n)f'(x_{n-1}) \ldots f'(x_1)||y_1 - x_1|.$$

Hence, we arrive at the following definition.

Definition 3.4 Let x_1 be an initial condition with orbit $\{x_1, x_2, x_3, \ldots\}$, i.e., $x_2 = f(x_1), x_3 = f(x_2) = f^2(x_1) \ldots x_n = f^{n-1}(x_1)$. The average separation rate per iteration is given by the Lyapunov number:

$$L(x_1) = \lim_{n\to\infty} |f'(x_1)f'(x_2) \ldots f'(x_n)|^{\frac{1}{n}}. \tag{3.21}$$

Assuming $f'(x_k) \neq 0$, for all k, and assuming also that the limit above exists, then the Lyapunov exponent of the orbit starting at x_1 is defined as

$$h(x_1) = \ln L(x_1),$$

which can be rewritten as

$$h(x_1) = \lim_{n\to\infty} \frac{1}{n} \sum_{k=1}^{n} \ln |f'(x_k)|. \tag{3.22}$$

Lyapunov numbers or exponents allow us to identify sensitive dependence on initial conditions. That is, if the Lyapunov number satisfies $L(x_1) > 1$ then it implies that nearby orbits will diverge from the original orbit $\{x_1, x_2, x_3, \ldots\}$. But $L(x_1) > 1$ also implies $h(x_1) > 0$. In practice, it is more common to use a positive Lyapunov exponent as a measure of sensitive dependence. Let's compute next the Lyapunov exponents for some special orbits.

Example 3.7 (*Equilibrium Points*) Let x_e be an equilibrium point of a general discrete model $x_{n+1} = f(x_n, \lambda)$. Assume x_1 to be an initial condition on the basin of attraction of x_e so that

$$x_n = f^n(x_1) \overrightarrow{n \to \infty} x_e.$$

Then the Lyapunov exponent of the orbit starting at x_1 satisfies

$$h(x_1) = \ln |f'(x_e)|.$$

Proof Continuity of $f'(x)$ implies

$$\lim_{n\to\infty} f'(x_n) = f'\left(\lim_{n\to\infty} x_n\right) = f'(x_e).$$

Continuity of ln implies

$$\lim_{n\to\infty} \ln |f'(x_n)| = \ln f'(x_e).$$

Thus, there must be an integer $N(\varepsilon)$ such that

$$|\ln |f'(x_k)| - \ln |f'(x_e)|| < \varepsilon, \quad \forall k \geq N(\varepsilon).$$

Then

$$\ln |f'(x_e)| - \varepsilon < \ln |f'(x_k)| < \ln |f'(x_e)| + \varepsilon, \quad \forall k \geq N(\varepsilon).$$

Considering the summation in Eq. (3.22), we get

$$\frac{1}{n}\sum_{k=1}^{n} \ln |f'(x_k)| = \underbrace{\frac{1}{n}\sum_{k=1}^{N(\varepsilon)} \ln |f'(x_k)|}_{A_n} + \underbrace{\frac{1}{n}\sum_{k=N(\varepsilon)+1}^{n} \ln |f'(x_k)|}_{B_n}.$$

Assume: $n > N(\varepsilon)$ and all terms in B_n to be ε-close to $\ln f'(x_e)$. Since there are $n - N(\varepsilon)$ terms in B_n we get

$$\left(\frac{n-N(\varepsilon)}{n}\right)\left[\ln|f'(x_e)|-\varepsilon\right] < B_n < \left(\frac{n-N(\varepsilon)}{n}\right)\left[\ln|f'(x_e)|+\varepsilon\right].$$

Adding A_n we get

$$A_n+\left(\frac{n-N(\varepsilon)}{n}\right)\left[\ln|f'(x_e)|-\varepsilon\right] < \frac{1}{n}\sum_{k=1}^{n}\ln|f'(x_k)| < A_n+\left(\frac{n-N(\varepsilon)}{n}\right)\left[\ln|f'(x_e)|+\varepsilon\right]. \tag{3.23}$$

Fixing ε fixes both $N(\varepsilon)$ and the summation in A_n, so that

$$\lim_{n\to\infty} A_n = 0.$$

Furthermore,

$$\lim_{n\to\infty}\frac{n-N(\varepsilon)}{n} = \lim_{n\to\infty}\frac{n+N(\varepsilon)}{n} = 1.$$

Letting $n \to \infty$ in Eq. (3.23), we get

$$\ln|f'(x_e)|-\varepsilon \le \lim_{n\to\infty}\frac{1}{n}\sum_{k=1}^{n}\ln|f'(x_k)| \le \ln|f'(x_e)|+\varepsilon.$$

Finally, letting $\varepsilon \to 0^+$ we find the Lyapunov exponent to be

$$h(x_0) = \lim_{n\to\infty}\frac{1}{n}\sum_{k=1}^{n}\ln|f'(x_k)| = \ln|f'(x_e)|.$$

□

We can now apply the previous result to a specific case.

Example 3.8 Consider the logistic map $x_{n+1} = rx_n(1-x_n)$. Let $r = 2.5$. We know from previous work that an equilibrium point is given by

$$x_e = 1 - \frac{1}{r} = 0.6.$$

Since $f(x) = rx(1-x)$ we get $f'(x) = r(1-2x)$ then $f'(x_e = 0.6) = -0.5$. The Lyapunov exponent of a nearby orbit starting at x_1 is

$$h(x_1) = \ln|0.5| = -0.6931,$$

which implies that nearby orbits converge, i.e., the equilibrium point $x_e = 0.6$ is stable.

Next, we consider periodic orbits.

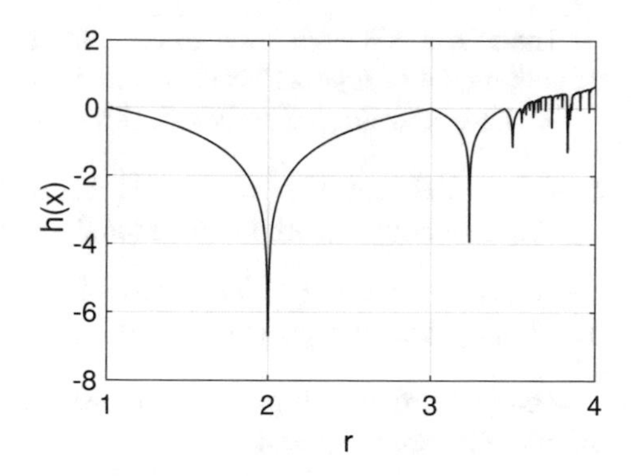

Fig. 3.15 Lyapunov exponents of the logistic growth model $x_{n+1} = rx_n(1 - x_n)$ over the interval $1 < r < 4$. Negative values are indicative of stability while positive values imply sensitive dependence on initial conditions, i.e., divergence of nearby orbits

Example 3.9 (*Periodic Orbits*) Let $\{x_1, x_2, \ldots, x_k\}$ be a period-k orbit. Then

$$h(x_1) = \frac{1}{k}\sum_{j=1}^{k} \ln|f'(x_j)|.$$

Proof

$$\begin{aligned}
h(x_1) &= \lim_{n\to\infty} \frac{1}{n}\left[\ln|f'(x_1)| + \ldots + \ln|f'(x_k)| + \right.\\
&\qquad \left.\ln|f'(x_1)| + \ldots + \ln|f'(x_k)| + \ldots\right]\\
&= \lim_{n\to\infty} \frac{1}{n}\left\{\frac{n}{k}\left[\ln|f'(x_1)| + \ldots + \ln|f'(x_k)|\right]\right\}\\
&= \lim_{n\to\infty} \frac{1}{k}\left[\ln|f'(x_1)| + \ldots + \ln|f'(x_k)|\right]\\
&= \frac{1}{k}\sum_{j=1}^{k} \ln|f'(x_j)|.
\end{aligned}$$

□

The following theorem says that transient behavior does not influence the Lyapunov exponent of an orbit.

Theorem 3.3 *Let $x_{n+1} = f(x_n, \lambda)$ be a general discrete model. Let $O(x_1) = \{x_1, x_2, x_3, \ldots\}$ be an orbit starting at x_1 such that $f'(x_k) \neq 0$, $\forall k$. Assume $O(x_1)$ converges asymptotically to a period-k orbit $O(y_1) = \{y_1, \ldots y_k\}$. Then the Lyapunov exponents of both orbits are the same, that is*

$$h(x_1) = h(y_1) = \frac{1}{k}\sum_{j=1}^{k} \ln|f'(y_j)|.$$

Figure 3.15 shows the Lyapunov exponents of the logistic map $x_{n+1} = rx_n(1 - x_n)$ over the interval $1 < r < 4$.

The MATLAB code used to compute the Lyapunov exponents of the logistic growth model of Fig. 3.15 is shown in the Appendix.

We have now arrived at the definition of chaotic behavior.

Definition 3.5 Let $x_{n+1} = f(x_n, \lambda)$ be a discrete model. Let $O(x_1) = \{x_1, x_2, x_3, \ldots\}$ be a bounded orbit of the model. the orbit $O(x_1)$ is said to be chaotic if

(i) $O(x_1)$ is not asymptotically periodic, and
(ii) The Lyapunov exponent satisfies $h(x_1) > 0$.

Thus, according to Fig. 3.15 the logistic map exhibits chaotic behavior for various values of r, close to $r = 4$.

Example 3.10 Consider the simple model $x_{n+1} = 2x_n \mod 1$. Let $x \neq 0.5$. Direct computations show that

$$h(x_1) = \lim_{n\to\infty} \frac{1}{n} \sum_{k=1}^{n} \ln|f'(x_k)| = \lim_{n\to\infty} \frac{1}{n} \sum_{k=1}^{n} \ln 2 > 0.$$

This relatively simple model represents multiplication by 2 over the bounded interval $[0 : 1]$. Yet, every orbit of the map that it is not asymptotically periodic is chaotic.

3.10 Exercises

Exercise 3.1 Most banks now use continuously compounded interest, as it makes programming the principal easier. The *compound interest formula* was given by

$$P_t = (1+i)^{kt} P_0,$$

where P_0 is the original principal, t is time in years, $i = r/k$ with r the annual interest rate and k the number of times interest is compounded per year, and P_t is the principal at time t. Continuously compounded interest is computed by taking the limit as $k \to \infty$. Use your limits from Calculus to show that

$$P_t \equiv P(t) = \lim_{k\to\infty} (1+i)^{kt} P_0 = P_0 e^{rt}.$$

Exercise 3.2 (*Annuity model*) Ordinary *annuity plans*, such as 401(k) savings plans, assume that an employee (and possibly employer) contribute regularly a fixed amount, d, into an account that earns compound interest. If we assume a fixed interest earned on the capital, then an *annuity model* is given by:

$$P_{n+1} = (1+i)P_n + d. \tag{3.24}$$

a. Assume an average monthly interest rate, $i = r/12$, where r is the annual rate. Find the solution of (3.24) for any P_0, i, and N, where N is the number of years of investing in the annuity.

b. Suppose that an employee starts an annuity, $P_0 = 0$ with a 401k plan that averages $r = 7.2\%$ annually. If this person wants to accrue \$500,000 after 20 years, how much does this employee need to contribute monthly to obtain this goal?

c. As noted in the text, the interest rates can vary over time and the principal can be tied to stock values. Write a brief description of how these ideas could be incorporated into the model.

Exercise 3.3 Consider the 1D map:

$$x_{n+1} = |x_n - 1|, \quad x \in \mathbb{R}.$$

(a) Find all fixed points and eventually fixed points.
(b) Find all the periodic points and eventually periodic points.
(c) What happens if $x \gg 1$? and $x \ll 1$? Hint: look at the cobweb diagram.

Exercise 3.4 Consider a discrete model of the form:

$$x_{n+1} = (1 + r)(x_n - b)^a + b, \tag{3.25}$$

where a, b, and r are constant parameters. Make the following substitution:

$$P_n = x_n - b,$$

into (3.25). The resulting simplified form has similarities to the *discrete Malthusian growth model*, (3.1). Follow the methods from that derivation to find an exact closed form solution for x_n, for all values of n, based on the initial condition x_0.

Exercise 3.5 (*The Allee effect*) The San Diego Zoo discovered that because their flamingo population was too small, it would not reproduce until they borrowed some from Sea World. Scientists have discovered that certain gregarious animals require a minimum number of animals in a colony before they reproduce successfully. This is called the Allee effect. Consider the following model for the population of a gregarious bird species, where the population xn, is given in thousands of birds:

$$x_{n+1} = x_n + \lambda x_n \left(1 - \frac{1}{16}(x_n - 6)^2\right). \tag{3.26}$$

(a) Find all fixed points and study their stability. Find the values of λ at which the fixed points bifurcate and classify the type of bifurcations.
(b) Let $\lambda = 0.2$. Draw two cobwebbing orbits. One with $x_0 = 4.0$ and one with $x0 = 4.3$. Explain the differences between the two orbits.

(c) Write a computer program to generate a bifurcation diagram. Make sure that positive and negative initial conditions are used. Compare the diagram with that of the logistic map. (Turn in the code and ONE plot of the bifurcation diagram).
(d) Give a brief biological description of what your results imply about this gregarious species of bird.

Exercise 3.6 (*Bistability in Population Genetics*) Consider the cubic map

$$x_{n+1} = \lambda x_n - x_n^3, \qquad -2 \le x_n \le 2, \quad 0 \le \lambda \le 3. \tag{3.27}$$

(a) Find all fixed points and study their stability. Find the values of λ at which the fixed points bifurcate and classify the type of bifurcations.
(b) Let $\lambda = 3$. Draw two cobwebbing orbits. One with $x_0 = 1.9$ and one with $x_0 = 2.1$. Explain the differences between the two orbits.
(c) Find a period-2 point as a function of λ. Hint: Use the fact that $f(x) = \lambda x - x^3$ is odd. Study the stability of this period-2 orbit, i.e., find the range of values of λ where the period-2 orbit is stable.
(d) Write a computer program to generate a bifurcation diagram. Make sure that positive and negative initial conditions are used. Compare the diagram with that of the logistic map. (Turn in the code and ONE plot of the bifurcation diagram).

Exercise 3.7 (*Epilepsy*) In this problem P_n represents the fraction of neurons of a large neural network that fire at time t_n. As a simple model of epilepsy, the dynamics of the network can be described by

$$P_{n+1} = 4CP_n^3 - 6CP_n^2 + (1 + 2C)P_n, \tag{3.28}$$

where C is a positive number and $0 \le P_n \le 1$.

(a) Find all fixed points and study their stability as a function of C.
(b) Graph P_{n+1} as a function of P_n for $C = 4$ and study the dynamics as $t \to \infty$ starting from an initial condition of $P_0 = 0.45$.
(c) Generate a bifurcation diagram and discuss the results.

Exercise 3.8 (*Gene and Neural Networks*) The following 1D map plays a role in the analysis of nonlinear models of gene and neural networks (Glass and Pasternack, 1978):

$$x_{n+1} = \frac{\alpha x_n}{1 + \beta x_n}, \tag{3.29}$$

where α and β are positive parameters and $x_n > 0$.

(a) Algebraically determine all fixed points.
(b) For each fixed point find the range of values of α and β for which it exists, indicate whether the fixed point is stable or unstable, and state whether the dynamics in the neighborhood of the fixed point are monotonic or oscillatory.

(c) Assume $\alpha = 1$ and $\beta = 1$. Sketch the graph of x_{n+1} as a function of x_n. Graphically (cobwebbing) iterate the equation starting with initial condition $x_0 = 10$. What happens as the number of iterates approaches ∞ ?

(d) Assume $\alpha = 1$ and $\beta = 1$. Algebraically determine x_{n+2} as a function of x_n, and x_{n+3} as a function of x_n. Based on these computations what is the algebraic expression for x_{n+k} ? What is the behavior of x_{n+k} as $k \to \infty$? This result should agree with the answer to part (c).

Exercise 3.9 Assume X to be a finite set with n points, $X = \{x_1, x_2, \ldots, x_n\}$. Assume also a discrete model of the form

$$x_{n+1} = f(x_n, \lambda), \quad f : X \to X.$$

Prove that every point of X is eventually periodic.

Exercise 3.10 An employee has been investing in a 401(k) plan for quite sometime, with about \$1,000 per quarter. Currently, the plan is worth \$100,000. The employee wishes to retire in 10 years from now. It is reasonable to assume that the account can earn about 6% annual interest compounded quarterly. (a) How much will be in the account at retirement time? (b) How much should each quarterly deposit be in order to accumulate \$400,000 by retirement time ?

Exercise 3.11 Consider the logistic map

$$x_{n+1} = rx_n(1 - x_n), \quad f : [0, 1] \to [0, 1].$$

Suppose $x_0 \in [0, 1]$. Obviously, if $x_0 = 0$ or $x_0 = 1$, then $x_n = f^n(x_0) = 0$, $n \geq 1$. Are these the only points that get mapped to 0 in finitely many steps ? Find a condition on $r \in [0, 4]$ that implies this so. What happens if your condition on r fails ?

Exercise 3.12 Consider the discrete model

$$x_{n+1} = ax_n + bx_n^2 + cx_n^3, \qquad x \in \mathbb{R}.$$

Discuss the stability properties of the equilibrium point $x_e = 0$ for $a < 1$, $a = 1$, and $a > 1$.

Exercise 3.13 Find the equilibrium points for each of the following discrete models:

(i) $x_{n+1} = \dfrac{x_n^3 + x_n}{2}, \quad x \in [-1, 1];$

(ii) $x_{n+1} = -\dfrac{x_n^3 + x_n}{2}, \quad x \in [-1, 1];$

(iii) $x_{n+1} = \dfrac{x_n(1 - x_n)}{2}, \quad x \in [0, 1].$

Exercise 3.14 Find a continuous function $f : (0, 1) \to (0, 1)$ for a discrete model, $x_{n+1} = f(x_n, \lambda)$, which does not have an equilibrium point.

Exercise 3.15 A model of population growth with harvesting can be written as

$$P_{n+1} = rP_n + k,$$

where r is the growth rate and k represents the constant level for harvesting. Suppose that the present size of a population is 5,500 and is growing by 10% per generation. Harvesting occurs at a constant rate of 400 per generation. (a) What will the size of the population be after 10 generations? (b) How many generations will it take for the population to reach 10,000?

Exercise 3.16 Consider again the population growth model with harvesting

$$P_{n+1} = rP_n + k.$$

Find a suitable substitution to convert this equation into a simpler form

$$x_{n+1} = ax_n.$$

Apply the solution to the Malthusian growth model to find a close form solution for x_n and, in turn, find one close form solution for P_n.

Exercise 3.17 Apply Newton's method to compute an approximate value for $\sqrt{2}$. Hint: consider the function $f(x) = x^2 - 2$ and solve $f(x) = 0$.

Exercise 3.18 Consider the *odd logistic model*

$$x_{n+1} = rx_n(1 - x_n^2), \quad f = rx(1 - x^2), \quad f : [0, 1] \to [0, 1].$$

Show that there is a period-doubling bifurcation of the trivial equilibrium point $x = 0$ at $r = -1$. To do this, you must verify that the conditions of the period doubling bifurcation theorem are satisfied.

Exercise 3.19 Consider the following discrete model

$$x_{n+1} = rx \cos x + cx^2 - 2x^3.$$

Describe the type of bifurcation that occurs at $r = 1$ in the cases (a) $c = 1$, (b) $c = 0$. For both cases, identify the type of bifurcation (using appropriate theorems) and sketch the bifurcation diagram, indicating stabilities.

Exercise 3.20 (*Tent Map*) Consider the tent map:

$$x_{n+1} = T(x_n) = \begin{cases} 2x_n, & x_n \leq 1/2 \\ 2(1 - x_n), & 1/2 \leq x_n \end{cases}$$

(a) Find all fixed points and period-2 orbits and determine their stability.
(b) Find a period-3 point $x_0 < 0.5$ such that $T(x_0) < 0.5$ and $T^2(x_0) > 0.5$.
(c) Show that the Tent map has infinitely many chaotic orbits.

Exercise 3.21 Consider again the tent map.

(a) Below is the graph of $T(x)$. Locate the fixed points and use Cobewebbing with $x_0 = 0.1$ to determine their stability.
(b) Find an expression for $T^2(x_n)$. Sketch the graph of $T^2(x)$ then determine the period-2 points and their stability using Cobwebbing.

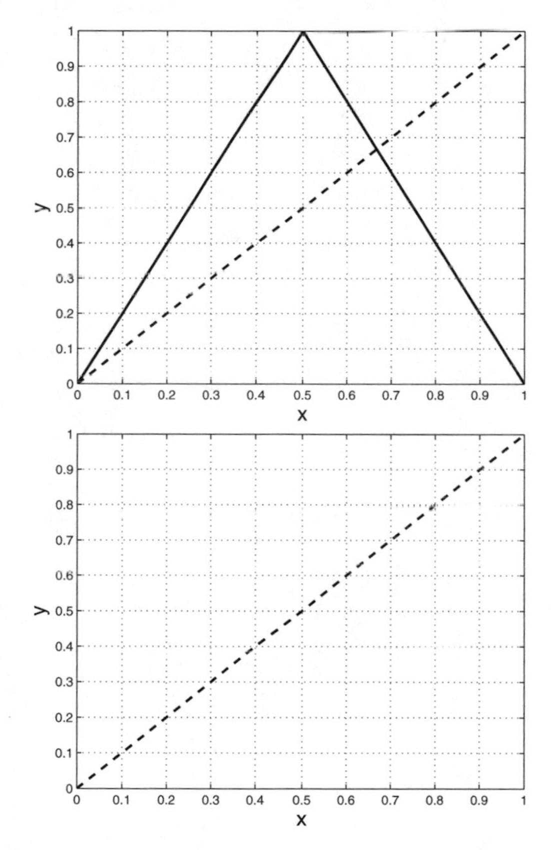

References

1. R.M. May, Deterministic models with chaotic dynamics. Nature **256**, 165–166 (1975)
2. K.T. Alligood, T.D. Sauer, J.A. Yorke, *Chaos: An Introduction to Dynamical Systems* (Springer, New York, 1996)
3. R.L. Devaney, *An Introduction to Chaotic Dynamical Systems*, 2nd edn. (Westview Press, Boulder, CO, 2003)

4. W.E. Ricker, *Handbook of Computations for Biological Statistics of Fish Populations*. Fisheries Research Board of Canada (1958)
5. M. Levi, *Why Cats Land on Their Feet: and 76 other Physical Paradoxes and Puzzles* (Princeton University Press, 2012)
6. R.M. May, Simple mathematical models with very complicated dynamics. Nature **261**, 459–467 (1975)

Chapter 4
Continuous Models

This chapter examines phenomena that change continuously, primarily in time. Examples are drawn from life sciences and engineering, including yeast populations, predator-prey interactions, laser beams, oscillations of a quartz crystal, sensors of magnetic fields, and pharmokinetic systems. These examples serve to illustrate that continuous models are characterized by the same type of long-term solution sets as the discrete ones, i.e., equilibrium and periodic points. The conditions for the existence and stability properties of these solution sets are revised. Qualitative methods, e.g., phase portraits, are also employed to study the behavior of continuous models.

4.1 Introduction

Continuous modeling typically employs Ordinary Differential Equations, or ODEs for short, to describe phenomena and data that vary continuously. In general, these models contain (one or many) state variables, which represent "point-measurements" of the phenomenon, such as the densities of two species, the number of photons in a laser beam, the voltage in an electronic circuit, or the magnetization state of a sensor. These state variables are the dependent variables of the ODE; however, the model can only contain one independent variable, typically time or a location in space.

4.2 Chemostat

A *chemostat* is a bioreactor in which a fresh medium (or nutrient) is added continuously, while culture liquid with left over microorganisms are also continuously removed at the same rate. This process ensures the culture volume to remain constant. The growth rate of the micororganisms is controlled by changing the rate at which

A. Palacios, *Mathematical Modeling*, Mathematical Engineering,
https://doi.org/10.1007/978-3-031-04729-9_4

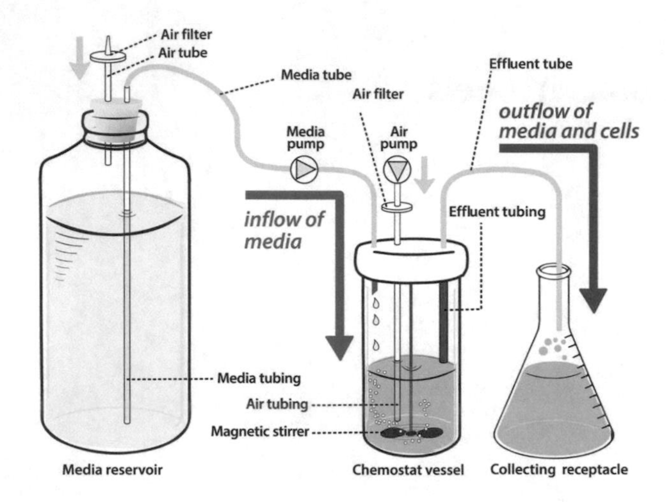

Fig. 4.1 Diagram of an open system describing a chemostat

the medium is added into the bioreactor. Chemostats are used in bioengineering research and industry aimed to accurately quantify the effect of nutrient concentration on differences in growth rates. In large breweries, for instance, chemostats are used to maintain consistency of product. This can be achieved by growing a particular strain of brewer's yeast inside a chemostat. Figure 4.1 shows the key elements of a chemostat.

It turns out that discrete models are not well suited to describe yeast grow because yeast grows continuously. However, we can modify the models from Chap. to include small time-steps, which in the limit leads to a continuous model in the form of an *Ordinary Differential Equation* or (ODE).

4.2.1 Continuous Model of Yeast Growth

Recall that a *discrete Malthusian growth model* has the form:

$$P_{n+1} = (1 + r)P_n, \tag{4.1}$$

where P_n is the population at time n and r is the per capita growth rate. Let $P_n \equiv P(t)$ and assume a time step of Δt, so $P_{n+1} \equiv P(t + \Delta t)$. If r is the per capita growth rate per unit time, then Eq. (4.1) becomes:

$$P(t + \Delta t) = (1 + r\Delta t)P(t) \quad \text{or} \quad \frac{P(t + \Delta t) - P(t)}{\Delta t} = rP(t),$$

Table 4.1 Monoculture yeast experiments for *Saccharomyces cerevisiae*

Time (hr)	Volume	Time (hr)	Volume	Time (hr)	Volume
0	0.37	18	10.97	38	12.77
1.5	1.63	23	12.5	42	12.87
9	6.2	25.5	12.6	45.5	12.9
10	8.87	27	12.9	47	12.7
18	10.66	34	13.27		

Taking the limit $\Delta t \to 0$, the left-hand side becomes the definition of the derivative, so this equation reduces to:

$$\boxed{\begin{aligned} \frac{dP}{dt} &= rP(t) \\ P(0) &= P_0, \end{aligned}} \tag{4.2}$$

which is the *continuous Malthusian growth model*. This is a *first order linear differential equation*, and its solution is:

$$P(t) = P_0 e^{rt}. \tag{4.3}$$

For the early stages of growth of *Saccharomyces cerevisiae*, we consider only the first 10 h (4 data points) of yeast data collected by Gause [1, 2], see Table 4.1. We then show two methods for estimating the growth rate, r.

The solution (4.3) contains two parameters, P_0 and r, that need to be fit. Using the linear least squares best fit to the logarithm of data

$$\ln(P(t)) = \ln(P_0) + rt,$$

we get the best continuous Malthusian growth model to be,

$$P(t) = 0.6045\, e^{0.2690\, t}.$$

Alternatively, we can use a nonlinear least squares best fit to the data by minimizing the sum of square errors:

$$J(P_0, r) = \sum_{i=1}^{4} \left(P_d(t_i) - P_0 e^{rt_i}\right)^2,$$

where t_i are the times in the data and $P_d(t_i)$ are the first 4 yeast volume data from Table 4.1. The Appendix A.3.1. contains the MatLab code for performing this non-

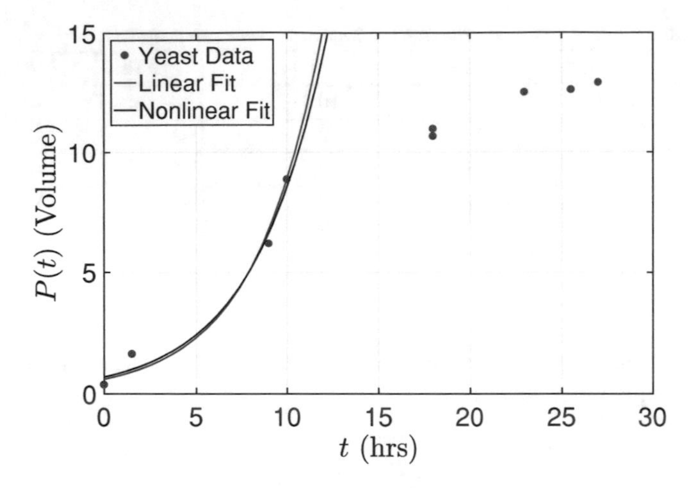

Fig. 4.2 A graph of the best fitting continuous Malthusian growth models through the first 10 h of the Gause data for *S. cerevisiae*

linear fit and finding the parameters. With these data, the nonlinear least squares best fit gives the best continuous Malthusian growth model,

$$P(t) = 0.6949\, e^{0.2511\, t},$$

which is similar to fit using the logarithm of the data. Figure 4.2 shows a graph of the second two parameter fits to the continuous Malthusian growth model.

4.2.2 Logistic Growth Model

The results of Fig. 4.2 indicate that a time-dependent Malthusian growth is clearly not a good fit, especially for larger values of time where the exponential growth of the Malthusian growth is way off the observed labeling off of the data set. In this section we derive a more appropriate model in the form of an *autonomous differential equation*:

$$\frac{dP}{dt} = f(P(t)) = f(P).$$

The *Malthusian growth model* represented by Eq. (4.2) contains the linear updating function $f(P) = rP$. Now, a Maclaurin series expansion of $f(P)$ is

$$f(P) = f(0) + f'(0)P + \frac{f''(0)}{2!}P^2 + \mathcal{O}\left(P^3\right),$$

where $\mathcal{O}\left(P^3\right)$ means order P^3.

The chemostat problem is a closed system. When the population is at *extinction*, the system is said to be at an *equilibrium point* $P_e = 0$. This equilibrium point yields, $f(0) = 0$. Since the *linear term*, see Eq. (4.2), originates from the *Malthusian growth*, we have

$$f'(0) = r.$$

Furthermore, we observe that the population growth rate declines for larger populations, so the second order term in the series expansion must be negative. This term is known as the *intraspecies competition*. Mathematically, this implies:

$$\frac{f''(0)}{2!} = -\frac{r}{M},$$

where r is from the Malthusian growth, and as we'll see later, M is the *carrying capacity*. Ignoring higher order terms of $f(p)$, we obtain the *logistic growth model*:

$$\boxed{\begin{aligned} \frac{dP}{dt} &= rP\left(1 - \frac{P}{M}\right), \\ P(0) &= P_0. \end{aligned}} \tag{4.4}$$

The logistic growth model or Eq. (4.4) can be solved using separation of variables:

$$\frac{dP}{P\left(1 - \frac{P}{M}\right)} = dt.$$

Expanding the left-hand side in terms of partial fractions:

$$\frac{1}{P\left(1 - \frac{P}{M}\right)} = \frac{1}{rP} + \frac{1}{Mr\left(1 - \frac{P}{M}\right)}.$$

Integrating:

$$\int_{P_0}^{P} \frac{1}{rP} dP + \int_{P_0}^{P} \frac{1}{Mr\left(1 - \frac{P}{M}\right)} dP = \int_{t_0}^{t} dt.$$

Assuming $t_0 = 0$, we arrive (after a little algebra) at the closed-form solution:

$$\boxed{P(t) = \frac{MP_0}{P_0 + (M - P_0)e^{-rt}}.} \tag{4.5}$$

Table 4.2 Monoculture yeast experiments for *Schizosaccharomyces kephir*

Time (hr)	Volume	Time (hr)	Volume	Time (hr)	Volume
9	1.27	42	2.73	87	5.67
10	1	45.5	4.56	111	5.8
23	1.7	66	4.87	135	5.83
25.5	2.33				

The solution given by Eq. (4.5) contains three parameters: P_0, r, and M, which must be fit to the yeast data. To do so, we consider the data set for em Saccharomyces cerevisiae, see Table 4.1, and a new data set for *Schizosaccharomyces kephir*, see Table 4.2.

The best unbiased fit to the data is obtained by minimizing the sum of square errors (SSE) between the data, $P_d(t_i)$, and the model, (4.5), where the SSE satisfies:

$$J(P_0, r, M) = \sum_{i=0}^{N} \left(P_d(t_i) - \frac{M P_0}{P_0 + (M - P_0)e^{-rt_i}} \right)^2.$$

Since the parameters appear nonlinearly in this formula, the minimization requires a computational nonlinear solver to obtain the best fitting parameters. Appendix A.3.1 has the MatLab code for performing this nonlinear least squares best fit and finding the parameters.

From the experimental data in Tables 4.1 and 4.2, the best fitting parameters for *S. cerevisiae* and *S. kephir* are given by:

$$P_0 = 1.2343, \ r = 0.25864, \ M = 12.7421,$$

and

$$P_0 = 0.67807, \ r = 0.057442, \ M = 5.8802$$

with least *SSE* = *4.9460* and *SSE* = *1.3850*, respectively. These results lead to the best fitting solutions:

$$P_{sc}(t) = \frac{12.742}{1 + 9.323\, e^{-0.2586t}} \quad \text{and} \quad P_{sk}(t) = \frac{5.880}{1 + 7.672\, e^{-0.05744t}}. \tag{4.6}$$

Figure 4.3 shows the graphs of the data with the best fitting logistic growth yeast models.

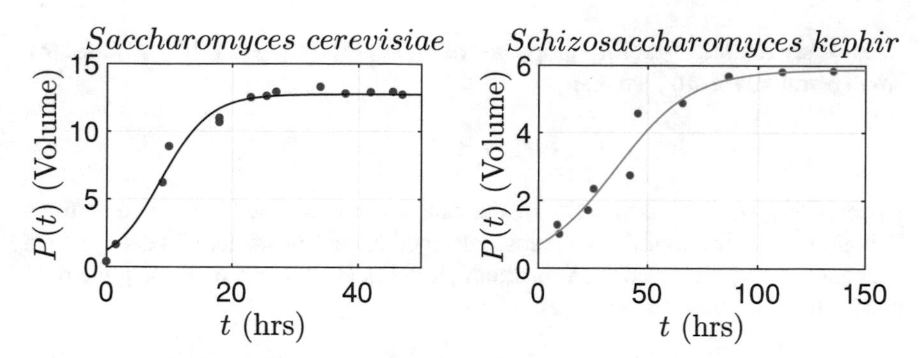

Fig. 4.3 The graph on the left shows the data and best fitting logistic growth model for *S. cerevisiae*, and the graph on the right is the same for *S. kephir*

4.3 Qualitative Analysis of Continuous Models

We now consider the *qualitative analysis* of a *continuous model*, described by a general autonomous ODE

$$\frac{dP}{dt} = f(P). \tag{4.7}$$

As it was the case of discrete models, the qualitative analysis seeks to understand the long-term behavior of solutions of Eq. (4.7), without having full knowledge of a closed-form solution. In a similar manner to discrete models, the analysis starts by computing the equilibrium points of the model. In the case of a continuous system, equilibrium points correspond to no change in the system. Thus, they are found by setting the derivative of P equal to zero:

$$f(P_e) = 0.$$

Observe that in the case of population dynamics this condition implies *no growth*. For instance, in the *logistic growth model*, (4.4), the equilibrium points satisfy:

$$rP_e\left(1 - \frac{P_e}{M}\right) = 0,$$

which yields two solutions:

$$P_{e_1} = 0, \qquad P_{e_2} = M.$$

The first equilibrium point corresponds to *extinction*, while the second one is essentially the carrying capacity of the environment. Notice the analogy of these two equilibria with those of the discrete model.

To study the local stability properties of an equilibrium point, let us consider a small perturbation, $\delta(t)$, so that

$$P(t) = P_e + \delta(t),$$

where $||\delta(t)|| \ll 1$. We now seek to understand how the perturbation evolves in time. If it grows then the equilibrium point is deemed locally unstable. If it decays, then the equilibrium point is stable. Substituting into Eq. (4.7), while expanding the right hand side in a Taylor series we get:

$$\frac{d\delta(t)}{dt} = f(P_e) + f'(P_e)\delta(t) + \mathcal{O}\left(\delta(t)^2\right). \tag{4.8}$$

Since P_e is an equilibrium, then $f(P_e) = 0$, and the linearization of the continuous model Eq. (4.7) becomes (after omitting higher order terms):

$$\frac{d\delta(t)}{dt} = f'(P_e)\delta(t). \tag{4.9}$$

We have then arrived at the following stability theorem.

Theorem 4.1 (Stability of 1D ODE) *Let $f(P)$ be differentiable, then the local stability of an equilibrium, P_e, for the one-dimensional ODE, (4.7), satisfies the following:*

- *If $f'(P_e) > 0$, then locally the solution grows exponentially (positive eigenvalue) and the equilibrium at P_e is unstable.*
- *If $f'(P_e) < 0$, then locally the solution decays exponentially (negative eigenvalue) and the equilibrium at P_e is stable.*
- *If $f'(P_e) = 0$, then more information must be obtained to determine the stability of the equilibrium at P_e.*

The proof of this important theorem is available in standard ODE texts, such as Guckheimer and Holmes [3] (Chap. 1), Strogatz [4] (Chap. 2), and Wiggins [5] (Chap. 1). It uses the Taylor's series, like (4.8), with the definition of stability to prove this local behavior. The last point is critical for changing behavior or bifurcation in an ODE, as some parameter changes and will be studied more in the next chapter.

Recall that the logistic growth model admits two *equilibria* are $P_{e_1} = 0$ and $P_{e_2} = M$. The function and its derivative satisfy:

$$f(P) = rP\left(1 - \frac{P}{M}\right), \quad \text{so} \quad f'(P) = r - \frac{2rP}{M},$$

where $r > 0$ is the Malthusian growth rate at low density and M is the carrying capacity. At the extinction equilibrium, $P_{e_1} = 0$, we have $f'(0) = r > 0$, which makes this equilibrium *unstable*. At the *carrying capacity equilibrium*, $P_{e_2} = M$, we have $f'(M) = -r < 0$, which makes this equilibrium *stable*. These results suggest that the differential equation governing the logistic growth model has an initial exponential growth before moving smoothly toward its carrying capacity.

4.3.1 Direction Fields and Phase Portraits in 1D

Consider the following *initial value problem*

$$\begin{aligned}\frac{dy}{dt} &= f(t, y),\\ y(t_0) &= y_0,\end{aligned} \tag{4.10}$$

Observe immediately that the function $f(t, y)$ is the *slope of the solution*, which can be easily found by computing on a planar grid for (t, y). Available programs, e.g., *dfield*, typically generate arrows or small lines showing the direction of the solution. The *direction field* or *slope field* is this *graphical representation* in the y versus t plane with arrows showing the direction of the solution. Solutions of Eq. (4.10) follow paths defined by the direction field.

Figure 4.4 shows the direction field for the best fitting logistic growth model for *S. cerevisiae*. It has small lines indicating the slope of the solution in (t, P) and plots various solutions using different initial conditions. The graph clearly shows all solutions approaching toward the carrying capacity of 12.742.

Existence and uniqueness of solutions to the differential Eq. (4.10) is established through the following theorem.

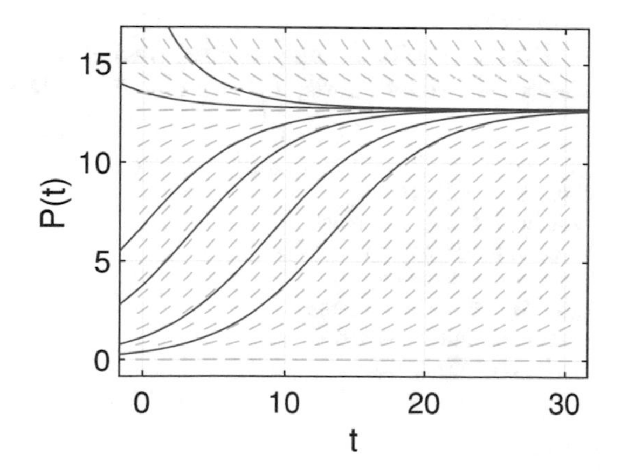

Fig. 4.4 The direction field (computed using the MATLAB code *dfield*) for the logistic growth model for *S. cerevisiae*

Theorem 4.2 (Existence and Uniqueness) *If f and $\partial f/\partial y$ are continuous in a rectangle $R : |t - t_0| \leq a, |y - y_0| \leq b$, then there is some interval $|t - t_0| \leq h \leq |a|$ in which there exists a unique solution $y = \phi(t)$ of the initial value problem (4.10).*

Theorem 4.2 guarantees that solution trajectories do not cross on a *graph*. Thus, geometrically it is easy to trace the direction of the solution from any starting point in the *direction field*.

The *direction field* of the *autonomous differential equation*,

$$\frac{dy}{dt} = f(y), \tag{4.11}$$

in the $y - t$ plane is constant for each value of y. At *equilibria*, y_e, the slope in the direction field is zero, $f(y_e) = 0$, represented by horizontal arrows. Between equilibria, the direction field has only slopes with the same sign, $f(y) < 0$ or $f(y) > 0$. It follows that solutions *monotonically* go toward or away from equilibria. Thus, the *qualitative behavior* of the autonomous differential equation is captured in a *1D-line* or *1D-phase portrait*, where equilibria are marked and solution directions are noted with arrows pointing right or left.

To create a 1D-phase portrait for the differential equation, (4.11), one graphs $f(y)$. The projection of $f(y)$ onto the y-axis provides the 1D-phase portrait, where equilibria are marked by circles with $f(y_e) = 0$. Arrows to the right are drawn for $f(y) > 0$ and to the left for $f(y) < 0$. When arrows of the phase portrait point toward an equilibrium, then it is *stable* and is indicated with a *solid circle*. When arrows of the phase portrait point away from an equilibrium, then it is *unstable* and is indicated with an *open circle*. When arrows of the phase portrait go in the same direction through an equilibrium, then it is *semi-stable* and is indicated with a *half open circle*.

The function for the logistic growth model is a parabola pointing down. Figure 4.5 shows the two equilibria, $P_{e_1} = 0$ and $P_{e_2} = M$ with the function being positive

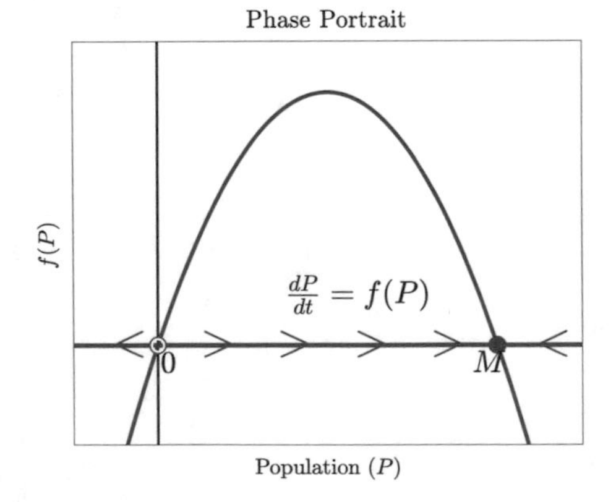

Fig. 4.5 The 1D phase portrait of the logistic growth model is the P-axis in this figure. From the arrows on this axis it shows the unstable extinction equilibrium and stable carrying capacity

between these equilibria. Since the arrow between these equilibria moves to the right, it is easy to see that the extinction equilibrium, $P_{e_1} = 0$, is *unstable*. The carrying capacity, $P_{e_2} = M$, is *stable*, as the arrows to the right and left of this equilibrium point toward the this point. It follows that any positive initial condition results in the solution approaching in time to the *carrying capacity*.

4.3.2 Stable Manifold Theorem

In this section we extend the qualitative analysis of one-dimensional models to higher dimensional systems. Let us start with an *n-dimensional linear ODE:*

$$\dot{x} = Ax, \qquad x \in \mathbb{R}^n, \tag{4.12}$$

where the matrix A has n eigenvalues with n (generalized) eigenvectors. Analytical and computational methods to solve this type of equations can be found in Brauer and Nohel [6] and Guckenheimer and Holmes [3]. Here, we provide a summary. The *fundamental solution set* of the original model Eq. (4.12) is given by:

$$\Phi(t) = e^{At},$$

which leads to the unique solution

$$\phi_t(x_0) = x(x_0, t) = e^{At}x_0.$$

The function $\Phi(t)$ is called the fundamental solution because it generates a *flow*: $e^{At}x_0 : \mathbb{R}^n \to \mathbb{R}^n$, which gives *all* the solutions to Eq. (4.12). Specifically, the *linear subspaces* spanned by the *eigenvectors* of A are *invariant* under the *flow*, $\phi_t(x_0) = e^{At}x_0$. The *eigenspaces* of A are *invariant subspaces* for the flow, $\phi_t(x_0) = e^{At}x_0$.

The subspaces spanned by the eigenvectors are divided into three classes:

1. The *stable subspace*, $E^s = \text{span}\{v^1, \ldots, v^{n_s}\}$,
2. The *unstable subspace*, $E^u = \text{span}\{u^1, \ldots, u^{n_u}\}$,
3. The *center subspace*, $E^c = \text{span}\{w^1, \ldots, w^{n_c}\}$,

where $v^1, \ldots, v^{n_s}$ are the n_s (generalized) eigenvectors whose eigenvalues have *negative real parts*, $u^1, \ldots, u^{n_u}$ are the n_u (generalized) eigenvectors whose eigenvalues have *positve real parts*, and $w^1, \ldots, w^{n_c}$ are the n_c (generalized) eigenvectors whose eigenvalues have *zero real parts*. Clearly, $n_s + n_u + n_c = n$, and the names reflect the behavior of the flows on the particular subspaces with those on E^s exponentially decaying, E^u exponentially growing, and E^c doing neither.

Let us now consider the equivalent nonlinear autonomous problem

$$\dot{x} = f(x), \qquad x \in \mathbb{R}^n, \qquad x(0) = x_0. \tag{4.13}$$

Existence of Solutions
There exist unique solutions to the *nonlinear system* in some small neighborhood of $t = 0$ near x_0 provided adequate smoothness of f. A qualitative analysis follows similar ideas and methods to those of one-dimensional models. That is, we start by computing *fixed points* or *equilibria* of Eq. (4.13) by solving $f(x_e) = 0$, which may be nontrivial and allow only numerical solutions.

Stability
Assume that x_e is a *fixed point* of Eq. (4.13), then to characterize the stability properties of the equilibrium, x_e, one examines the *linearization* at x_e, which is given by the system:

$$\dot{\delta(t)} = Df(x_e)\delta(t), \qquad \delta(t) \in \mathbb{R}^n, \tag{4.14}$$

where $Df = [\partial f_i/\partial x_j]$ is the *Jacobian matrix* of the first partial derivatives of $f = [f_1(x_1, \ldots, x_n), f_2(x_1, \ldots, x_n), \ldots, f_n(x_1, \ldots, x_n)]^T$ and $\delta(t)$ is a small perturbation of the equilibrium, i.e., $x = x_e + \xi$ with $\xi \ll 1$. Since Eq. (4.14) is also a linear system of the form given by Eq. (4.12), then the *linearized flow map* near x_e is given by:

$$D\phi_t(x_e)\delta = e^{tDf(x_e)}\delta. \tag{4.15}$$

Ideally, one would like to decompose the space of flows at least locally (near a fixed point) into the behaviors similar to the ones observed for the linear system. Figure 4.6a provides a cartoon of the manifold extension for (4.13), decomposing the flows into the *stable subspace*, W^s, the *unstable subspace*, W^u, and the *center subspace*, W^c. However, unlike the linear system, nonlinearities dominate when $Df(x_e)$ has zero or purely imaginary eigenvalues, W^c, so the primary stability theorems examine flows only for *hyperbolic fixed points*. For hyperbolic fixed points, the Hartman-Grobman Theorem [3, 7] guarantees local behavior near x_e of (4.13) is similar to the linearization (4.14), which is illustrated in Fig. 4.6b.

Definition 4.1 (*Hyperbolic Fixed Point*) When $Df(x_e)$ has no eigenvalues with **zero real part**, x_e is called a *hyperbolic* or *nondegenerate fixed point*.

It follows that for the *nonlinear ODE* the behavior can only be defined locally, so one defines the *local stable and unstable manifolds*.

Definition 4.2 (*Local Stable and Unstable Manifold*) Define the *local stable and unstable manifolds* of the *fixed point*, x_e, $W^s_{loc}(x_e)$, $W^u_{loc}(x_e)$, as follows:

- $W^s_{loc}(x_e) = \{x \in U | \phi_t(x) \to x_e \text{ as } t \to \infty, \text{ and } \phi_t(x) \in U \text{ for all } t \geq 0\}$,

- $W^u_{loc}(x_e) = \{x \in U | \phi_t(x) \to x_e \text{ as } t \to -\infty, \text{ and } \phi_t(x) \in U \text{ for all } t \leq 0\}$,

where $U \subset \mathbb{R}^n$ is a neighborhood of the *fixed point*, x_e.

These *invariant manifolds*, $W^s_{loc}(x_e)$ and $W^u_{loc}(x_e)$, provide nonlinear analogues of the flat stable and unstable eigenspaces, E^s and E^u of the linear problem. The primary theorem is the *Stable Manifold Theorem*, which shows that $W^s_{loc}(x_e)$ and $W^u_{loc}(x_e)$ are tangent to the eigenspaces, E^s and E^u.

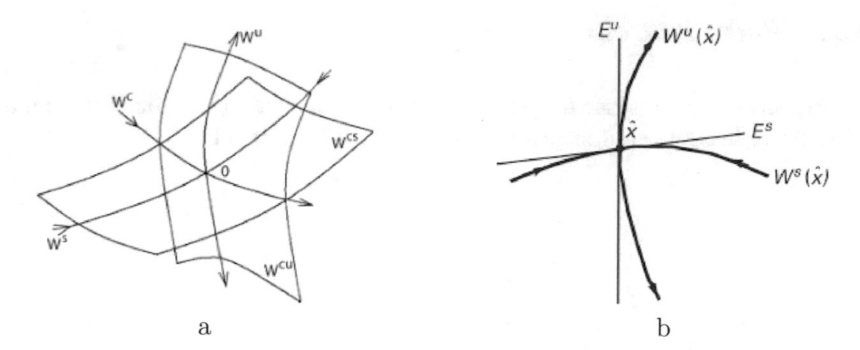

Fig. 4.6 **a** This figure illustrates the local manifolds for a nonlinear system (4.13) about an equilibrium at 0 with $W^s_{loc}(x_e)$ stable, $W^u_{loc}(x_e)$ unstable, and $W^c_{loc}(x_e)$ neither. **b** This figure shows only the stable and unstable manifolds of the Stable Manifold Theorem, where $\hat{x}$ is a hyperbolic fixed point

Theorem 4.3 (Stable Manifold Theorem) *Suppose that $\dot{x} = f(x)$ has a hyperbolic fixed point, x_e. Then there exist local stable and unstable manifolds, $W^s_{loc}(x_e)$ and $W^u_{loc}(x_e)$, of the same dimensions, n_s and n_u, as those of the eigenspaces, E^s and E^u, of the linearized system and tangent to E^s and E^u at x_e. $W^s_{loc}(x_e)$ and $W^u_{loc}(x_e)$ are as smooth as the function, f.*

The proof of this theorem is found in several texts on ODEs [7, 8] and plays a central role in studying continuous dynamical systems. As noted earlier, this theorem avoids discussion about a *center manifold* being tangent to E^c, confining the results to hyperbolic fixed points. However, the *center manifold* often relates to studies in *bifurcation theory*, which is covered in the next chapter.

The *local invariant manifolds* can be extended to global analogues, which can have profound effects on the behavior of the ODE. The *global stable manifold*, W^s, follows points in $W^s_{loc}(x_e)$ flow backwards in time:

$$W^s(x_e) = \bigcup_{t \leq 0} \phi_t(W^s_{loc}(x_e)).$$

The *global unstable manifold*, W^u, follows points in $W^u_{loc}(x_e)$ flow forward in time:

$$W^u(x_e) = \bigcup_{t \geq 0} \phi_t(W^u_{loc}(x_e)).$$

Existence and uniqueness ensures that two stable (unstable) manifolds of distinct fixed points, x_{1e}, x_{2e}, cannot intersect. However, intersections of stable and unstable manifolds of distinct fixed points or the same fixed point can occur. These intersections are often the source of complex dynamics, such as chaos.

4.3.3 Phase Portraits

In this section we introduce the *phase portrait* as a method to visualize the solution set of the following two-dimensional autonomous ODE model:

$$\frac{dx_1}{dt} = f_1(x_1, x_2), \tag{4.16}$$

$$\frac{dx_2}{dt} = f_2(x_1, x_2).$$

Such portraits are generated by projecting solution trajectories, $(t, x(t), y(t))$ onto the (x, y) *phase plane* with arrows (small lines) showing the direction of flows. Mathematical software such as Maple, Pplane, MATLAB, and XPPAUT, can produce these portraits with built-in routines to generate phase portraits.

According to the Stable Manifold Theorem in 2D, there are three generic cases:

- *Stable (or Unstable) Node*, where the equilibrium has two negative (positive) eigenvalues.
- *Stable (or Unstable) Focus or Spiral*, where the equilibrium has two complex eigenvalues with negative (positive) real parts.
- *Saddle Node*, where the equilibrium has one negative and one positive eigenvalue.

Figure 4.7 shows some representative examples from 2D linear ODEs of the form (4.12). Since they are generic cases, it means that they occur most frequently in 2D models. A single model can produce many of these portraits simply by varying a single parameter, which is usually known as the *bifurcation parameter*.

A *center manifold* corresponds to the particular case where an equilibrium of Eq. (4.13) has eigenvalues with zero real part. When this happens in a model, the system experiences the highest sensitivity since small perturbations can lead to drastically different behavior. The 2D system (4.12) with a zero eigenvalue results in the *degenerate* case where there is a *line of equilibria*. This reflects the boundary between problems with stable or unstable nodes and ones with a saddle node. When the 2D system (4.12) has purely imaginary eigenvalues, then the phase portrait is concentric

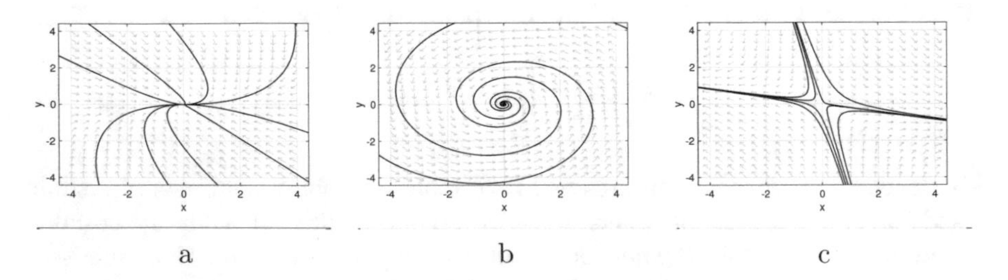

Fig. 4.7 **a** This phase portrait shows a generic *stable node*. **b** This phase portrait shows a generic *unstable focus or spiral*. **c** This phase portrait shows a generic *saddle node*

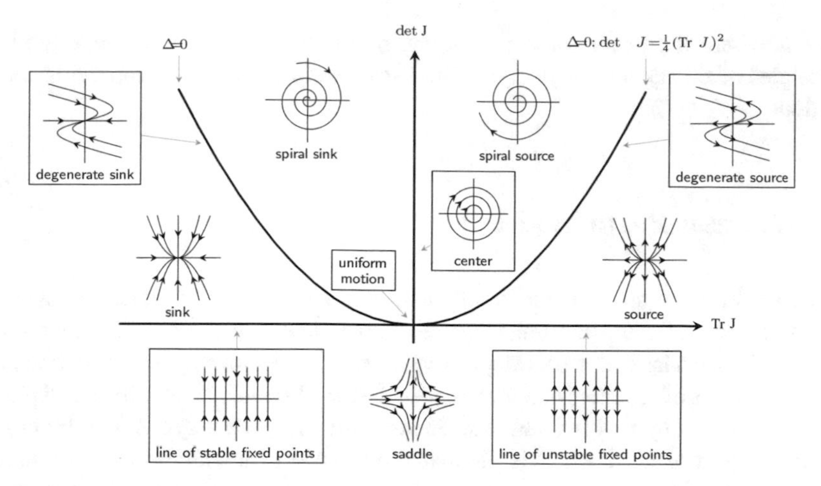

Fig. 4.8 Trace-Determinant stability plane

ellipses about the equilibrium. This is the boundary between the equilibrium being asymptotically stable (stable focus) and being unstable (unstable spiral). There are a number of other *nongeneric* cases, such as equal eigenvalues, but they have minimal effects on the behavior of the mathematical model, but have other interesting mathematical behaviors.

All the different possibilities of real and complex eigenvalues for the 2D linearized ODE (4.16) are controlled by the values of $\text{tr}(J)$ and $\det(J)$, where $J = Df(x_e)$ from (4.14). Letting $\Delta = (\text{tr}(J))^2 - 4\det(J)$, we can use the $\text{tr}(J)$ and $\det(J)$ to visualize all possible stability properties through the "trace-determinant stability plane" of Fig. 4.8. Many references [9–11] provide a complete set of possible 2D phase portraits for (4.12).

The qualitative analysis of (4.16) continues geometrically by the introduction of *nullclines* to better visualize the behavior in the 2D phase plane. Nullclines of the nonlinear system show geometrically where equilibria exist and divide the region into sections with similar flows.

Definition 4.3 (*Nullclines*) *Nullclines* for (4.16) are the curves in the 2D phase plane when either

$$f_1(x_1, x_2) = 0 \qquad \text{or} \qquad f_2(x_1, x_2) = 0.$$

The curve for $f_1(x_1, x_2) = 0$ shows all points in the phase plane where solutions are vertical or trajectory flows are only in the x_2-direction. Similarly, when $f_2(x_1, x_2) = 0$, solutions are horizontal or trajectory flows are only in the x_1-direction.

Geometrically, equilibria occur where the separate nullclines intersect, since $f_1(x_1, x_2) = f_2(x_1, x_2) = 0$. Furthermore, the nullclines divide the phase plane into regions where the flows for the solutions of (4.16) are monotonic in x_1 and x_2. Examples of this geometric analysis are discussed in the mixed competition model section.

4.4 A Laser Beam Model

The term **laser** is an acronym for **l**ight **a**mplification by **s**timulated **e**mission of **r**adiation and is a device that emits light through optical amplification. The theoretical foundations for a laser were developed by Einstein on electromagnetic radiation [12]. A laser consists of an active material or *gain medium* bound by two mirrors, all placed inside a cavity. Figure 4.9 shows a *solid-state laser*, where a synthetic ruby crystal acts as the *gain medium*, which is stimulated by flashes generated from a high-voltage power supply. These photons stimulate electrons in ruby atoms to a higher energy level. Subsequently, the electrons return to their ground state releasing coherent light through *spontaneous emission of radiating energy*, which is readily focused to a narrow beam.

The emitted photons travel at the speed of light in the medium and may collide with already excited atoms, producing more photons in a *stimulated emission* process. This process yields multiple photons, amplifying the light. The mirrors allow photons to continue bouncing back and forth. But a partial mirror (labeled 95% in Fig. 4.9) allows some photons to escape in a very concentrated beam of powerful laser light. The first ruby laser was built in 1960 by Maiman at Hughes Research Laboratories [13], based on theoretical work by Townes and Schawlow [14].

A mathematical model derived by Haken [15] has the form

$$\frac{dn}{dt} = GnN - kn, \tag{4.17}$$

where $n(t)$ is the number of photons in the laser field, $G > 0$ is a parameter that measures the gain in the medium, $N(t)$ is the number of excited atoms, and $k > 0$ is the typical lifetime of a photon. An excited atom emits a photon, dropping to a lower energy level. This is modeled by

$$N(t) = N_0 - \alpha n,$$

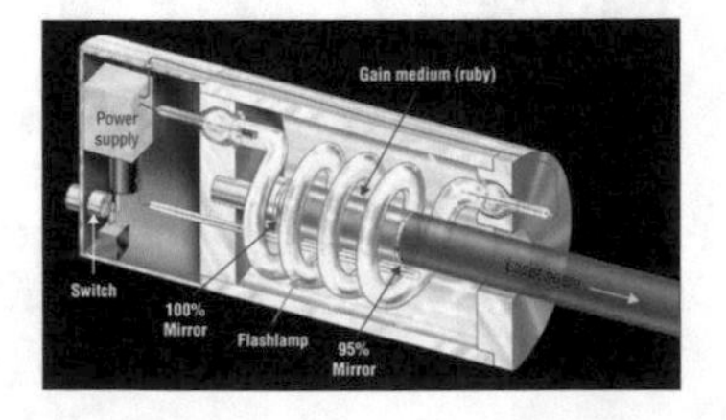

Fig. 4.9 A solid-state laser, also known as a ruby laser. *Source* Wikipedia

where N_0 is the number of excited atoms, which are assumed constant in the absence of the pump. This relation indicates that the number of excited atoms decreases linearly, rate $\alpha > 0$, with the number of photons in the field. Substituting $N(t)$ into Eq. (4.17) yields a mathematical model for the laser beam

$$\frac{dn}{dt} = (GN_0 - k)n - \alpha G n^2, \tag{4.18}$$

which is a quadratic model, like the logistic growth model (4.4). It follows that the qualitative analysis of these models are the same.

4.4.1 Equilibria for Laser

Equilibria of the laser model (4.18) satisfy:

$$(GN_0 - k)n - \alpha G n^2 = 0, \qquad \text{so} \qquad n_{e_1} = 0 \text{ or } n_{e_2} = \frac{GN_0 - k}{\alpha G}.$$

The first equilibrium point, $n_{e_1} = 0$, corresponds to the absence of photons being stimulated, i.e., there is no stimulated emission of energy and the laser system behaves as a standard light. When $N_0 > k/G$, the second equilibrium point, n_{e_2}, corresponds to sufficient stimulus of energy, leading to a laser beam.

Consider the quadratic function

$$f(n) = (GN_0 - k)n - \alpha G n^2, \qquad \text{where} \qquad f'(n) = (GN_0 - k) - 2\alpha G n.$$

This produces a parabola pointing down for all positive parameters. When $N_0 > k/G$, then $n_{e_2} = \frac{GN_0-k}{\alpha G} = M > 0$ produces the equivalent 1D-phase portrait of Fig. 4.5, where n_{e_2} is a stable equilibrium (with n_{e_1} unstable). When $N_0 = k/G$, the two equilibrium points merge together (threshold), and the vertex of the parabola occurs at the origin (semi-stable). When $N_0 < k/G$, there is insufficient stimulus energy and the parabola intersects the negative axis for n_{e_2}, becoming unstable (with n_{e_1} stable). Linearization at $n_{e_1} = 0$ gives

$$f'(0) = GN_0 - k,$$

and stability of this equilibrium depends on the sign of $f'(0)$, negative being stable.

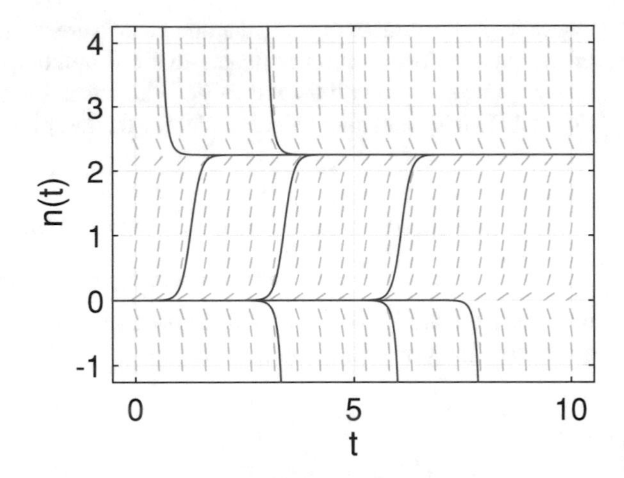

Fig. 4.10 Direction field for the mathematical model (4.18) of a solid-state laser device. Parameters are: the gain, $G = 0.1$; the number of excited atoms, $N_0 = 100$; the typical lifetime of a photon, $k = 1$; the rate atoms are lost, $\alpha = 40$. The two equilibria are $n_{e_1} = 0$ (unstable) and $n_{e_2} = (GN_0 - k)/(\alpha G)$ (stable)

4.4.2 *Visualization of Laser Model*

This quadratic laser model has isomorphic graphs to the logistic growth model, so the laser model direction field is similar to Fig. 4.4. Figure 4.10 shows the direction field for the model (4.18), which is readily generated by software packages, such as Maple's `DEplot` or the MatLab's `dfield` [16]. The solid lines in this figure show representative solution curves from numerical integration of the model.

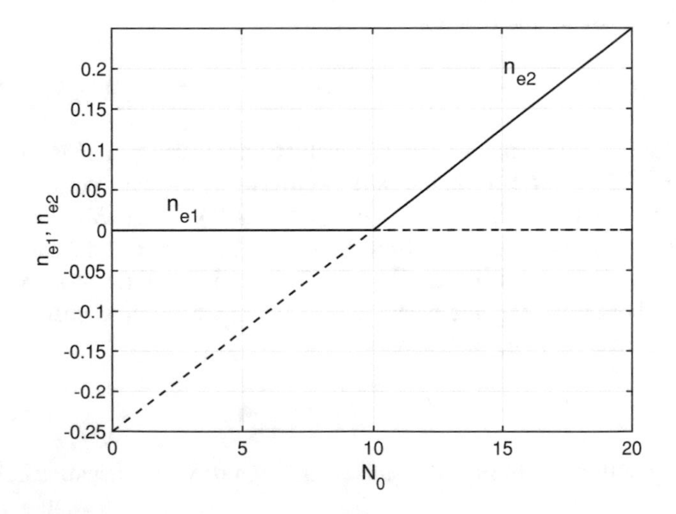

Fig. 4.11 Bifurcation diagram of a laser beam model as N_0 varies. Other parameters are the same as in Fig. 4.10. The zero equilibrium, n_{e1}, is stable (solid line) when $N_0 < k/G$ and unstable for $N_0 > k/G$. The nontrivial (laser state) equilibrium state, n_{e2}, is unstable when $N_0 < k/G$ and stable for $N_0 > k/G$. At $N_0 = k/G$ a change of stability occurs in what is known as a *transcritical bifurcation*

The equilibria, n_{e_1} and n_{e_2}, vary with the parameters in the model. The solution trajectories are basic S-shaped curves which vary in steepness with the parameters. However, it is the equilibria and their stability that are most significant in the qualitative behavior. Figure 4.11 shows n_{e_1} and n_{e_2} as functions of the parameter N_0 and gives the their stability. This type of plot is known as a *bifurcation diagram.* Solid lines indicate stable equilibrium points, while dashed lines depict unstable equilibrium points. Chapter will present greater details on the basic principles of bifurcation theory.

4.5 Two Species Competition Model

Let $x(t)$ and $y(t)$ represent two species assumed to evolve according to the logistic growth model, which we write in polynomial form:

$$\frac{dx}{dt} = a_1 x - a_2 x^2$$
$$\frac{dy}{dt} = b_1 y - b_2 y^2,$$

where $a_i > 0$ and $b_i > 0$. Each of these species satisfies a solution of the form given by Eq. (4.5). If we now include the negative effects of *interspecies competition*, the previous set of equations yields a *two species competition model*:

$$\begin{aligned} \frac{dx}{dt} &= a_1 x - a_2 x^2 - a_3 xy = f_1(x, y), \\ \frac{dy}{dt} &= b_1 y - b_2 y^2 - b_3 yx = f_2(x, y), \end{aligned} \tag{4.19}$$

where $a_i > 0$ and $b_i > 0$. This time the system of ODEs does not have, however, an analytical solution. Nevertheless, we can infer the behavior of the solutions from a qualitative analysis of the model Eq. (4.19).

4.5.1 Qualitative Analysis

We start the analysis by recognizing that, typically, two possible outcomes can arise as a result of the two species competing for resources:

1. *Competitive Exclusion*—one species out competes the other and, as it reaches a stable equilibrium, it becomes the only survivor.
2. *Coexistence*—both species coexist while reaching a mutually stable equilibrium.

Existence of Equilibrium Solutions
Analysis begins by calculating the equilibrium points(x_e, y_e), of the system. This is done by solving $f_i(x_e, y_e) = 0, i = 1, 2$. That is

$$\begin{aligned}a_1x_e - a_2x_e^2 - a_3x_ey_e &= 0,\\ b_1y_e - b_2y_e^2 - b_3x_ey_e &= 0.\end{aligned}$$

Solving this system of equations, simultaneously, we find four equilibrium points. One equilibrium point is the trivial equilibrium

$$P_1(x_e, y_e) = (0, 0),$$

which corresponds to both species becoming extinct. In addition, there are two more equilibrium points that correspond to *carrying capacity equilibria*:

$$P_2(x_e, y_e) = \left(\frac{a_1}{a_2}, 0\right) \quad \text{and} \quad P_3(x_e, y_e) = \left(0, \frac{b_1}{b_2}\right),$$

where y_e is at *carrying capacity* (when y survives) and x_e is at *carrying capacity* (when x survives), respectively. The fourth equilibrium point corresponds to *coexistence* of the two species, which satisfies:

$$P_4(x_e, y_e) = \left(\frac{a_1b_2 - a_3b_1}{a_2b_2 - a_3b_3}, \frac{a_2b_1 - a_1b_3}{a_2b_2 - a_3b_3}\right),$$

where, generically, it can be assumed that $a_2b_2 \neq a_3b_3$. Technically, either x_e or y_e can be negative, depending on the choice of parameters. Nevertheless, we consider only the biologically meaningful case in which the parameter values yield $x_e > 0$ and $y_e > 0$.

Stability Analysis
The local stability properties of the equilibrium points can be inferred from the linearization of Eq. (4.21). This requires that we examine the spectrum of eigenvalues of the Jacobian matrix evaluated at each of the equilibria:

$$J(x_e, y_e) = \begin{pmatrix} a_1 - 2a_2x_e - a_3y_e & -a_3x_e \\ -b_3y_e & b_1 - 2b_2y_e - b_3x_e \end{pmatrix}. \tag{4.20}$$

For the trivial equilibrium point we get:

$$J(0, 0) = \begin{pmatrix} a_1 & -0 \\ 0 & b_1 \end{pmatrix}.$$

Similarly, the Jacobian matrix evaluated at the second and third equilibrium points are:

$$J\left(\frac{a_1}{a_2}, 0\right) = \begin{pmatrix} -a_1 & -a_3\frac{a_1}{a_2} \\ 0 & b_1 - \frac{b_3 a_1}{a_2} \end{pmatrix}, \quad J\left(0, \frac{b_1}{b_2}\right) = \begin{pmatrix} a_1 - \frac{a_3 b_1}{b_2} & 0 \\ -b_3\frac{b_1}{b_2} & -b_1 \end{pmatrix}.$$

The spectrum of eigenvalues at these three equilibrium points is:

$$\begin{aligned} P_1(0,0): &\quad \lambda_1 = a_1, \quad \lambda_2 = b_1, \\ P_2\left(\tfrac{a_1}{a_2}, 0\right): &\ \lambda_1 = -a_1,\ \lambda_2 = \frac{b_1 a_2 - b_3 a_1}{a_2}, \\ P_3\left(0, \tfrac{b_1}{b_2}\right): &\ \lambda_1 = -b_1,\ \lambda_2 = \frac{a_1 b_2 - a_3 b_1}{b_2}. \end{aligned}$$

$$\begin{aligned} P_1(0,0): &\quad \lambda_1 = a_1, \quad \lambda_2 = b_1, \\ P_2\left(\tfrac{a_1}{a_2}, 0\right): &\ \lambda_1 = -a_1,\ \lambda_2 = \frac{b_1 a_2 - b_3 a_1}{a_2}, \\ P_3\left(0, \tfrac{b_1}{b_2}\right): &\ \lambda_1 = -b_1,\ \lambda_2 = \frac{a_1 b_2 - a_3 b_1}{b_2}. \end{aligned}$$

Since all parameters are positive, it follows that the extinction equilibrium point is unstable. From a biological standpoint, this result implies that competition cannot result in both species disappearing. One or both must survive. Thus, if $b_1 a_2 < b_3 a_1$, then both eigenvalues of $J(P_2)$ are negative, so P_2 is a stable sink. Alternative, if $b_1 a_2 > b_3 a_1$, then one eigenvalue is positive and one is negative, so P_2 is unstable, a saddle point. Similarly, if $a_1 b_2 < a_3 b_1$ then P_3 is a stable sink. Otherwise, if $a_1 b_2 > a_3 b_1$ then P_3 is unstable, a saddle point.

The eigenvalues at the fourth equilibrium point depend on all of the parameters a_i and b_i, so each problem has to be studied on a case by case basis.

Example 4.1 Consider the competition model given by:

$$\begin{aligned} \frac{dx}{dt} &= 0.1\,x - 0.01\,x^2 - 0.02\,xy, \\ \frac{dy}{dt} &= 0.2\,y - 0.03\,y^2 - 0.04\,xy. \end{aligned}$$

The four equilibrium points are found to be:

$$P_1 = (0,0), \quad P_2 = (10,0), \quad P_3 = \left(0, \frac{20}{3}\right), \quad P_4 = (2,4).$$

The linearization yields the following Jacobian matrix:

$$J(X_e, Y_e) = \begin{pmatrix} 0.1 - 0.02X_e - 0.02Y_e & -0.02X_e \\ -0.04Y_e & 0.2 - 0.06Y_e - 0.04X_e \end{pmatrix} \tag{4.21}$$

Evaluating Eq. (4.21) at the extinction and carrying capacity equilibria, we get:

$$J(0,0) = \begin{pmatrix} 0.1 & 0 \\ 0 & 0.2 \end{pmatrix}, \quad J(10,0) = \begin{pmatrix} -0.1 & -0.2 \\ 0 & -0.2 \end{pmatrix}, \quad J\left(0, \tfrac{20}{3}\right) = \begin{pmatrix} -\frac{0.1}{3} & 0 \\ -\frac{0.8}{3} & -0.2 \end{pmatrix}.$$

The form of all these matrices implies that the eigenvalues can be read from the diagonal. The extinction equilibrium has two positive eigenvalues, giving an unstable node, which is expected as low populations should grow in a Malthusian manner. Both carrying capacity equilibria have two negative eigenvalues, giving stable nodes. Since the monocultures satisfy logistic growth, we expect at least one eigenvalue to be negative with its eigenvector pointed along the axis. This population exhibits competitive exclusion with the other eigenvalue being negative, making these equilibria asymptotically stable. One population approaches its carrying capacity and the other species goes extinct.

At the coexistence equilibrium, $(x_e, y_e) = (2, 4)$, the Jacobian matrix satisfies:

$$J(2,4) = \begin{pmatrix} -0.02 & -0.04 \\ -0.16 & -0.12 \end{pmatrix},$$

which has eigenvalues $\lambda_1 = 0.0243$ and $\lambda_2 = -0.1643$ giving a saddle node. Figure 4.12 shows the solutions near this coexistence equilibrium split between

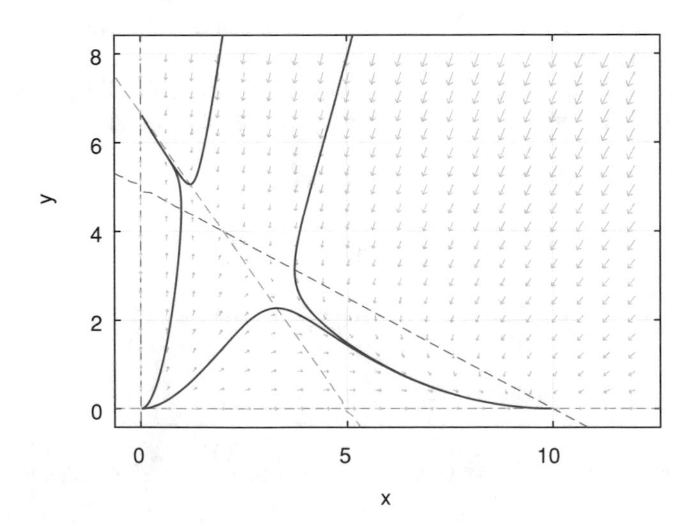

Fig. 4.12 This graph shows the direction field and nullclines (dashed lines) for Example 1 and demonstrates competitive exclusion. Solutions tend to the carrying capacities of either species X or Y

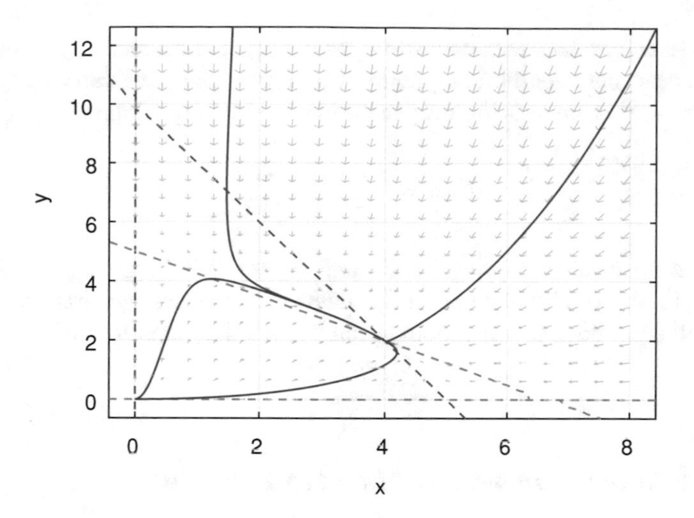

Fig. 4.13 This graph shows the direction field and nullclines for Example 2 and demonstrates cooperative coexistence. Solutions tend to the coexistence equilibrium

ones going to either of the stable equilibria. The stable manifold of this equilibrium forms a *separatrix* dividing the region into solutions going to each of the *competitive exclusion* equilibria. Note that all of these equilibria in this example are one of the generic cases of the Stable Manifold Theorem.

Example 4.2 Consider the competition model given by:

$$\frac{dx}{dt} = 0.1\,x - 0.02\,x^2 - 0.01\,xy, \tag{4.22}$$
$$\frac{dy}{dt} = 0.2\,y - 0.04\,y^2 - 0.03\,xy.$$

The linear analysis of this system is performed like the previous example by finding the Jacobian matrix for the right side of (4.22) and evaluating it at the equilibria. The Jacobian matrix satisfies:

$$J(X_e, Y_e) = \begin{pmatrix} 0.1 - 0.04X_e - 0.01Y_e & -0.01X_e \\ -0.03Y_e & 0.2 - 0.08Y_e - 0.03X_e \end{pmatrix} \tag{4.23}$$

The extinction equilibrium has exactly the same linear analysis to the previous example. At the carrying capacity equilibria, $(X_e, Y_e) = (5, 0)$ or $(X_e, Y_e) = (0, 5)$, we evaluate (4.23) and obtain:

$$J(5, 0) = \begin{pmatrix} -0.1 & -0.05 \\ 0 & 0.05 \end{pmatrix} \quad \text{and} \quad J(0, 5) = \begin{pmatrix} 0.05 & 0 \\ -0.15 & -0.2 \end{pmatrix}$$

which have eigenvalues with opposite signs giving saddle nodes. Figure 4.13 shows solutions near these equilibria asymptotically approaching the carrying capacity of X or Y on the axes, but moving away in the interior of the quadrant. When $(X_e, Y_e) = (4, 2)$,

$$J(4,2) = \begin{pmatrix} -0.02 & -0.02 \\ -0.12 & -0.18 \end{pmatrix},$$

which has eigenvalues $\lambda_1 = -0.0062$ and $\lambda_2 = -0.1938$ giving a stable node. Figure 4.13 shows all solutions in the interior of the first quadrant approaching this stable equilibria, so when both species exist, they tend toward a *cooperative* equilibria.

4.5.2 Fitting a Competition Model to Yeast Data

Gause [1, 2] also performed a third set of experiments, which combines yeast cultures to examine competition between the species. Two repetitions were done of each experiment, and data were combined and shifted to match described conditions. Table 4.3 shows the results.

In this section we attempt to fit a competition model to the data shown in Table 4.3. Thus, we need to estimate six unknown parameters, a_i and b_i, $i = 1, 2, 3$. and two initial conditions, x_0 and y_0. We consider additional assumptions in order to reduce the number of parameters that need to be fit. For instance, since in the absence of the other yeast species, assuming the same experimental conditions, the competition model (4.19) should match the monoculture logistic models given by Eq. (4.6). This assumption implies that the rate constants, a_1, a_2, b_1, and b_2, derive from fitting the monoculture logistic growth data of Tables 4.1 and 4.2. That is:

$$a_1 = 0.25864, \quad a_2 = 0.020298, \quad b_1 = 0.057442, \quad \text{and} \quad b_2 = 0.0097687.$$

Table 4.3 Competition yeast experiments with *S. cerevisiae* and *S. kephir*

Time (hr)	0	1.5	9	10	18	18	23
Vol (*S. cerevisiae*)	0.375	0.92	3.08	3.99	4.69	5.78	6.15
Vol (*S. kephir*)	0.29	0.37	0.63	0.98	1.47	1.22	1.46
Time (hr)	25.5	27	38	42	45.5	47	
Vol (*S. cerevisiae*)	9.91	9.47	10.57	7.27	9.88	8.3	
Vol (*S. kephir*)	1.11	1.225	1.1	1.71	0.96	1.84	

In this way, we have already reduced the total number of fitting unknowns to four: two parameters, a_3 and b_3, which represent the interspecies competition coefficients, and two initial conditions, x_0 and y_0. To estimate these four unknowns, we minimize the following SSE formula:

$$J(a_3, b_3, x_0, y_0) = \sum_{i=0}^{N} \left(\left(x_d(t_i) - x(t_i; a_3, x_0) \right)^2 + \left(y_d(t_i) - y(t_i; b_3, y_0) \right)^2 \right)^2, \tag{4.24}$$

where $x_d(t_i)$ and $y_d(t_i)$ are the data for *S. cerevisiae* and *S. kephir*, respectively. The corresponding solutions to Eq. (4.19) at times t_i are $x(t_i; a_3, x_0)$ and $y(t_i; b_3, y_0)$ where these numerical solutions depend on a_3, x_0, b_3, and y_0. The algorithm is initiated with a reasonable guess to the parameters and initial conditions. A numerical ODE solver is applied to the model Eq. (4.19) and inserted into the SSE formula (4.24). Finally a nonlinear minimizing program is used on the SSE to find the best fitting parameters and initial conditions.

The Appendix A.3.2 provides details for implementation of this computer algorithm with this two yeast species competition model. The MatLab code gives the best fitting interspecies competition parameters for the competition model with:

$$a_3 = 0.057011 \quad \text{and} \quad b_3 = 0.0047576$$

and initial conditions:

$$x(0) = 0.41095 \quad \text{and} \quad y(0) = 0.62578.$$

The least sum of square errors is 9.312. We then arrive at the following best fit competition model for Gause mixed culture data:

$$\begin{aligned} \frac{dx}{dt} &= 0.25864\,x - 0.020298\,x^2 - 0.057011\,xy, \\ \frac{dy}{dt} &= 0.057442\,y - 0.0097687\,y^2 - 0.0047576\,xy. \end{aligned} \tag{4.25}$$

The best fitting competition model is readily simulated and compared to the Gause mixed culture data.

Figure 4.14 shows numerical simulations of the fitted competition model with the data.

The best fit model Eq. (4.25) has four equilibrium points: extinction, $P_1(0, 0)$, carrying capacity, $P_2(12.742, 0)$ or $P_3(0, 5.8802)$, and coexistence, $P_4(10.257, 0.88482)$. A linearization analysis (left as an exercise) yields the following spectrum of eigenvalues of the Jacobian matrix:

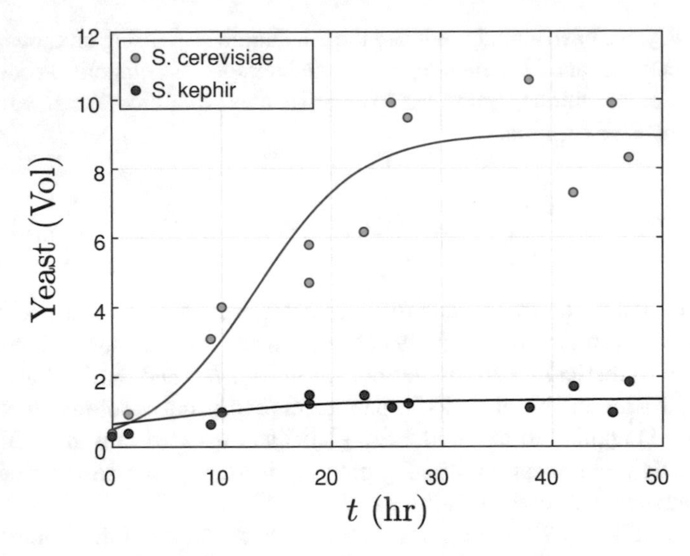

Fig. 4.14 Simulations of a two-species competition model fitted for *S. cerevisiae* and *S. kephir* mixed culture data

$$
\begin{aligned}
&P_1(0,0): && \lambda_1 = 0.25864, && \lambda_2 = 0.05744,\\
&P_2(12.742,0): && \lambda_1 = -0.25864, && \lambda_2 = -0.00318,\\
&P_3(0,5.8802): && \lambda_1 = -0.0766, && \lambda_2 = -0.0574,\\
&P_4(10.257,0.88482): && \lambda_1 = -0.2199, && \lambda_2 = 0.00301.
\end{aligned}
$$

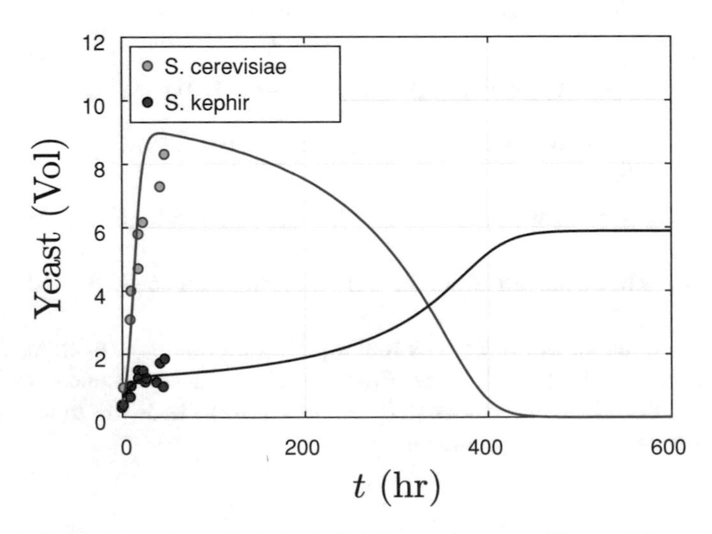

Fig. 4.15 Long-term numerical simulations of a two-species competition model reveal that, over long periods of time, species *S. cerevisiae* dies out while *S. kephir* survives

It follows that P_1 is unstable (as expected), $J(P_2)$ and $J(P_3)$ have both negative eigenvalues, so they are stable. $J(P_4)$ has one negative eigenvalue and one positive eigenvalue, so P_4 is a saddle point. The eigenvector associated with λ_1 is tangent to the one-dimensional *separatrix*, which connects to the extinction equilibrium separating the first quadrant into regions, which are attracted to one of the two carrying capacity equilibria, where one species goes extinct.

Figure 4.15 shows a longer term series simulation of the best fitting competition model Eq. (4.25). The culture of *S. cerevisiae* initially grows very rapidly because the saddle node has a fairly strong negative eigenvalue, attracting solutions toward the coexistence equilibrium. However, it eventually dies out. On the other hand, the culture of *S. kephir* grows slower but, in the long-term, it possesses a competitive advantage, which is very common among similar species.

4.6 Predator-Prey Model

In the 1920s, Vito Volterra began to study the population of sharks, which typically prey on small fish such as sardines and tuna. It can be said that Volterra's work, and book [17], *A Mathematical Theory of the Struggle for Life* led to the beginning of the field of Mathematical Ecology. Since sharks are, obviously, predators, and sardines or tuna are prey, their encounters are known as *predator-prey* interactions. Nowadays, many other species are known to interact with one another in a predator-prey manner. For instance, lynx are highly specialized predators that depend on one type of prey, hares, for survival. In this section, we introduce *predator-prey models* as a mean to describe the life-cycle of many animals as they compete for survival.

4.6.1 *Sharks and Food Fish*

Figure 4.16 showcases the predator-prey fluctuations between populations of sharks and those of tuna.

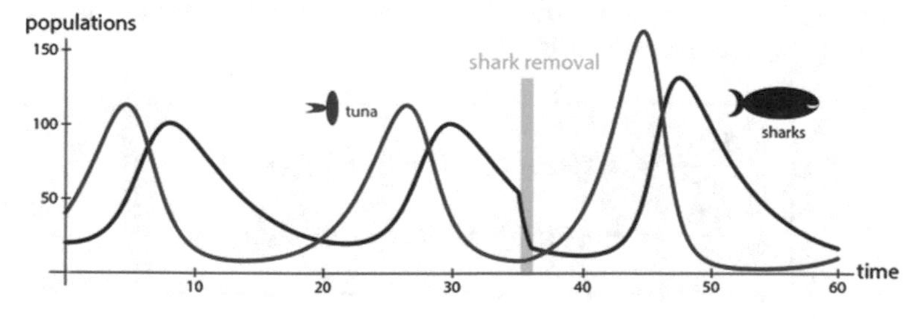

Fig. 4.16 Shark and tuna interactions [18]

Our goal is to derive a mathematical model that describes these interactions. Let us choose tuna as the prey, and make two modeling assumptions.

1. Population of sardines or tuna are usually kept down exclusively by sharks.
2. Population of sharks is at the limit of its food supply, and is kept in check by the lack of tuna.

4.6.2 *Lotka-Volterra Model*

The first version of the *predator-prey model* was published by Lotka [19, 20]. In that work, Lotka was able to show that his model can reproduce the periodic cycle that is commonly observed in the interaction of the species, as is shown in Fig. 4.16. A few years later, Volterra [21] introduced a similar model to explain the fluctuations in the population of fish, e.g., sharks. Nowaday, the combined efforts have led to the *Lotka-Volterra predator-prey model*, which can be derived based on first principles.

Let $x(t)$ represent the population of food fish (tuna), and let $y(t)$ represent the number of predators (sharks). Predation is modeled by assuming random contact between the species in proportion to their populations with a fixed percentage of those contacts resulting in death of the prey species. This is modeled by a negative term, $-a_2x(t)y(t)$. Thus, the fish growth model is:

$$\frac{dx(t)}{dt} = a_1x(t) - a_2x(t)y(t).$$

The primary growth for the shark population depends on adequate nutrients from predation on food fish. This growth rate is similar to the death rate for the fish population, $b_2y(t)x(t)$. Without food fish, the shark population declines in proportion to its own population, $-b_1y(t)$. Thus, the simplified growth model for the shark population is:

$$\frac{dy(t)}{dt} = -b_1y(t) + b_2y(t)x(t).$$

Combining these last two sets of equations, we arrive at the following predator-prey model for sharks and food fish:

$$\begin{aligned} \frac{dx(t)}{dt} &= a_1x(t) - a_2x(t)y(t) = f_1(x, y), \\ \frac{dy(t)}{dt} &= -b_1y(t) + b_2y(t)x(t) = f_2(x, y). \end{aligned} \tag{4.26}$$

Existence of Equilibrium Points

We look for equilibrium points, (x_e, y_e), by solving, simultaneously, $f_1(x, y) = 0$

and $f_2(x, y) = 0$, which yields two equilibrium points: P_1 which corresponds to *extinction* and P_2, which corresponds to *coexistence*:

$$P_1(x_e, y_e) = (0, 0) \quad \text{and} \quad P_2(x_e, y_e) = \left(\frac{b_1}{b_2}, \frac{a_1}{a_2}\right),$$

It is interesting that the nonzero equilibrium for fish, x_e, depends only on the parameters governing the shark population, while the nonzero shark equilibrium, L_e, depends only on the parameters governing the fish population.

Stability

The local stability properties of the equilibrium points can be obtained by studying the eigenvalues of the Jacobian matrix:

$$J = \begin{pmatrix} a_1 - a_2 y_e & -a_2 y_e \\ b_2 y_e & -b_1 + b_2 x_e \end{pmatrix}.$$

The spectrum of eigenvalues of the Jacobian matrix is:

$$\begin{aligned} &P_1(0,0): && \lambda_1 = a_1, && \lambda_2 = -b_1, \\ &P_2(12.742, 0): && \lambda_1 = 0 + i\sqrt{a_1 b_1}, && \lambda_2 = 0 - i\sqrt{a_1 b_1}. \end{aligned}$$

It follows that the trivial equilibrium, $P_1(x_e, y_e) = (0, 0)$, is a saddle node with solutions exponentially growing along the x-axis and decaying along the y-axis. The second equilibrium, $P_2(x_e, y_e)$, is a center, which suggests that the solution of the predator-prey model contains a family of infinitely many periodic solutions parameterized only by initial conditions. This last solution produces a *structurally unstable model*. The model is structurally unstable because small perturbations from the nonlinear terms could result in the solution either spiraling toward or away from the equilibrium or possibly a completely different trajectory. Figure 4.17 shows numerical simulations of the predator-prey model Eq. (4.26).

The orbits are periodic, so the solutions are integrated for one period to determine the average value of the solutions. Assume a period of T, then the average populations of fish and sharks satisfy:

$$\bar{x} = \frac{1}{T}\int_0^T x(t)dt \quad \text{and} \quad \bar{y} = \frac{1}{T}\int_0^T y(t)dt.$$

From Eq. (4.26), we write:

$$\begin{aligned} \frac{1}{T}\int_0^T \frac{x'(t)}{x(t)} dt &= \frac{1}{T}\int_0^T (a_1 - a_2 y(t))\, dt, \\ \frac{1}{T}\ln(x(t))\Big|_0^T &= a_1 - \frac{a_2}{T}\int_0^T y(t)dt. \end{aligned}$$

The left hand side above is **zero** because $x(T) = x(0)$ from the assumption of periodicity. This gives the average shark population:

$$\frac{1}{T}\int_0^T y(t)dt = \bar{y} = \frac{a_1}{a_2}.$$

An almost identical argument gives the average fish population:

$$\frac{1}{T}\int_0^T x(t)dt = \bar{x} = \frac{b_1}{b_2}.$$

It follows that the average population around any periodic orbit is given by the equilibrium value:

$$(\bar{x}, \bar{y}) = \left(\frac{b_1}{b_2}, \frac{a_1}{a_2}\right).$$

The Lotka-Volterra model (4.26) is structurally unstable because of the center node. However, the equilibrium is robust because all periodic orbits have the same mean, the equilibrium.

4.7 Method of Averaging

The *method of averaging* is a useful tool in dynamical systems, where time-scales in a differential equation are separated between a fast oscillation and slower behavior. The fast oscillations are averaged out to allow the determination of the qualitative behavior of an averaged dynamical system. The averaging method dates from perturbation problems that arose in celestial mechanics, when Lagrange [22] formulated the gravitational three-body problem as a perturbation of the two-body problem. The

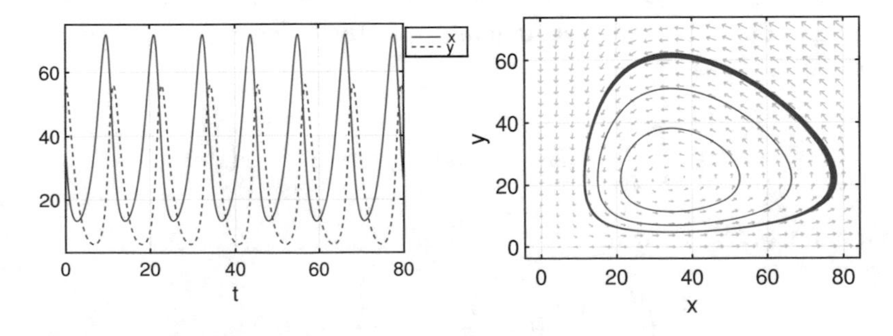

Fig. 4.17 Numerical simulations of the predator-prey model associated with the interaction between sharks and tuna. (Left) Time series solution. (Right) Phase portrait. Parameters are: $a_1 = 0.453$, $a_2 = 0.0205$, $b_1 = 0.79$, $b_2 = 0.0229$

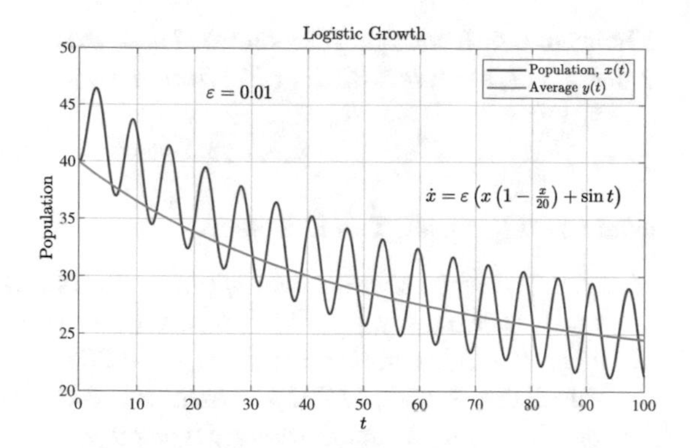

Fig. 4.18 The logistic growth model with seasonal variation in growth rate is shown with the averaged solution

validity of this method waited until Fatou [23] proved some of the asymptotic results. Significant results, including Krylov-Bogoliubov [24], followed in the 1930s, making averaging methods important classical tools for analyzing nonlinear oscillations.

The method of averaging is applicable to systems of the form:

$$\dot{x} = \varepsilon f(x, t, \varepsilon), \qquad x \in U \subset \mathbb{R}^n, \qquad \varepsilon \ll 1, \tag{4.27}$$

where $f : \mathbb{R}^n \times \mathbb{R} \times \mathbb{R}^+ \to \mathbb{R}^n$ is C^r, $r \geq 1$, bounded on bounded sets, and T-periodic in t; U is bounded and open. The associated autonomous averaged system is given by:

$$\dot{y} = \frac{\varepsilon}{T} \int_0^T f(y, t, 0) dt = \varepsilon \bar{f}(y). \tag{4.28}$$

The averaged system (4.28) is chosen to be easier to study, yet its properties should reflect the dynamics of (4.27).

Example 4.3 Consider the logistic growth model with some seasonal variation:

$$\dot{x} = \varepsilon \left(x \left(1 - \frac{x}{M} \right) + \sin(\omega t) \right), \qquad x \in \mathbb{R}, \quad 0 < \varepsilon \ll 1.$$

When the seasonal growth is averaged over its period, then the averaged equation satisfies:

$$\dot{y} = \varepsilon y \left(1 - \frac{y}{M} \right), \qquad y \in \mathbb{R}.$$

Figure 4.18 shows a simulation of the seasonal growth model with the averaged logistic growth model. The solution $x(t)$ shows complicated dynamics. However, when the oscillations are removed, the solution $y(t)$ reduces to the standard logistic growth model where a stable equilibrium occurs at $y_e = M$ with an unstable equilibrium at $y_e = 0$. The solution $x(t)$ simply oscillates about the solution $y(t)$.

The Method of Averaging is stated in the following theorem.

Theorem 4.4 (Averaging Theorem) *There exists a C^r change of coordinates $x = y + \varepsilon w(y, t, \varepsilon)$ under which (4.27) becomes*

$$\dot{y} = \varepsilon \bar{f}(y) + \varepsilon^2 f_1(y, t, \varepsilon),$$

where f_1 is of period T in t. Moreover,

1. *If $x(t)$ and $y(t)$ are solutions of (4.27) and (4.28) based at x_0, y_0, respectively, at $t = 0$, and $|x_0 - y_0| = \mathcal{O}(\varepsilon)$, then $|x(t) - y(t)| = \mathcal{O}(\varepsilon)$ on a time scale $t \sim \frac{1}{\varepsilon}$.*
2. *If p_0 is a hyperbolic fixed point of (4.28) then there exists $\varepsilon_0 > 0$ such that, for all $0 < \varepsilon \leq \varepsilon_0$, (4.27) possesses a unique hyperbolic periodic orbit $\gamma_\varepsilon(t) = p_0 + \mathcal{O}(\varepsilon)$ of the same stability type as p_0.*
3. *If $x^s(t) \in W^s(\gamma_\varepsilon)$ is a solution of (4.27) lying in the stable manifold of the hyperbolic periodic orbit $\gamma_\varepsilon = p_0 + \mathcal{O}(\varepsilon)$, $y^s(t) \in W^s(p_0)$ is a solution of (4.28) lying in the stable manifold of the hyperbolic fixed point p_0 and $|x^s(0) - y^s(0)| = \mathcal{O}(\varepsilon)$, then $|x^s(t) - y^s(t)| = \mathcal{O}(\varepsilon)$ for $t \in [0, \infty)$. Similar results apply to solutions lying in the unstable manifolds on the time interval $t \in (-\infty, 0]$.*

More theoretical details and proofs for this theorem are found in standard texts of ODEs, such as Guckenheimer and Holmes (Chap. 4 [3]), Hale (Chap. 5 [25]), Sanders et al. [26]. Information on the stable manifold and hyperbolic fixed points are covered in Sect. 4.3.2. This theorem states that an equilibrium point of the averaged Eq. (4.28) corresponds to a periodic orbit of the original model (solution $x(t)$).

The near unitary transformation, $x = y + \varepsilon w(y, t, \varepsilon)$, has w being T-periodic in t. We show some details for the formulation of $w(y, t, \varepsilon)$ and $f_1(y, t, \varepsilon)$. A Taylor expansion of (4.27) gives this perturbation problem in *standard form*:

$$\dot{x} = \varepsilon f^{(1)}(x, t) + \varepsilon^2 f^{(2)}(x, t, \varepsilon),$$

with $f^{(1)}$ and $f^{(2)}$ being T-periodic, $x \in \mathbb{R}^n$, and $\varepsilon \ll 1$. Decomposing $f^{(1)}$ into its mean and oscillating components gives:

$$f^{(1)}(x, t) = \bar{f}^{(1)}(x) + \hat{f}^{(1)}(x, t).$$

Differentiating the near-identity transformation gives:

$$\left[I + \varepsilon D_y w\right] \dot{y} + \varepsilon \frac{\partial w}{\partial t} = \varepsilon \bar{f}^{(1)}(y + \varepsilon w) + \varepsilon \hat{f}^{(1)}(y + \varepsilon w, t) + \mathcal{O}\left(\varepsilon^2\right),$$

so

$$\dot{y} = \varepsilon \left[I + \varepsilon D_y w\right]^{-1} \left[\bar{f}^{(1)}(y + \varepsilon w) + \hat{f}^{(1)}(y + \varepsilon w, t) - \frac{\partial w}{\partial t} + \mathcal{O}(\varepsilon)\right].$$

We take w to be the antiderivative of the oscillatory part $\hat{f}^{(1)}$, $\frac{\partial w}{\partial t} = \hat{f}^{(1)}$, then using ε-expansions, we obtain:

$$\dot{y} = \varepsilon \bar{f}(y) + \varepsilon^2 \left[D_y f(y,t,0) w(y,t,0) - D_y w(y,t,0) \bar{f}^{(1)}(y) \right] + \mathcal{O}\left(\varepsilon^3\right),$$

where $f_1(y,t,\varepsilon)$ is defined to be the ε terms with order greater than or equal to two. The proof is completed using analytical tools, like Gronwall's inequality, to prove the solutions remain $\mathcal{O}(\varepsilon)$ close to each other.

4.7.1 Quasilinear ODE and Lagrange Standard Form

Many mathematical models begin as linear systems of ODEs, which are perturbed by some small nonlinearity. This is observed in the models of several classic oscillatory systems, which are studied in more detail later in this chapter.

Consider the initial value problem:

$$\dot{x} = A(t)x + \varepsilon g(x,t,\varepsilon), \qquad x(0) = x_0, \tag{4.29}$$

where $x \in \mathbb{R}^n$, $A(t)$ is a continuous $n \times n$ matrix function and $g(x,t,\varepsilon)$ is a sufficiently smooth function of t and x. When $\varepsilon = 0$, Eq. (4.29) is a first order linear system of differential equations. We assume that $\Phi(t)$ is the *fundamental matrix solution* [6] of the unperturbed system ($\varepsilon = 0$), and $y(t)$ satisfies $y(0) = x_0$ and becomes part of comoving (Lagrangian) coordinates with

$$x = \Phi(t)y, \qquad \text{so} \qquad \dot{x} = \dot{\Phi}(t)y + \Phi(t)\dot{y}.$$

Since $x(t)$ solves the perturbed system above, we have

$$\dot{\Phi}(t)y + \Phi(t)\dot{y} = A(t)\Phi(t)y + \varepsilon g(\Phi(t)y,t,\varepsilon),$$

or

$$\Phi(t)\dot{y} = \left(A(t)\Phi(t) - \dot{\Phi}(t)\right)y + \varepsilon g(\Phi(t)y,t,\varepsilon).$$

(Note if A is constant, $\Phi(t) = e^{At}$.) Since $\Phi(t)$ is the fundamental matrix solution of the unperturbed system, so $\dot{\Phi}(t) = A(t)\Phi(t)$, it follows that:

$$\Phi(t)\dot{y} = \varepsilon g(\Phi(t)y,t,\varepsilon), \qquad \text{equivalently} \qquad \dot{y} = \varepsilon \Phi^{-1}(t) g(\Phi(t)y,t,\varepsilon).$$

This equation is said to have the *Lagrange standard form* and can be written without loss of generality as

$$\dot{y} = \varepsilon f(y,t,\varepsilon),$$

which is the same form as our weakly nonlinear ODE given by (4.27). This version of the standard form will be used in the next section.

4.8 Linear and Nonlinear Oscillators

This section presents an overview of mathematical models for linear and nonlinear oscillators. Linear oscillators are well studied and have their natural frequencies independent of initial conditions and amplitude of oscillation. Thus, the oscillatory behavior of a solution is similar for initial conditions close to each other. This is a simplifying property that is broken by nonlinearity. A nonlinear oscillator can have solutions, which depend on amplitude and that are sensitive in their long-term behavior to initial conditions. This richness in solution dynamics make the study of nonlinear oscillators a challenge.

4.8.1 Linear Oscillators

A general model for a linear oscillator can be written in the form

$$\mathcal{L}(x) = 0,$$

where $\mathcal{L}$ is some linear operator acting on the state variable x. In fact, a linear differential equation of the form

$$a_2 \frac{d^2x}{dt^2} + a_1 \frac{dx}{dt} + a_0 x(t) = F(t) \tag{4.30}$$

is used as a "universal" model of a linear oscillator system. In this formulation, $F(t)$ represents an externally applied driving force. Figure 4.19 shows a spring-mass system that is usually studied in basic mechanics [27, 28] and a first course in ODEs [9].

Newton's second law of motion [27, 28] states that the mass of an object, m, times its acceleration is equal to the sum of all forces acting on it. The object in Fig. 4.19 is the mass, m, with its position being x, where it is assumed that the mass

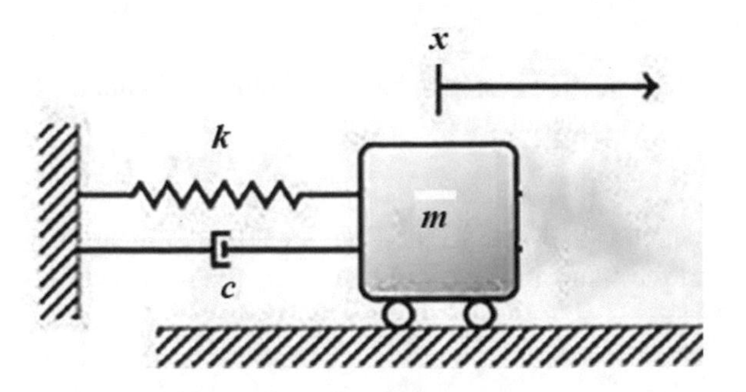

Fig. 4.19 Mass-spring-damping oscillator system

is in equilibrium at $x = 0$. We consider three forces acting on m. The first force is a Hooke's law spring, which dictates that the force the spring exerts on the mass is proportional to the distance from equilibrium and given by

$$F_s(x) = -kx,$$

where k is a material-dependent constant that defines the stiffness of the spring.

The second force comes from friction or viscous damping. Typically, this is modeled as being proportional to the velocity of the mass, so

$$F_f = -cv = -c\frac{dx}{dt}.$$

The third force considered is an external force, $F(t)$. Combining these in Newton's Second Law give:

$$m\frac{d^2x}{dt^2} = F_s + F_f + F(t) = -kx - c\frac{dx}{dt} + F(t),$$

which readily becomes Eq. (4.30), with $a_2 = m$, $a_1 = c$, and $a_0 = k$.

When there is no external force ($F(t) = 0$) and frictional forces are ignored ($c = 0$), this spring-mass model reduces to the *simple harmonic oscillator*:

$$\boxed{m\frac{d^2x}{dt^2} + kx = 0.} \tag{4.31}$$

This second order ODE has eigenvalues:

$$\lambda = \pm i\sqrt{\frac{k}{m}} = \pm i\omega_0,$$

which has a solution of the form:

$$x(t) = A\cos(\omega_0 t + \varphi),$$

where A represents the amplitude of the oscillations, φ is the phase, and ω_0 is the undamped angular frequency. The period of the motion, T, and frequency, f, are

$$T = \frac{2\pi}{\omega_0} \quad \text{and} \quad f = \frac{1}{T}.$$

Note that the harmonic oscillator has only one frequency and that the period of the motion is the inverse of the frequency. Furthermore, the frequency does not depend on

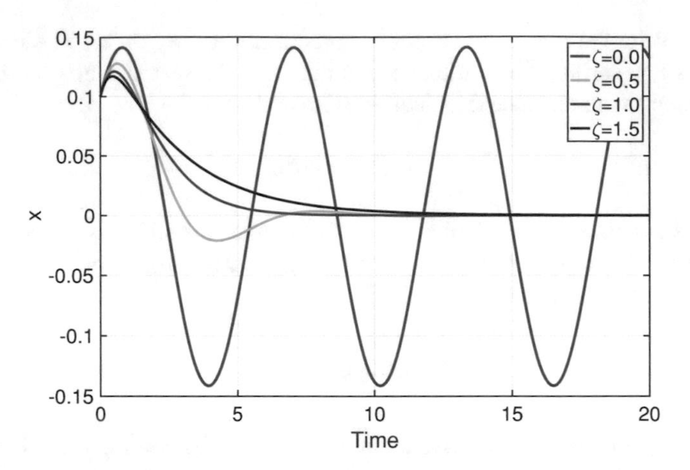

Fig. 4.20 Modes of oscillation of a linear oscillator model (4.32)

the amplitude or initial conditions, but depends on the spring stiffness and magnitude of the mass.

In practical applications, there is friction or damping, so $c \neq 0$. This gives the *damped harmonic oscillator*, $m\ddot{x} + c\dot{x} + kx = 0$, which can be written in the form:

$$\boxed{\frac{d^2x}{dt^2} + 2\zeta\omega_0\frac{dx}{dt} + \omega_0^2 x = 0,} \tag{4.32}$$

where

$$\zeta = \frac{c}{2\sqrt{mk}}$$

is known as the damping ratio and $\omega_0 = \sqrt{\frac{k}{m}}$ is defined as before. This model has eigenvalues:

$$\lambda = \left(-\zeta \pm \sqrt{\zeta^2 - 1}\right)\omega_0.$$

With all positive coefficients in the characteristic equation, the eigenvalues must all have negative real parts. This results in the **three** classic cases for the damped harmonic oscillator, Eq. (4.32). Figure 4.20 illustrates the behavior of these cases for the damped harmonic oscillator and adds the harmonic oscillator ($\zeta = 0$).

(i) *Overdamped*: If $\zeta > 1$, then the eigenvalues are negative with $\lambda_1 < \lambda_2 < 0$, which gives a solution of the form:

$$x(t) = c_1 e^{\lambda_1 t} + c_2 e^{\lambda_2 t},$$

where c_1 and c_2 are arbitrary constants. Thus, the solution returns to its equilibrium state $x = 0$ without oscillating with larger damping ratios returning faster to the equilibrium state.

(ii) *Critically damped*: If $\zeta = 1$, then $\lambda = -\zeta\omega_0$ (repeated eigenvalue), which gives a solution of the form:

$$x(t) = c_1 e^{-\zeta\omega_0 t} + c_2 t e^{-\zeta\omega_0 t}.$$

Oscillations are still not possible, and the system converges to the trivial equilibrium, $x = 0$.

(iii) *Underdamped*: If $0 < \zeta < 1$, then the eigenvalues are two complex conjugate roots $\lambda = -\zeta\omega_0 \pm i\omega_0\sqrt{1-\zeta^2}$, which has a solution of the form:

$$x(t) = Ae^{-\zeta\omega_0 t}\cos\left(\omega_0\sqrt{1-\zeta^2}t + \varphi\right).$$

The mass oscillates, but the oscillations die out and the system converges to the trivial equilibrium, $x = 0$.

This section concludes with the application of an external driving force, so that the *driven harmonic oscillator* model satisfies:

$$\frac{d^2x}{dt^2} + 2\zeta\omega_0\frac{dx}{dt} + \omega_0^2 x = \frac{F(t)}{m}. \tag{4.33}$$

Specifically, we study the effects of applying a sinusoidal force of the form

$$F(t) = F_0\cos(\omega t).$$

This second order linear nonhomogeneous ODE model is readily solved with the *method of undetermined coefficients* [9] (or alternately by *variation of parameters* [9]). The resulting particular solution, $x_p(t)$, satisfies:

$$x_p(t) = -\frac{(\omega^2-\omega_0^2)\frac{F_0}{m}}{(\omega^2-\omega_0^2)+(2\zeta\omega_0\omega)^2}\cos(\omega t) + \frac{(2\zeta\omega_0\omega)\frac{F_0}{m}}{(\omega^2-\omega_0^2)+(2\zeta\omega_0\omega)^2}\sin(\omega t),$$

which with a little bit of algebra, can be written in amplitude-phase form

$$x_p(t) = \frac{F_0}{m\omega\sqrt{\left(1-\frac{\omega_0^2}{\omega^2}\right)^2 + (2\zeta\omega_0)^2}}\cos(\omega t - \phi), \tag{4.34}$$

where the phase ϕ is given by

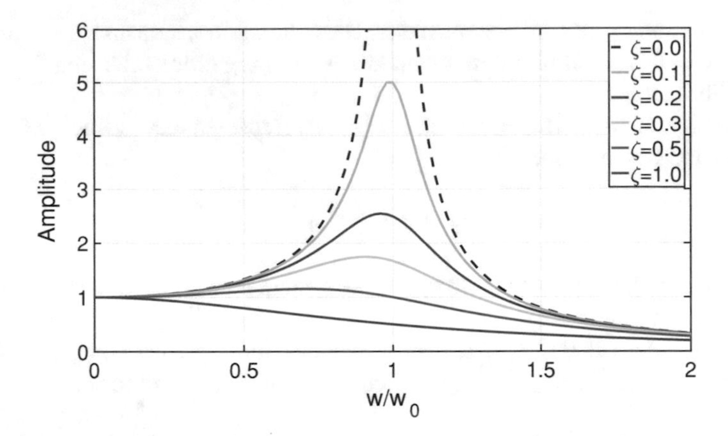

Fig. 4.21 Amplitude response in the steady-state solution of a sinusoidally forced linear oscillator model (4.33)

$$\phi = \arctan\left(-\frac{2\zeta\omega_0\omega}{(\omega^2 - \omega_0^2)}\right).$$

The general solution, $x_g(t)$, to the forced linear oscillator Eq. (4.33) is expressed as

$$x_g(t) = x_h(t) + x_p(t),$$

where $x_h(t)$ is the homogeneous solution of the unforced system, $F_0 = 0$, which is governed by the damping ratio. This solution is known as the *transient solution* because in all three cases where $\zeta > 0$, $x_h(t)$ asymptotically approaches the trivial equilibrium. For this reason, $x_p(t)$ is known as the *steady-state solution*. Figure 4.21 shows the amplitude of the steady-state solution as a function of the relative frequency ω/ω_0, with various values of the damping ratio ζ. Observe that as ζ becomes smaller, the peak in the amplitude response becomes larger. The location of the peak approaches the relative frequency $\omega/\omega_0 = 1$, where the denominator is minimized. When $\omega/\omega_0 = 1$ and $\zeta = 0$, the undamped harmonic oscillator is in *resonance* , and the particular solution, $x_p(t)$, is unbounded.

4.8.2 Conversion to a System of Differential Equations

Section 4.3.2 examines the first order linear system of ODEs given by Eq. (4.12) and shows how the stability properties relate to the eigenvalues of the matrix A. An nth-order scalar linear differential equation of the form

$$\frac{d^n y}{dt^n} + a_{n-1}\frac{d^{n-1}y}{dt^{n-1}} + \ldots + a_1\frac{dy}{dt} + a_0 y(t) = 0, \tag{4.35}$$

is readily transformed to $\dot{x} = Ax$ with $x \in \mathbb{R}^n$ using a standard technique [9]. The following sequence of substitutions is introduced:

$$x_1 = y, \quad x_2 = \frac{dy}{dt}, \quad \ldots, \quad x_n = \frac{d^{n-1}y}{dt^{n-1}},$$

so

$$\begin{aligned}
&\dot{x}_1 = x_2, \quad \dot{x}_2 = x_3, \quad \ldots, \quad \dot{x}_{n-1} = x_n, \\
&\dot{x}_n = -a_0 x_1 - a_1 x_2 - \ \ldots \ -a_{n-1} x_n.
\end{aligned}$$

It follows that

$$A = \begin{pmatrix} 0 & 1 & 0 & \ldots & 0 \\ 0 & 0 & 1 & \ldots & 0 \\ \vdots & \ddots & \ddots & \ddots & \vdots \\ 0 & \ldots & 0 & \ldots & 1 \\ -a_0 & -a_1 & -a_2 & \ldots & -a_{n-1} \end{pmatrix}.$$

Since (4.35) and (4.12) with A above are equivalent ODEs, the solutions of the characteristic equation of (4.35) match the eigenvalues of A. It follows that the behavior of the fundamental solution of (4.12) produces the same behavior as the solution to (4.35) based on the Stable Manifold Theorem (Theorem 4.3).

Example 4.4 Consider the unforced linear oscillator model:

$$\frac{d^2y}{dt^2} + 2\zeta\omega_0 \frac{dy}{dt} + \omega_0^2 y = 0. \tag{4.36}$$

Let

$$x_1 = y, \qquad x_2 = \frac{dy}{dt},$$

where x_1 is the position and x_2 is the velocity. Differentiating gives:

$$\begin{aligned}
\dot{x}_1 &= \frac{dy}{dt} = x_2, \\
\dot{x}_2 &= \frac{d^2y}{dt^2} = -2\zeta\omega_0 \frac{dy}{dt} - \omega_0^2 y = -2\zeta\omega_0 x_2 - \omega_0^2 x_1.
\end{aligned}$$

In matrix form with $x = (x_1, x_2)^T$, this system is written:

$$\dot{x} = Ax = \begin{pmatrix} 0 & 1 \\ -\omega_0^2 & -2\zeta\omega_0 \end{pmatrix} x.$$

The characteristic equation for A is easily seen to satisfy:

$$\lambda^2 + 2\zeta\omega_0\lambda + \omega_0^2 = 0,$$

which matches the characteristic equation for (4.32).

The equilibrium for this system is $x_e = (x_{1e}, x_{2e})^T = (0, 0)^T$. Since the eigenvalues have negative real parts, the Stable Manifold Theorem 4.3 shows that only the stable linear subspace, E^s, exists for this example, implying all solutions asymptotically approach $(x_{1e}, x_{2e})^T$. With the techniques of Sect. 4.3.3, one can produce the 2D phase portrait for this example with the axes being the position, x_1, and velocity, x_2. The overdamped case produces a stable node, the critically damped case gives a stable improper node, and the underdamped case produces a stable focus or spiral, where all cases show the solution trajectories approaching the equilibrium or origin.

Mathematical models for nonlinear oscillators are, in general, much less amenable to analysis. Finding closed form solutions for nonlinear ODEs is rare, occurring only for special cases. However, there exist many ideas and methods from dynamical systems theory for obtaining qualitative behavior of these nonlinear ODEs. Below we use several classic nonlinear oscillators to illustrate key mechanisms and mathematical tools for studying these oscillatory behaviors.

4.8.3 Duffing Oscillator

The Duffing oscillator is a well-known example of a nonlinear oscillator that serves as a model for a periodically forced elastic beam. The canonical model is given by

$$\boxed{\ddot{x} + \delta\dot{x} + \beta x + \alpha x^3 = \gamma\cos\omega t,} \tag{4.37}$$

where x represents the deflections of the beam from a zero equilibrium, δ describes damping strength, β and α are, respectively, the linear and nonlinear elastic properties of the beam, and $\gamma\cos\omega t$ represents a sinusoidal forcing term of strength γ. Mathematically, this model differs from the forced linear harmonic oscillator (4.33) with its cubic term; however, this can significantly affect the dynamics.

Figure 4.22 illustrates the experimental setup. The beam hangs fixed from one end within a rigid frame, while the other end is deflected by the attraction force of two magnets that are held fixed within the rigid frame. The entire rigid frame is subject to a periodically driving force of amplitude γ and frequency w. The case of $\gamma > 0$ renders the model equations *nonautonomous*, and it can lead to significantly more complicated behavior, e.g., chaos.

Unforced Duffing Oscillator

The unforced Duffing equation with no damping, i.e., $\gamma = 0$ and $\delta = 0$, satisfies:

$$\ddot{x} + \beta x + \alpha x^3 = 0.$$

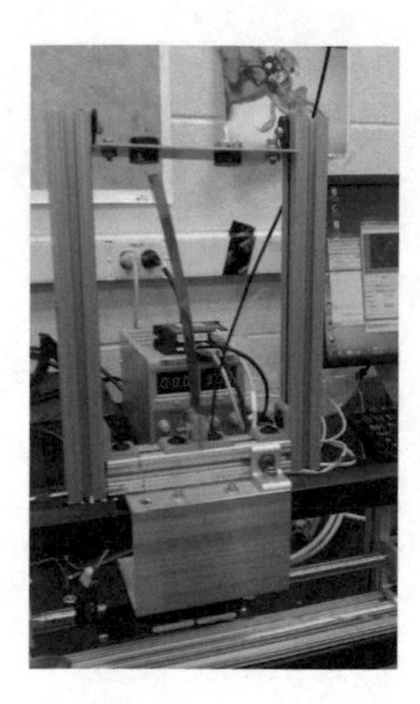

Fig. 4.22 Duffing oscillator. *Source* Georgia Tech

If we multiply by the velocity term, $\dot{x}$, we obtain the total derivative:

$$\dot{x}(\ddot{x} + \beta x + \alpha x^3) = \frac{d}{dt}\left(\tfrac{1}{2}\dot{x}^2 + \tfrac{1}{2}\beta x^2 + \tfrac{1}{4}\alpha x^4\right) = 0.$$

Integration shows that the term in parenthesis must be constant, C, so

$$H(x, \dot{x}, t) = \tfrac{1}{2}\dot{x}^2 + \tfrac{1}{2}\beta x^2 + \tfrac{1}{4}\alpha x^4 = C.$$

In *Hamiltonian mechanics*, a classical physical system is defined by a set of canonical coordinates, (x, y). Hamilton's equations define the time evolution of the system by

$$\dot{x} = \frac{\partial H}{\partial y} \quad \text{and} \quad \dot{y} = -\frac{\partial H}{\partial x},$$

where the Hamiltonian, $H(x, y, t)$, often corresponds to the total energy of the system, which in a closed system is the sum of the kinetic and potential energy. By taking the canonical coordinates to be the position, x, and velocity, $y = \dot{x}$, with $H(x, y, t) = H(x, y)$ defined by:

$$H(x, y) = \tfrac{1}{2}y^2 + \tfrac{1}{2}\beta x^2 + \tfrac{1}{4}\alpha x^4, \tag{4.38}$$

we see that

$$\dot{x} = \frac{\partial H}{\partial y} = y \quad \text{and} \quad \dot{y} = \ddot{x} = -\frac{\partial H}{\partial x} = -\beta x - \alpha x^3,$$

so is a Hamiltonian function. Figure 4.23 shows the energy potential, $H(x, y, t)$, for an unforced Duffing oscillator with schematic solution trajectories and where equilibria occur at the minima of these surfaces.

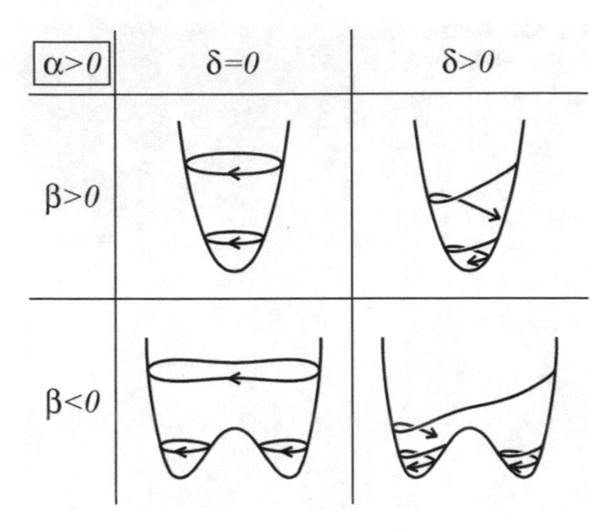

Fig. 4.23 Energy potential, $H(x, y, t)$, for an unforced Duffing oscillator with schematic solution trajectories. In the absence of damping, $\delta = 0$, energy is conserved, and the system is Hamiltonian with $H(x, y, t) = H(x, y)$. In the presence of damping, $\delta > 0$, energy is dissipated, so $\dot{H}(x, y, t) < 0$ and the system settles into an equilibrium state

For no damping, $\delta = 0$, energy is conserved. The Hamiltonian function is not explicitly dependent of time, and solutions of the unforced Duffing Eq. (4.37) exist on level curves of $H(x, y)$ and are oscillatory. When $\beta > 0$ the Hamiltonian function is a single-well potential, which leads to oscillations around a single equilibrium (see Fig. 4.24a). For $\beta < 0$, the Hamiltonian is a double-well potential function, and the system oscillates around two different equilibria (see Fig. 4.24c).

With damping, $\delta > 0$, energy is dissipated, and the Hamiltonian function is time dependent, $H(x, y, t)$, and satisfies:

$$\frac{dH}{dt}(x, y, t) = -\delta\dot{x}^2 = -\delta y^2 \leq 0. \tag{4.39}$$

This implies that solution trajectories decrease along the surface of $H(x, y, t)$ until they converge to an equilibrium. For $\beta > 0$, there is only one equilibrium point, but for $\beta < 0$, there are three equilibria (see Fig. 4.24b and d).

The qualitative behavior of the solutions begins with finding the equilibria. With $\dot{x} = y$, the unforced Duffing oscillator (4.37) is written:

$$\begin{aligned} \dot{x} &= y \\ \dot{y} &= -\delta y - \beta x - \alpha x^3. \end{aligned} \tag{4.40}$$

Equilibria occur when $y_e = 0$ and $\beta x_e + \alpha x_e^3 = 0$, so

$$x_e = 0, \pm\sqrt{-\beta/\alpha}.$$

For $\beta/\alpha \geq 0$, only the trivial equilibrium, $(x_e, y_e) = (0, 0)$, exists, while for $\beta/\alpha < 0$, there are **three** equilibria.

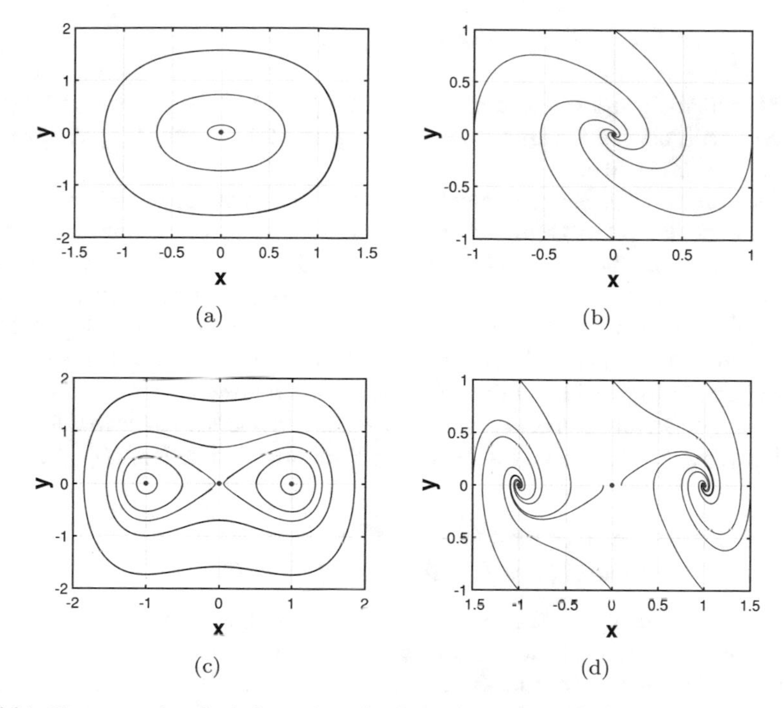

Fig. 4.24 Phase portraits of solution trajectories for an unforced Duffing oscillator with parameter $\alpha > 0$. In **a** $\beta > 0$ and $\delta = 0$, which yields a family of periodic solutions around the trivial equilibrium $(0, 0)$. In **b** $\beta > 0$ and $\delta > 0$, thus damping causes the motion to decay towards zero. In **c** $\beta < 0$ and $\delta = 0$, leads to two equilibria and families of periodic oscillations. In **d** $\beta < 0$ and $\delta > 0$, in which case damping causes again decay of solution trajectories towards one of two nontrivial equilibria. Which one is observed depends on initial conditions

To determine the local stability of the equilibria, we linearize (4.40) and compute its Jacobian matrix:

$$DF(x_e, y_e) = \begin{pmatrix} 0 & 1 \\ -\beta - 3\alpha x_e^2 & -\delta \end{pmatrix}.$$

The characteristic equation is given by:

$$\lambda^2 + \delta\lambda + \beta + 3\alpha x_e^2 = 0.$$

For $\alpha > 0$ with $\beta > 0$ and $\delta > 0$, only the trivial equilibrium, $x_e = 0$, exists, and both eigenvalues have negative real part (stable node), so all solutions are locally asymptotically stable. For $\alpha > 0$ with $\beta < 0$ and $\delta > 0$, the trivial equilibrium, $x_e = 0$, has eigenvalues with opposite signs (saddle node), so solutions are unstable. However, the nontrivial equilibria, $x_e = \pm\sqrt{-\beta/\alpha}$, have the characteristic equation:

$$\lambda^2 + \delta\lambda - 2\beta = 0,$$

which has both eigenvalues with negative real part (stable nodes), so again all solutions are locally asymptotically stable. These arguments agree with the phase portraits seen in Fig. 4.24b and d.

If $\delta = 0$ and $\beta > 0$, then the eigenvalues for $x_e = 0$ are purely imaginary, so linear analysis does not give stability though this case does result in a neutrally stable equilibrium. Similarly, for $\alpha > 0$ with $\beta < 0$ and $\delta = 0$, the nontrivial equilibria have purely imaginary eigenvalues and other means of analysis are required to show these equilibria are neutrally stable. These properties are observed in Fig. 4.24a, c. Chapter studies the richness in the behavior changes for this example as parameters change, introducing the ideas of *pitchfork* and *Hopf bifurcations*.

A Hamiltonian function or energy function can provide a valuable tool for obtaining global results for some ODEs. When the Hamiltonian function, $H(x, y, t)$ satisfies:

(i) $H(x, y, t) = 0$ if and only if $(x, y) = (0, 0)$,
(ii) $H(x, y, t) > 0$ and $\dot{H}(x, y, t) < 0$ for $(x, y) \neq (0, 0)$,

then it is a *Lyapunov function* [5, 25] and theorems show that $(x_e, y_e) = (0, 0)$ is globally asymptotically stable. For the unforced Duffing oscillator with $\alpha > 0$ and $\beta > 0$, the Hamiltonian function, (4.38), clearly satisfies (i). From the derivative of $H(x, y, t)$ given in (4.39) and $\delta > 0$, Condition (ii) also holds. The theory of Lyapunov functions gives the unique equilibrium, $(x_e, y_e) = (0, 0)$, is globally asymptotically stable for the unforced Duffing oscillator.

Weakly Forced Duffing Oscillator
A weakly forced Duffing oscillator [3] is derived from model (4.37) and written:

$$\boxed{\ddot{x} + \omega_0^2 x = \varepsilon(\gamma \cos\omega t - \delta\dot{x} - \alpha x^3),} \tag{4.41}$$

where $\omega_0^2 = \beta$, which with $x_1 = x$, $x_2 = \dot{x}$, is written in system form:

$$\begin{aligned} \dot{x}_1 &= x_2 \\ \dot{x}_2 &= -\omega_0^2 x_1 + \varepsilon(\gamma\cos\omega t - \delta x_2 - \alpha x_1^3). \end{aligned} \tag{4.42}$$

We could use the method from Sect. 4.7.1 to transform (4.42) into Lagrange standard form. However, the cubic term complicates the analysis, so a related transformation, the *van der Pol transformation*, which is based on the frequency of the forcing function, is applied with the assumption that the natural frequency, ω_0 is close to the forcing frequency, $\omega_0^2 = \omega^2 + \varepsilon\Omega$. This invertible transformation is given by:

$$\begin{pmatrix} x_1 \\ x_2 \end{pmatrix} = \Phi(t)\begin{pmatrix} y_1 \\ y_2 \end{pmatrix}, \qquad \text{where} \qquad \Phi(t) = \begin{pmatrix} \sin\omega t & \cos\omega t \\ \omega\cos\omega t & -\omega\sin\omega t \end{pmatrix}.$$

Note that replacing ω with ω_0 above gives a fundamental solution to (4.42) with $\varepsilon = 0$, which could be used in creating the Lagrange standard form.

With the van der Pol transformation, it follows that

$$\begin{pmatrix} \dot{y}_1 \\ \dot{y}_2 \end{pmatrix} = \Phi^{-1}(t) \left[-\dot{\Phi}(t) \begin{pmatrix} y_1 \\ y_2 \end{pmatrix} + \begin{pmatrix} x_2 \\ -\omega_0^2 x_1 + \varepsilon(\gamma \cos \omega t - \delta x_2 - \alpha x_1^3) \end{pmatrix} \right],$$

$$\begin{pmatrix} \dot{y}_1 \\ \dot{y}_2 \end{pmatrix} = \frac{\varepsilon}{\omega} \begin{bmatrix} (-\Omega x_1 + \gamma \cos \omega t - \delta x_2 - \alpha x_1^3) \cos \omega t \\ (\Omega x_1 - \gamma \cos \omega t + \delta x_2 + \alpha x_1^3) \sin \omega t \end{bmatrix}, \tag{4.43}$$

where $x_1 = y_1 \sin \omega t + y_2 \cos \omega t$ and $x_2 = \omega(y_1 \cos \omega t - y_2 \sin \omega t)$. This ODE in $(y_1, y_2)^T$ has the form of Eq. (4.27), which allows the Theorem of Averaging 4.4.

Expanding the right hand side of Eq. (4.43) with our expressions for x_1 and x_2, we integrate over the period of the forcing function, $2\pi/\omega$, letting y_1 and y_2 be treated as constant. The averaging process eliminates all of the trigonometric functions and reduces the averaged system to:

$$\dot{y}_1 = \frac{\varepsilon}{2\omega} \left[-\Omega y_2 + \gamma - \omega \delta y_1 - \frac{3\alpha}{4}(y_1^2 + y_2^2) y_2 \right] + \mathcal{O}\left(\varepsilon^2\right), \tag{4.44}$$

$$\dot{y}_2 = \frac{\varepsilon}{2\omega} \left[\Omega y_1 - \omega \delta y_2 + \frac{3\alpha}{4}(y_1^2 + y_2^2) y_1 \right] + \mathcal{O}\left(\varepsilon^2\right).$$

Sys. (4.44) is rewritten in polar coordinates with $y_1 = r \cos \phi$ and $y_2 = r \sin \phi$, so $r^2 = y_1^2 + y_2^2$ and $\phi = \arctan(y_2/y_1)$. This yields the following system:

$$\dot{r} = \frac{\varepsilon}{2\omega} [\gamma \cos \phi - \omega \delta r] + \mathcal{O}\left(\varepsilon^2\right), \tag{4.45}$$

$$r\dot{\phi} = \frac{\varepsilon}{2\omega} \left[\Omega r + \frac{3\alpha}{4} r^3 - \gamma \sin \phi \right] + \mathcal{O}\left(\varepsilon^2\right).$$

Equilibria are found numerically by setting the right hand side of Sys. (4.45) to zero and solving for (r, ϕ). The results are shown in Fig. 4.25 for various values of the parameter α. When $\alpha = 0$, the amplitude response exhibits a peak near the resonance condition $\omega = \omega_0$, as the nonlinear term vanishes, and the Duffing oscillator behaves as a forced linear oscillator. The case $\alpha > 0$ corresponds to a hardening beam, so the amplitude response curves to the right. The opposite case, $\alpha < 0$, corresponds to a softening beam, where the amplitude response curves towards the left.

4.8.4 *Van der Pol Oscillator*

Another classic example of a nonlinear oscillator is the van der Pol oscillator [29]. This oscillator contains a nonlinear damping term, and it originally modelled the

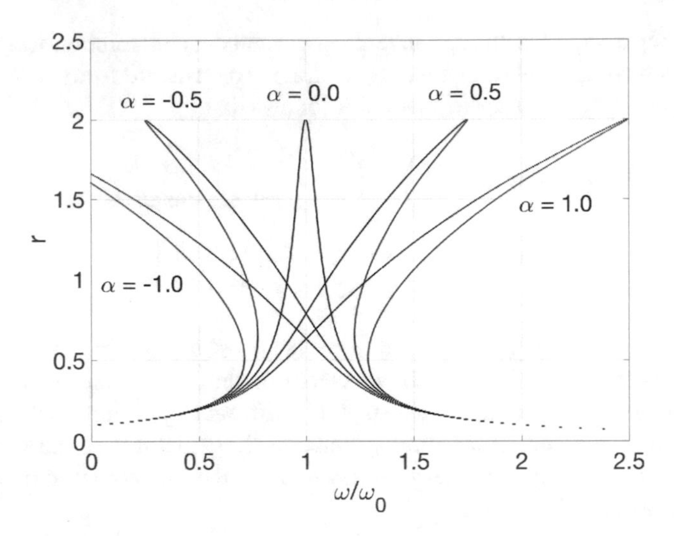

Fig. 4.25 Amplitude and frequency response of a weakly forced Duffing oscillator. Black lines indicate stable equilibrium points, while red lines depict unstable equilibrium points. Parameters are: $\omega_0 = 1$, $\varepsilon\delta = 0.2$ and $\varepsilon\gamma = 2.5$

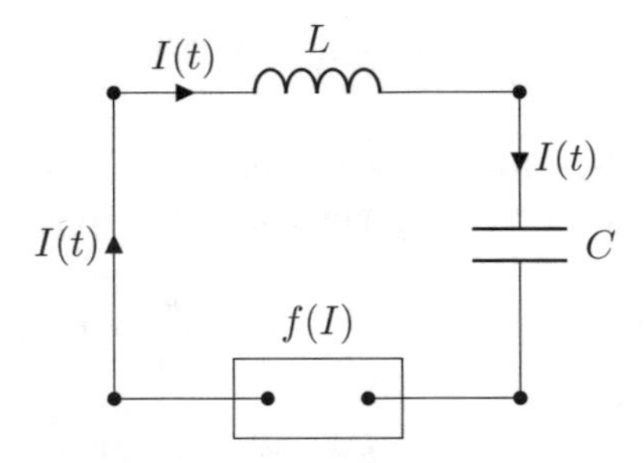

Fig. 4.26 The van der Pol oscillator serves as a model for a tetrode multivibrator circuit, whose current-voltage characteristic, $V = f(I)$, resembles a cubic function

"tetrode multivibrator" circuit used in early commercial radios [30]. In addition, its behavior simulates a tunnel diode in an electric circuit and has modelled other types of natural oscillatory systems, such as neurons. The circuit is seen in Fig. 4.26.

The mathematical model for the van der Pol oscillator is:

$$\boxed{\ddot{x} - \varepsilon(1 - x^2)\dot{x} + x = 0,} \tag{4.46}$$

which has the nonlinear damping term, $-\varepsilon(1 - x^2)\dot{x}$. Letting $\dot{x} = y$, (4.46) becomes the system:

$$\begin{aligned} \dot{x} &= y \\ \dot{y} &= -x + \varepsilon(1 - x^2)y \end{aligned} \tag{4.47}$$

Small Damping

As in Sect. 4.8.3, we apply the *van der Pol transformation*. This invertible transformation is given by:

$$\begin{pmatrix} x \\ y \end{pmatrix} = \Phi(t) \begin{pmatrix} u \\ v \end{pmatrix}, \quad \text{where} \quad \Phi(t) = \begin{pmatrix} \sin t & \cos t \\ \cos t & -\sin t \end{pmatrix}.$$

Since $\Phi(t)$ gives a fundamental solution for the linearization of (4.47), it is equivalent to the method from Sect. 4.7.1 transforming (4.47) into Lagrange standard form. It follows that

$$\begin{pmatrix} \dot{u} \\ \dot{v} \end{pmatrix} = \Phi^{-1}(t) \left[-\dot{\Phi}(t) \begin{pmatrix} u \\ v \end{pmatrix} + \begin{pmatrix} y \\ -x + \varepsilon(1 - x^2)y \end{pmatrix} \right],$$

$$\begin{pmatrix} \dot{u} \\ \dot{v} \end{pmatrix} = \varepsilon \begin{bmatrix} (1 - x^2)y \cos t \\ -(1 - x^2)y \sin t \end{bmatrix}, \tag{4.48}$$

where $x = u \sin t + v \cos t$ and $y = u \cos t - v \sin t$). This ODE in $(u, v)^T$ has the form of Eq. (4.27), which allows the Theorem of Averaging 4.4.

The right hand side of Sys. (4.48) is expanded with the expressions for x and y, and we average over $t \in [0, 2\pi]$, letting u and v be treated as constant. The averaging process eliminates all of the trigonometric functions and reduces the averaged system to:

$$\dot{u} = \frac{\varepsilon}{8} \left(4 - (u^2 + v^2)\right) u + \mathcal{O}\left(\varepsilon^2\right), \tag{4.49}$$

$$\dot{v} = \frac{\varepsilon}{8} \left(4 - (u^2 + v^2)\right) v + \mathcal{O}\left(\varepsilon^2\right).$$

Sys. (4.49) is rewritten in polar coordinates with $u = r \cos \phi$ and $v = r \sin \phi$, so $r^2 = u^2 + v^2$ and $\phi = \arctan(v/u)$. This yields the amplitude-phase equations:

$$\dot{r} = \frac{\varepsilon}{8} r(4 - r^2) + \mathcal{O}\left(\varepsilon^2\right), \tag{4.50}$$

$$\dot{\phi} = 0 + \mathcal{O}\left(\varepsilon^2\right).$$

Ignoring $\mathcal{O}\left(\varepsilon^2\right)$ terms, we see the equilibria of Sys. (4.49) occur when the amplitude equation, which is decoupled from the phase equation, satisfies $\dot{r} = 0$. Since $r \geq 0$, it follows that there are two equilibria, $r_{e_1} = 0$ and $r_{e_2} = 2$. In fact, the equilibrium $r_{e_2} = 2$ corresponds to a *limit cycle* of the model Eq. (4.47). A limit cycle is a periodic solution that appears isolated, i.e., not part of a family of solutions, in phase space.

If we define

$$f(r) = \frac{\varepsilon}{8} r(4 - r^2), \quad \text{where} \quad f'(r) = \frac{\varepsilon}{8}(4 - 3r^2),$$

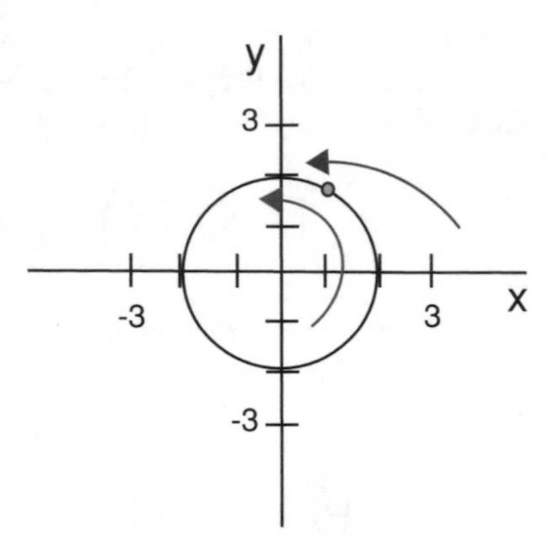

Fig. 4.27 Stable limit cycle oscillation in a weakly damped van der Pol oscillator

then stability of the trivial equilibrium and of the limit cycle are easily studied. Evaluating at $r_{e_1} = 0$ and $r_{e_2} = 2$, we find

$$f'(0) = \frac{\varepsilon}{2} > 0, \qquad f'(2) = -\varepsilon < 0.$$

It follows that the trivial equilibrium, $r_{e_1} = 0$, is unstable, while the limit cycle, $r_2 = 2$, is stable. Figure 4.27 shows a simulation of the model Sys. 4.47, where we observe the stable cycle and solution trajectories approaching the limit cycle.

The equation in r from (4.50) without the $\mathcal{O}\left(\varepsilon^2\right)$ term is solved explicitly by either the separation of variables or Bernoulli's method. Consider

$$\dot{r} = \frac{\varepsilon}{8} r(4 - r^2), \qquad \text{with} \qquad r(0) = r_0,$$

then its solution is given by

$$r(t) = \frac{2r_0}{\sqrt{r_0^2 + (4 - r_0^2)e^{-\varepsilon t}}} = 2 + \left(1 - \frac{4}{r_0^2}\right) e^{-\varepsilon t} + \mathcal{O}\left(e^{-2\varepsilon t}\right).$$

As expected, $r \to 2$ as $t \to \infty$, confirming the stability of the limit cycle. This shows that the average method yields the amplitude of the oscillations remains constant (slow variation), while moding out the fast sinusoidal oscillations.

Large Damping

In the van der Pol Eq. (4.46), the case of large damping occurs when $\varepsilon \gg 1$. Using Liénard's transformation,

$$y = \frac{\dot{x}}{\varepsilon} - x + \frac{x^3}{3},$$

we differentiate this transformation and use the van der Pol Eq. (4.46) to obtain the following system:

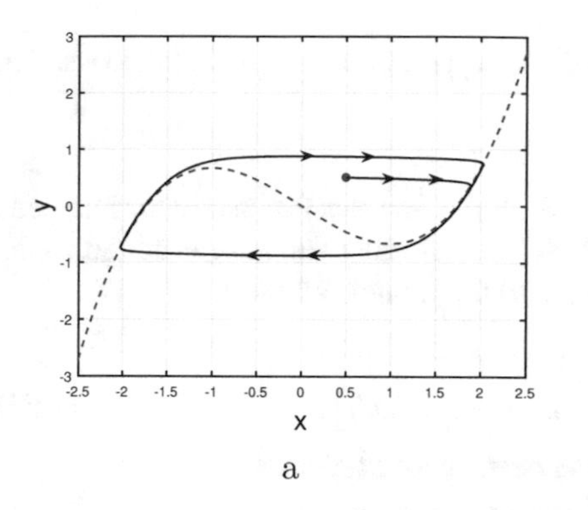

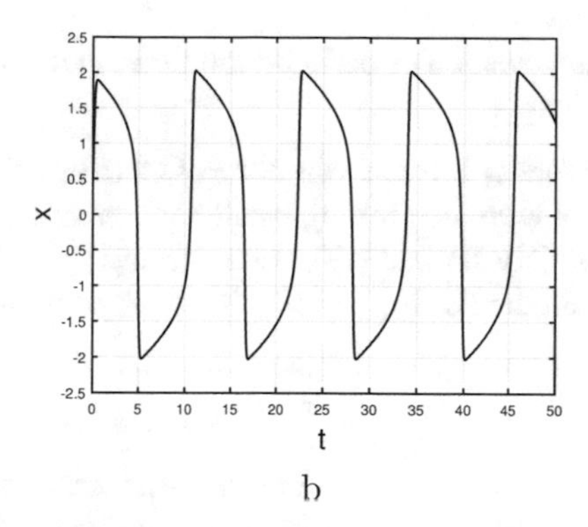

Fig. 4.28 **a** Phase portrait of a van der Pol oscillator with large damping, i.e., $\varepsilon \gg 1$. When a solution trajectories approaches the cubic nullcline (dashed curve), $y = x^3/3 - x$, the motion is very slow but it quickly jumps horizontally once it reaches either of the local minima or maxima of the nullcline. These type of oscillations where we have a slow buildup followed by a quick discharge are called "relaxation oscillations". **b** Time series solution of the system versus time. The characteristic "relaxation oscillations" are seen in this plot

$$\begin{aligned}\dot{x} &= \varepsilon\left(y - \tfrac{1}{3}x^3 + x\right), \\ \dot{y} &= -\frac{x}{\varepsilon}.\end{aligned} \tag{4.51}$$

The curve $y = \frac{x^3}{3} - x$ is the nullcline for $\dot{x}$ and serves to separate fast motion from slow motion in phase space. Since $\varepsilon \gg 1$, then the relation $|\dot{x}| >> |\dot{y}| = \mathcal{O}(1/\varepsilon)$ holds for most x and y. This implies that for most x and y, solution trajectories move fast in the horizontal direction and very slow in the vertical direction. However, when a solution enters the region where $|y - (\frac{x^3}{3} - x)| = \mathcal{O}\left(1/\varepsilon^2\right)$, then $\dot{x}$ and $\dot{y}$ are comparable because both of them are $\mathcal{O}(1/\varepsilon)$. This implies that solution trajectories move slowly along the curve $y = \frac{x^3}{3} - x$, then rapidly after they eventually exit from this region. See Fig. 4.28. This combination of slow and fast motion leads to what is known as a *relaxation oscillation*. A time series of the solution trajectory around the limit cycle is shown in Fig. 4.28, and the fast and slow motion is readily seen.

The stability of the trivial equilibrium is studied by computing the Jacobian matrix of Sys. (4.51), where

$$J(x_e, y_e) = \begin{bmatrix} \varepsilon(1 - x_e^2) & \varepsilon \\ -\frac{1}{\varepsilon} & 0 \end{bmatrix}.$$

At $(x_e, y_e) = (0, 0)$, the eigenvalues of the Jacobian matrix are

$$\lambda = \tfrac{1}{2}\left(\varepsilon \pm \sqrt{\varepsilon^2 - 4}\right),$$

which shows the trivial equilibrium is unstable for $\varepsilon > 0$. When $\varepsilon = 0$, the eigenvalues are purely imaginary, $\lambda = \pm i$, which is indicative of a *Hopf bifurcation* and the creation of periodic solutions. Details of this type of bifurcation are discussed in Chap. .

Weakly Forced van der Pol Oscillator
As with the Duffing oscillator (see Sect. 4.8.3), the van der Pol oscillator has been widely studied with weak forcing [3, 31]. A weak forcing function is added to the van der Pol Eq. (4.46) with weak damping, and the model satisfies:

$$\boxed{\ddot{x} + x = \varepsilon\left[(1 - x^2)\dot{x} + \gamma\cos(\omega t)\right].} \tag{4.52}$$

With $x_1 = x$ and $x_2 = \dot{x}_1$, Eq. (4.52) is written in system form:

$$\begin{aligned}\dot{x}_1 &= x_2,\\ \dot{x}_2 &= -x_1 + \varepsilon\left[(1 - x_1^2)x_2 + \gamma\cos(\omega t)\right].\end{aligned}$$

Following the techniques of Sect. 4.8.3 and applying the *van der Pol transformation*, the invertible transformation satisfies:

$$\begin{pmatrix} x_1 \\ x_2 \end{pmatrix} = \Phi(t)\begin{pmatrix} y_1 \\ y_2 \end{pmatrix}, \quad \text{where} \quad \Phi(t) = \begin{pmatrix} \sin\omega t & \cos\omega t \\ \omega\cos\omega t & -\omega\sin\omega t \end{pmatrix}.$$

With the van der Pol transformation and the assumption that the natural frequency is close to the forcing frequency, $\omega^2 = 1 + \varepsilon\Omega$, it follows that

$$\begin{aligned}\begin{pmatrix} \dot{y}_1 \\ \dot{y}_2 \end{pmatrix} &= \Phi^{-1}(t)\left[-\dot{\Phi}(t)\begin{pmatrix} y_1 \\ y_2 \end{pmatrix} + \begin{pmatrix} x_2 \\ -x_1 + \varepsilon\left[(1 - x_1^2)x_2 + \gamma\cos(\omega t)\right] \end{pmatrix}\right],\\ \begin{pmatrix} \dot{y}_1 \\ \dot{y}_2 \end{pmatrix} &= \frac{\varepsilon}{\omega}\begin{bmatrix} (\Omega x_1 + (1 - x_1^2)x_2 + \gamma\cos(\omega t))\cos\omega t \\ (-\Omega x_1 - (1 - x_1^2)x_2 - \gamma\cos(\omega t))\sin\omega t \end{bmatrix},\end{aligned} \tag{4.53}$$

where $x_1 = y_1\sin\omega t + y_2\cos\omega t$ and $x_2 = \omega(y_1\cos\omega t - y_2\sin\omega t)$. This ODE in $(y_1, y_2)^T$ has the form of Eq. (4.27), so allows the Theorem of Averaging 4.4.

Expanding the right hand side of Sys. (4.53) with our expressions for x_1 and x_2, we integrate over the period of the forcing function, $2\pi/\omega$, letting y_1 and y_2 be treated as constant. The averaging process eliminates all of the trigonometric functions and reduces the averaged system to:

$$\begin{aligned}\dot{y}_1 &= \frac{\varepsilon}{2\omega}\left[\Omega y_2 + \gamma + \omega y_1\left(1 - \frac{y_1^2 + y_2^2}{4}\right)\right] + \mathcal{O}\left(\varepsilon^2\right),\\ \dot{y}_2 &= \frac{\varepsilon}{2\omega}\left[-\Omega y_1 + \omega y_2\left(1 - \frac{y_1^2 + y_2^2}{4}\right)\right] + \mathcal{O}\left(\varepsilon^2\right).\end{aligned} \tag{4.54}$$

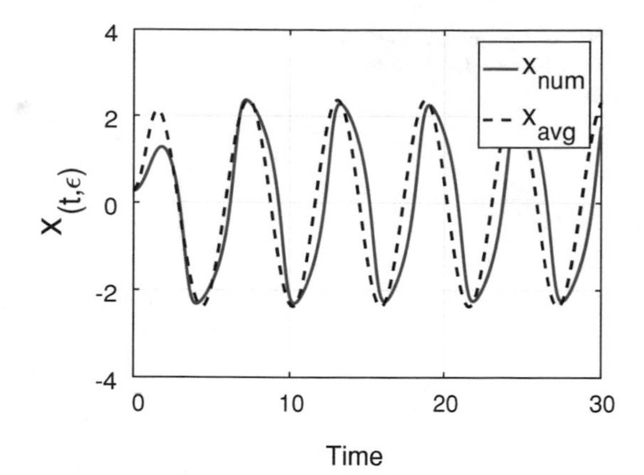

Fig. 4.29 Comparison of numerical integration of a weakly forced van der Pol oscillator with a solution obtained through the method of averaging

Figure 4.29 shows a comparison of solutions of the weakly forced van der Pol oscillator obtained by numerical integration of Sys. (4.53) with those obtained via averaging, Sys. (4.54). The MATLAB code that was written to generate these simulations can be found in the Appendix.

4.8.5 The FitzHugh-Nagumo Model

In the introduction of this book the Hodgkin-Huxley Eq. () gave a representative PDE model, describing the dynamics of an action potential in space and time along a giant squid axon. Their model [32–34], based on detailed experiments, continues to generate novel research from the precision of their electrophysiological measurements and careful description of the biophysics of neural excitation. However, their model remains too complex for detailed analysis and implementation in problems, such as collections of neurons.

Numerous researchers have developed simplified models that capture the qualitative properties of neuronal excitation and action potential propagation with simpler models. FitzHugh [35, 36] proposed an early model, inspired by the van der Pol oscillator (see Sect. 4.8.4), which has some of the neural qualitative behaviors and has the general form

$$\begin{aligned}\frac{dV}{dt} &= f(V) - W + I \\ \frac{dW}{dt} &= \frac{1}{c}(V + a - bW),\end{aligned} \tag{4.55}$$

where V is the voltage across the membrane cell and W is a variable that models the recovery process of the membrane. $f(V)$ is usually a third degree polynomial,

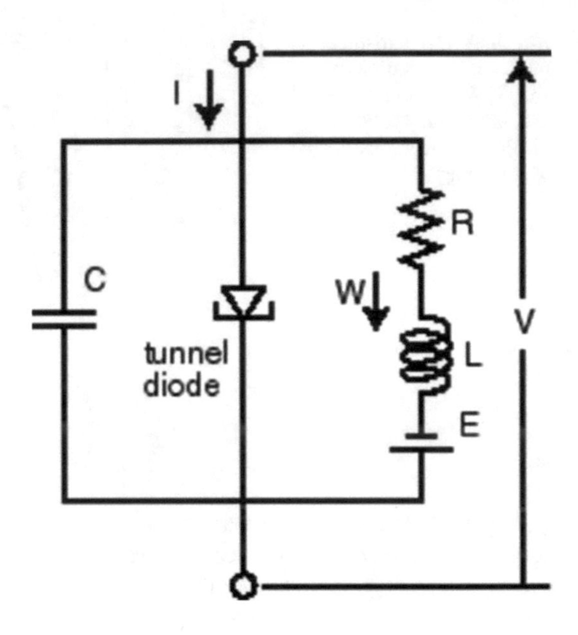

Fig. 4.30 Equivalent circuit of the FitzHugh-Nagumo model

and a, b and c are constant parameters. I represents the external stimulus current. Nagumo [37] proposed an equivalent circuit, see Fig. 4.30, which contains a tunnel diode for modeling the nonlinear response of the membrane current, a capacitor that models the membrane capacitance, a resistor that represents channel resistance, and an inductor. This model (and circuit) became known as the FitzHugh-Nagumo model.

This section analyzes a special form of the model:

$$\begin{aligned}\frac{dx}{dt} &= x - \tfrac{1}{3}x^3 - y + I \\ \frac{dy}{dt} &= \frac{1}{c}(x + a - by).\end{aligned} \tag{4.56}$$

As a model for neural excitation, the behavior of Sys. (4.56) needs to reproduce different responses from changing the external stimulus, I. In particular, a neuron responds to a weak stimulus by simply damping out. A larger stimulus results in an action potential with a rapid voltage increase followed by becoming quiescent. A sufficiently large stimulus results in repetitive spikes of action potentials.

The analysis of Eq. (4.56) begins with finding the equilibria, using a nullcline analysis similar to Sect. 4.5. The x-nullcline satisfies:

$$x - \tfrac{1}{3}x^3 - y + I = 0 \quad \text{or} \quad y = x - \tfrac{1}{3}x^3 + I,$$

while the y-nullcline satisfies:

$$\tfrac{1}{c}(x + a - by) = 0 \quad \text{or} \quad y = \frac{x + a}{b}.$$

Equilibria occur where these nullclines intersect, so depend on the parameters a, b, and I. The parameter I shifts the cubic x-nullcline vertically, and it is easily seen that this nullcline has a maximum slope of one at $x = 0$. Geometrically, it follows that if the linear y-nullcline has a slope greater than one, so $0 < b < 1$, then there is a unique equilibrium. This satisfies the equation:

$$x^3 + \frac{3(1-b)}{b}x + \frac{3(a - bI)}{b} = 0, \qquad 0 < b < 1, \tag{4.57}$$

which is most easily solved numerically.

As the FitzHugh-Nagumo model represents a neuron, when the neuron is at rest ($I = 0$), there should be a unique stable equilibrium. The stability of Eq. (4.56) depends on its linearization, and this system has the Jacobian matrix:

$$J(x_e) = \begin{pmatrix} 1 - x_e^2 & -1 \\ \frac{1}{c} & -\frac{b}{c} \end{pmatrix}.$$

At the unique equilibrium point, x_e, the characteristic equation for its eigenvalues, λ satisfies:

$$\lambda^2 - \text{Tr}(J)\lambda + \text{Det}(J) = 0, \tag{4.58}$$

where

$$\text{Tr}(J) = 1 - x_e^2 - \frac{b}{c}, \qquad \text{Det}(J) = \frac{b}{c}x_e^2 + \frac{1-b}{c}.$$

The trace-determinant stability plane of Fig. 4.8 shows that when $\text{Tr}(J) < 0$ and $\text{Tr}^2(J) - 4\text{Det}(J) < 0$, the equilibrium is a stable spiral sink, and when $\text{Tr}(J) > 0$, it becomes an unstable spiral source.

Consider the FitzHugh-Nagumo model (4.56) with the parameters $a = 0.7$, $b = 0.8$, and $c = 12.5$. We illustrate the response of this model from the resting state as various external stimuli, I are applied. The resting state from (4.57) gives the equilibrium:

$$x_e = -1.199 \quad \text{and} \quad y_e = -0.6243.$$

From the characteristic equation (4.58) with this equilibrium, the eigenvalues are:

$$\lambda = -0.2513 \pm 0.2119i,$$

giving a stable spiral.

We assume a small external stimulus with the neuron beginning at rest, $I = 0.15$, which shifts the equilibrium to:

$$x_e = -1.104 \quad \text{and} \quad y_e = -0.505.$$

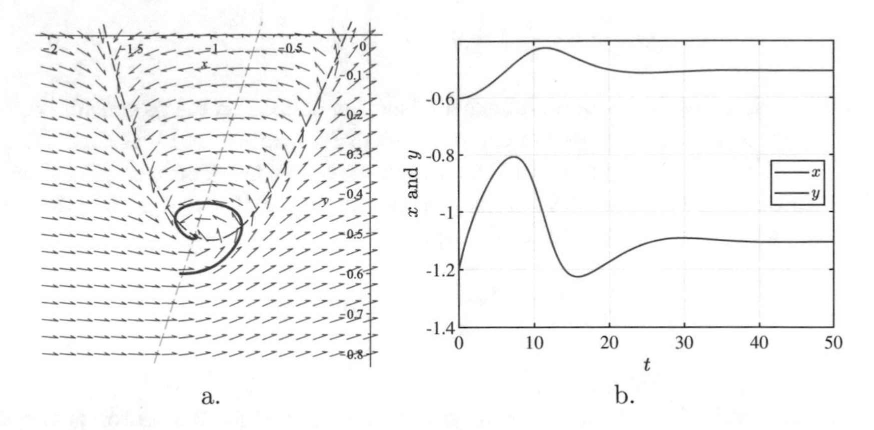

Fig. 4.31 FitzHugh-Nagumo model (4.56) is simulated with parameters $a = 0.7$, $b = 0.8$, and $c = 12.5$ and external stimulus $I = 0.15$. **a** gives the phase portrait of the simulation, including the x and y nullclines notably intersecting to the left of the minimum of the cubic. **b** gives the times series response to (4.56) and demonstrates simply decay of the stimulus by the neuron

The eigenvalues continue to have negative real parts at this equilibrium:

$$\lambda = -0.142 \pm 0.272i.$$

giving a stable spiral. Figure 4.31 shows the phase portrait and time series response to this small external stimulus. Starting at the resting phase, this neuronal response shows a basic damped oscillator settling at the equilibrium nearby.

Next a larger external stimulus to the neuron beginning at rest is employed with $I = 0.3$. This shifts the equilibrium to:

$$x_e = -0.993 \quad \text{and} \quad y_e = -0.367.$$

The eigenvalues continue to have negative real parts at this equilibrium:

$$\lambda = -0.0253 \pm 0.2802i,$$

implying a local stable spiral. Figure 4.32 shows the phase portrait and time series response to this larger external stimulus. Starting at the resting phase, this neuronal response has a very different trajectory as the action potential fires with a rapid increase in the voltage, which is followed by a recovery period with a voltage below the equilibrium. The solution trajectory follows close to the outer branches of the cubic or x-nullcline and eventually spiraling into the shifted equilibrium at the intersection of the nullclines. This is the behavior expected of a neuron firing once and returning to a resting state.

With still a larger external stimulus, $I = 0.35$, to the neuron beginning at rest, the equilibrium shifts to:

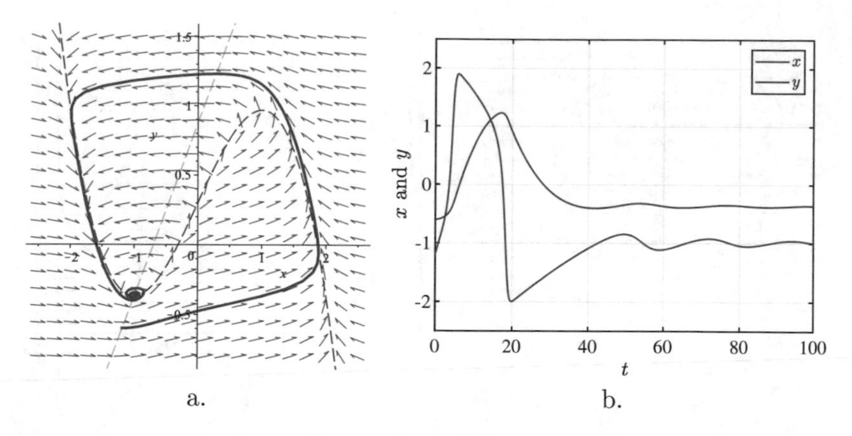

Fig. 4.32 FitzHugh-Nagumo model (4.56) is simulated with parameters $a = 0.7$, $b = 0.8$, and $c = 12.5$ and external stimulus $I = 0.3$. **a** gives the phase portrait of the simulation, including the x and y nullclines intersecting near the minimum of the cubic. **b** gives the times series response to (4.56) and shows a single action potential, where the solution trajectory jumps quickly to the positive descending branch of the cubic, decaying slowly before quickly following the negative descending branch of the cubic, and finally spiraling into the stable equilibrium

$$x_e = -0.951 \quad \text{and} \quad y_e = -0.314.$$

Now the eigenvalues have positive real parts at this equilibrium:

$$\lambda = 0.0153 \pm 0.2715i.$$

locally giving an unstable spiral. Figure 4.33 shows the phase portrait and time series response to this external stimulus. This response is very similar to the van der Pol relaxation oscillator (see Sect. 4.8.4). From the resting state, the voltage of the neuron rapidly increases to the positive descending branch of the cubic or x-nullcline with the speed depending on c. The solution trajectory more slowly decays in the x-direction until after the relative maximum of the cubic. Subsequently, the voltage rapidly decays below the equilibrium, then slowly increases along the negative descending branch of the cubic (refractory period). After passing the minimum of the cubic, the trajectory roughly repeats this process, leading to an infinite sequence of action potentials as seen in Fig. 4.33b. Many neurons have been shown to exhibit regular periodic spiking, qualitatively making this model behavior appropriate though somewhat different from the actual waveform of neuronal voltage.

This section shows that the FitzHugh-Nagumo model (4.56) provides some of the qualitative behavior of the Hodgkin-Huxley experiments on the giant squid axon when stimulated by an external stimulus, so provides a relatively simple model for studying neural behavior. However, the voltage response of this model fails to capture many of the details of the voltage waveform observed in experiments, which often require modeling the actual molecular channels in the neuronal membrane. However,

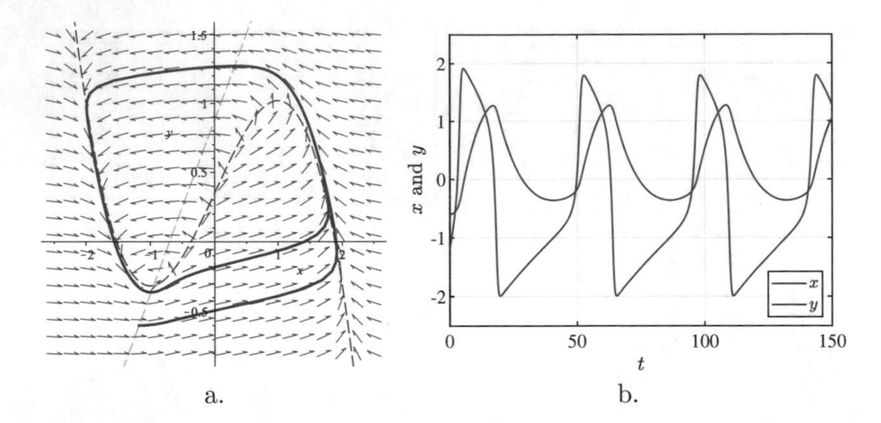

Fig. 4.33 FitzHugh-Nagumo model (4.56) is simulated with parameters $a = 0.7$, $b = 0.8$, and $c = 12.5$ and external stimulus $I = 0.3$. **a** gives the phase portrait of the simulation, including the x and y nullclines again intersecting to the left of the minimum of the cubic. **b** gives the times series response to (4.56)

Eq. (4.56) does illustrate the significant change of behavior, which occurs as the external stimulus increases from a quiescent state to a stable single action potential to a series of action potential spikes. These are observed by the linear analysis and the roots of the characteristic equation shifting from negative to positive real part, which is also known as a *Hopf bifurcation* and will be studied in more detail in Chap. . For our example above, this critical change in stability occurs at $I_c = 0.3313$.

4.9 Crystal Oscillators

A crystal is a solid in which the constituent atoms, molecules, or ions are packed in a regularly ordered, repeating pattern extending in all three spatial dimensions [38, 39]. Almost any object made of an elastic material could be used like a crystal, with appropriate transducers, since all objects have natural resonant frequencies of vibration. For example, steel is very elastic and has a high speed of sound. It was often used in mechanical filters before quartz. The resonant frequency depends on size, shape, elasticity, and the speed of sound in the material. High-frequency crystals are typically cut in the shape of a simple, rectangular plate. Low-frequency crystals, such as those used in digital watches, are typically cut in the shape of a tuning fork. For applications not needing very precise timing, a low-cost ceramic resonator is often used in place of a quartz crystal.

When a crystal of quartz is properly cut and mounted, it can be made to distort in an electric field by applying a voltage to an electrode near or on the crystal. This property is known as electrostriction or inverse piezoelectricity. When the field is removed, the quartz will generate an electric field as it returns to its previous shape, and this can

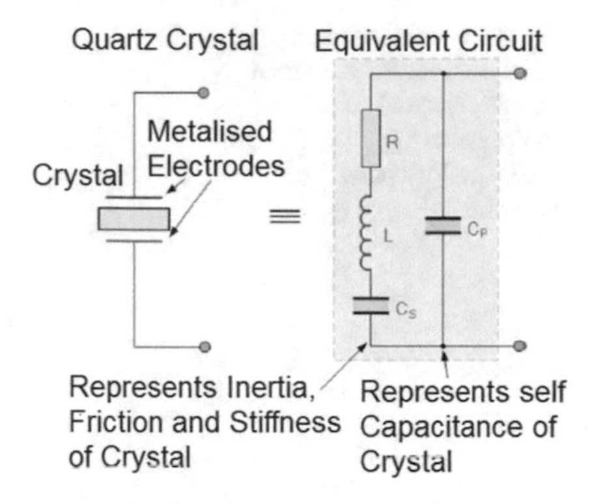

Fig. 4.34 (Left) Schematic of a quartz crystal. (Right) A quartz crystal behaves like circuit composed of an inductor, capacitor and resistor, so it oscillates with a precise resonant frequency when it is subjected to an electric field

generate a voltage. The result is that a quartz crystal behaves like a circuit composed of an inductor, capacitor and resistor, with a precise resonant frequency [40], see Fig. 4.34.

Quartz has the further advantage that its elastic constants and its size change in such a way that the frequency dependence on temperature can be very low. The specific characteristics will depend on the mode of vibration and the angle at which the quartz is cut (relative to its crystallographic axes). Therefore, the resonant frequency of the plate, which depends on its size, will not change much, either. This means that a quartz clock, filter or oscillator will remain accurate. For critical applications the quartz oscillator is mounted in a temperature-controlled container, called a crystal oven, and can also be mounted on shock absorbers to prevent external mechanical vibrations.

4.9.1 Two-Mode Oscillator Model

The crystal oscillator circuit sustains oscillation by taking a voltage signal from the quartz resonator, amplifying it, and feeding it back to the resonator. The rate of expansion and contraction of the quartz is the resonant frequency, and is determined by the cut and size of the crystal. When the energy of the generated output frequencies matches the losses in the circuit, an oscillation can be sustained. The frequency of the crystal is slightly adjustable by modifying the attached capacitances. A varactor, a diode with capacitance depending on applied voltage, is often used in voltage-controlled crystal oscillators, VCO.

The analog port of the VCO chip is modeled by a nonlinear resistor R^-, see Fig. 4.35, that obeys the voltage-current relationship

$$v(i) = -ai + bi^3,$$

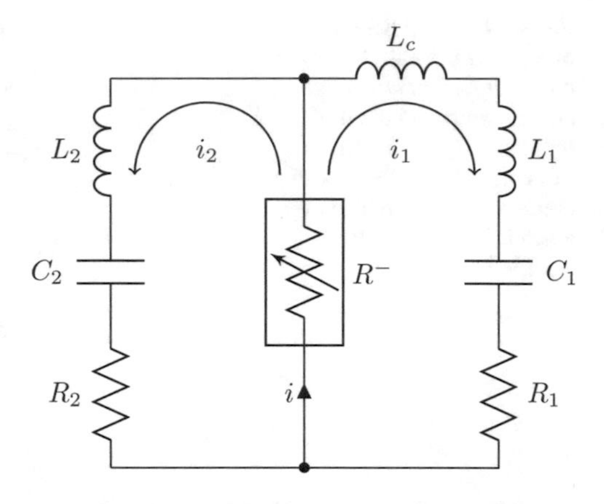

Fig. 4.35 Two-mode crystal oscillator circuit. A second set of spurious RLC components (R_2, L_2, C_2) are introduced by parasitic elements

where a and b are constant parameters. A major reason for the wide use of crystal oscillators is their high Q factor. This is a dimensionless parameter that indicates how underdamped an oscillator is. For a crystal oscillator, it can be defined as the ratio of the resonant frequency with respect to the half-power bandwidth, i.e., the bandwidth over which the power of vibration is greater than half the power at the resonant frequency. Higher Q indicates that the oscillations die out more slowly. A typical Q value for a quartz oscillator ranges from 104 to 106, compared to perhaps 102 for an LC oscillator. The maximum Q for a high stability quartz oscillator can be estimated as $Q = 1.6 \times 107/f$, where f is the resonance frequency in megahertz.

The inductance of the leads connecting the crystal to the VCO port is represented by L_c. In addition, parasitic elements can be represented by a series resonator (L_2, C_2, R_2) connected in parallel with the nonlinear resistor. The resulting circuit, depicted in Fig. 4.35, forms a two-mode resonator model. To derive a mathematical model for the time-evolution of the circuit we employ Kirchoff's Law of Circuits [9].

4.9.2 *Kirchoff's Law of Circuits*

The German physicists Gustav Kirchhoff introduced in 1845 two laws that serve to model the current and voltage evolution of electrical circuits, although they can also be used in the frequency domain to model networks.

Kirchhoff's First Law of Current. The algebraic sum of currents, I_k, in a network of conductors meeting at a point is zero. Mathematically speaking, this law implies that

$$\sum_{k=1}^{n} I_k = 0.$$

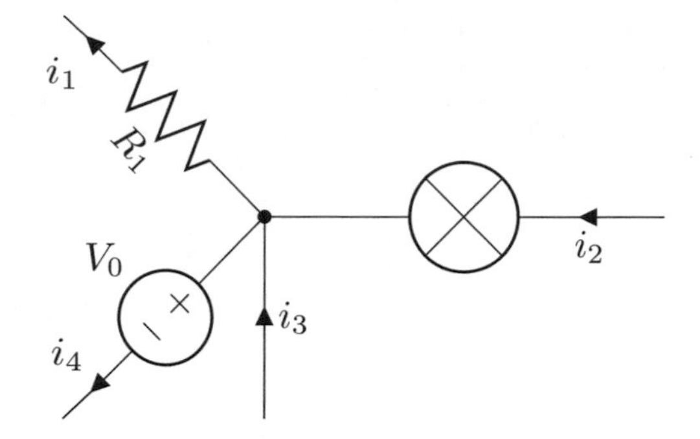

Fig. 4.36 Kirchoff's current law: The algebraic sum of currents in a network of conductors meeting at a point is zero

Figure 4.36 below illustrates the first law. The sum of the currents flowing through the resistor, R_1, the voltage source V_0, must be equal to the current, i_2, flowing through the mixer plus the current, i_3, entering the common node. That is:

$$i_2 + i_3 = i_1 + i_4.$$

Kirchhoff's Second Law of Voltage. The directed sum of voltages, V_k, around any closed loop is zero. That is,

$$\sum_{k=1}^{n} V_k = 0.$$

Using the RLC circuit diagram of Fig. 4.37 as a reference, this law implies that the sum of the voltage across each of the circuit elements, resistor, R, capacitor, C, and inductance, L, plus the voltage provided by the source V_0 must add up to zero. That is

$$V_0 + V_R + V_L + V_C = 0,$$

where V_R, V_L and V_C, are, respectively, the individual voltages across the resistor, inductor, and capacitor.

Now, the individual voltages across the components of the circuit are computed as follows. The voltage across the resistor is calculated using Ohm's law: $V_R = RI(t)$, where R is the resistor and $I(t)$ is the current flowing across it. The voltage, V_L, across the inductor, L, is determined by Faraday's law of induction: $V_L = L\frac{dI(t)}{dt}$. Finally, the voltage, V_C, across the capacitor, C, is given by $V_C = \frac{q(t)}{C}$, where $q(t)$ is the charge store across the capacitor. Kirchhoff's second law of voltage leads to the following relation among the voltages:

$$-V_0 = RI + \frac{q}{C} + L\frac{dI}{dt}. \tag{4.59}$$

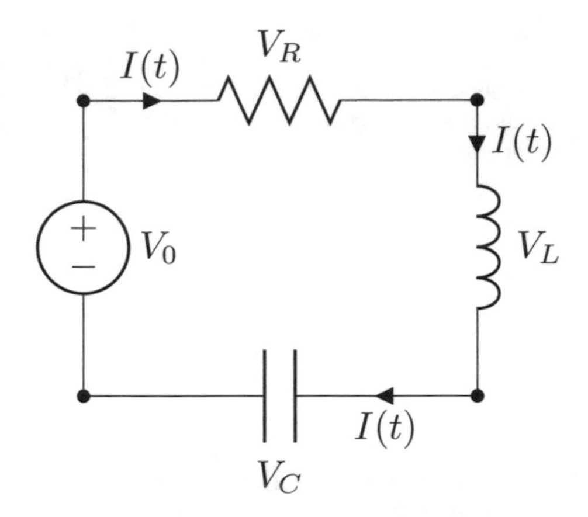

Fig. 4.37 Kirchoff's voltage law: The directed sum of the voltages around any closed loop is zero

Letting $V(t) = -V_0$, and since $I(t) = \frac{dq(t)}{dt}$, we can differentiate every term of Eq. (4.59) to derive a mathematical model for the RLC circuit in terms of $I(t)$, as follows

$$\boxed{L\frac{d^2I}{dt^2} + R\frac{dI}{dt} + \frac{1}{C}I = \frac{dV}{dt}.} \tag{4.60}$$

This model Eq. (4.60) is valid for both types of voltages, a DC voltage and an AC voltage. In the DC case, V_0 is constant so the right-hand side of Eq. (4.60) becomes zero, leading to a homogenous, linear, second order ODE model. In the AC case, we can consider a periodic voltage source, with amplitude, A, and frequency, ω, i.e., of the form, $V(t) = A \sin \omega t$, which yields a non-homgeneous, linear, second order ODE model. A critical observation is that in both cases, DC and AC voltage, the model Eq. (4.60) is of the same form as the universal model of a linear oscillator system, which was introduced earlier through Eq. (4.30). All the different cases, and solutions, that were obtained earlier also apply to Eq. (4.60), with a_0, a_1 and a_2 replaced by $1/C$, R, and L, respectively, and $F(t) = A \sin \omega t$.

Crystal Oscillator. Let's consider now the circuit that serves as a model for the crystal oscillator, see Fig. 4.35. Following on the previous example, we can apply Kirchhoff's voltage law to each RLC loop in the crystal oscillator circuit to get

$$\begin{aligned} (L_1 + L_c)\frac{d^2i_1}{dt^2} &+ R_1\frac{di_1}{dt} + \frac{1}{C_1}i_1 = -\frac{dv(t)}{dt} \\ L_2\frac{d^2i_2}{dt^2} &+ R_2\frac{di_2}{dt} + \frac{1}{C_2}i_2 = -\frac{dv(t)}{dt}, \end{aligned} \tag{4.61}$$

where $v(i(t)) = -ai + bi^3$, in which i is the current circulating through the nonlinear resistor R^-. From Kirchhoff's first law of current, we have

$$i = i_1 + i_2.$$

Thus, $v(i(t))$ becomes: $v(i(t)) = -a(i_1 + i_2) + b(i_1 + i_2)^3$. Differentiating $v(i(t))$, with respect to t, we get

$$\frac{dv}{dt} = \left(-a + 3b(i_1 + i_2)^2\right)\left(\frac{di_1}{dt} + \frac{di_2}{dt}\right).$$

Substituting $\frac{dv}{dt}$ into Eq. (4.61) yields the following governing equations for the crystal oscillator circuit:

$$\boxed{L_j \frac{d^2 i_j}{dt^2} + R_j \frac{di_j}{dt} + \frac{1}{C_j} i_j = \left[a - 3b(i_1 + i_2)^2\right]\left[\frac{di_1}{dt} + \frac{di_2}{dt}\right],} \tag{4.62}$$

where $j = 1, 2$ and L_c has been included in L_1.

4.9.3 *Averaging*

Next we apply the technique to convert the two-mode crystal oscillator model (4.62) to a system of differential equations. Since the model equations (4.62) are actually two 2nd-order differential equations, we would need to introduce two sets of variables, one for current i_1 and one for current i_2, as follows. Letting

$$x_1 = i_1,\ x_2 = \frac{di_1}{dt},\ x_3 = i_2,\ x_4 = \frac{di_2}{dt},$$

and $\omega_{0j}^2 = 1/L_j C_j$ and $X = [x_1, x_2, x_3, x_4]^T$, the model equations (4.62) can be rewritten as

$$\frac{dX}{dt} = F(X) \equiv AX + \mathcal{N}(X), \tag{4.63}$$

where

$$A = \begin{bmatrix} 0 & 1 & 0 & 0 \\ -\omega_{01}^2 & \dfrac{a - R_1}{L_1} & 0 & \dfrac{a}{L_1} \\ 0 & 0 & 0 & 1 \\ 0 & \dfrac{a}{L_2} & -\omega_{02}^2 & \dfrac{a - R_2}{L_2} \end{bmatrix}, \quad \mathcal{N}(X) = \begin{bmatrix} 0 \\ \dfrac{-3b}{L_1}(x_1 + x_3)^2(x_2 + x_4) \\ 0 \\ \dfrac{-3b}{L_2}(x_1 + x_3)^2(x_2 + x_4) \end{bmatrix}.$$

The terms AX and $\mathcal{N}(X)$ represent the linear and nonlinear terms, respectively, which, together, govern the behavior of the two-mode crystal oscillator.

In what follows we assume nonresonance conditions among ω_{01} and ω_{02}, so that there are no nonzero integers p and q for which $\dfrac{\omega_{01}}{\omega_{02}} = \dfrac{p}{q}$. Using the invertible van der Pol transformation

$$\begin{pmatrix} x_1 \\ x_2 \\ x_3 \\ x_4 \end{pmatrix} = \Phi(t) \begin{pmatrix} y_1 \\ y_2 \\ y_3 \\ y_4 \end{pmatrix},$$

where

$$\Phi(t) = \begin{pmatrix} \cos\omega_1 t & -\sin\omega_1 t & 0 & 0 \\ -\omega_1 \sin\omega_1 t & -\omega_1 \cos\omega_1 t & 0 & 0 \\ 0 & 0 & \cos\omega_2 t & -\sin\omega_2 t \\ 0 & 0 & -\omega_2 \sin\omega_2 t & -\omega_2 \cos\omega_2 t \end{pmatrix}.$$

Equation (4.63) can be rewritten as

$$\frac{dY}{dt} = \varepsilon F(Y, t), \tag{4.64}$$

where $Y = [y_1, \ldots, y_4]^T$ and the derivatives are explicitly given by

$$\begin{aligned}
\dot{y}_1 &= \frac{1}{\omega_1}\left\{ \Omega_1\big[y_1 \cos(\omega_1 t) - y_2 \sin(\omega_1 t)\big] - \mathcal{N}_1 \right\} \sin(\omega_1 t) \\
\dot{y}_2 &= \frac{1}{\omega_1}\left\{ \Omega_1\big[y_1 \cos(\omega_1 t) - y_2 \sin(\omega_1 t)\big] - \mathcal{N}_1 \right\} \cos(\omega_1 t) \\
\dot{y}_3 &= \frac{1}{\omega_2}\left\{ \Omega_2\big[y_3 \cos(\omega_2 t) - y_4 \sin(\omega_2 t)\big] - \mathcal{N}_2 \right\} \sin(\omega_2 t) \\
\dot{y}_4 &= \frac{1}{\omega_2}\left\{ \Omega_2\big[y_3 \cos(\omega_2 t) - y_4 \sin(\omega_2 t)\big] - \mathcal{N}_2 \right\} \cos(\omega_2 t),
\end{aligned}$$

with $\omega_{01}^2 - \omega_1^2 = \varepsilon\Omega_1$ and $\omega_{02}^2 - \omega_2^2 = \varepsilon\Omega_2$. Averaging over the periods $T_1 = 2\pi/\omega_1$ and $T_2 = 2\pi/\omega_2$ we arrive at the simplified equation

$$\frac{dY}{dt} = \varepsilon\bar{F}(Y) \equiv \varepsilon\bar{A}Y + \varepsilon\bar{\mathcal{N}}(Y), \tag{4.65}$$

where $\bar{F}(Y) = \lim\limits_{T\to\infty} \dfrac{1}{T}\displaystyle\int_0^T F(Y, t)dt$ and

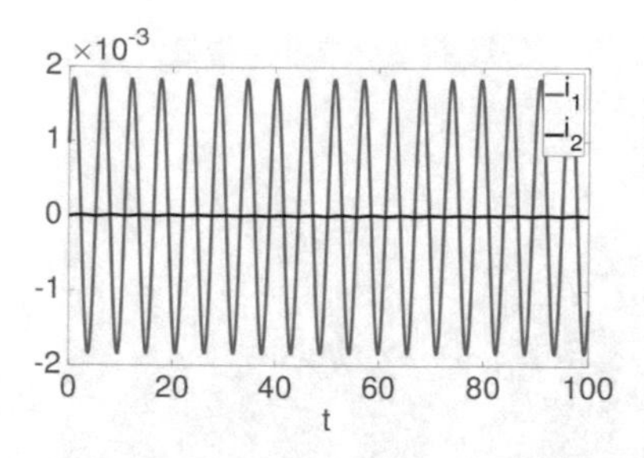

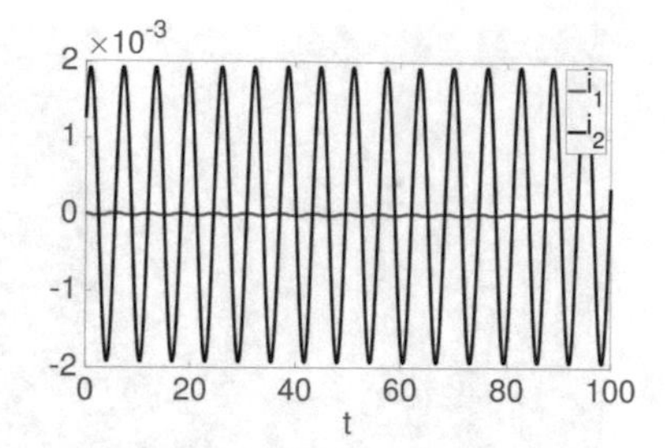

Fig. 4.38 Time series solutions of a two-mode crystal oscillator model (4.62). (Left) Stable Mode 1 $(i_1(t), 0)$, and (right) Stable Mode 2 $(0, i_2(t))$. Parameter values are: $R_1 = 30.9\Omega$, $R_2 = 181.1\Omega$, $L_1 = 5.2E-04H$, $L_2 = 2.6E-04H$, $C_1 = 1.0E-13F$, $C_2 = 2.5E-14F$, $a = 939$, $b = 3E08$

$$\bar{A} = \begin{bmatrix} \frac{a-R_1}{2L_1} & -\frac{\Omega_1}{2\omega_1} & 0 & 0 \\ \frac{\Omega_1}{2\omega_1} & \frac{a-R_1}{2L_1} & 0 & 0 \\ 0 & 0 & \frac{a-R_2}{2L_2} & -\frac{\Omega_2}{2\omega_2} \\ 0 & 0 & \frac{\Omega_2}{2\omega_2} & \frac{a-R_2}{2L_2} \end{bmatrix}, \quad \bar{\mathcal{N}}(X) = -\frac{3b}{8} \begin{bmatrix} \frac{y_1}{L_1}[y_1^2 + y_2^2 + 2(y_3^2 + y_4^2)] \\ \frac{y_2}{L_1}[y_1^2 + y_2^2 + 2(y_3^2 + y_4^2)] \\ \frac{y_3}{L_2}[y_3^2 + y_4^2 + 2(y_1^2 + y_2^2)] \\ \frac{y_4}{L_2}[y_3^2 + y_4^2 + 2(y_1^2 + y_2^2)] \end{bmatrix}.$$

Equation (4.65) can be interpreted as a representation of the original model Eq. (4.63) with respect to two rotating frames of reference, one rotating with speed ω_1 and one with speed ω_2. Observe that in this rotating coordinate system, $y_1 = y_2 = 0$ and $y_3 = y_4 = 0$ are two invariant subspaces so that the two modes of oscillation of the crystal decouple from one another. Figure 4.38 shows the two modes of oscillations, in the variables x_1 and x_3, decoupled via the method of averaging. Later on in Chap. we will employ crystal oscillators as building blocks to design a model for a precision timing device through networks of interconnected crystals.

4.10 Fluxgate Magnetometer

We have already seen examples of various systems, natural and artificial ones, that exhibit oscillatory behavior, i.e., cyclic behavior that repeats at regular intervals. Additional examples include: the rhythmic light pulses of fireflies [41, 42], see Fig. 4.39, the electrical activity of neuron cells that make up central pattern generators in biological systems [43–49], see Fig. in the introduction, the patterns of lights produced by arrays of coupled lasers [50, 51], voltage variations in modern communication systems [52, 53], the growth and decay of population sizes between competing species [54–56], bubble formation and evolution in fluidization and mixing processes [57], and variations in phase and current in arrays of Josephson junctions [58–61] in quantum physics.

Fig. 4.39 Complex interactions among fireflies can lead them to coordinate the rhythmic flashing lights produced by each individual firefly. Collectively, the swarm can then achieve synchronization and oscillate in unison. *Source* National Geographic

In the absence of noise, the underlying cyclic fluctuations in a given nonlinear system can arise from individual units that oscillate on their own, also known as *endogenous* or *self-excited* oscillators, or from *exogenous* units that oscillate only when they are externally driven or coupled together. *Circadian rhythms* , which regulate the daily cycle of many living organisms, plants, and animals, for instance, are endogenously generated. In fact, the first endogenous circadian oscillation to be observed was the movement of the leaves of *Mimosa pudica*, a plant studied by the French scientist Jean-Jacques dÒetous de Mairan.[1]

In addition, bistability—the property that allows a system to rest in either of two states—underlies the basic oscillatory behavior of many other natural and artificial systems. States may include typical invariant sets, such as equilibrium points, periodic and quasi-periodic solutions, and chaotic attractors. In the absence of an external stimulus, the state variable $x(t)$ of a bistable system will relax to one of the invariant sets, and it will remain in that state unless it is switched or forced to another state. It is in this sense that the system exhibits "memory." Which invariant set the system will relax to depends typically on the set of initial conditions.

Next we introduce the concept of bistability in more detail and, later on, we will use it to derive a mathematical model for a fluxgate magnetometer.

[1] *Source: Wikepedia* https://en.wikepedia.org.

4.10.1 Bistability

All bistable systems employ some form of energy source as the underlying principle that allows them to switch between states. The source of energy is due typically through external forcing or through the coupling mechanism. For instance, dynamic sensors [62–66], operate as exogeneous oscillators with nonlinear input-output characteristics, often corresponding to a bistable potential energy function of the form

$$\frac{dx}{dt} = -\nabla U(x), \tag{4.66}$$

where $x(t)$ is the state variable of the natural system or artificial device, e.g., magnetization state, and U is the bistable potential function. Examples include: fluxgatemagnetometers [67, 68], ferroelectric sensors [69], and mechanical sensors, e.g., acoustic transducers made with piezoelectric materials. Later on in this chapter we will derive a mathematical model for a fluxgate magnetometer and show how bistability plays a critical role.

To get insight, consider Fig. 4.40(top), which illustrates the case of a double-well potential function

$$U(x) = -ax^2 + bx^4,$$

whose minima are located at $\pm x_m$ and the height of the potential barrier between the two minima is labeled by U_0.

Without an external excitation (periodic forcing or noise), the state point $x(t)$ of the exogenous oscillator described by Eq. (4.66) will rapidly relax to one of two stable attractors, which correspond to the minima of the potential energy function $U(x)$. In the presence of an external periodic forcing term $f(t)$, with frequency ω, the state variable in $U(x + f(t))$ can be induced to oscillate periodically (with a well-defined waveform) between its two stable attractors $-x_m$ and $+x_m$, as is illustrated in Fig. 4.40(bottom). The forcing term is also known as *biasing signal* in the engineering literature.

In biological systems, bistability is a key feature for understanding and engineering cellular functions such as: storing and processing information by the human brain during the decision-making process [70]; regulation of the cell cycle [71, 72]; sporulation, which controls the timing and dynamics of dramatic responses to stress [73]; design and construction of synthetic toggle switches [74]; and in gene regulatory networks responsible for embryonic stem cell fate decisions [75]. In chemical systems bistability is central to the analysis of relaxation kinetics [76]. In mechanical systems, bistable mechanisms are commonly employed in the design and fabrications of Micro-Electro-Mechanical-Systems (MEMS) versions of relays, valves, clips, and threshold switches [77, 78]. In electronics, hysteresis and bistability are combined to design and fabricate Schmitt trigger circuits, which convert analog input signals

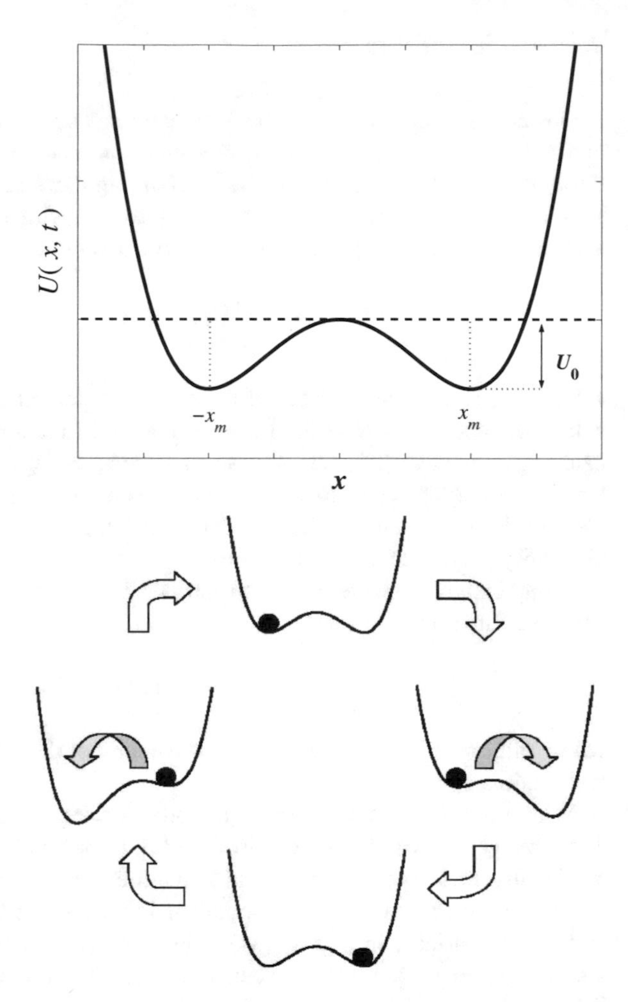

Fig. 4.40 (Top) Bistable Potential $U(x) = -ax^2 + bx^4$. (Bottom) Switching between wells of a potential function can be achieved by a sufficiently large biasing signals (or noise) greater than the potential barrier

to digital output signals [79]. In neuroscience, bistability is at the core of Hopf models [80–82], which describe the input/output response of neurons through differential equations of the form

$$\tau_i \frac{dV_i}{dt} = -V_i + g(V_i), \tag{4.67}$$

where τ_i is a suitable time constant that controls how quickly unit(neuron) i responds to a stimulus, V_i is the output (typically voltage) of unit i, and g is the activation function, which normally represents a saturation nonlinearity property of neurons.

Let's now discuss a specific example of a nonlinear device that is governed by stability–a fluxgate magnetometer.

Table 4.4 Magnetic fields measurements for different objects vary according to distance and size of each individual object. Units appear in nano-Tesla nT

Ferrous Components	"Near"	"Far"
Ship (100 tons)	100 ft ($300 - 700$ nT)	1000 ft ($0.3 - 0.7$ nT)
Train Engine	500 ft ($5 - 200$ nT)	1000 ft ($1 - 50$ nT)
Automobile (1 ton)	30 Feet (40 nT)	100 ft (1 nT)
Rifle	5 ft ($10 - 50$ nT)	10 ft ($2 - 10$ nT)
Screwdriver (5")	5 ft ($5 - 10$ nT)	10 ft ($0.5 - 1$ nT)

4.10.2 What is a Fluxgate Magnetometer

A fluxgate magnetometer is a relatively simple device that is built to measure magnetic fields produced by certain materials [83–86]. In the biomedical field, for instance, research with magnetic tracers has lead scientists to consider using fluxgate sensors to study the mechanical activity of the large intestine [87]. Other potential applications include remote sensing [88], geological explorations of the deep ocean [89], vehicle guidance in agriculture [90], and traffic control [91]. Magnetic materials such as iron, cobalt, and steel, contain tiny subatomic regions of magnetism called domains. When these domains align the result is a magnetic field. Now in any ferrous object the magnetic lines of force or flux are greatest at the ends of a magnet or dipole. Of course, magnetic forces vary according to size, shape, and orientation of the object. Consider a simple rifle. This rifle can be approximated by a magnetic dipole and has its own variations of magnetic lines of force. These lines of force influence the Earth's magnet forces which cause a change in the Earth's ambient local magnetic field near the rifle. This change is commonly know as an anomaly. Fluxgate magnetometers can measure this magnetic field anomalies. The amount of measurable change in an anomaly force depends on the size and distance from the device. A basic idea of some of the typical measurements can be seen in Table 4.4, which was originally produced by Breiner [92]. Today's highly specialized fluxgate devices boast laboratory sensitivity levels as low as $10pT/\sqrt{Hz}$.

4.10.3 How Does a Fluxgate Magnetometer Work?

In its most basic form, a classical fluxgate sensor consist of two detection coils wound around a ferromagnetic core in opposite directions to one another, as is illustrated in Fig. 4.41. The *excitation coil* serves to drive the core into saturation [93], so it can oscillate, between two stable magnetization states. The *pick-up coil* serves to record the oscillations. Then an external field or signal can be detected by processing its effects on the input-put response of the oscillating signal. We explain this in more detail.

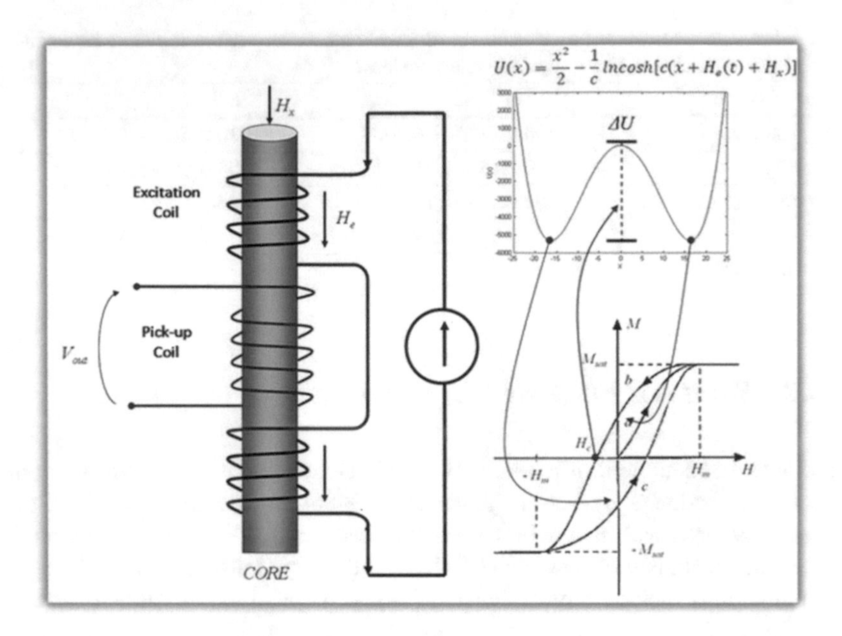

Fig. 4.41 Schematic configuration of a traditional fluxgate magnetometer as proposed by Forster [94]. *Source* University of Catania http://www.measurement.dees.unict.it/rtd_fluxgate.html

The underlying principle of operation of the fluxgate magnetometer is based on the concept of bistability discussed earlier on. The core exhibits *hysteresis* in its input-output response to an external magnetic field. That is, it can switch between its two (assumed to be stable) magnetization states when an external field H_x is applied. In practice, the coercive field ΔU (roughly the deterministic switching threshold between the stable states of an energy function $U(x)$) can be quite high, so that the device might show little response to a target signal H_x of amplitude far smaller than the energy barrier height. Hence, the *standard detection method* consists of applying a known time-periodic (usually taken to be triangular or sinusoidal) bias signal, $H_e(t)$, of very large amplitude, to periodically drive the core between its two stable magnetization states, see Fig. 4.41.

To detect a small target signal (dc or low-frequency), the spectral-based [85, 86, 95, 96], readout mechanism is employed. In the absence of a target signal, i.e., $H_x = 0$, the power spectral density contains only the odd harmonics of the bias frequency ω. But when $H_x > 0$, the potential energy function $U(x + H_e(t) + H_x)$ is skewed, resulting in the appearance of even harmonics; the response at the second harmonic 2ω is then used to detect and quantify the target signal, as is shown in Fig. 4.42.

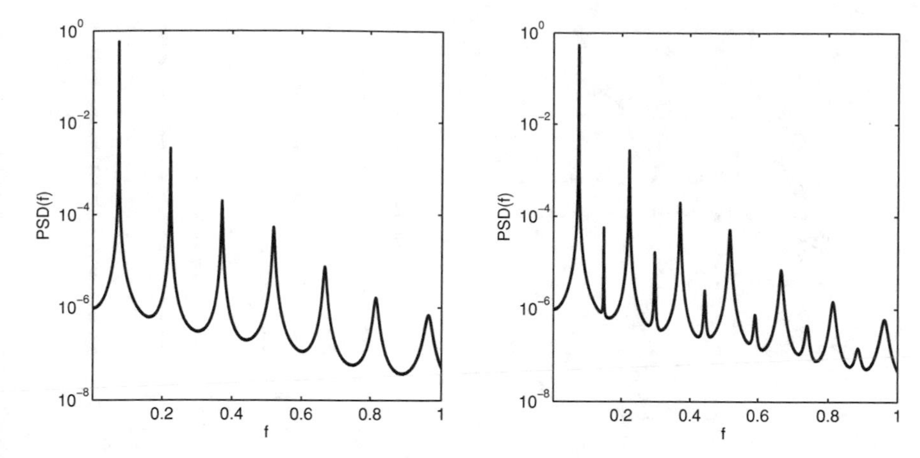

Fig. 4.42 Power Spectrum Decomposition of the oscillations in a bistable overdamped system subject to a periodic forcing. (Left) In the absence of an external signal the PSD shows only even harmonics. (Right) In the presence of an external signal, however, the PSD exhibits odd as well as even harmonics. Typically, the first even harmonic is used as a detection mechanism

4.10.4 Modeling Single-Core Dynamics

To derive the governing equations for a fluxgate magnetometer, we start with Landau's theory [97] on the physics of *phase transitions*, which allows to formulate a mathematical model in the form of a continuous differential equation for the average magnetization state of a fluxgate device.

We consider a single-core fluxgate with a two-coil structure (a primary coil and a secondary coil) wound around a suitable magnetic core, as is depicted in Fig. 4.41. The magnetization of the core is governed by the excitation field H_e produced in the primary coil and the core is composed of a ferromagnetic material with the characteristic "sharp" input-output hysteresis loop, corresponding to a bistable potential energy function, which underpins the system dynamics; the minima of this potential energy function correspond to the two (stable) steady magnetization states, see Fig. 4.40. In order to reverse the core magnetization, a suprathreshold excitation field is required. Here, the "threshold" represents the minimum field required to switch the saturation of the material. Mathematically, it corresponds to the inflection point(s) in the potential function. With an alternating excitation (or bias) magnetic field H_e, the output voltage V_o at the secondary coil will be alternating and symmetric in time. The presence of an external "target" magnetic field H_x will break this symmetry and the resulting temporal asymmetry can be used to monitor the target field amplitude.

A simple way to model the ferromagnetic dynamics is through an Ising-type model. We assume the core to be composed of a set of atomic magnets, called "spins", arranged on a regular lattice that represents the crystal structure of the core [98]. Thermal fluctuations tend to disrupt the orientation of the spins while spin interactions tend to align the spins with each other. When the temperature T exceeds

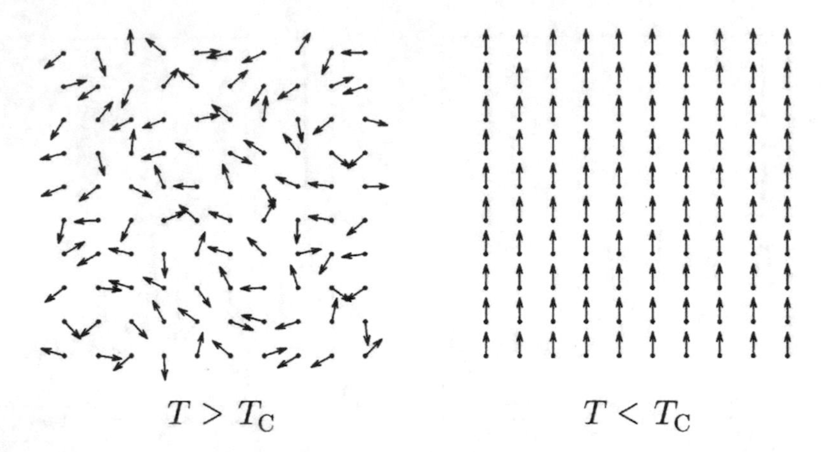

Fig. 4.43 (Left) A ferromagnet in a paramagnet state, well above the critical temperature T_c the magnetic spins are randomly organized. (Right) Below the critical temperature, a majority of the spins are uniformly organized in one of two states, "up" or "down". *Source Introduction to the Theory of Neural Computation* [101]

a critical temperature T_c, called the Curie temperature, the system exhibits a *phase transition* [99, 100] from a *paramagnet state* with little magnetization properties to a *ferromagnetic state*, where magnetization is uniform, see Fig. 4.43.

A particularly useful simplification is to consider spin 1/2 magnetic materials, so that only two distinct directions are possible: "spin up" ($S_i = +1$) and "spin down" ($S_i = -1$), where S_i is the state variable that describes the orientation of the spin found at lattice i. Then the average magnetic field $\langle h_i \rangle$ at spin S_i is determined by adding the average contributions from all neighboring spins S_j and from any external h^{ext} applied field through:

$$\langle h_i \rangle = \sum_{j \to i} w_{ij} \langle S_j \rangle + h^{ext}, \tag{4.68}$$

where w_{ij} is the coupling strength of the influence of spin S_j on S_i. The applied field h_i can induce the magnetic spin to switch back and forth between its two states $+1$ and -1. The actual switching mechanism can be modeled by an activation function:

$$\langle S_i \rangle = \tanh(\beta \langle h_i \rangle), \tag{4.69}$$

where the parameter β is related to the temperature T through $\beta = 1/(k_B T)$, with k_B being Boltzmann's constant. Substituting Eq. (4.68) into Eq. (4.69), shows that the *average magnetization* is given by

$$\langle S_i \rangle = \tanh(\beta \sum_{j \to i} w_{ij} \langle S_j \rangle + \beta h^{ext}). \tag{4.70}$$

Our interest is in the ferromagnetic state, in which $\langle S_i \rangle = \langle S \rangle$. Assuming identical coupling strengths $w_{ij} = 1/N$, where N is the number of spins, we find a single equation for the average magnetization:

$$\langle S \rangle = \tanh(\beta \langle S \rangle + \beta h^{ext}). \tag{4.71}$$

An extension of this last equation with continuous updating of the average magnetization state leads to the following simple model of the fluxgate core dynamics

$$\tau \frac{dx}{dt} = -x + \tanh\left(\frac{x + h^{ext}}{T}\right), \tag{4.72}$$

where τ is a relaxation parameter, $x = \langle S \rangle$, $k_B = 1$ so that $\beta = 1/T$ is a pseudo-temperature parameter, and $h = h^{ext}$. There is a close analogy of this model with those of artificial Hopfield neural networks [102]. The saturation nonlinearity of the tanh function, for instance, is equivalent to the activation function that controls the response of individual neurons. There is also an analogy with the *energy function* introduced by Hopfield [102] in neural network theory. In our case,

$$\tau \frac{dx}{dt} = -\frac{\partial U}{\partial x}(x, t),$$

where

$$U(x, t) = \frac{x^2}{2} - \frac{1}{c} \ln \cosh\left(c\,(x + H_e(t) + H_x)\right),$$

where $c = 1/T$, $H_e(t) = h^{ext}$. Observe that this is the same potential function that appears in Fig. 4.41. Using this notation, the model equation for a single-core fluxgate magnetometer becomes

$$\boxed{\tau \frac{dx}{dt} = -x + \tanh(c(x + H_e(t) + H_x)).} \tag{4.73}$$

Figure 4.44 shows a visualization of the vector field for the model equation (4.73), generated with the aid of the MATLAB software package *dfield* [16]. Solid lines are representative solution curves obtained by numerically integrating the model equations with a variety of initial conditions. The external field has been set to $H_x = 0$ and the biasing signal has also been set to $H_e = 0$, while the temperature related parameter $c = 3$ corresponds to the case of a ferromagnetic material. Observe that for this choice of parameters the model exhibits two stable equilibrium points $x_e = \pm 1$ and one unstable $x_e = 0$.

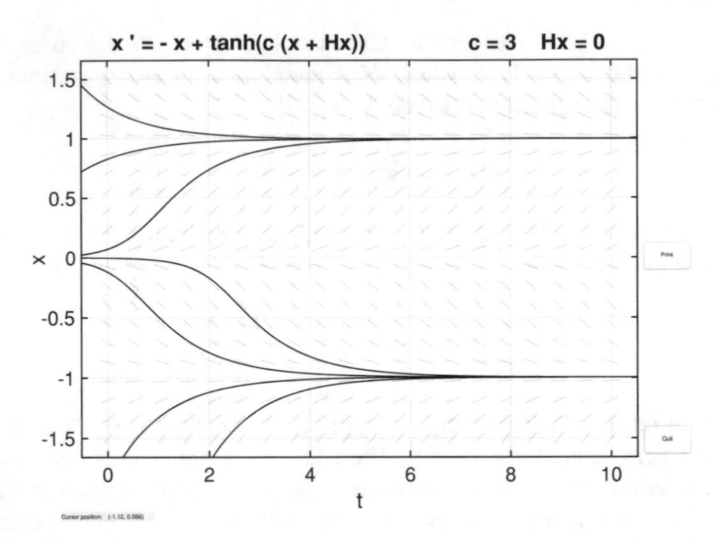

Fig. 4.44 Visualization of the vector field generated by the mathematical model that governs the magnetization state of a fluxgate magnetometer. Parameters are: $c > 1$ determines a ferromagnetic material while $0 < c < 1$ defines a paramagnetic material; $H_e(t)$ is the external biasing signal that drives the system to oscillate; H_x is the target or external field to be detected

In fact, solving

$$-x + \tanh(c(x + H_x)) = 0,$$

confirms that $x_e = \{-1, 0, 1\}$ are all indeed equilibrium points. To determine their stability properties we set $f(x) = -x + \tanh(c(x + H_x))$ and compute the derivative:

$$f'(x) = -1 + c\,\mathrm{sech}^2(c(x + H_x)).$$

Direct substitution yields $f'(x = 0) = c - 1$, so that $x_e = 0$ is unstable whenever $c > 1$. This is always the case in the ferromagnetic regime since it requires $c > 1$. Similarly, for the nonzero equilibrium points, we make use of the fact that $0 < \mathrm{sech}\, c < 1$, so that $f'(x_e = \pm 1) < 0$. It follows that the other two equilibrium points $x_e = \pm 1$ are stable. These calculations confirm the stability properties that are inferred from direct observation of Fig. 4.44.

We now set $H_x = 0.5$ and generate once again the vector field visualization shown in Fig. 4.45. This time only one equilibrium $x_e = +1$ appears to be stable.

A more convenient method to visualize what is happening with the equilibrium points when H_x varies is to plot simultaneously the functions $\tanh(c\ (x + H_x))$ and x. The intersection of the graphs of these two functions correspond to the location of the equilibrium points. Thus, when $H_x = 0$, Fig. 4.46(left) shows three intersection points which correspond to the tree equilibria $x_e = \{-1, 0, 1\}$. But when $H_x = 0.5$

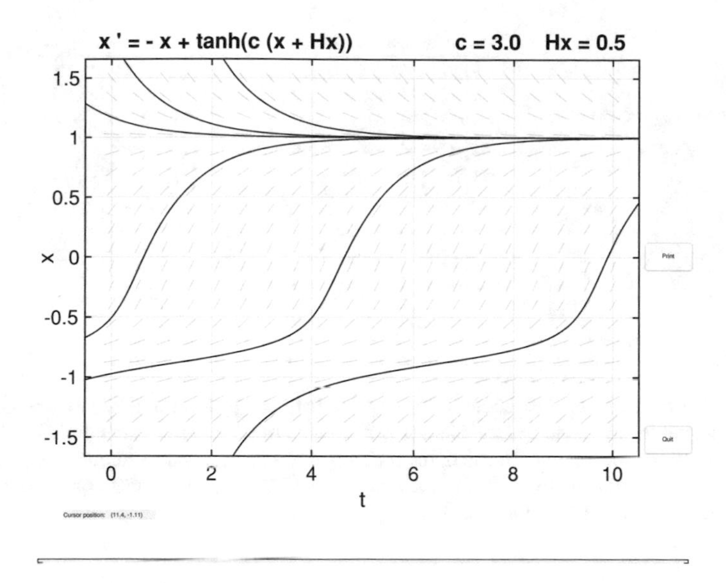

Fig. 4.45 Same as Fig. 4.44 with a nonzero target field H_x

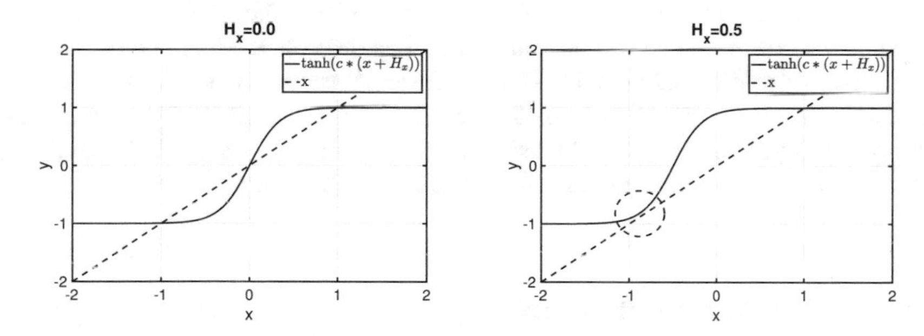

Fig. 4.46 Equilibrium points of the model of a fluxgate magnetometer are the intersection points of the graphs of $\tanh(c\,(x + H_x))$ and x. (Left) For $H_x = 0$ there are two nonzero stable equilibrium points $x_e = \pm 1$ and one unstable at $x_e = 0$. For $H_x \neq 0$, there is a threshold value at which the negative equilibrium and the zero one annihilate each other via a saddle-node bifurcation

there is only one $x_e = 1.0$. In fact, something similar happens when $H_x = -0.5$, but now the negative equilibrium $x_e = -1.0$ appears. This last case is not plotted for brevity.

Thus, as H_x increases continuously from zero the equilibria at -1 and at 0 move towards each other until they collide with one another at a critical threshold value of H_x. Past this critical value these two equilibrium points disappear. In reverse order, i.e., decreasing now H_x towards zero, a Taylor series expansion shows that the scenario under which the two equilibrium points re-appear is via a saddle-node bifurcation. This task is left as an exercise for the reader. Geometrically, the effect of continuously varying the external field H_x between positive and negative values is

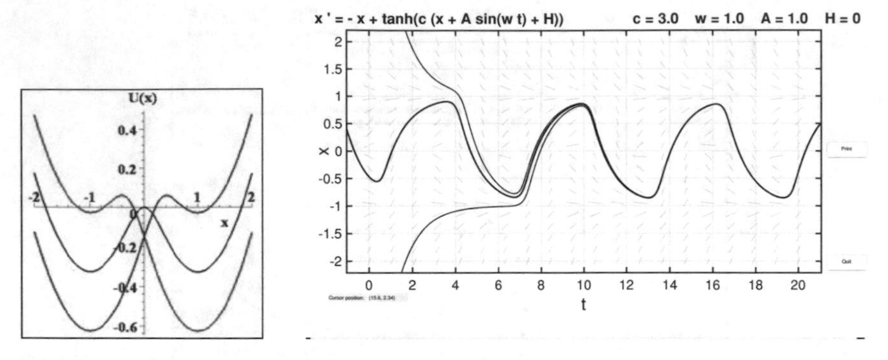

Fig. 4.47 (Left) Effect of continuously varying H_x between positive and negative values in the fluxgate magnetometer model is equivalent to periodically rocking the potential function $U(x)$ that governs its behavior. (Right) In practice, the rocking effect is achieved by applying a periodically varying external field of the form $H_e(t) = A\ \sin(w\,t)$, which leads to sustained oscillations. In this example, $H_x = 0$, $A = 1.0$ and $w = 1.0$

equivalent to periodically "rocking" the potential function that governs the behavior of the magnetometer. This rocking effect, in turn, induces the system to oscillate. In practice, the coercive field can be quite high, thus the system has to be periodically perturbed by a biasing signal of the form $H_e(t) = A\ \sin(w\,t)$ in order for the system to sustain its oscillations. Figure 4.47 illustrates the sustained oscillations that can be achieved through a sinusoidal biasing signal. In this example the external field is set to $H_x = 0$ but the same effect is achieved for positive target fields.

4.11 Compartmental Models

Many natural and artificial systems are often modeled by decomposing them into a number of interacting subsystems, or *compartments*. Each compartment contains a series of *states*, which are typically interconnected (or coupled) to other states from other compartments. This subdivision into a discrete number of states make *compartmental models* highly popular, and useful to model systems with large number of components. Examples include: large population dynamics, such as epidemiology and sociology, ecology, chemical kinetics, i.e., pharmokinetics. In this section, we illustrate the use of compartmental models for the ongoing COVID pandemic. First, we introduce some general concepts.

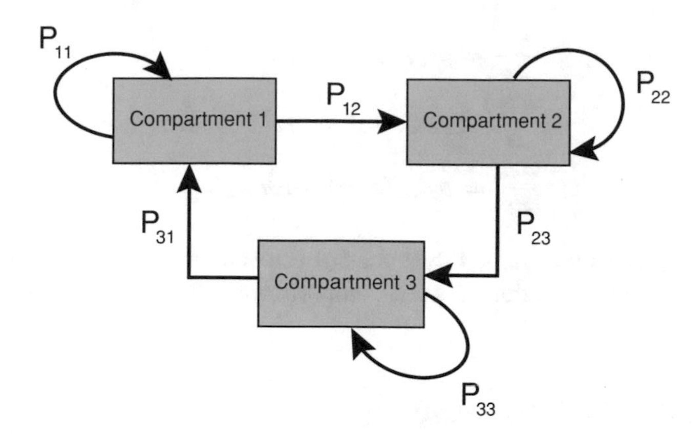

Fig. 4.48 Representative example of a three-compartmental model

4.11.1 General Setting

Let N denote the total size of the system, e.g., number of species or population size. Assume the population is then divided into $k \geq 2$ qualitatively distinct classes or compartments. Also, let x_i represent the number of individuals, n_i, in compartment i at time t, so that $x_i \in \mathbb{R}^{n_i}$, represents the state variable in compartment i, where $i = 1, \ldots k$. Then,

$$X(t) = [x_1(t), x_2(t), \ldots, x_n(t)]^T,$$

represents the state of the entire population. The phase space is $X \subset \mathbb{R}^N$, where $N = n_1 + n_2 + \ldots + n_k$.

We also consider the following modeling assumptions.

1. Let p_{ij} be the transition rate that defines the probability that an individual in compartment i will move to compartment j.
2. Transitions among individuals are made independently throughout all compartments. This means that each individual in each compartment has the same likelihood of transitioning.

Compartmental models, and their modeling assumptions, are commonly visualized through flow diagrams, with blocks representing compartments, and arrows representing interconnections or transport of states from one compartment to another. Figure 4.48 illustrates a representative example of a three-compartmental model.

Under the configuration shown in Fig. 4.48, a compartmental model can be written in the following form:

$$\begin{aligned}\frac{dx_1}{dt} &= p_{11}f_1(x_1) + p_{31}h(x_3, x_1)\\ \frac{dx_2}{dt} &= p_{22}f_2(x_2) + p_{12}h(x_1, x_2)\\ \frac{dx_3}{dt} &= p_{33}f_3(x_3) + p_{23}h(x_2, x_3),\end{aligned} \tag{4.74}$$

where f_i represents the internal dynamics of each compartment i, and $h(x_j, x_i)$ is the interconnectivity function between compartment j and compartment i.

4.11.2 COVID-19 Modeling

Throughout history, humans have faced the challenges of serious diseases and pandemics. Some examples include: the Spanish flu of 1918, the Swine flu of 2009, and most recently, the COVID-19 pandemic of 2020. Scientists all over the world have developed mathematical models aimed at describing the spatio-temporal evolution of infectious diseases. The hope is for the models to guide leaders to create preventive measures and implement effective policy to eradicate the diseases.

The first compartmental model that forecasted the progress of an epidemic was introduced by Kermack and McKendrick in 1927 [103]. The model contains three compartments: S, I, ad R, or SIR for short. S represents the number of susceptible individuals, while I is the number of infected cases, and R is the number of recovered individuals. This partition leads to a compartmental model in the form of a system of three differential equations describing the interactions between compartments. The most common form of a SIR compartmental model is

$$\begin{aligned}\frac{dS}{dt} &= -\beta\tfrac{S}{N}I\\ \frac{dI}{dt} &= \beta\tfrac{S}{N}I - \gamma I\\ \frac{dR}{dt} &= \gamma I,\end{aligned} \tag{4.75}$$

where β denotes the infectious rate, $1/\gamma$ is the average latent time, and N is the total (assumed to be constant) population size, so that $N = S(t) + I(t) + R(t)$.

In a recent attempt to model the dynamics of the COVID-19 virus, Dashtbali and Mirzaie [104] extend the SIR to create a compartmental model that is tailored to the characteristics of COVID-19, as observed empirically. The model contains 8 compartments that are account for individuals that are: (S) Susceptible, (M) Semi-susceptible, (E) Exposed, (I) Infected, (H) Hospitalized, (R) Recovered, (D) Dead, and (V) Vaccinated. The system of ODEs, written in dimensionless variables, becomes:

$$\begin{aligned}
\frac{dS}{dt} &= -(1-\alpha)\sigma(t)SI - \alpha S \\
\frac{dM}{dt} &= -\sigma(t)MI \\
\frac{dE}{dt} &= (1-\alpha)\sigma(t)SI + \sigma MI - E \\
\frac{dI}{dt} &= E - I \\
\frac{dH}{dt} &= I - H \\
\frac{dR}{dt} &= H(t) \\
\frac{dD}{dt} &= H(t) \\
\frac{dV}{dt} &= \alpha S(t),
\end{aligned} \tag{4.76}$$

where α accounts for vaccine coverage, the function $\sigma(t)$ describes the impact of social distancing, and $N = S + E + I + H + R + D + V$ is the total population

Table 4.5 Parameter estimation for a 7-compartment model for COVID-19

Parameter	Germany	Italy	Belgium	Egypt	Nigeria	Japan
N (million)	80	60	11.5	102	195	126.5
σ	0.5	0.56	0.54	0.47	0.57	0.55
α	0.2–0.6	0.2–0.6	0.2–0.6	0.2–0.6	0.2–0.6	0.2–0.6

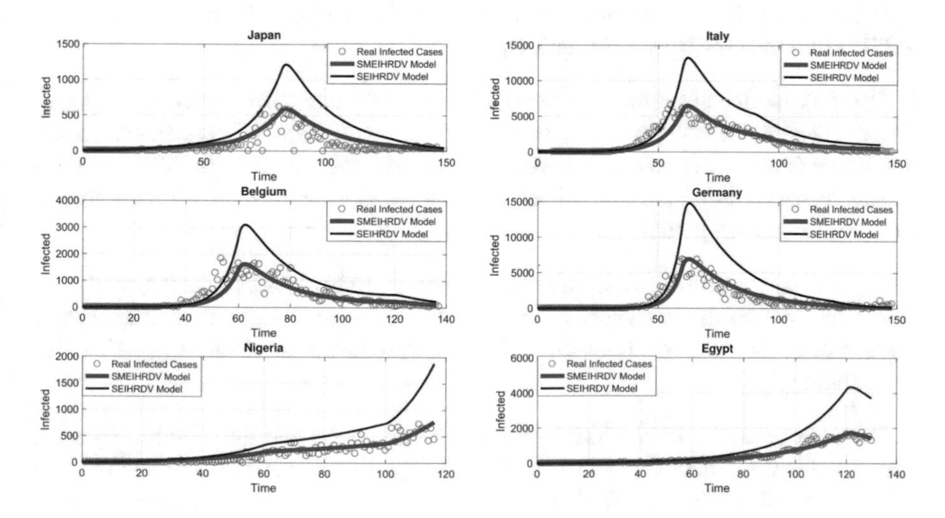

Fig. 4.49 Numerical simulations of a three-compartmental model for COVID-19 [104]

size. Table 4.5 lists the parameters estimated based on the spread of the COVID-19 virus across seven different countries.

Figure 4.49 shows results of numerical simulations of the 8-compartment model Eq. (4.76), carried out by Dashtbali and Mirzaie [104]. In one case, the compartment associated with the number of semi-susceptible individuals is employed, and in the other case it is omitted. Their simulations show that including the number of semi-susceptible individuals produces significantly more accurate results.

4.12 Exercises

Exercise 4.1 Logistic Growth for U. S. Population: The logistic growth model satisfies the ODE:

$$\frac{dP}{dt} = rP\left(1 - \frac{P}{M}\right), \qquad P(0) = P_0,$$

where $P(t)$ is the population at time t, r is the Malthusian growth rate, M is the carrying capacity, and P_0 is the initial population. The table below gives the population of the U. S. during the twentieth century (with the population given in millions).

Year	Census	Year	Census	Year	Census
1900	76.2	1940	132.2	1980	226.5
1910	92.2	1950	150.7	1990	248.7
1920	106.0	1960	179.3	2000	281.4
1930	122.8	1970	203.3	2010	308.7

a. Solve the ODE and use the populations in 1900, 1930, and 1960 with $t = 0$ at 1900 to find the constants r, M, and P_0.

b. Use this model to predict the population in 1990 and 2010. Taking the actual census data as the best predicted value, find the percent error between the model and the census data. (Use percent error as the signed relative error or $100\times$(model - census)/census.) What is the predicted limiting population for the U.S. according to this logistic growth model?

c. Repeat Parts a and b, using the populations in 1900, 1940, and 1980 with $t = 0$ at 1900 to find the constants r, M, and P_0. How much does this change your parameter values and how does this affect your predicted populations (1990, 2010, and limiting)?

d. Use your general solution to the logistic growth model and find a nonlinear least squares fit to the census data from 1900 to 1980 (with $t = 0$ at 1900) to find the best fitting constants r, M, and P_0. How much does this change your parameter values and how does this affect your predicted populations (1990, 2010, and limiting)?

e. Create a graph showing the data and the **3** models found above. Describe how well the models fit the data. List both strengths and weaknesses of using the logistic growth model and the different fits performed above.

Exercise 4.2 The *logistic growth model* satisfies the ODE:

$$\frac{dP}{dt} = rP\left(1 - \frac{P}{M}\right), \qquad P(0) = P_0,$$

where $P(t)$ is the population at time t, r is the Malthusian growth rate, M is the carrying capacity, and P_0 is the initial population. Microbial populations often fit this model very well. The bacterium *Staphylococcus aureus*, a fairly common pathogen, can cause food poisoning. Below are data from one experiment[2], where a normal strain is grown using control conditions and the optical density (OD_{650}) is measured to determine an estimate of the number of bacteria in the culture.

t (hr)	OD_{650}	t (hr)	OD_{650}	t (hr)	OD_{650}
0	0.032	2.0	0.170	4.0	0.309
0.5	0.039	2.5	0.229	4.5	0.327
1.0	0.069	3.0	0.261	5.0	0.347
1.5	0.110	3.5	0.288		

Solve the ODE and use the data above to find the best fitting constants r, M, and P_0, i.e., a nonlinear least squares fit to the data. Create a graph showing the data and the best fitting model and describe how well the model fits the data.

Exercise 4.3 Remifentanil is an opioid drug that acts very quickly and is rapidly degraded by esterases. A bolus injection of 15 μ/kg was administered to a patient and the plasma concentration was measured over a period of time. Data from the experiment is given in the table below.

t	R	t	R	t	R
1	53.15	7	5.56	25	0.76
2	28.25	8	3.96	30	0.57
3	18.82	10	3.02	40	0.39
4	11.19	12	2.2	60	0.23
5	8.93	15	1.72	90	0.13
6	7.13	20	1.17		

where t is in min and R is the concentration of remifentanil in ng/ml.

a. Perform a least squares fit of a quadratic (3-parameter) model to the remifentanil concentration in the plasma,

$$Q(t) = q_0 + q_1 t + q_2 t^2.$$

[2] Data from the laboratory of Anca Segall at San Diego State University collected by Carl Gunderson, 1998.

Write the equation for this best-fitting model. Also, consider a 3-compartment model for disappearance of remifentanil given by

$$C(t) = a_1 e^{-\lambda_1 t} + a_2 e^{-\lambda_2 t} + a_3 e^{-\lambda_3 t}.$$

Suppose that other measurements suggest that the *rapid distribution phase* has $\lambda_1 = 0.6$, the *middle distribution phase* has $\lambda_2 = 0.15$, and the *terminal/elimnation phase* has $\lambda_3 = 0.025$. Perform a least squares fit to find the 3-parameters, a_1, a_2, and a_3, and write the formula with the best-fitting parameters in your model. Write the least sum of square errors for each model. Create a graph of the data and these two models using a logarithmic scale for the vertical axis (semilog graph). Which model appears to fit the data best?

b. Perform the **exponential peeling** procedure separating out the data over the ranges $t \in [0, 7]$, $t \in [8, 25]$, and $t \in [30, 90]$. Give the best fitting exponential models (using a linear least squares fit to the logarithms of the concentrations) over these ranges, where you start with the *terminal/elimination phase* and subtract each model from the data before computing the next exponential fit. Give the complete 3-compartment model using this exponential peeling procedure. Include the sum of square errors for this model.

c. Start with the 6 parameters from the exponential peeling (3 coefficients and 3 exponents) and use MatLab's `fminsearch`(Nonlinear Least Squares fit) to find the best fitting parameters for the 3-compartment model. This procedure fails with a direct least squares fit of the model to the data, so perform a least squares best fit of the logarithm of the model to the logarithm of the data. Write this best fitting model with its parameters and give the sum of square errors. Write a brief paragraph comparing and contrasting the models in the entire problem and giving your opinion on which ones are best from both how they fit the data and the computational effort. Create a semilog graph including the data, the 3-compartment model from Part a, the exponential peeling model from Part b, and the Nonlinear least squares best fit model.

Exercise 4.4 Consider the following 2D problem:

$$L\frac{d^2 I}{dt} + R\frac{dI}{dt} + \frac{1}{C} I = 0, \tag{4.77}$$

where $L, C > 0$ and $R \geq 0$ are parameters.

(a) Transform the equation to a 2D linear system.
(b) Show that the $(0, 0)$ equilibrium is asymptotically stable if $R > 0$ and neutrally stable if $R = 0$.
(c) Classify the $(0, 0)$ equilibrium depending on whether $R^2C - 4L$ is positive, negative or zero. Sketch the phase portrait for each case.

Exercise 4.5 Consider the weak forced Duffing oscillator.

(a) Show the details of the van der Pol transformation, converting Duffing's equation (4.42) into Sys. (4.43).
(b) Show the details of applying the Method of Averaging to (4.43) in order to obtain Sys. (4.44).
(c) Show the details of transforming the averaged system (4.44) in coordinates y_1 and y_2 into Sys. (4.45) in polar coordinate.

Exercise 4.6 The spruce budworm is a serious pest in eastern Canada, where it attacks the leaves of the balsam fir tree. When an outbreak occurs, the budworms can defoliate and kill most of the fir trees in the forest in about four years. In 1978 Ludwig proposed the following model to explain the outbreak of budworms:

$$\frac{dN}{dT} = RN\left(1 - \frac{N}{K}\right) - P(N) \quad \text{with} \quad P(N) = \frac{BN^2}{A^2 + N^2} \tag{4.78}$$

where $A, B, K, R > 0$. The parameter R represents growth rate and K the carrying capacity. The term $P(N)$ represents the death rate due to predators (cf. birds).

(a) Plot a typical graph for $P(N)$ and give an ecological explanation for its shape as well as the meaning of A and B.
(b) By rescaling time: $t = \tau T$ and population: $n = \eta N$, show that (4.78) can be rewritten as

$$\frac{dn}{dt} = rn\left(1 - \frac{n}{k}\right) - \frac{n^2}{1 + n^2}. \tag{4.79}$$

Find τ, η, k and r as a function of the original parameters A, B, K, R.
(c) Show that (0,0) is an equilibrium and determine its stability. Show that the remaining equilibria of (4.79) happen at the intersection points of $f(n) = \frac{n}{1+n^2}$ and $g(n) = r\left(1 - \frac{n}{k}\right)$. For a fixed value of k, sketch $f(n)$ and the 5 possibilities for $g(n)$ such that, for increasing r, you have (i) 1, (ii) 2, (ii) 3, (iv) 2, and (v) 1 intersections between f and g.
(d) For each of the 5 cases in (c), label the fixed points, sketch the phase line and give a detailed description of what you expect from the dynamics of the population. (Do not attempt to find an explicit form for the fixed points.)

Exercise 4.7 Consider the following 2D linear system of differential equations

$$\begin{cases} \dfrac{dx}{dt} = ax + by \\ \dfrac{dy}{dt} = cx + dy \end{cases}$$

for the following cases:

$$\begin{array}{lllll} \text{(a)} & a = 1, & b = 3, & c = 1, & d = -1 \\ \text{(b)} & a = 4, & b = -3, & c = 1, & d = 0 \\ \text{(c)} & a = -1, & b = 2, & c = -2, & d = -1 \end{array}$$

(a) Find the solution for $(x(t), y(t))$ using eigenvalues/eigenvectors.
(b) Determine the stability of the origin.
(c) Sketch the phase plane with isoclines and eigenvectors.

Exercise 4.8 In several situations, due to nonlinearities, it is not possible, or too cumbersome, to find an explicit form for the equilibrium points. Nonetheless, in those cases the Jacobian matrix tends to have the simple form:

$$A = \begin{pmatrix} a & b \\ c & d \end{pmatrix} \qquad a, b, c, d \in \mathbb{R} \tag{4.80}$$

(a) Using *only* the signs of the trace and the determinant of A what can you tell about the stability of the equilibrium point?
(b) Using *only* the signs of a, b, c, d what can you tell about the stability of the equilibrium point?

Exercise 4.9 A mathematical model that describes the motion of a swinging mass-spring system is given by the following system of differential equations

$$\begin{aligned} \frac{Ld^2\theta}{dt^2} + 2\frac{dL}{dt}\frac{d\theta}{dt} &= -g\sin(\theta), \\ \frac{d^2L}{dt^2} - L\left[\frac{d\theta}{dt}\right]^2 &= g\cos(\theta) - \frac{k}{m}(L - L_0), \end{aligned} \tag{4.81}$$

where L_0 is the unstretched length of the spring, m is the mass of the object, and k is the spring constant.

(a) Transform (4.81) to a first-order system of differential equations and find all equilibrium points.
(b) Characterize the stability of each equilibrium point. Find the characteristic equation and determine the eigenvalues.

(c) Find the period of motion in the L direction and in the θ direction.

Exercise 4.10 Mathematical models for nonlinear oscillators are usually written in the following form:

$$\ddot{x} + \omega^2 x = \varepsilon g(x, \dot{x}, t, \varepsilon), \quad x(t_0) = a_1, \quad \dot{x}(t_0) = a_2.$$

Find a suitable transformation to write the equation above in standard form for the method of averaging.

Exercise 4.11 Apply the method of averaging to study *Mathieu's equation*:

$$\ddot{x} + (1 + 2\varepsilon \cos(2t))x = 0.$$

Exercise 4.12 Consider the equation:

$$\ddot{x} + x = \varepsilon \dot{x}^2 \cos t.$$

Transform this equation to standard form and then apply the method of averaging to study its solution. Plot the solution of the original equation obtained by numerical integration and the solution produced by averaging. Compare the two solutions.

Exercise 4.13 Consider the following 1D model

$$\dot{x} = \varepsilon x \sin^2(t).$$

(a) Integrate the model equation to find an exact solution.
(b) Apply the method of averaging to find an approximate analytical solution.
(c) Compare the exact solutions against the averaged solution.

Exercise 4.14 Consider Mathieu's equation

$$\frac{d^2u}{dt^2} + (\delta + \varepsilon \cos 2t)u = 0. \tag{4.82}$$

(a) Use the generalized method of averaging to determine the equations describing the slow variations in the amplitude and the phase.
(b) Write a MATLAB (or equivalent software) program to numerically integrate (4.82). Graph (in one single plot) the averaged solution and the numerical solution. Explain the results.

Exercise 4.15 Consider a simplified version of the nonlinear Mathieu's equation

$$\ddot{x} + (1 + \varepsilon \cos t)x + \varepsilon \beta x^3 = 0.$$

Determine periodic solutions via averaging.

Exercise 4.16 Consider a bucket with a whole in the bottom. If at a given time you see the bucket empty, can you figure out when (if ever) it was full? Explain your answer.

Let $h(t)$ represent the height of the water remaining in the bucket at time t. The differential equation that models the leaky bucket is:

$$\text{(Initial Value Problem)} \begin{cases} \dfrac{dh}{dt} = -k\sqrt{h} \\ h(0) = 1 \quad \text{(full bucket)}, \end{cases} \tag{4.83}$$

where k is an arbitrary constant that controls how fast the bucket becomes empty. Perform the following tasks.

(a) Use *separation of variables* to find the explicit analytical solution to the initial value problem in terms of t_e (time when the bucket becomes empty).
(b) Verify that your solution satisfies the Initial Value Problem.
(c) Sketch the solution in the $x - t$ plane for different values of t_e.
(d) Explain why there are infinitely many solutions of the differential equation.

Exercise 4.17 Consider the model

$$\dot{x} = \varepsilon(1 - 2\cos\theta)$$
$$\dot{\theta} = x,$$

where $x \in \mathbb{R}$, $x(0) = x_0$, $\theta \in \S^1$, and $\theta(0) = \theta_0$.

Show that the error between the exact and the averaged solution grows like $(\Delta\theta, \Delta x) \approx (\varepsilon t^2, \varepsilon t)$.

Exercise 4.18 In response to glucose, β-cells of the pancreatic islet secrete insulin, which causes the increase use or uptake of glucose in target tissues such as muscle, liver, and adipose tissue. When blood levels of glucose decline, insulin secretion stops, and the tissues begin to use their energy stores instead. Interruption of this control system results in diabetes. It is believed that electrical bursting plays an important role in the release of insulin from the cell. A possible mechanism for generating bursting behavior is through the FitzHugh-Nagumo model

$$\begin{cases} \dfrac{dx}{dt} = y - x^3 + 3x^2 + I - z \\ \dfrac{dy}{dt} = 1 - 5x^2 - y \\ \dfrac{dz}{dt} = r\left[s\left(x + \dfrac{1+\sqrt{5}}{2}\right) - z\right], \end{cases} \tag{4.84}$$

where I is an input current applied to the cell, and r and s are positive parameters.

(a) Assume $I = 0$, $z = 0$ and $r = 0$. Observe that the first two equations decouple from the last equation. Find *all* equilibrium points of the first two equations.
(b) Determine the local stability properties of *all* equilibrium points in part (a), and sketch the phase plane (xy-plane) solutions using eigenvalues, eigenvectors,, and nullcline curves.
(c) Repeat (a) and (b) with $I = 0.4$, $I = 2$, $I = 4$, and various values of z. Explain how the phase plane solutions change as I and z change.
(d) Write a MATLAB program to integrate the full system of equations with the same values of input current: $I = 0.4$, $I = 2$, and $I = 4$. For each value of I, produce the following graphs: (1) Time series of $x(t)$, $y(t)$, and $z(t)$, all in the same plot; (2) Phase plane diagrams: x versus y, x versus z, and y versus z, all in the same plot.
(e) Write a brief explanation of your results relevant to the biological interpretation of the model.

Exercise 4.19 Michael Crichton in the *Andromeda Strain* (1969) states that "A single cell of the bacterium *E. coli* would, under ideal circumstances, divide every twenty minutes... It can be shown that in a single day, one cell of *E. coli* could produce a super-colony equal in size and weight to the entire planet Earth". A single *E. coli* has a volume of about 1.7 μm^3. The diameter of the Earth is 12,756 km, so assuming it is a perfect sphere, determine how long it takes for an ideally growing colony (Malthusian growth) of *E. coli* (doubling every 20 min) to equal the volume of the Earth.

Exercise 4.20 Consider a modified version of the van der pol model

$$\frac{d^2x}{dt^2} + \mu(x^2 - 1)\frac{dx}{dt} + x = a \tag{4.85}$$

where μ and a are constant parameters.

(a) Convert the second order ODE to a first-order system of ODE's.
(b) Analytically, calculate *all* equilibrium points and study their stability.
(c) Find the curves in (μ, a) parameter-space at which the eigenvalues of the linearized Jacobian matrix J are purely imaginary, which is equivalent to the condition: $trace(J) = 0$ and $det(J) > 0$. This locus of points is called a Hopf bifurcation.

(d) Sketch a diagram in the (μ, a) plane illustrating the change in stability of each equilibrium point when both μ and a change. Sketch phase portraits of different types of behaviors.

Exercise 4.21 Compute all equilbrium points of the best fitted two-species competition model Eq. (4.25) and perform a linear stability analysis. Write your conclusions about the competition of the two species.

Exercise 4.22 Use the analysis of the Lotka-Volterra or Predator-Prey model to explain why the application of DDT to control scale insects in the citrus industry led to excessive application of the pesticide, which in turn resulted in many environment problems highlighted in the classic book that really started the environmental movement, Rachel Carlson's *Silent Spring* (referring to the devastation of the bird populations). The citrus groves were rapidly invaded by scale insects, which caused tremendous destruction until lady bugs, the scale insect's natural predator, were imported as a control. Give at least two reasons why use of pesticides result in escalating use of pesticides that further put farmers in debt and only enrich the chemical industry. Use modeling methods to show some smarter way to control an agricultural pest.

Exercise 4.23 The Colpitts oscillator is a parallel nonlinear LC circuit that was designed to be an almost sinusoidal oscillator [105]. Figure 4.50 shows a schematic circuit diagram containing a feedback circuit built with a single bipolar junction transistor, T (circle in the middle), which acts as a gain element. The feedback occurs because the output of the transistor is connected to its input in a feedback loop that contains a resonant network made up of an inductor, L, and a pair of capacitors, C_1 and C_2. The input-output characteristics of the transistor are responsible for the nonlinear behavior of the entire circuit.

A distinguishing feature of the Colpitts oscillator is that the feedback for the active device is taken from a voltage divider made of two capacitors in series across the inductor. This design leads to a circuit with high quality factor, which in turn yields better frequency stability. Additionally, these type of nonlinear oscillators can also perform in a large frequency bandwidth, ranging from a few Hertz all the way up to the gigahertz range.

In this exercise you will derive the model equations of a Colpitts oscillator. The state variables that determine the evolution of the circuit at any time are: the transistor collector voltage, V_{C_1}, and the voltage across the second capacitor, V_{C_2}, and the current, I_L in the inductor L. The modeling process requires the following steps.

a. In this first part, you will derive the governing equation for the voltage, V_{C_1}, across the capacitor C_1. The current-voltage relation, $I_{C_1} - V_{C_1}$, across the capacitor, C_1, indicates that

$$C_1 \frac{dV_{C_1}}{dt} = I_{C_1}.$$

Using Kirchhoff's first law of currents, show that

$$I_{C_1} = I_L - \alpha I_E,$$

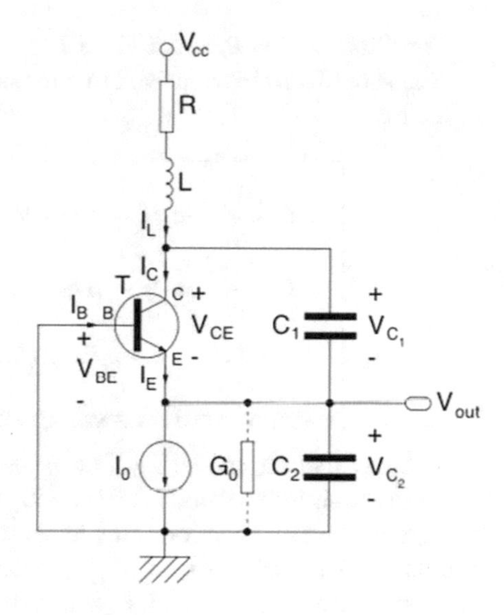

Fig. 4.50 Schematic circuit diagram for a Colpitts oscillator

where I_E is the current in the base-emitter, B-E, given by

$$I_E(V_{BE}) = \frac{I_S}{\alpha} \exp\left(\frac{V_{BE}}{V_T}\right),$$

in which α is the common-base forward short-circuit current gain of the transistor, I_S is the saturation current of the B-E junction, V_{BE} is the voltage across the B-E junction, and $V_T = k_b T/q$ is the thermal voltage with k_b being the Boltzmann constant, T is the absolute temperature expressed in Kelvin degrees, and q is the electron charge.

b. Now you will do something similar for obtaining an evolution equation for the variations of the voltage, V_{C_2}, across the second capacitor C_2. The current-voltage relation, $I_{C_2} - V_{C_2}$, across the capacitor, C_2, indicates that

$$C_1 \frac{dV_{C_1}}{dt} = I_{C_2}.$$

Using Kirchhoff's first law of currents, show that

$$I_{C_2} = I_L - \alpha I_0 - G_0 V_{C_2} + (1 - \alpha) I_E(-V_{C_2}),$$

where G_0 is a parasitic conductance G_0.

c. Now, you will derive an equation for the voltage, V_L, across the inductor. Apply Kirchhoff's second law of voltage to show that

$$V_{CC} = V_L + V_{C_1} + V_{C_2} + V_{R_L}.$$

Substitute $V_{R_L} = RI_L$ and $V_L = L\frac{dI_L}{dt}$, to write a differential equation for $\frac{dI_L}{dt}$.

You should have now arrived to a mathematical model for the Colpitts oscillator in the form:

$$\begin{aligned} C_1\frac{dV_{C_1}}{dt} &= I_L - \alpha I(-V_{C_2}) \\ C_2\frac{dV_{C_2}}{dt} &= I_L - \alpha I_0 - G_0 V_{C_2} + (1-\alpha)I_E(-V_{C_2}) \\ L\frac{dI_L}{dt} &= -V_{C_1} - V_{C_2} - RI_L + V_{CC}. \end{aligned} \tag{4.86}$$

d. Show that the model Eq. (4.86) has one unique equilibrium, $(V^*_{C_1}, V^*_{C_2}, I^*_L)$. Write the equilibrium in terms of $I_E(-V^*_{C_2})$. The value, $|V^*_{C_2}|$ represents the threshold voltage of the B-E junction. This leads to two modes of operation of the Colpitts oscillator. One where, $V_{BE} > |V^*_{C_2}|$, in which the transistor conducts, i.e., $I_E > 0$. And one where, $V_{BE} < |V^*_{C_2}|$, in which the transistor is cut off, i.e., $I_E \approx 0$.

e. Derive a dimensionless version of the model Eq. (4.86) as follows. Shift the origin of the phase-space coordinates to the equilibrium point $(V^*_{C_1}, V^*_{C_2}, I^*_L)$ of Eq. (4.86) by applying the transformation:

$$x_1 = \frac{V_{C_1} - V^*_{C_1}}{V_T}, \quad x_2 = \frac{V_{C_2} - V^*_{C_2}}{V_T}, \quad x_3 = \frac{I_L - I^*_L}{I_0},$$

and by re-scaling time by

$$\tau = \omega_0 t, \qquad \omega_0 = \frac{1}{\sqrt{L\dfrac{C_1C_2}{C_1+C_2}}},$$

where ω_0 represents the resonant frequency of the circuit. Show that the dimensionless model takes the form

$$\begin{aligned} \frac{dx_1}{d\tau} &= \frac{g}{Q(1-\kappa)}\left[-\alpha\left(e^{-x_2}-1\right)+x_3\right] \\ \frac{dx_2}{d\tau} &= \frac{g}{Q\kappa}\left[(1-\alpha)\left(e^{-x_2}-1\right)+x_3\right] - Q_0(1-\kappa)x_2 \\ \frac{dx_3}{d\tau} &= -\frac{Q\kappa(1-\kappa)}{g}(x_1+x_2) - \frac{1}{Q}x_3, \end{aligned} \tag{4.87}$$

where the parameters κ, Q_0, Q and g are defines as

$$\kappa = \frac{C_2}{C_1+C_2}, \quad Q_0 = G_0\omega_0 L, \quad Q = \frac{\omega_0 L}{R}, \quad g = \frac{I_0 L}{V_T R(C_1+C_2)}.$$

An advantage of these dimensionless coordinates is that the subspace $x_2 < 0$ is associated with the active or conductive mode of the transistor, while $x_2 > 0$ corresponds to the cut-off mode.

f. We can further reduce the number of parameters in the dimensionless model by assuming an ideal current bias, where $G_0 \to 0$, which, in turn, implies that $Q_0 \to 0$. If we neglect the base current of the transistor, so that $\alpha = 1$, we can obtain an ideal model in the form

$$\begin{aligned} \frac{dx_1}{d\tau} &= \frac{g}{Q(1-\kappa)}\left[1 - e^{-x_2} + x_3\right] \\ \frac{dx_2}{d\tau} &= \frac{g}{Q\kappa}x_3 \\ \frac{dx_3}{d\tau} &= -\frac{Q\kappa(1-\kappa)}{g}(x_1 + x_2) - \frac{1}{Q}x_3. \end{aligned} \tag{4.88}$$

In this model the trivial equilibrium (0, 0, 0) corresponds to the equilibrium point, $(V_{C_1}^*, V_{C_2}^*, I_L^*)$. Calculate the Jacobian matrix of the linearization of Eq. (4.88) about the trivial equilibrium point (0, 0, 0). Compute the eigenvalues of the characteristic polynomial associated with the Jacobian matrix. Perform computer simulations to show that oscillations do not exist when $g < 1$, while stable oscillations exist for $g > 1$, independently of the parameter Q.

Exercise 4.24 Crystal Oscillator. In this exercise we will explore an alternative derivation to the averaged equations for a crystal oscillator.

(a) Let us start with a *dimensionless derivation* of the model equations. Re-scale time by $t = \sqrt{L_1 C_1}\tau$. Let $\Omega_1^2 = 1$, $\Omega_2^2 = \frac{L_1}{L_2}\frac{C_1}{C_2}$, $L_r = \frac{L_1}{L_2}$, and $\varepsilon = \sqrt{\frac{C_1}{L_1}}$, and relabel τ as time t. Then show that Eq. (4.62) becomes:

$$\begin{aligned} \frac{d^2 i_1}{dt^2} + \Omega_1^2 i_1 &= \varepsilon\left\{-R_1\frac{di_1}{dt} + \left[a - 3b(i_1 + i_2)^2\right]\left[\frac{di_1}{dt} + \frac{di_2}{dt}\right]\right\} \\ \frac{d^2 i_2}{dt^2} + \Omega_2^2 i_2 &= \varepsilon L_r\left\{-R_2\frac{di_2}{dt} + \left[a - 3b(i_1 + i_2)^2\right]\left[\frac{di_1}{dt} + \frac{di_2}{dt}\right]\right\}. \end{aligned} \tag{4.89}$$

(b) Show that the invertible transformations

$$\begin{aligned} i_1 &= x_1 \cos\phi_1; \quad i_1' = -\Omega_1 x_1 \sin\phi_1; \\ i_1'' &= \Omega_1 x_1' \sin\phi_1 - \Omega_1^2 x_1 \cos\phi_1 - \Omega_1 x_1 \psi_1' \cos\phi_1; \\ i_2 &= x_2 \cos\phi_2; \quad i_2' = -\Omega_2 x_2 \sin\phi_2; \\ i_2'' &= \Omega_2 x_2' \sin\phi_2 - \Omega_2^2 x_2 \cos\phi_2 - \Omega_2 x_2 \psi_2' \cos\phi_2; \\ \phi_1 &= \Omega_1 t + \psi_1; \quad \phi_2 = \Omega_2 t + \psi_2, \end{aligned} \tag{4.90}$$

allow us to rewrite Eq. (4.89) in the following form.

$$\begin{bmatrix} \mathbf{x}' \\ \boldsymbol{\psi}' \\ \boldsymbol{\phi}' \end{bmatrix} = \begin{bmatrix} 0 \\ 0 \\ {}^{0} \end{bmatrix} + \varepsilon \begin{bmatrix} \mathbf{X}^{[1]}(\mathbf{x}, \boldsymbol{\phi}, \varepsilon) \\ {}^{[1]}(\mathbf{x}, \boldsymbol{\phi}, \varepsilon) \\ 0 \end{bmatrix}, \tag{4.91}$$

where $\mathbf{x} = (x_1, x_2)$, $\boldsymbol{\phi} = (\phi_1, \phi_2)$, $\boldsymbol{\psi} = (\psi_1, \psi_2)$, ${}^{0} = (\Omega_1, \Omega_2)$, $\mathbf{X}^{[1]} = (X_1^{[1]}, X_2^{[1]})$ and ${}^{[1]} = (\Omega_1^{[1]}, \Omega_2^{[1]})$. Explicitly:

$$\begin{aligned}
X_1^{[1]} &= \frac{1}{\Omega_1}\{R_1\Omega_1 x_1 \sin\phi_1 + \\
&\quad [a - 3b(x_1\cos\phi_1 + x_2\cos\phi_2)^2][-\Omega_1 x_1 \sin\phi_1 - \Omega_2 x_2 \sin\phi_2]\}\sin\phi_1 \\
X_2^{[1]} &= \frac{L_r}{\Omega_2}\{R_2\Omega_2 x_2 \sin\phi_2 + \\
&\quad [a - 3b(x_1\cos\phi_1 + x_2\cos\phi_2)^2][-\Omega_1 x_1 \sin\phi_1 - \Omega_2 x_2 \sin\phi_2]\}\sin\phi_2 \\
\Omega_1^{[1]} &= \frac{1}{\Omega_1 x_1}\{R_1\Omega_1 x_1 \sin\phi_1 + \\
&\quad [a - 3b(x_1\cos\phi_1 + x_2\cos\phi_2)^2][-\Omega_1 x_1 \sin\phi_1 - \Omega_2 x_2 \sin\phi_2]\}\cos\phi_1 \\
\Omega_2^{[1]} &= \frac{L_r}{\Omega_2 x_2}\{R_2\Omega_2 x_2 \sin\phi_2 + \\
&\quad [a - 3b(x_1\cos\phi_1 + x_2\cos\phi_2)^2][-\Omega_1 x_1 \sin\phi_1 - \Omega_2 x_2 \sin\phi_2]\}\cos\phi_2
\end{aligned}$$

(c) Observe that now the first two equations in (4.91) are in standard form [26]. Apply the method of averaging over the phase variables to obtain:

$$\begin{bmatrix} \mathbf{x}' \\ \boldsymbol{\psi}' \\ \boldsymbol{\phi}' \end{bmatrix} = \begin{bmatrix} 0 \\ 0 \\ \boldsymbol{\Omega}^{0} \end{bmatrix} + \varepsilon \begin{bmatrix} \bar{\boldsymbol{X}}^{[1]}(\mathbf{x}, \boldsymbol{\phi}) \\ \bar{\boldsymbol{\Omega}}^{[1]}(\mathbf{x}, \boldsymbol{\phi}) \\ 0 \end{bmatrix}, \tag{4.92}$$

where

$$\begin{aligned}
\bar{\boldsymbol{X}}^{[1]}(\mathbf{x}, \boldsymbol{\phi}) &= \frac{1}{(2\pi)^2}\int_{\mathbf{T}^2} \mathbf{X}^{[1]}(\mathbf{x}, \boldsymbol{\phi}, 0)\, d\phi_1\, d\phi_2 \\
\bar{\boldsymbol{\Omega}}^{[1]}(\mathbf{x}, \boldsymbol{\phi}) &= \frac{1}{(2\pi)^2}\int_{\mathbf{T}^2} \boldsymbol{\Omega}^{[1]}(\mathbf{x}, \boldsymbol{\phi}, 0)\, d\phi_1\, d\phi_2.
\end{aligned}$$

Simplify and show that $\bar{\boldsymbol{\Omega}}^{[1]}(\mathbf{x}, \boldsymbol{\phi}) = (0, 0)$, i.e., $\psi_1' = \psi_2' = 0$, so that the averaged system (4.92) can be re-written as

$$
\begin{aligned}
x_1' &= \varepsilon(a - R_1)x_1 - \varepsilon\frac{3b}{4}\left(x_1^2 + 2x_2^2\right)x_1 \\
x_2' &= \varepsilon L_r(a - R_2)x_2 - \varepsilon L_r\frac{3b}{4}\left(x_2^2 + 2x_1^2\right)x_2 \\
\phi_1' &= \Omega_1 \\
\phi_2' &= \Omega_2.
\end{aligned} \tag{4.93}
$$

References

1. G. F. Gause, Experimental studies on the struggle for existence. I. Mixed populations of two species of yeast. J. Exp. Biol. **9**, 389 (1932)
2. G.F. Gause, *Struggle for Existence* (Hafner, New York, 1934)
3. J. Guckenheimer, P.J. Holmes, *Nonlinear Oscillations* Dynamical Systems and Bifurcations of Vector Fields. (Springer, New York, 1993)
4. S.H. Strogatz, *Nonlinear Dynamics and Chaos* (Persus, Reading, MA, 1994)
5. S. Wiggins, *Introduction to Applied Nonlinear Dynamical Systems* (Springer, New York, 1990)
6. F. Brauer, J.A. Nohel, *The Qualitative Theory of Ordinary Differential Equations: An Introduction* (W. A. Benjamin, New York, 1969)
7. P. Hartman, *Ordinary Differential Equations* (Wiley, New York, 1964)
8. L. Perko, *Differential Equations and Dynamical Systems*, 3rd edn. (Springer, New York, 2001)
9. W.E. Boyce, R.C. Diprima, D.B. Meade, *Elementary Differential Equations and Boundary Value Problems*, 11th edn. (Wiley, 2017)
10. J.R. Brannan, W.E. Boyce, *Differential Equations: An Introduction to Modern Methods and Applications*, 3rd edn. (Wiley, 2015)
11. J.M. Mahaffy, *Math 337—Elementary Differential Equations*, Lecture Notes, 01 Apr. 2020. https://jmahaffy.sdsu.edu/courses/f15/math337/beamer/LinSys2B-04.pdf
12. A. Einstein, Zur quantentheorie der strahlung. Physikalische Zeitschrift **18**, 121–128 (1917)
13. T.H. Maiman, Stimulated optical radiation in ruby. Nature **187**(4736), 493–494 (1960)
14. C.H. Townes, *A Century of Nature: Twenty-one Discoveries that Changed the World*, Chapter the First Laser (University of Chicago Press, 2003), pp. 107–112
15. H. Haken, *Synergetics* (Springer, Berlin, 1983)
16. J.C. Polking, *Dfield and Plane*
17. V. Volterra, *Lecons sur la Théorie Mat'ematique de la Lutte pour la Vie* (Gauthier-Villars, 1931)
18. A. Garfinkel, J. Shevtsov, Y. Guo, *Modeling Life* (Springer, 2017), pp. 1–68
19. A.J. Lotka, Contribution to the theory of periodic reaction. J. Phys. Chem. **14**, 271–274 (1910)
20. A.J. Lotka, *Elements of Physical Biology* (Williams and Wilkins, Baltimore, 1925)
21. V. Volterra, Fluctuations in the abundance of species, considered mathematically. Nature **118**, 558–560 (1926)
22. J-L. Lagrange, *Mécanique analytique*, 4th edn. (Gauthier-Villars et fils, Paris, 1788)
23. P. Fatou, Sur le mouvement d'un système soumis à des forces à courte période. Bull. de la Soc. Math. de France **56**, 98–139 (1928)
24. N.M. Krylov, N.N. Bogoliubov, *New Methods of Nonlinear Mechanics in their Application to the Investigation of the Operation of Electronic Generators* (United Scientific and Technical Press, Moscow, 1934)
25. J.K. Hale, *Ordinary Differential Equations* (Wiley-Interscience, New York, 1969)
26. J.A. Sanders, F. Verhulst, J. Murdock, *Averaging Methods in Nonlinear Dynamical Systems* (Springer, 2nd edn., 2007)

27. R. Feynman, R. Leighton, M. Sands, *The Feynman Lectures on Physics, Volume*. The Feynman Lectures Website, vol. 1, Online edn. (Caltech, 2013)
28. D. Halliday, R. Resnick, J. Walker, *Fundamentals of Physics*, 10th edn. (Wiley, New York, 2013)
29. B. Van der Pol, On "relaxation-oscillations". London, Edinburgh, and Dublin Philos. Mag. J. Sci. Ser. **7**(2), 978–992 (1926)
30. S. Strogatz, From kuramoto to crawford: exploring the onset of synchronization in populations of coupled oscillators. Physica D **143**, 1–20 (2000)
31. J. Hale, *Ordinary Differential Equations* (Dover Publications, 2009)
32. A.L. Hodgkin, A.F. Huxley, A quantitative description of membrane current and its application to conduction and excitation in nerve. J. Physiol. London **117**, 500–544 (1952)
33. A.L. Hodgkin, A.F. Huxley, Propagation of electrical signals along giant nerve fibres. Proc. R. Soc. Lond. B Biol. Sci. **140**, 177–183 (1952)
34. A.L. Hodgkin, A.F. Huxley, Currents carried by sodium and potassium ions through the membrane of the giant axon of loligo. J. Physiol. **116**, 449–472 (1952)
35. R. FitzHugh, Thresholds and plateaus in the hodgkin-huxley nerve equations. J. Gen. Physiol. **43**, 867–869 (1960)
36. R. FitzHugh, Impulses and physiological states in theoretical models of nerve membrane. Biophys. J. **1**, 445–466 (1961)
37. J. Nagumo, S. Arimoto, S. Yoshizawa, An active pulse transmission line simulating nerve axon. Proc. IRE. **50**, 2061–2070 (1962)
38. A.K. Poddar, U.L. Rohde, *Crystal Oscillators* (Wiley Encyclopedia and Electronics Engineering, 2012), pp. 1–38
39. J. wang, R. Wu, J. Du, T. Ma, D. Huang, W. Yan, The nonlinear thickness-shear ovibrations of quartz crystal plates under a strong electric field, in *IEEE International Ultrasonics Symposium Proceedings*, vol. 10.1109 (IEEE, 2011), pp. 320–323
40. Marrison Warren, The evolution of the quartz crystal clock. Bell Syst. Tech. J. **27**, 510–558 (1948)
41. J. Buck, E. Buck, Mechanism of rhythmic synchronous flashing of fireflies: fireflies of south-east ASIA may use anticipatory time-measuring in synchronizing their flashing. Science **159**, 1319–1327 (1968)
42. G.B. Ermentrout, J. Rinzel, Beyond a pacemaker's entrainment limit: phase walk-through. Am. J. Physiol. **246**, 102–106 (1984)
43. P.L. Buono, M. Golubitsky, A. Palacios, Heteroclinic cycles in rings of coupled cells. Physica D **143**, 74–108 (2000)
44. A. Cohen, S. Rossignol, S. Grillner, (eds.), *Neural Control of Rhythmic Movements in Vertebrates*, New York (Wiley, 1988)
45. M. Golubitsky, I.N. Stewart, P.-L. Buono, J.J. Collins, A modular network for legged locomotion. Physica D **115**, 56–72 (1998)
46. N. Kopell, G.B. Ermentrout, Coupled oscillators and the design of central pattern generators. Math. Biosci. **89**, 14–23 (1988)
47. N. Kopell, G.B. Ermentrout, Phase transitions and other phenomena in chains of oscillators. SIAM J. Appl. Math. **50**, 1014–1052 (1988)
48. A.H. Cohen, S. Rossignol, S. Grillner (eds.), *Systems of Coupled Oscillators as Models of Central Pattern Generators*, New York. (Wiley, 1988)
49. X. Wang, J. Rinzel, Alternating and synchronous rhythms in reciprocally inhibitory model neurons. Neural Comput. **4**, 84–97 (1992)
50. W. Rappel, Dynamics of a globally coupled laser model. Phys. Rev. E **49**, 2750–2755 (1994)
51. K. Wiesenfeld, C. Bracikowski, G. James, R. Rajarshi, Observation of antiphase states in a multimode laser. Phys. Rev. Lett. **65**(14), 1749–1752 (1990)
52. L. Pecora, T.L. Caroll, Synchronization in chaotic systems. Phys. Rev. Lett. **64**, 821–824 (1990)
53. C.W. Wu, L.O. Chua, A unified framework for synchronization and control of dynamical systems. Int. J. Bifurc. Chaos **4**(4), 979–998 (1994)

54. R.M. May, Biological populations with no overlapping generations: stable points, stable cycles, and chaos. Science **186**, 645–647 (1974)
55. R.M. May, Biological population obeying difference equations: stable points, stable cycles, and chaos. J. Theor. Biol. **51**, 511–524 (1975)
56. R.M. May, Simple mathematical models with very complicated dynamics. Nature **261**, 459–467 (1975)
57. J.S. Halow, E.J. Boyle, C.S. Daw, C.E.A. Finney, *PC-based, near real-time, 3-dimensional simulation of fluidized beds* (Fluidization IX Durango, Colorado, 1998)
58. D.G. Aronson, M. Golubitsky, M. Krupa, Coupled arrays of josephson junctions and bifurcation of maps with s_n symmetry. Nonlinearity **4**, 861–902 (1991)
59. E. Doedel, D. Aronson, H. Othmer, The dynamics of coupled current-biased Josephson junctions: Part I. IEEE Trans. Circuits Syst. **35**(7), 0700–0810–0700–0817 (1988)
60. E. Doedel, D. Aronson, H. Othmer, The dynamics of coupled current-biased Josephson junctions: Part II. Int. J. Bifurc. Chaos **1**(1), 51–66 (1991)
61. P. Hadley, M.R. Beasley, K. Wiesenfeld, Phase locking of Josephson-junction series arrays. Phys. Rev. B **38**, 8712–8719 (1988)
62. W. Göpel, J. Hesse, J.N. Zemel, *Sensors A Comprehensive Survey*. Micro and Nanosensor Technology, vol. 8 (VCH Verlagsgesellschaft, Weinheim, 1995)
63. J.M. Janicke, *The Magnetic Measurement Handbook* (Magnetic Research Press, NJ, 1994)
64. J.E. Lenz, A review of magnetic sensors. Proceedings of the IEEE **78**, 973–989 (1990)
65. E. Ramsden, Measuring magnetic fields with fluxgate sensors. *Sensors*, pp. 87–90 (1994)
66. P. Ripka, Noise and stability of magnetic sensors. J. Magn. Magn. Mater. **157–158**, 424–427 (1996)
67. W. Bornhofft, G. Trenkler, Sensors, a comprehensive survey, in *Magnetic Sensors*, ed. by W. Gopel, J. Hesse, J. Zemel, vol. 5 (VCH, 1989), pp. 152–165
68. W. Geyger, *Nonlinear Magnetic Control Devices* (McGraw Hill, New York, 1964)
69. A. Barone, G. Paterno (eds.), *Physics and Applications of the Josephson Effect* (Wiley, New York, 1982)
70. C.M. Ajo-Franklin, D.A. Drubin, J.A. Eskin, E.P.S. Gee, D. Landgraf, I. Phillips, P.A. Silver, Rational design of memory in eukaryotic cells. Genes Dev. **21**, 2271–2276 (2007)
71. J.J. Tyson, K. Chen, B. Novak, Network dynamics and cell physiology. Nature Rev. Mol. Cell Biol. **2**, 908–916 (2001)
72. M.J. Solomon, Hysteresis meets the cell cycle. Acad. Sci. USA **100**(3), 771–772 (2003)
73. C.A. Voigt, D.M. Wolf, A.P. Arkin, Bacillus subtilis sin operon. Genetics **169**, 1187–1202 (2005)
74. T.S. Gardner, C.R. Cantor, J.J. Collins, Construction of a genetic toggle switch in escheria coli'. Nature **403**, 339–342 (2000)
75. V. Chickarmane, C. Troein, U.A. Nuber, H.M. Sauro, C. Peterson, Transcriptional dynamics of the embryonic stem cell switch. PLoS Comput. Biol. **2**(9), e123 (2006)
76. T. Wilhelm, The smallest chemical reaction system with bistability. BMC Syst. Biol. **3**, 90–98 (2009)
77. J. Qiu, J.H. Lang, A curved-beam bistable mechanism. J. Microelectromechanical Syst. **13**(2), 137–146 (2004)
78. M. Hoffmann, P. Kopka, E. Voges, All-silicon bistable micromechanical fiber switch based on advanced bulk micromachining. J. Sel. Top. Quantum Electron **5**(1), 46–51 (1999)
79. O.H. Schmitt, A thermionic trigger. J. Sci. Inst. **15**, 24–26 (1938)
80. M.A. Cohen, S. Grossberg, Absolute stability of global pattern formation and parallel memory storage by competitive neural networks. IEEE Trans. Syst. Man Cybern. **13**, 815–826 (1983)
81. J.J. Hopfield, Neural networks and physical systems with emergent collective computational abilities. Proc. Nat. Acad. Sci. USA **79**, 2554–2558 (1982)
82. J.J. Hopfield, Neurons with graded response have collective computational properties like those of two-state neurons. Proc. Nat. Acad. Sci. USA **81**, 3088–3092 (1984)
83. R. Koch, J. Deak, G. Grinstein, Fundamental limits to magnetic-field sensitivity of flux-gate magnetic-field sensors. Appl. Phys. Lett. **75**(24), 3862–3864 (1999)

84. D. Robbes, C. Dolabdjian, S. Saez, Y. Monfort, G. Kaiser, P. Ciureanu, Highly sensitive uncooled magnetometers: state of the art. superconducting magnetic hybrid magnetometers, an alternative to squids? IEEE Trans. Appl. Supercond. **11**(1), 629–634 (2001)
85. P. Ripka, Review of fluxgate sensors. Sens. Actuators A **33**, 129–141 (1996)
86. P. Ripka, New directions in fluxgate sensors. J. Magn. Magn. Mater. **215–216**, 735–739 (2000)
87. A. Ferreira, A.A.O. Carneiro, E.R. Moraes, R.B. Oliveira, O. Baffa, Study of the magnetic content movement present in the large intestine. J. Magn. Magn. Mater. **283**, 16–21 (2004)
88. R.D. Gupta, *Remote Sensing Geology* (Springer, Berlin, Heidelberg, 2003)
89. P. Herring, *The Biology of the Deep Ocean* (Oxford University Press, New York, 2002)
90. E.R. Benson, T.S. Stombaugh, N. Noguchi, J.D. Will, J.F. Reid, An evaluation of a geomagnetic direction sensor for vehicle guidance in predision agriculture applications. *Annual International Meeting Orlando* (1998)
91. F. Kaluza, A. Grúger, H. Grúger, New and future applications of fluxgate sensors. Sens. Actuators A **106**, 48–51 (2003)
92. S. Breiner, *Application Manual for Portable Magnetometers* (Geometrics, 2190 Fortune Drive San Jose, CA 95131, 1999)
93. R. Noble, Fluxgate magnetometry. *Electonics World*, pp. 726–732, Sept. 1991
94. F. Forster, A method for the measurement of dc field differences and its application to nondestructive testing. Nondestruct. Test **13**, 31 (1955)
95. F. Primdahl, Fluxgate magnetometers, in *Bibliography of Fluxgate Magnetometers*, vol. 41 (Publications of the Earth Physics Branch, 1970)
96. P. Ripka, Advances in fluxgate sensors. Sens. Actuators A **106**, 8–14 (2003)
97. L. Landau, On the theory of phase transitions. Zh. Eksp. Teor. Fiz. **7**, 1932 (1937)
98. F. Brailsford, *Magnetic Materials* (Wiley, New York, 1951)
99. G. Bertotti, *Hystersis in Magnetism* (Academic Press, San Diego, CA, 1998)
100. H.E. Stanley, *Introduction to Phase Transitions and Critical Phenomena* (Oxford University Press, Oxford, 1971)
101. J. Hertz, A. Krogh, R.G. Palmer, *Introdution to the Theory of Neural Computation* (Addison-Wesley Co., New York, 1991). Santa Fe Institute, Studies in the Sciences of Complexity
102. J.J. Hopfield, Neural networks and physical systems with emergent collective computational abilities, in *Proceedings of the National Academy of Sciences*, USA, pp. 2554–2558 (1982)
103. W.O. Kermack, A.G. McKendrick, A contribution to themathematical theory of epidemics. Proc. Roy. Soc. London **115**(772), 700–721 (1927)
104. M. Dashtbali, M. Mirzaie, A compartmental model tat predicts the effect of social distancing and vaccination on controlling covid-19. Sci. Rep. **11**, 8191 (2021)
105. O. de Feo, G.M. Maggio, M.P. Kennedy, The colpitts oscillator: families of periodic solutions and their bifurcations. Int. J. Bifurc. Chaos **10**(5), 935–958 (2000)

Chapter 5
Bifurcation Theory

Bifurcation theory is the study of the changes in the number of solutions and the type of solutions to models as parameters are varied. The term *bifurcation* was first introduced by Henri Poincaré in 1885 [1] to describe the relatively subtle changes in stationary points of a model. However, there are situations in a system where sudden and "catastrophic" changes occur, such as the collapse of the Tacoma Narrows Bridge [2], see Fig. 5.1.

René Thom introduced *catastrophe theory* [3, 4] as a special case of bifurcation theory or *singularity theory*. In particular, Thom identified equilibria with the minimum of a potential function to study the changes in behavior in *gradient systems*.

Chapters 3 and 4 illustrated various models, using discrete and continuous dynamical systems. These models were shown to exhibit a variety of behaviors, and the mathematical analyses centered on examining the equilibria and linearization about these equilibria to determine stability of the model system. The models are connected to real world data by varying the kinetic parameters in the systems. This chapter explores the changes that can occur to the behavior of the state variables when the parameters vary. The study of those changes is based on *bifurcation theory*, and it includes both, discrete and continuous dynamical systems.

This chapter explores a variety of discrete and continuous models, where changes in parameters result in different types of behavior reflected in the state variables. We present a number of examples, then introduce fundamental definitions of several generic types of bifurcations. Readers interested in a more in-depth treatment of bifurcation theory are encouraged to study Guckenheimer and Holmes [5], Wiggins [6], and Chow and Hale [7] and more specific studies in pattern formation theory by Hoyle [8] and Murray [9].

A. Palacios, *Mathematical Modeling*, Mathematical Engineering,
https://doi.org/10.1007/978-3-031-04729-9_5

Fig. 5.1 Collapse of the Tacoma Narrows bridge. *Source*: Wikipedia

5.1 Examples and Phase Portraits

Mathematically, a bifurcation occurs if the *phase portrait* of the model system changes its topological structure as one or more system parameters are varied (see Sect. 4.3.3). Algebraic systems have the form

$$f(\mathbf{x}, \boldsymbol{\lambda}) = 0, \quad \mathbf{x} \in \mathbb{R}^n, \ \boldsymbol{\lambda} \in \mathbb{R}^p, \ f(x) \in \mathbb{R}^m,$$

where $\lambda \in \mathbb{R}^p$ is a vector of parameters (with p components) for tuning the model. These are called *bifurcation parameters*, and the point at which changes occur is called the *bifurcation point*. Discrete models (see Chap. 3) with parameters are written:

$$\mathbf{x}_{n+1} = f(\mathbf{x}_n, \boldsymbol{\lambda}). \tag{5.1}$$

Parameterized systems of Ordinary Differential Equations (ODEs) (see Chap. 4) have the form:

$$\frac{d\mathbf{x}}{dt} = f(\mathbf{x}, \boldsymbol{\lambda}), \tag{5.2}$$

while a system of Partial Differential Equations (PDEs) (see Chap. 8)

$$\frac{\partial \mathbf{u}(\mathbf{x}, t)}{\partial t} = F(\mathbf{u}(\mathbf{x}, t), \boldsymbol{\lambda}). \tag{5.3}$$

The mathematical analyses of these models for fixed parameters were examined in Sects. 3.4 and 4.3. This chapter studies qualitative changes as $\boldsymbol{\lambda}$ varies. For example, a change in λ in (5.1) or (5.2) might result in the loss of stability of a *fixed point* of (5.1) or an *equilibrium* of (5.2). Below are some illustrative examples before we present some important definitions and theorems.

The discrete logistic growth model was examined in some detail in Sect. 3.8. That example showed that a simple scalar quadratic map exhibited a process called period doubling as only the parameter for growth rate was increased. For a range of values the growth rate caused the output of the logistic growth model to become *chaotic* and appear random. The example below examines a discrete polynomial map in two dimensions with five-fold symmetry and displays a high degree of complexity.

Example 5.1 (*Sand Dollar*) Discrete models with symmetry can exhibit long-term behavior, in the form of *attractors*, with pattern-forming characteristics appearing in the time-average of a chaotic dynamical system. In this example, we explore, throughout numerical simulations, the pattern forming characteristics, and changes in the attractor, of a symmetric discrete dynamical system. But first, let us discuss briefly the role of symmetry.

Symmetry is a geometrical concept that refers to the set of transformations that leave an object unchanged. In a mathematical model, those transformations typically form a *group of symmetries*, while the object itself is the mathematical equation that defines the model. Consider a discrete dynamical system of the form

$$z_{n+1} = f(z_n, \lambda), \tag{5.4}$$

where $z_n \in \mathbf{C}^m$ and $\lambda \in \mathbf{C}^p$ is a vector of parameters. Let Γ be a set of invertible linear transformations of the vector space $\mathbb{R}^n$ into itself. This is equivalent to saying that Γ is a matrix group, i.e., closed under multiplication. The mathematical model (5.4) is said to have Γ-symmetry if

$$f(\gamma z_n, \lambda) = \gamma f(z_n, \lambda),$$

for all $z \in \mathbb{R}^m$ and for all $\gamma \in \Gamma$. For the case of $\Gamma = \mathbf{D}_N$-symmetry, where $\mathbf{D}_N$ is the orthogonal group of symmetries of an N-gon, the standard action of the group, $\mathbf{D}_N$ is: $\gamma_1 z = \bar{z}$, $\gamma_2 z = e^{2\pi i/N} z$, where $z \in \mathbf{C}$. In this action, γ_1 represents a reflection across the real-axis, while γ_2 describes a cyclic rotation along the unit circle. It can be shown that the general form of a $\mathbf{D}_N$ discrete model under the group $\mathbf{D}_N = \{\gamma_1, \gamma_2\}$ has the form:

$$\boxed{z_{n+1} = p(u, v, \mu) z_n + q(u, v, \lambda) \bar{z}_n^{N-1},} \tag{5.5}$$

where p and q are polynomial functions of $u = |z_n|^2$ and $v = z_n^N + \bar{z}_n^N$, and μ is a vector of parameters. This means that the right-hand side of Eq. (5.5) satisfies $f(\gamma z_n, \lambda) = \gamma f(z_n, \lambda)$, where $\gamma = \gamma_1$ or $\gamma = \gamma_2$. A specific discrete model with $\mathbf{D}_5$-symmetry is

$$z_{n+1} = (\lambda + \alpha u + \beta v)\, z_n + \gamma \bar{z}_n^4, \tag{5.6}$$

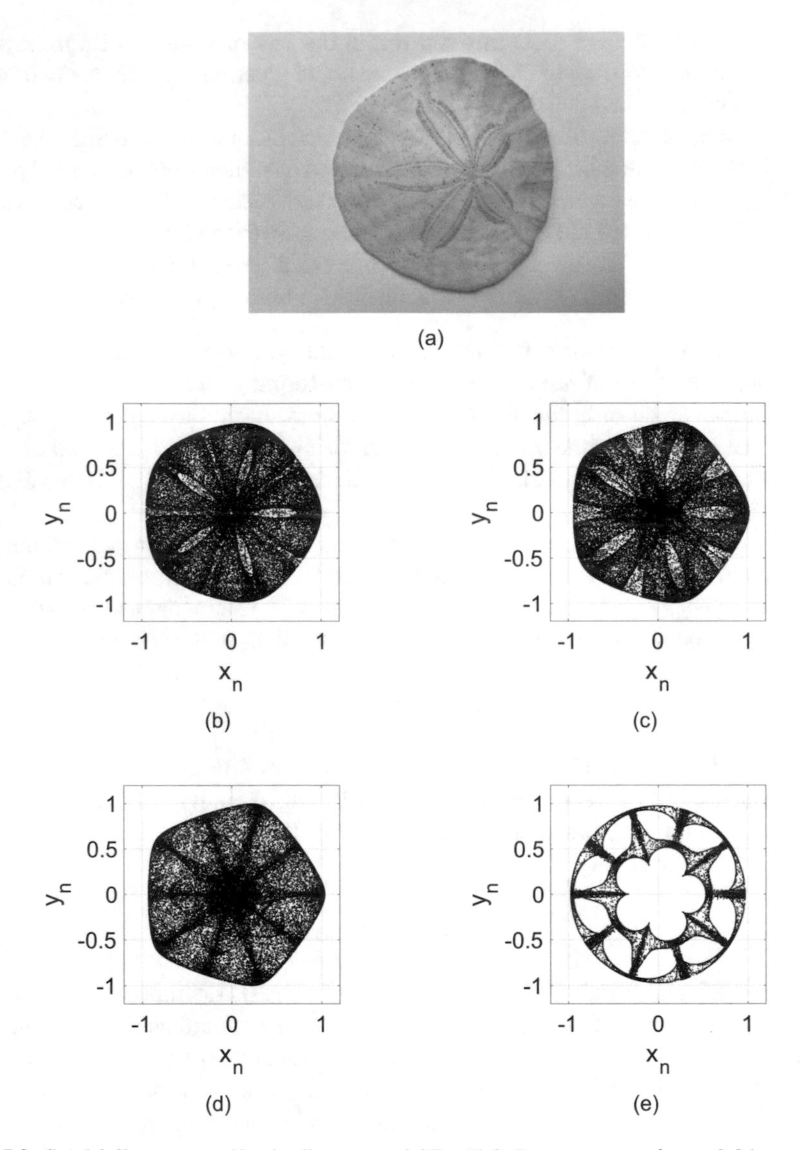

Fig. 5.2 Sand dollar generated by the discrete model Eq. (5.6). Parameters are: $\lambda = -2.34, \alpha = 2.0$. Subfigures **b–d** showcase the variations in the long-term attractor as the following parameters vary: **a** $\beta = 0.2, \gamma = 0.1$; **b** $\beta = 0.4, \gamma = 0.1$; **c** $\beta = 0.2, \gamma = 0.25$; **d** $\beta = 0.2, \gamma = -0.08$

where $u = |z_n|^2$, and $v = z_n^5 + \bar{z}_n^5$. The long-term behavior of this model leads to a pattern forming system, which is illustrated in Fig. 5.2.

Example 5.2 (*Euler's Beam Column*) Euler's column buckling experiment, illustrated in Fig. 5.3, consists of an elastic beam subjected to a compressive force. With

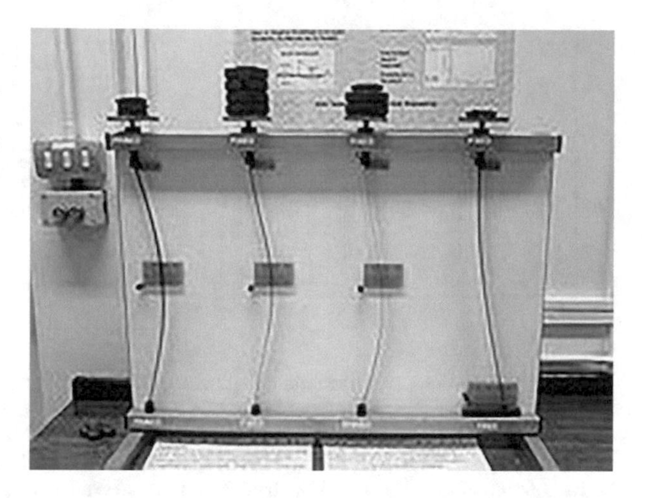

Fig. 5.3 Euler beam experiment. An elastic beam is subjected to a compressive force. Upon reaching a certain threshold value of the compressive force, the trivial solution, unbuckled state, loses stability and a buckled state, right or left, emerges through a pitchfork bifurcation. Source: Wikipedia

a critical compressive force (bifurcation point) the beam deforms into one of two buckled stationary states, either to the right or to the left. Which state appears depends on model-dependent features, such as material imperfections or thermal fluctuations. According to Bernoulli-Euler beam theory, [10, 11] a mathematical model for the angle, $\theta(t)$, between the undeformed rod and the tangent of the deformed rod with hinged boundary conditions is

$$E I\theta''(x) + P \sin \theta(x) = 0, \qquad \theta(0) = 0, \quad \theta(L) = 0, \quad 0 < x < L, \tag{5.7}$$

where x is the material coordinate, E is the elastic modulus, I is moment of inertia, P is the compressive force, and L is the length of the beam.

Equation (5.7) is a second order nonlinear boundary value problem with several parameters. This model is studied with a *reduced order model*, using a procedure called Lyapunov-Schmidt reduction [12]. (Details are discussed in Chap. 10.) The reduced order model satisfies:

$$\frac{dx}{dt} = \lambda x - x^3. \tag{5.8}$$

This is a *phenomenological model*, where the original material-related parameters E, I, and P have been "washed away," leaving one parameter, λ. This parameter idealizes the effect of the compressive force and the buckling phenomenon. When $\lambda = 0$ (absence of the compressive force), there is only one solution to the algebraic equation $f(x, \lambda) = \lambda x - x^3 = 0$, so only one equilibrium, $x_e = 0$. However, when $\lambda > 0$, there are three equilibria, $x_e = 0, \pm\sqrt{\lambda}$, where nonzero equilibria correspond to the two possible buckled states. Figure 5.4 illustrates the changes in the phase

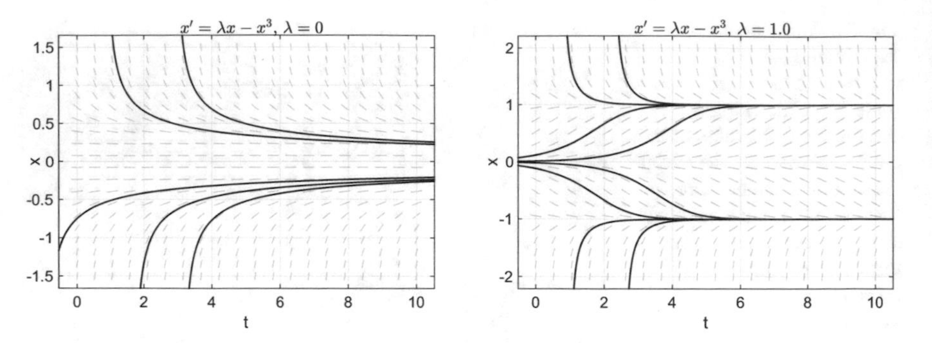

Fig. 5.4 Bifurcations in a phenomenological model for the buckling phenomenon of the Euler beam experiment where, λ represents the effects of a compressive force. (Left) In the absence of such force, the beam remains in its resting state, so solution trajectories in the phase portrait asymptotically converge towards zero. (Right) When $\lambda > 0$, the force causes the beam to buckle to the right or left ($x_e = \pm\sqrt{\lambda}$). Which of these two states appears depends on initial conditions of integration. In the experiments, the buckled state is determined by material imperfections

portrait of the reduced order model (5.8). Each curve in the diagram corresponds to a *solution trajectory* from numerically integrating (5.8) with a particular initial condition. Observe that when $\lambda > 0$, the zero solution (unbuckled state) exists, but is unstable. This is because $x_e = 0$ represents an ideal state in which the beam simply collapses. It is possible for this state to occur but material imperfections make it highly unlikely.

The example above is a scalar ODE with a single parameter, and the bifurcation occurs as the parameter varies and the number of equilibria changes. Chapter 4 showed the example of the *Duffing oscillator* (Sect. 4.8.3), where varying the elastic properties affects the number of equilibria, while varying the damping strength changes the qualitative behavior of the solution trajectories. The next example explores a chemical system of two ODEs, where parameter changes alter qualitative behavior between stable and oscillatory behavior.

Example 5.3 (*The Brusselator Model*) The Brusselator model was originally proposed by Prigogine and Nicolis [13] for describing an *autocatalytic reaction* in the form:

$$\alpha \xrightarrow{k_1} X, \qquad 2X + Y \xrightarrow{k_2} 3X,$$

$$\beta + X \xrightarrow{k_3} Y + D, \qquad X \xrightarrow{k_4} E,$$

where α and β are reactants (substrates), D and E the final products, and X and Y are the autocatalytic reactants. The kinetic constants are $k_i, i = 1 \ldots 4$.

Let $[\cdot]$ denote concentrations of a chemical species, then the chemical reactions above are readily converted into ODEs for any of the species using the *law of mass action*. The law of mass action, based on the frequency of molecular collisions, states that the rate of change of the concentration of a substance is proportional with rate, k_i, to the product of the concentrations of the different chemical species in the reactions that contain the substance with the sign depending on whether the substance is produced or lost.

The resulting ODEs for $[X]$ and $[Y]$ are found to be:

$$\frac{d[X]}{dt} = k_1[\alpha] + k_2[X]^2[Y] - k_3[\beta][X] - k_4[X]$$

$$\frac{d[Y]}{dt} = -k_2[X]^2[Y] + k_3[\beta]X,$$

where the concentrations $\alpha > 0$ and $\beta > 0$ are assumed to be constant. Letting $k_3 = k_4$ and $k_1k_3/k_2 = k_3^3/k_2^2 = 1$ and nondimensionalizing the above equations leads to the more familiar form of the Brusselator model given by

$$\boxed{\begin{aligned} \frac{dX}{dt} &= X^2Y - (1+\beta)X + \alpha, \\ \frac{dY}{dt} &= -X^2Y + \beta[X], \end{aligned}} \tag{5.9}$$

where α and β are positive real constants, while $X, Y \in \mathbb{R}$ represent the dimensionless concentrations of the two original reactants.

The Brusselator model (5.9) has a unique equilibrium at $(X_e, Y_e) = (\alpha, \beta/\alpha)$. Figure 5.5 shows some representative trajectories of the 2D phase portraits. For $\alpha = 1$ and $\beta = 1.7$, the equilibrium is stable, as nearby solutions asymptotically approach it. Increasing β to 3, while holding α fixed, makes the equilibrium become unstable and the solution trajectories now approach a stable limit cycle. This is an example of a bifurcation with a transition from a stable equilibrium to a stable periodic solution. The mechanism that underlies the changes to oscillatory behavior is known as a Hopf bifurcation, which is detailed later in Sect. 5.5.

5.2 Conditions for Bifurcations

The examples above show that bifurcations occur when parameters vary in a dynamical system, which result in changes in the number of equilibria or the qualitative behavior near the equilibria. Most models are nonlinear, and a complete study of the types of behavior over all model parameters is practically impossible. However, we are often interested in specific phenomena where there are changes occurring for a limited range of conditions reflected in the model parameters. This chapter studies which types of transitions are likely for particular forms of models, such as changing the number of equilibria or the appearance of a periodic solution. These bifurca-

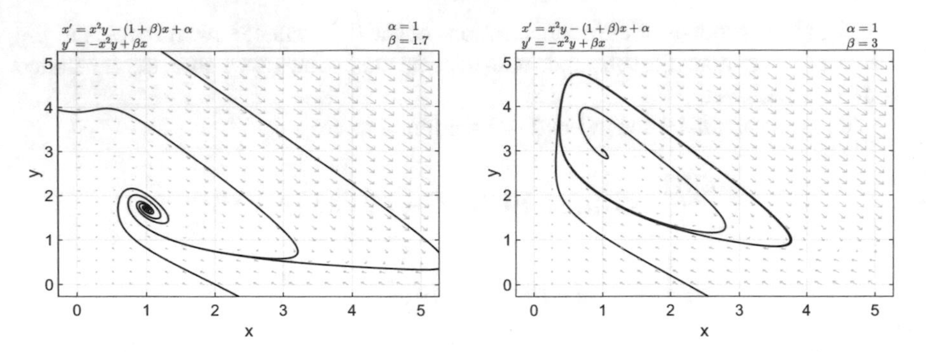

Fig. 5.5 Phase portraits of the Brusselator model (5.9) illustrate a Hopf bifurcation, where changes in β result in the qualitative behavior difference of a stable equilibrium point (left) and a stable periodic solution (right). Parameters are: (Left) $\alpha = 1.0$, $\beta = 1.7$. (Right) $\alpha = 1.0$, $\beta = 3.0$

tions occur generically in discrete models, ODEs, and more complicated models, like PDEs.

Though a bifurcation may require variations in multiple parameters, for simplicity we begin by examining a scalar parameter, $\lambda \in \mathbb{R}$. The Implicit Function Theorem [14] for a scalar satisfies the following:

Theorem 5.1 (Implicit Function Theorem for $\mathbb{R}^2$) *Consider a continuously differentiable function, $f : \mathbb{R}^2 \to \mathbb{R}$ and a point $(x_0, \lambda_0) \in \mathbb{R}^2$ so that $f(x_0, \lambda_0) = c$. If $\frac{\partial f}{\partial x}(x_0, \lambda_0) \neq 0$, then there is a neighborhood of (x_0, λ_0) so that whenever x is sufficiently close to x_0, there is a unique λ so that $f(x(\lambda), \lambda) = c$. Moreover, this assignment makes x a continuous function of λ.*

This theorem generalizes to higher dimensions and even function spaces. Figure 5.6 illustrates geometrically the results of the Implicit Function Theorem for $\mathbb{R}^2$, where we take $c = 0$. The figure shows that around a small neighborhood of $(0, 0)$, small changes in the parameter λ lead to smooth changes in the location of the x.

If we consider $x \in \mathbb{R}$ in the ODE (5.2), then equilibria are found by solving the algebraic equation:

$$f(x, \lambda) = 0, \qquad x,\ f(x) \in \mathbb{R}. \tag{5.10}$$

As noted in Chap. 4, finding equilibria can be a difficult problem. However, the Implicit Function Theorem states that if there exists (x_0, λ_0) satisfying (5.10), so that x_0 is an equilibrium and if $\frac{\partial f}{\partial x}(x_0, \lambda_0) \neq 0$, then in a neighborhood of x_0, there exists a continuous function $x(\lambda)$ with $f(x(\lambda), \lambda) = 0$. It follows that locally the equilibrium changes smoothly in λ with $x(\lambda)$ defining a curve of equilibria parametrized by λ, so no bifurcation occurs to a different number of equilibria. From the Implicit Function Theorem, it follows that a bifurcation at (x_0, λ_0) leading to a different number of equilibria requires the condition:

$$\frac{\partial f}{\partial x}(x_0, \lambda_0) = 0.$$

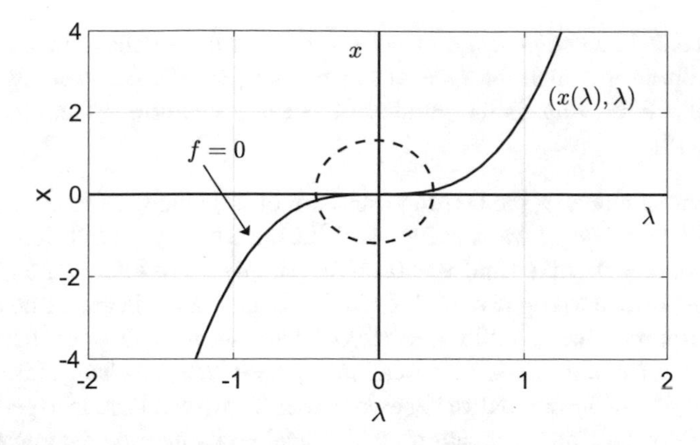

Fig. 5.6 The Implicit Function Theorem implies that a solution $x(\lambda)$ to the algebraic equation $f(x, \lambda) = 0$ exists in a neighborhood of a known solution $(0, 0)$. To have a bifurcation the Implicit Function Theorem must fail

Note that if we define $g(x, \lambda) = f(x, \lambda) - x$ in Eq. (5.1), then a similar analysis applies to discrete models.

We formally define the following bifurcation condition.

Definition 5.1 (*Bifurcation Condition*) Consider a mathematical model in the form of an Ordinary Differential Equation

$$\frac{dx}{dt} = f(x, \lambda), \qquad x, \lambda \in \mathbb{R}$$

Assume (x_0, λ_0) to be an equilibrium. To have bifurcations in the number of equilibria of the model the following equations must be solved simultaneously

$$\boxed{f(x, \lambda) = 0 \quad \text{and} \quad f_x(x, \lambda) = 0,} \tag{5.11}$$

where f and f_x are evaluated at (x_0, λ_0).

5.3 Codimension of a Bifurcation

We examine how many parameters must be simultaneously varied in the model to observe a change. The number of parameters that must be varied to observe a change in a system's behavior, so that a bifurcation occurs, is known as the *codimension of a bifurcation*. More precisely, the codimension is defined as follows.

Definition 5.2 The *codimension of a bifurcation* is the smallest dimension of the parameter space in which the type of bifurcation persists. Equivalently, the *codimension of a bifurcation* is the number of equality conditions that characterize a bifurcation [5].

Consider the model of the Duffing oscillator (4.37)which has four parameters α, β, γ and δ. The phase portraits in Fig. 4.24 illustrate two types of bifurcations. It is assumed that $\alpha > 0$ is fixed and $\gamma = 0$, so the parameters of interest are (β, δ). As β varies from positive to negative with δ fixed, the qualitative behavior changes from one equilibrium to three equilibria (*pitchfork bifurcation*). Since this only uses one parameter, it is a codimension 1 bifurcation at $\beta = 0$. When δ changes from positive to zero with $\beta > 0$, the qualitative behavior varies from a stable node to periodic solutions (Hamiltonian Hopf bifurcation), which again only has one parameter varying, so is a codimension 1 bifurcation at $\delta = 0$.

The codimension of a particular bifurcation is the minimum number of parameters that are varied for some model. Whenever one observes a specific qualitative change or bifurcation in a model, then necessarily that change results from the same number of parameters of all related generic models undergoing that type of bifurcation. This allows one to study the changes in behavior in a model under a common underlying mechanism.

5.4 Codimension One Bifurcations in Discrete Systems

Consider discrete models of the form:

$$\mathbf{x}_{n+1} = f(\mathbf{x}_n, \lambda), \tag{5.12}$$

where $\mathbf{x} \in \mathbb{R}^n$, $\lambda \in \mathbb{R}$ is the bifurcation parameter, and $f : \mathbb{R}^n \times \mathbb{R} \to \mathbb{R}^n$ is a smooth map with respect to both $\mathbf{x}$ and λ. Since $\lambda \in \mathbb{R}$, it follows that any bifurcations are codimension 1, which is the concentration of our study in this section.

Let $J(\mathbf{x}, \lambda)$ be the Jacobian matrix for f. A *local bifurcation* occurs at $(\mathbf{x}_e, \lambda_c)$ if $J(\mathbf{x}_e, \lambda_c)$ has an eigenvalue, σ with $|\sigma| = 1$. If $\sigma = 1$, then the bifurcation involves equilibria or *fixed points* and the associated types are saddle-node, transcritical, or pitchfork bifurcations. If $\sigma = -1$, then the bifurcation involves the emergence of period-doubling solutions. Otherwise, if a pair of complex eigenvalues crosses the unit circle, so $|\sigma_{1,2}| = 1$, then the bifurcation is a Hopf bifurcation for discrete models and is known as a Neimark-Sacker bifurcation.

5.4.1 Continuability

For a discrete model (5.12), the existence of a bifurcation is closely connected to the concept of continuability and uses information from the Implicit Function Theorem 5.1.

Definition 5.3 (*Continuability*) Let $f : \mathbb{R}^n \times \mathbb{R} \to \mathbb{R}^n$ be a smooth map on $\mathbb{R}^n \times \mathbb{R}$. Assume $\mathbf{x}_e$ to be an equilibrium or *fixed point*, such that

$$f(\mathbf{x}_e, \lambda) = \mathbf{x}_e. \tag{5.13}$$

Then the equilibrium $\mathbf{x}_e(\lambda)$ is *locally continuable* if it lies on a continuous path in λ.

A bifurcation in a discrete system (5.12) occurs when the equilibrium, $\mathbf{x}_e(\lambda)$, can no longer be uniquely continued as λ varies. This is formalized in the following theorem.

Theorem 5.2 *Consider the discrete model (5.12), where $f : \mathbb{R} \times \mathbb{R} \to \mathbb{R}$. Let x_e be an equilibrium, satisfying (5.13). If $\frac{\partial f}{\partial x}(x_e, \lambda_c) \neq +1$, then (x_e, λ_c) is locally continuable.*

The proof of this theorem follows from applying the Implicit Function Theorem 5.1 to $g(x, \lambda) = f(x, \lambda) - x$. It is readily extended to higher dimensions where $\mathbf{x} \in \mathbb{R}^n$ with generalizations of Theorem 5.1. This theorem shows that for a bifurcation to occur at an equilibrium, x_e, then $\frac{\partial f}{\partial x}(x_e, \lambda_c) = \pm 1$ for some λ_c. We explore several models and various types of codimension one bifurcations that occur in discrete systems (5.12).

5.4.2 Saddle-Node Bifurcation

The logistic growth model (3.8) is commonly used for population studies and often forms the basis for agencies managing animal populations. Consider an extension of the logistic growth model, where a term is included for constant harvesting, h, so

$$p_{n+1} = p_n + rp_n\left(1 - \frac{p_n}{M}\right) - h = f(p_n, h). \tag{5.14}$$

This model is appropriate for some species growing logistically and at each discrete time period a constant population (hunting or fishing limits), h, is removed.

Figure 5.7 shows simulations of the logistic growth model with constant harvesting (5.14) where $r = 0.2$, $M = 100$, and various values of h. When $h = 0$, then we observe the model approaching its natural carrying capacity, M. As h increases, this largest equilibrium decreases, and as expected, if the constant removal of the species is sufficiently high, then the species goes extinct. A serious weakness in this model

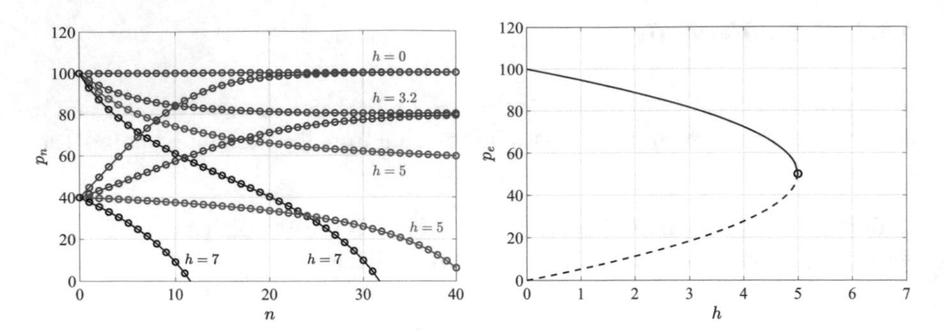

Fig. 5.7 The figure on the left shows simulations of the logistic growth model with constant harvesting (5.14) where $r = 0.2$, $M = 100$, and various values of h. Across the range of h there is a distinct change in behavior between a stable steady state to populations going extinct. The right figure shows a bifurcation diagram illustrating the equilibria as h varies with a bifurcation point at $h = 5$. The equilibria along the solid line are stable equilibria, while those on the dashed line are unstable

is that as populations get small, then harvesting a constant number becomes difficult, so is unrealistic. In fisheries management, the regulatory agency wants to set the limit of fish caught to be sustainable, but allow adequate catch for the fishermen to make their living, which often are conflicting goals.

The equilibria for (5.14) are found solving the quadratic, $p_e = f(p_e, h)$, so

$$p_e = \tfrac{M}{2} \pm \sqrt{\tfrac{M^2}{4} - \tfrac{Mh}{r}},$$

which are the two equilibria, provided two real solutions exist. When $h = 0$, we get the results from Chap. 3 with $p_e = 0$, the unstable extinction equilibrium, and $p_e = M$, the stable carrying capacity equilibrium. The right graph in Fig. 5.7 shows the continuous change in these equilibria as h varies. When $h = rM/4$, then the two equilibria coalesce at $p_e = M/2$, so it is a *bifurcation.*

For $h > rM/4$, no real solutions exist, so there are no equilibria for the model (5.14). This bifurcation is called a *saddle-node* or *blue sky* bifurcation. For this logistic growth model with constant harvesting (5.14), the larger equilibrium is stable, while the smaller equilibrium is unstable. Note that

$$\frac{\partial f}{\partial p}(p, h) = 1 + r - \frac{2rp}{M}, \qquad \text{so} \qquad \frac{\partial f}{\partial p}(M/2, rM/4) = 1.$$

Thus, this example satisfies the conditions for a bifurcation at $p_e = M/2$.

Saddle-node bifurcations are generic among codimension-one bifurcations in discrete systems. The following theorem provides the conditions for the existence (and direction) of a saddle-node bifurcation.

Theorem 5.3 (Saddle Node Bifurcation) *Let* $f : \mathbb{R} \times \mathbb{R} \to \mathbb{R}$ *with*

$$f(x_e, \lambda_c) = x_e \quad and \quad f'(x_e, \lambda_c) = 1, \qquad \text{(bifurcation condition)}$$
$$A = \frac{\partial f}{\partial \lambda} \neq 0 \quad and \quad D = \frac{\partial^2 f}{\partial x^2} \neq 0, \qquad \text{(nondegeneracy condition)}$$

where all derivatives are evaluated at the bifurcation point (x_e, λ_c). Then two curves of fixed points emanate from (x_e, λ_c). The new fixed points exist for $\lambda > \lambda_c$, if $DA < 0$, and for $\lambda < \lambda_c$, if $DA > 0$. The upper branch of fixed points is stable and the lower one is unstable, if $D < 0$, and the stabilities are reversed if $D > 0$.

The proof of this theorem relies on the Implicit Function Theorem 5.1 and details are found in [15]. We apply this theorem to the logistic growth model with constant harvesting (5.14), where the bifurcation conditions were shown above at $(p_e, h_c) = (M/2, rM/4)$. The nondegeneracy conditions for this model are

$$A = \frac{\partial f}{\partial h} = -1 \neq 0 \quad \text{and} \quad D = \frac{\partial^2 f}{\partial p^2} = -\frac{2r}{M} \neq 0.$$

It follows that $DA = 2r/M > 0$, which implies the fixed point emanate to the left with $h < h_c$, as seen in the right diagram in Fig. 5.7. In addition, with $D < 0$, Theorem 5.3 gives the upper branch of equilibria is stable, while the lower branch is unstable as depicted in Fig. 5.7.

Example 5.4 Consider the generic discrete model given by:

$$x_{n+1} = \lambda - x_n^2.$$

Equilibria are found by solving $x_e = \lambda - x_e^2$ or $(x_e + \frac{1}{2})^2 = \lambda + \frac{1}{4}$, so

$$x_e = -\tfrac{1}{2} \pm \sqrt{\lambda + \tfrac{1}{4}}.$$

It follows that $\lambda < -\frac{1}{4}$ has no real solutions, so a *saddle node bifurcation* occurs at $\lambda_c = -\frac{1}{4}$ with $x_e = -\frac{1}{2}$. Since

$$\tfrac{\partial f}{\partial x}(x, \lambda) = -2x, \qquad \text{it follows} \qquad \tfrac{\partial f}{\partial x}(x_e, \lambda_e) = 1.$$

Thus, this example satisfies the bifurcation condition of Theorem 5.3. Checking the degeneracy conditions, we find

$$A = \tfrac{\partial f}{\partial \lambda}(x_e, \lambda_e) = 1, \qquad \text{and} \qquad D = \tfrac{\partial^2 f}{\partial x^2}(x_e, \lambda_e) = -2.$$

Since $DA = -2 < 0$ and $D < 0$, Theorem 5.3 states that new fixed points emanate to the right, $\lambda > -\frac{1}{4}$, and the upper branch of fixed points are stable, while the lower branch is unstable. Figure 5.8 shows the bifurcation diagram for this example.

The Continuability Theorem 5.2 shows that the stable branch of equilibria has continuous stable equilibria, $x_e(\lambda)$, provided $\left|\frac{\partial f}{\partial x}(x_e, \lambda)\right| < 1$. For this example, this

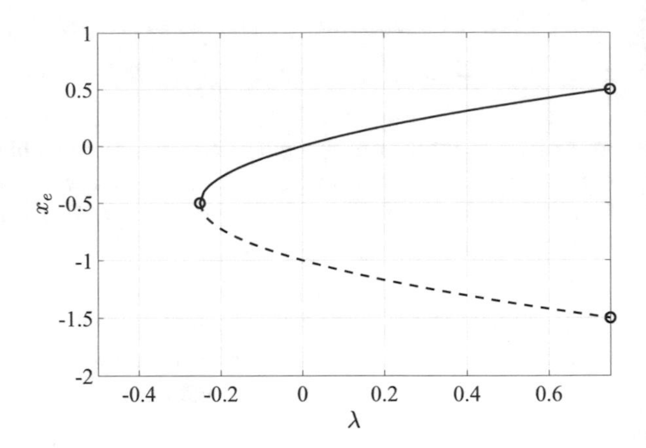

Fig. 5.8 This figure presents the bifurcation diagram for Example 5.4. This saddle-node bifurcation at the leftmost circle has stable equilibria with solid lines and unstable equilibria with dashed lines. The other circles denote additional bifurcation points, where period doubling occurs

stable branch satisfies:

$$\left|\frac{\partial f}{\partial x}(x_e, \lambda)\right| = |-2x_e| = |1 - \sqrt{4\lambda + 1}| < 1.$$

Solving this inequality gives $-\frac{1}{4} < \lambda < \frac{3}{4}$, which is the domain for this branch of stable equilibria. A similar argument with the Continuability Theorem 5.2 gives the same domain for the unstable branch of equilibria seen in Fig. 5.8. At $\lambda_b = \frac{3}{4}$, another bifurcation occurs, a *period doubling* bifurcation, which is studied later in this chapter.

5.4.3 Transcritical

The previous section introduced the logistic growth model with constant harvesting (5.14), which is based on controlled hunting or fishing that removes a fixed number of animals in a given time period. A related logistic growth model considers proportional harvesting, such as when fish are caught by nets, so the removal is proportional to the existing population. This model satisfies the discrete model:

$$p_{n+1} = p_n + rp_n\left(1 - \frac{p_n}{M}\right) - hp_n = f(p_n, h), \tag{5.15}$$

where h presents the intensity of the proportional harvesting. This model better represents how hunting and fishing occurs when there are lower densities of the species. This model fails for larger populations, as there are limits to how much can be harvested for a number of reasons, such as limited markets or storage in a boat.

Figure 5.9 shows simulations of the logistic growth model with proportional harvesting (5.15) where $r = 0.2$, $M = 100$, and various values of h. As before when $h = 0$, the model approaches its natural carrying capacity, M. As h increases,

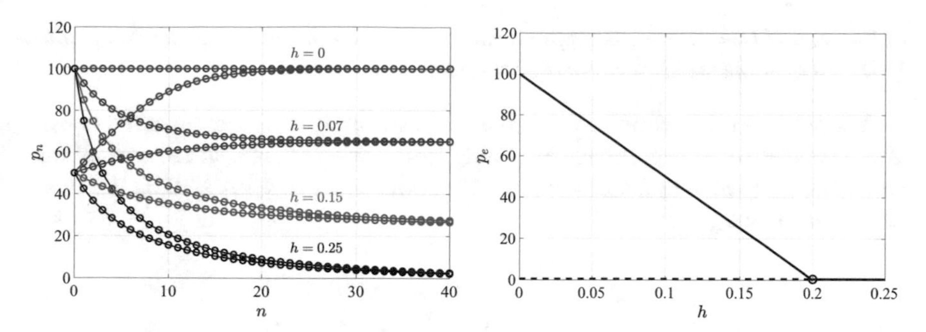

Fig. 5.9 The figure on the left shows simulations of the logistic growth model with proportional harvesting (5.15) where $r = 0.2$, $M = 100$, and various values of h. Across the range of h there is a linear decline in the carrying capacity of the stable steady state population until it goes extinct. The right figure shows a bifurcation diagram illustrating the equilibria as h varies with a bifurcation point at $h = 0.2$. The equilibria along the solid lines are stable equilibria, while those on the dashed line are unstable

this largest equilibrium decreases linearly, since this is equivalent to proportionally decreasing the Malthusian growth rate, r. When the proportional removal of the species matches the birth rate, then the species goes extinct.

The equilibria for (5.15) are found by solving the quadratic, $p_e = f(p_e, h)$, so

$$p_e = 0 \qquad \text{or} \qquad p_e = M - \tfrac{Mh}{r}.$$

For populations the second equilibrium must be non-negative, so $h < r$. As before, when $h = 0$, there is the unstable extinction equilibrium, $p_e = 0$, and the stable carrying capacity equilibrium, $p_e = M$. The right graph in Fig. 5.9 shows the linear decline of the stable carrying capacity with h, while the extinction equilibrium remains unstable until $h_c = r$. Here a bifurcation occurs and subsequently extinction, $p_e = 0$, becomes the stable equilibrium. This bifurcation is called a *transcritical bifurcation*. Note that $\frac{\partial f}{\partial p}(p, h) = 1 + r - h - \frac{2rp}{M}$, so $\frac{\partial f}{\partial p}(0, r) = 1$. Thus, this example satisfies the bifurcation conditions that $f(0, r) = 0$ and $\frac{\partial f}{\partial p}(0, r) = 1$.

Transcritical bifurcations are another generic codimension-one bifurcation in discrete systems. The following theorem provides the conditions for the existence and stability of a transcritical bifurcation.

Theorem 5.4 (Transcritical Bifurcation) *Let* $f : \mathbb{R} \times \mathbb{R} \to \mathbb{R}$ *and let*

$$f(x_e, \lambda_c) = x_e \text{ and } \tfrac{\partial f}{\partial p}(x_e, \lambda_c) = 1, \qquad \textit{(bifurcation conditions)}$$
$$A = \frac{\partial f}{\partial \lambda} = 0, \quad D = \frac{\partial^2 f}{\partial x^2} \neq 0, \quad \text{and} \quad E = \left(\frac{\partial^2 f}{\partial x \partial \lambda}\right)^2 - \frac{\partial^2 f}{\partial x^2}\frac{\partial^2 f}{\partial \lambda^2} \neq 0,$$

where all derivatives are evaluated at the bifurcation point, (x_e, λ_c). *Then there are two curves of fixed points in a neighborhood of* (x_e, λ_c), *which intersect transversely* *at (x_e, λ_c). If $D < 0$ then the upper branch is stable and the lower one is unstable. If $D > 0$ then the stability properties are reversed.*

The proof of this theorem are found in [15]. We apply this theorem to the logistic growth model with proportional harvesting (5.15), where the bifurcation conditions were shown above at $(p_e, h_c) = (0, r)$. Computing the quantities A, D, and E at the bifurcation point gives:

$$A = \frac{\partial f}{\partial h} = -p, \qquad \text{so} \quad A = 0,$$

$$D = \frac{\partial^2 f}{\partial p^2} = -\frac{2r}{M}, \qquad \text{so} \quad D < 0,$$

$$E = \left(\frac{\partial^2 f}{\partial h \partial p}\right)^2 - \frac{\partial^2 f}{\partial p^2}\frac{\partial^2 f}{\partial h^2} = (-1)^2 + \left(\frac{2r}{M} \cdot 0\right), \qquad \text{so} \quad E = 1.$$

These quantities satisfy the hypotheses of Theorem 5.4, showing that this model does have a *transcritical bifurcation* at $(p_e, h_c) = (0, r)$. Furthermore, with $D < 0$, Theorem 5.4 implies the upper branch of equilibria is stable, while the lower branch is unstable as shown in Fig. 5.9.

Example 5.5 Consider the logistic growth model

$$x_{n+1} = \lambda x_n(1 - x_n).$$

We know from previous chapters that this model has two fixed points $x_{e_1} = 0$ and $x_{e_2} = 1 - \frac{1}{\lambda}$. At $\lambda = 1$, $x_{e_2} = 0$, so that the two branches meet. Then the condition $f'_{\lambda_c}(x_e) = 1$ is satisfied when $(x_e, \lambda_c) = (0, 1)$. Direct calculations at $(x_e, \lambda_c) = (0, 1)$ yield

$$A = x(1 - x) = 0, \quad D = -2, \quad E = 1 - 2x = 1.$$

Hence, by Theorem 5.4, a transcritical bifurcation occurs at $(x_e, \lambda_c) = (0, 1)$. Furthermore, since $D < 0$ then the upper branch $x_{e_2} = 1 - \frac{1}{\lambda}$ is stable while the lower branch $x_{e_1} = 0$ is unstable.

5.4.4 Pitchfork Bifurcation

J.J. Hopfield [16, 17] proposed a connection between *Ising models* and networks of neurons, which can be modeled using a mean-field theory approach. The approach is similar to the one used earlier on in Chap. 4 for modeling a fluxgate magnetometer. In fact, it leads to a very similar model for a neuron, even though the two systems are completely different–a common occurrence in mathematical modeling. What Hopfield suggested was that neurons could be modeled as a lattice of N binary "spins", S_i, $i = 1, \ldots, N$, in which each spin can be in one of two states: "spin up" or $S_i = +1$ and "spin down" or $S_i = -1$. The average state of the neuron, u_i, at spin

S_i, is found by adding the average contributions from all neighboring spins S_j and from any external input, u^{ext}, through

$$\langle u_i \rangle = \sum_{j \to i} J_{ij} \langle S_j \rangle + u^{ext}, \tag{5.16}$$

where J_{ij} is the coupling strength of the influence of spin S_j on spin S_i. The average state of the neuron, u_i, can, in turn, induce a response on spin S_i, to switch back and forth between its two states $+1$ and -1. The actual switching mechanism can be modeled by an activation function:

$$\langle S_i \rangle = \tanh(\omega \langle u_i \rangle). \tag{5.17}$$

where the parameter ω is related to the temperature T through $\omega = 1/(k_B T)$, with k_B being Boltzmann's constant. Substituting Eq. (5.16) into Eq. (5.17), leads to an *average neuron response* given by

$$\langle S_i \rangle = \tanh(\omega \sum_{j \to i} J_{ij} \langle S_j \rangle + \omega u^{ext}). \tag{5.18}$$

Our interest is in the regime in which the neurons acquire an uniform polarization, so that $\langle S_i \rangle = \langle S \rangle$. Assuming identical coupling strengths $J_{ij} = 1/N$, where N is the number of spins, we find a single equation for the average polarization of the neurons

$$\langle S \rangle = \tanh(\omega \langle S \rangle + \omega u^{ext}). \tag{5.19}$$

An extension of this last equation with discrete updating of the average polarization state leads to the following simple model [16–18] of N neurons.

$$x_{n+1} = \tanh\left(\omega x_n + u\right) = f(x), \tag{5.20}$$

where $x \in \mathbb{R}$ is the state variable that represents the polarization of the neurons, u is an external input or applied stimulus, ω is a self-feedback parameter, which is assumed to be constant. This model is often used as a basic unit to forming larger neural networks, also known as Hopfield networks or recurrent neural network (RNN), for AI (Artificial Intelligence). We wish to highlight that the RNN model Eq. (5.20) is, essentially, the same as the model that governs the response of a fluxgate magnetometer, see Eq. (4.73). This is the result of both systems being governed by similar underlying principles, in spite of neurons and fluxgate magnetometers being unrelated to one another.

The model Eq. (5.20) actually represents a codimension-two problem since it involves two independent parameters, ω and u. In this example, we consider, however, the special case of no external stimulus, so that $u = 0$, which renders the problem codimension-one. The feedback-parameter, ω, is then treated as the main bifurcation parameter.

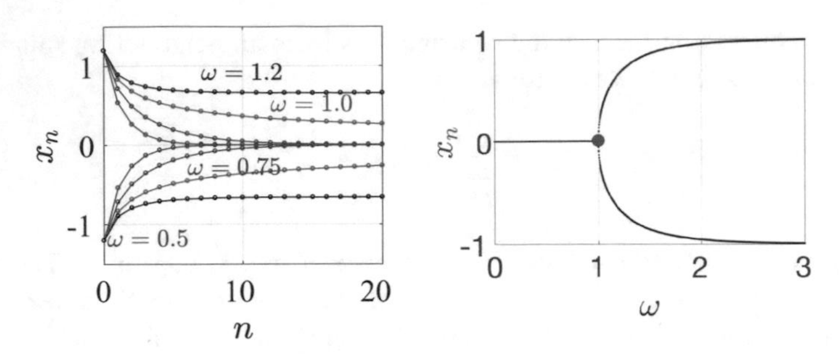

Fig. 5.10 (Left) simulations of the RNN model (5.20) for various values of ω. For the range $0 < \omega < 1$, the solution of the model settles into lack of activity, represented by the stable trivial fixed point. For values of ω in the range $\omega > 1$, the solution settles into a nontrivial steady state. (Right) bifurcation diagram illustrating the changes in fixed points as the feedback parameter ω varies. The fixed points along the solid line are stable states, while those on the dashed line are unstable

Figure 5.10 (left) shows simulations of the RNN model for various values of the parameter ω. The simulations show that for small values of ω, in the interval $0 < \omega < 1$, the neuron has no activity, so it settles into the trivial fixed point $x = 0$. But for values in the range $\omega > 1$, the zero fixed point loses stability and two new branches of non trivial fixed points emerge.

The fixed points of the RNN model are found by solving the equation

$$\tanh(\omega x) = x,$$

which leads to

$$x_{e_1} = 0, \qquad x_{e_{2,3}} = \pm\sqrt{\frac{\omega - 1}{\omega}}.$$

Observe that the trivial fixed points exist for all values of the parameter ω, while the last two fixed points, $x_{e_{2,3}}$, exist only when $\omega > 1$. These last two fixed points merge with the trivial fixed point, x_{e_1}, when $\omega = 1$. Furthermore, the plots in Fig. 5.10 suggest that the fixed points, $x_{e_{2,3}}$, are always stable whenever they exist, while the trivial fixed point changes from stable to unstable as ω changes from $\omega < 1$ to $\omega > 1$. This type of transition, in which two new stable fixed points appear on one side of the parameter range, while a third fixed point exchanges stability, is known as *pitchfork bifurcation*.

Direct calculations show that

$$\frac{\partial f}{\partial x} = \omega\big(1 - \tanh(\omega x)\big).$$

Then, $\frac{\partial f}{\partial \omega}(x_{e_1}) = \omega\big(1 - \frac{\omega-1}{\omega}\big) = 1$. It follows that this example satisfies the bifurcation condition $f(0, \omega = 1) = 0$ and $\frac{\partial f}{\partial x}(0, \omega = 1) = 1$.

The following theorem lists the conditions and provides details for calculating the direction of the *pitchfork bifurcation*.

Theorem 5.5 (Pitchfork Bifurcation)
Let $f_\lambda : \mathbb{R} \to \mathbb{R}$ and let

$$f_{\lambda_c}(x_e) = x_e, \qquad f'_{\lambda_c}(x_e) = 1 \qquad \textit{(bifurcation condition)}$$
$$A = \frac{\partial f}{\partial \lambda} = D = \frac{\partial^2 f}{\partial x^2} = 0, \quad E = \frac{\partial^2 f}{\partial x \partial \lambda} \neq 0, \quad F = \frac{\partial^3 f}{\partial x^3} \neq 0,$$

where all derivatives are evaluated at the bifurcation point $(\lambda_c, x_e) = (0, 0)$. Then there is a branch of fixed points which passes through (x_e, λ_c) transverse to $\lambda = 0$. These fixed points are unstable in $\lambda < 0$ and stable in $\lambda > 0$ if $E < 0$, with stabilities reversed if $E > 0$. A second branch of fixed points bifurcates from (x_e, λ_c) tangential to $\lambda = 0$ into $\lambda > 0$ if $EF < 0$ or into $\lambda < 0$ if $EF > 0$. This second branch of fixed points is stable if it exists in $\lambda < 0$ and $E < 0$ or in $\lambda > 0$ and $E > 0$ (a supercritical bifurcation). Otherwise it is unstable (a subcritical bifurcation).

The proof of this theorem is found in [15]. We apply this theorem to the RNN model Eq. (5.20), where the bifurcation conditions were shown above at $(x_{e_1}, \omega_c) = (0, 1)$. Computing the quantities A, D, E, and F gives:

$$\begin{aligned}
A &= \frac{\partial f}{\partial \omega} &&= x\big(1 - \tanh^2(\omega x)\big) \\
D &= \frac{\partial^2 f}{\partial x^2} &&= \mp 2\sqrt{\tfrac{\omega-1}{\omega}} \\
E &= \frac{\partial^2 f}{\partial x \partial \omega} &&= 1 - \tanh^2(\omega x) - 2\omega x \tanh(\omega x)\big(1 - \tanh^2(\omega x)\big) \\
F &= \frac{\partial^3 f}{\partial x^3} &&= -2\omega \pm 4\omega(\omega - 1)
\end{aligned}$$

Evaluating these derivatives at the critical bifurcation point $(x_{e_1}, \omega_c) = (0, 1)$ we get

$$A = 0, \qquad D = 0, \qquad E = 1 - 2x^2 = 1, \qquad F = -2.$$

These quantities satisfy the hypotheses of Theorem 5.5, showing that this model does exhibit a *pitchfork bifurcation* at $(x_{e_1}, \omega_c) = (0, 1)$. Furthermore, since $E > 0$ the branch of fixed points $x_{e_{2,3}}$ is stable in $\omega > \omega_c = 1$. Since $E > 0$ and $F < 0$, Theorem 5.5 implies that the second branch of fixed points $x_{e_{2,3}}$ is stable in $\omega > \omega_c = 1$. Thus, the bifurcation is supercritical, as is shown in Fig. 5.10 (right).

Example 5.6 Consider the cubic map

$$\boxed{x_{n+1} = \lambda x_n - x_n^3.}$$

This model has two fixed points:

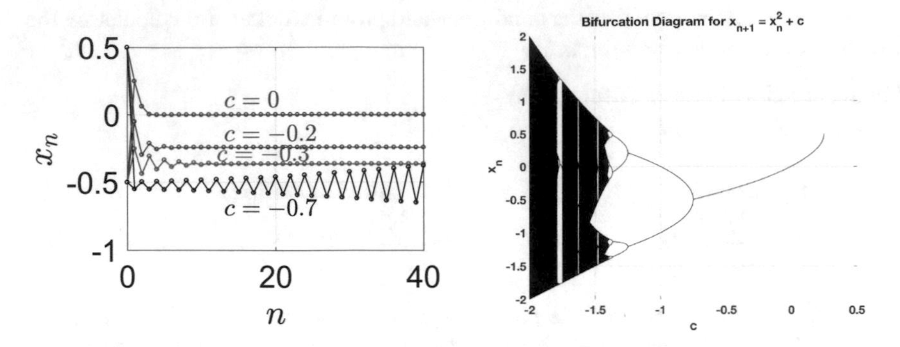

Fig. 5.11 (Left) simulations of the quadratic model (5.21) for various values of c. (Right) Bifurcation diagram as a function of the varying parameter c. Across the range $-3/4 < c < 1/4$ all solution trajectories converge to a nontrivial fixed point. When c is closer to $c = -3/4$ solutions tend to oscillate

$$x_{e_1} = 0, \qquad x_{e_2} = \pm\sqrt{\lambda - 1}.$$

At $\lambda = 1$, $x_{e_2} = 0$, so that the two branches meet. Then the condition $f'_{\lambda_c}(x_e) = 1$ is satisfied when $(x_e, \lambda_c) = (0, 1)$. Direct calculations at $(x_e, \lambda_c) = (0, 1)$ yield

$$A = x = 0, \quad D = -6x = 0, \quad E = 1, \quad F = -6.$$

Hence, by Theorem 5.5, a pitchfork bifurcation occurs at $(x_e, \lambda_c) = (0, 1)$. Furthermore, since $E > 0$ the branch of zero fixed points $x_{e_1} = 0$ is stable in $\lambda < \lambda_c = 1$ and unstable in $\lambda > \lambda_c = 1$. $E > 0$ also implies that the second branch of fixed points $x_{e_2} = \pm\sqrt{\lambda - 1}$ is stable in $\lambda > \lambda_c = 1$. Thus the pitchfork bifurcation is supercritical.

5.4.5 Period Doubling Bifurcation

The quadratic map

$$\boxed{x_{n+1} = x_n^2 + c,} \tag{5.21}$$

where c is a real-valued constant, has been used in the analysis of systems of semiconductor lasers [19], as it represents a phenomenological model of a chaotic system that can be synchronized through coupling.

Figure 5.11 (left) shows simulations of the quadratic model for various values of the parameter c. The simulations reveal that if $-3/4 < c < 1/4$ then solution trajectories approach a nontrivial fixed point. For values of c closer to $c = -3/4$ the solutions start to oscillate.

Let us examine the model in more detail. Equilibrium points are found by solving

$$x^2 + c = x,$$

which yields two solutions

$$x_{e1,2} = \frac{1}{2} \pm \frac{1}{2}\sqrt{1-4c}.$$

Observe that these solutions exist only when $c \leq 1/4$. To study their stability, we compute the derivative of $f(x) = x^2 + c$, which satisfies $f'(x) = 2x$. For the first equilibrium point, $x_{e1} = \frac{1}{2} + \frac{1}{2}\sqrt{1-4c}$, to be stable, it must satisfy

$$|f'(x_{e1})| < 1, \quad \text{or} \quad -1 < \frac{1}{2} + \frac{1}{2}\sqrt{1-4c} < 1,$$

but this inequality has no solution. Thus, x_{e1} is always unstable. For the second equilibrium point, $x_{e2} = \frac{1}{2} - \frac{1}{2}\sqrt{1-4c}$, to be stable, we get

$$|f'(x_{e2})| < 1, \quad \text{or} \quad -1 < \frac{1}{2} - \frac{1}{2}\sqrt{1-4c} < 1,$$

which is satisfied when c is in the interval

$$-\frac{3}{4} < c < \frac{1}{4}.$$

A bifurcation occurs when $|f'(x_{e1,2})| = 1$. That is,

$$|1 \pm \sqrt{1-4c}| = 1,$$

which is satisfied when $c = 1/4$ or when $c = -3/4$. This result suggests that two transitions occur. One at $c = 1/4$ and one at $c = -3/4$. Indeed, the first transition, at $c = 1/4$, corresponds to a saddle-node bifurcation and its analysis is left as an exercise.

The second transition is associated with the onset of period-2 oscillations. Period-2 points are found by solving

$$f^2(x) = (x^2 + c)^2 + c = x.$$

Expanding we get

$$(x^2 - x + c)(x^2 + x + 1 + c) = 0.$$

The first quadratic polynomial yields the equilibrium points. The second quadratic function leads to the period-2 points, mainly

$$x_{1,2} = \frac{1}{2} \pm \frac{1}{2}\sqrt{-3-4c}.$$

Observe that period-2 points exist only when $c < -3/4$. Figure 5.11 (right) is a bifurcation diagram for the quadratic map. The diagram helps visualize the sequence of changes in solution types as the bifurcation parameter c varies.

We now show that the second bifurcation at $c = -3/4$ corresponds to a *period-doubling bifurcation*.

The following theorem lists the conditions and provides details for calculating the direction of the *pitchfork bifurcation*.

Theorem 5.6 (Period-Doubling Bifurcation) *Let $f_\lambda : \mathbb{R} \to \mathbb{R}$ and let*

$$f_{\lambda_c}(x_e) = x_e, \qquad f'_{\lambda_c}(x_e) = -1 \qquad \textit{(bifurcation condition)}$$
$$A = 2\frac{\partial^2 f}{\partial\lambda\partial x} + \frac{\partial f}{\partial\lambda}\frac{\partial^2 f}{\partial x^2} \neq 0, \quad D = \frac{1}{2}\left(\frac{\partial^2 f}{\partial x^2}\right)^2 + \frac{1}{3}\frac{\partial^3 f}{\partial x^3} \neq 0,$$

where all derivatives are evaluated at the bifurcation point $(\lambda_c, x_e) = (0, 0)$. Then a curve of periodic points of period two bifurcates from (λ_c, x_e) into $\lambda > 0$ if $AD < 0$ or $\lambda < 0$ if $AD > 0$. The fixed point from which these solutions bifurcate is stable in $\lambda > 0$ and unstable in $\lambda < 0$ if $A > 0$, with the signs of λ reversed if $A < 0$. The bifurcating cycle of period two is stable if it coexists with an unstable fixed point and vice versa. The bifurcation is supercritical if the bifurcating solution of period two stable and subcritical otherwise.

The proof of this theorem are found in [15]. We apply this theorem to the quadratic Eq. (5.21), where the bifurcation conditions were shown above at $(x_{e_2}, c) = (x_{e_2}, -3/4)$. Computing the quantities A and D, gives:

$$A = 2\frac{\partial^2 f}{\partial\lambda\partial x} + \frac{\partial f}{\partial\lambda}\frac{\partial^2 f}{\partial x^2} = 0 + 1 \times 2 = 2, \qquad \text{so} \quad A > 0,$$
$$D = \frac{1}{2}\left(\frac{\partial^2 f}{\partial x^2}\right)^2 + \frac{1}{3}\frac{\partial^3 f}{\partial x^3} = 2x^2 + \frac{2}{3}, \qquad \text{so} \quad D > 0.$$

Since $AD > 0$ then the period-doubling branch bifurcates into $c < -3/4$. And since $A > 0$ the branch of fixed points x_{e_2} is stable when $c > -3/4$ and unstable otherwise. Figure 5.11 (right) shows a bifurcation diagram of the solutions of the quadratic model as a function of the parameter c. The diagram illustrates period-doubling cascades starting from the nontrivial equilibrium point x_{e_2}.

Example 5.7 Consider the logistic growth model

$$x_{n+1} = \lambda x_n(1 - x_n).$$

We know from previous chapters that this model has two fixed points $x_{e_1} = 0$ and $x_{e_2} = 1 - \frac{1}{\lambda}$. At $\lambda = 3$, $x_{e_2} = 2/3$, and the bifurcation condition $f'_{\lambda_c}(2/3) = -1$ is

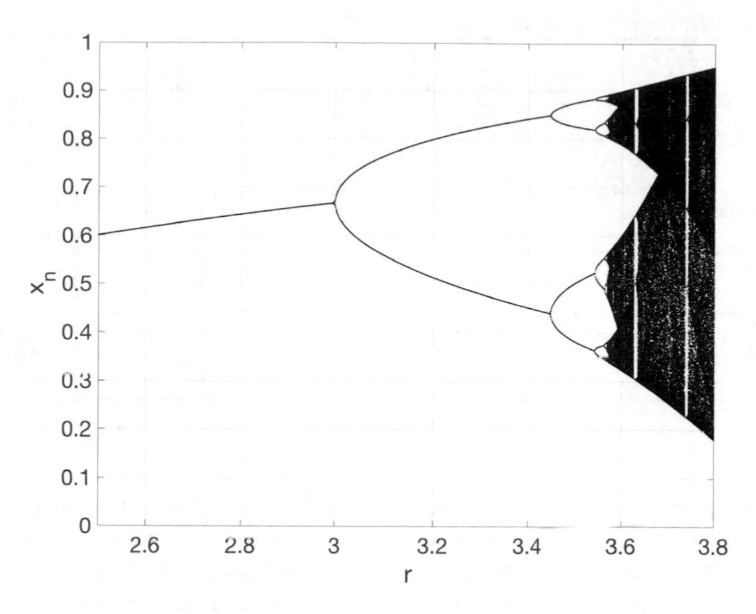

Fig. 5.12 Period-doubling bifurcations in the logistic model $x_{n+1} = rx_n(1 - x_n)$

satisfied. Thus, we set $(x_e, \lambda_c) = (2/3, 3)$. Direct calculations at $(x_e, \lambda_c) = (2/3, 3)$ yield

$$A = 2(1 - 2x) - 2\lambda x(1 - x) = -2, \quad D = 2\lambda^2 - \frac{2}{3} = \frac{52}{3}.$$

Hence, by Theorem 5.6, a period doubling bifurcation occurs at $(x_e, \lambda_c) = (2/3, 3)$. Furthermore, since $AD < 0$ then the period-doubling branch bifurcates into $\lambda > \lambda_c = 3$. And since $A < 0$ the branch of fixed points $x_{e_2} = 1 - \frac{1}{\lambda}$ is stable when $\lambda < \lambda_c = 3$ and unstable otherwise. Figure 5.12 shows a bifurcation diagram of the solutions of the logistic model as a function of the parameter r. The diagram illustrates period-doubling cascades starting from the nontrivial equilibrium point x_{e_2}.

Table 5.1 summarizes all the cases of codimension one bifurcations that we have discussed so far. All derivatives are evaluated at the bifurcation points (x_e, λ_c).

In addition, another nondegeneracy condition for the transcritical bifurcation is $f_{x\lambda}^2 - f_{xx} f_{\lambda\lambda} \neq 0$ and for the period doubling: $2 f_{xx} + f_\lambda f_{xx} \neq 0$ and $\frac{1}{2} f_{xx}^2 + \frac{1}{2} f_{xxx} \neq 0$.

5.4.6 *Neimark-Sacker Bifurcation*

There is also an equivalent mechanism to that of the Hopf bifurcation in discrete models. It's called the *Neimark-Sacker bifurcation* (NS). In a similar fashion to the Hopf

Table 5.1 Classification of codimension one bifurcations in discrete models

Bifurcation	$\frac{\partial f}{\partial x}$	$\frac{\partial f}{\partial \lambda}$	$\frac{\partial^2 f}{\partial x^2}$	$\frac{\partial^2 f}{\partial x \partial \lambda}$	$\frac{\partial^3 f}{\partial x^3}$	Normal form
Saddle-Node (fold)	1	$\neq 0$	$\neq 0$			$x_{n+1} = \lambda + x_n + x_n^2$
Transcritical	1	0	$\neq 0$	$\neq 0$		$x_{n+1} = \lambda x_n - x_n^2$
Pitchfork	1	0	0	$\neq 0$	$\neq 0$	$x_{n+1} = \lambda x_n - x_n^3$
Period-doubling	-1	$\neq 0$	$\neq 0$			$x_{n+1} = -(1 + \lambda)x_n + x_n^3$

bifurcation, the NS bifurcation occurs when a pair of complex-valued eigenvalues, $\sigma_{1,2}$, cross the unit circle. This can be defined as $|\sigma_{1,2}| = 1$ or $\sigma_1 \bar{\sigma}_1 = 1$, and $\sigma_1 \in \mathbf{C}$. The bifurcation can also be either supercritical or subcritical. In the former case, a stable focus loses its stability as a parameter is varied with the consequent emergence of a stable cycle or quasi-cycle–typically known as closed invariant curves in the literature. In the latter case of a subcritical NS bifurcation, a stable focus enclosed by an unstable closed curve loses its stability with the consequent disappearance of the closed invariant curve as a parameter is varied. We discuss both cases next, following closely the presentation in [20].

The normal form for a NS bifurcation problem takes the form

$$\begin{bmatrix} x_1 \\ x_2 \end{bmatrix} \mapsto (1+\lambda) \begin{bmatrix} \cos\theta & -\sin\theta \\ \sin\theta & \cos\theta \end{bmatrix} \begin{bmatrix} x_1 \\ x_2 \end{bmatrix} + (x_1^2 + x_2^2) \begin{bmatrix} \cos\theta & -\sin\theta \\ \sin\theta & \cos\theta \end{bmatrix} \begin{bmatrix} d & -b \\ b & d \end{bmatrix} \begin{bmatrix} x_1 \\ x_2 \end{bmatrix} \quad (5.22)$$

where λ is the distinguished bifurcation parameter, $\theta = \theta(\lambda)$, $b = b(\lambda)$, $d = d(\lambda)$, $d(0) \neq 0$, and $0 < \theta(0) < \pi$. Observe that $(x_1, x_2) = (0, 0)$ is an equilibrium point for all values of λ. The first term contains linear terms and the second one the nonlinear part. Thus, the Jacobian matrix is

$$(df) = (1+\lambda) \begin{bmatrix} \cos\theta & -\sin\theta \\ \sin\theta & \cos\theta \end{bmatrix},$$

whose eigenvalues are $\sigma(\lambda) = (1+\lambda)e^{\pm\theta i}$. Thus, at $\lambda = 0$, in particular, the zero equilibrium is nonhyperbolic due to a complex-conjugate pair of eigenvalues on the unit circle, i.e., $|\sigma(0)| = 1$.

We can rewrite the normal forms in complex coordinates by setting $z = x_1 + x_2 i$ and $d_1 = d + bi$, which yields

$$z \mapsto e^{\theta i} z(1 + \lambda + d_1 |z|^2).$$

If we let $\mu = \mu(\lambda) = (1+\lambda)e^{\theta(\lambda)i}$ and $c_1 = c_1(\lambda) = d_1(\lambda)e^{\theta(\lambda)i}$ we arrive at the complex version of the normal form

$$\boxed{z_{n+1} = (\mu + c_1|z_n|^2)z_n} \tag{5.23}$$

Theorem 5.7 (Kuznetsov [20])

Consider the complex version of the normal forms (5.23) for the Neimark-Sacker bifurcation. Let $d(0) = Re\{e^{-\theta_0 i}c(0)\}$ be the first Lyapunov coefficient. If $d(0) \neq 0$ then

(a) If $d(0) < 0$, then the normal form has a fixed point at the origin, which is asymptotically stable for $\lambda \leq 0$ (weakly at $\lambda = 0$) and unstable for $\lambda > 0$. Thus, the bifurcation is classified as a supercritical Neimark-Sacker bifurcation. Moreover, there is a unique and stable closed invariant curve that exists for $\lambda > 0$ and has radius $\mathcal{O}(\sqrt{\lambda})$.

(b) If $d(0) > 0$, then the normal form has a fixed point at the origin, which is asymptotically stable for $\lambda < 0$ and unstable for $\lambda \geq 0$ (weakly at $\lambda = 0$). Thus, the bifurcation is classified as a supercritical Neimark-Sacker bifurcation. Moreover, there is a unique and unstable closed invariant curve that exists for $\lambda < 0$ and has radius $\mathcal{O}(\sqrt{-\lambda})$.

We can also rewrite the normal form equations in polar coordinates by letting $z_n = \rho_n e^{\varphi_n i}$, which (after some simplifications) yields

$$\begin{aligned} \rho_{n+1} &= \rho_n(1 + \lambda + d(\lambda)\rho_n^2) + \ldots \text{ h.o.t} \\ \varphi_{n+1} &= \varphi_n + \theta(\lambda) + \ldots \text{ h.o.t.} \end{aligned} \tag{5.24}$$

Observe that the amplitude equations for ρ decouple from the phase equations φ. This decoupling facilitates the calculation of fixed points and their stability properties. First of all, $\rho_{e_1} = 0$ is always a fixed point for all values of λ. If we let

$$f(\rho) = \rho(1 + \lambda + d(\lambda)\rho^2),$$

then $f'(\rho_{e_1} = 0) = 1 + \lambda$ implies that $\rho_{e_1} = 0$ is stable if $\lambda < 0$ an unstable if $\lambda > 0$.
Additionally, for $\lambda > 0$ there is a nonzero fixed-point

$$\rho_{e_2} = \sqrt{-\frac{\lambda}{d(\lambda)}}.$$

which corresponds to a periodic orbit of Eq. (5.23). Similarly, direct computations show $f'(\rho_{e_2}) = 1 - 2\lambda$. Thus, ρ_{e_2} is stable whenever $\lambda > 0$. Under this scenario, the NS bifurcation is classified as supercritical.

Figure 5.13 illustrates both the supercritical and subcritical Neimark-Sacker bifurcations.

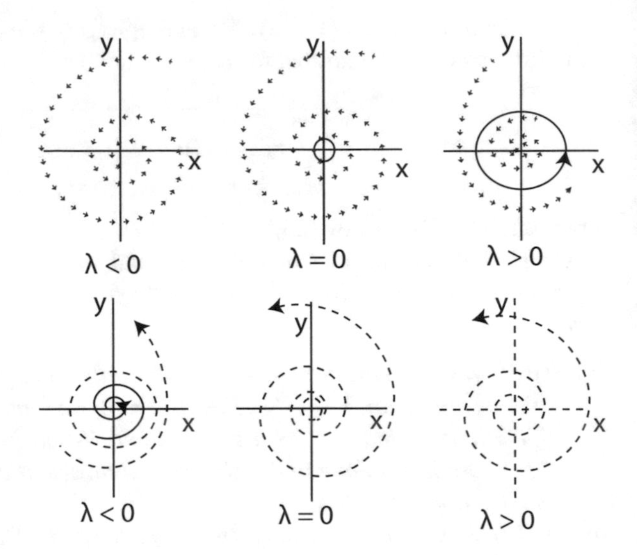

Fig. 5.13 (Top) Supercritical and (bottom) Subcritical Neimark-Sacker bifurcation

Example 5.8 Consider the following discrete model of the interaction between a predator P and a prey N

$$\begin{aligned} N_{t+1} &= rN_t e^{-bP_t} \\ P_{t+1} &= N_t(1 - e^{-aP_t}), \end{aligned} \tag{5.25}$$

where $a, b, r > 0$. This model makes he simplifying assumption that the predator can consume the prey without limit. In fact, the special case $a = b$ can lead to unbounded growth in solutions. We do not consider that case in here. Hone et al. [21] showed that a nondimensionalized version of the model takes the form

$$\begin{aligned} x_{t+1} &= rx_t e^{-y_t} \\ y_{t+1} &= x_t(1 - e^{-ay_t}), \end{aligned} \tag{5.26}$$

which reduces the total number of positive parameters from three to two, a and r. Neither one of these remaining parameters can be removed by rescaling.

Since x and y represent the size of two populations, prey and predator, respectively, we will only consider solutions on the first quadrant of the phase space, where $x, y \geq 0$. On this quadrant, the only fixed points are: $(x_{e_1}, y_{e,1}) = (0, 0)$ and (x_{e_2}, y_{e_2}), where

$$x_{e_2} = \frac{r^a \log r}{r^a - 1}, \qquad y_{e_2} = \log r.$$

The Jacobian matrix of the linearized system around the (x_{e_2}, y_{e_2}) fixed point is

$$J = \begin{bmatrix} 1 & -x_{e_2} \\ 1 - r^{-a} & a x_{e_2} r^{-a} \end{bmatrix}.$$

The characteristic polynomial for J is $\sigma^2 - \text{tr}(J)\sigma + \det(J) = 0$, where

$$\text{tr}(J) = 1 + \frac{\log(1+v)}{v}, \quad \det(J) = \left(\frac{1}{a} + \frac{1}{v}\right) \log(1+v),$$

where $v = r^a - 1$. Now, recall the condition for the NS bifurcation: $\sigma_1 \bar{\sigma}_1 = 1$ and $\sigma_1 \in \mathbf{C}$. This is also equivalent to $\det(J) = 1$, which yields, in solving for $a(v)$:

$$a_c(v) = \left(\frac{1}{\log(1+v)} - \frac{1}{v}\right)^{-1}.$$

Direct (and tedious) calculations show that along $a_c(v)$, $\Delta = \text{tr}^2(J) - \det(J) < 0$, so that the eigenvalues are indeed complex. Consequently, at $a = a_c(v)$ the nontrivial fixed point $(x_{e,2}, y_{e,2})$ loses stability in a Neimark-Sacker bifurcation. An analysis of the direction of bifurcation is deferred as an exercise. Figure 5.14 illustrates the closed invariant curve that emerges via the NS bifurcation. The simulations were performed in MATLAB, see Appendix.

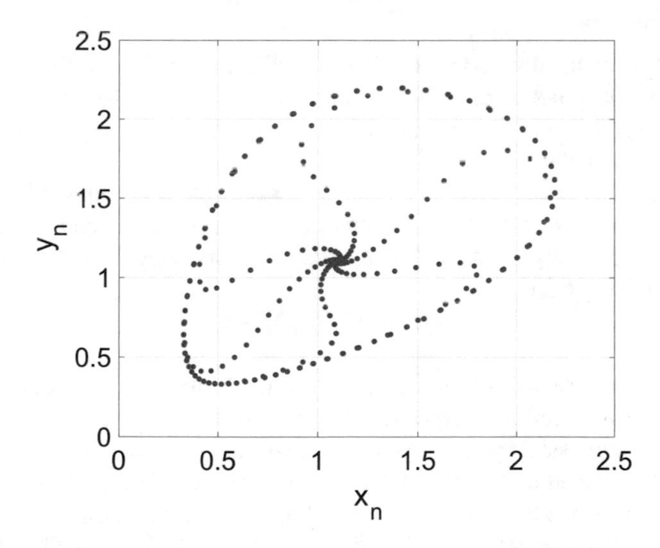

Fig. 5.14 Representative orbit of the discrete model (5.26) showing the closed invariant curve that emerges via a Neimark-Sacker bifurcation. Parameters are: $a = 40$ and $r = 3$. Initial condition: $(1.1, 1.1)$, number of iterates $N = 40$

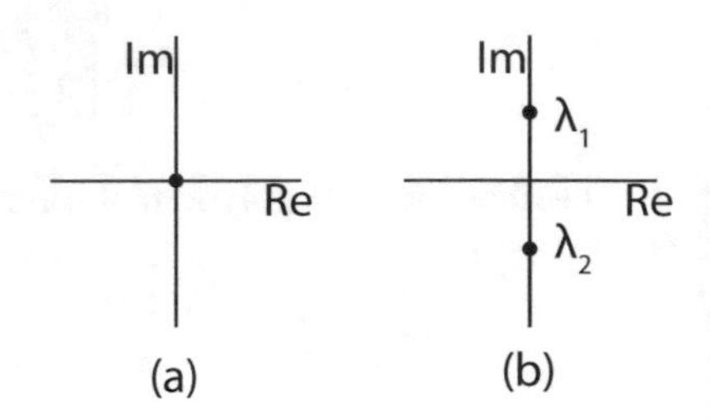

Fig. 5.15 Critical eigenvalues associated with codimension one bifurcations in continuous models

5.5 Codimension One Bifurcations in Continuous Systems

In the case of continuous models, there are two critical values of the eigenvalues that can lead to codimension one bifurcations, either $\lambda = 0$ or $\lambda = 0 \pm \tau i$. These two cases are illustrated in Fig. 5.15.

These two critical values of the eigenvalues associated with codimension-one bifurcations can lead to four common mechanisms for changes in behavior in continuous models when one single parameter is varied:

- Saddle node bifurcation
- Transcritical bifurcation
- Pitchfork bifurcation
- Hopf bifurcation

The first three involve changes between equilibrium points while the fourth one involves the emergence of periodic oscillations. All four cases are the simplest cases of bifurcations that one can encounter in a model because it requires that we only change one parameter.

We introduced bifurcations in continuous models first because it turns out that much of the analysis of the bifurcations in discrete systems, such as (5.12), can be inferred from the analysis on continuous models by applying the theory discussed in Sect. 5.5 on the system

$$g(\mathbf{x}, \lambda) = f(\mathbf{x}, \lambda) - \mathbf{x}.$$

In this section we will discuss each one of the bifurcations scenarios described above. The presentation will be motivated by a corresponding example of a mathematical model of a physical system. Each model will involve a distinguished parameter λ that must be varied and a critical value λ_c where the bifurcation occurs. Without loss of generality, we can assume $\lambda_c = 0$. Similarly, we can assume there is a non-hyperbolic equilibrium at $\mathbf{x}_e = \mathbf{0}$. Together, $(\mathbf{x}_e, \lambda_c) = (\mathbf{0}, 0)$ will define the location of the bifurcation.

5.5.1 Saddle-Node Bifurcation

We now return to our running example, the fluxgate magnetometer, first introduced in Chap. 4. Recall the mathematical model from Eq. (4.73):

$$\tau \frac{dx}{dt} = -x + \tanh(c(x + H_e(t) + H_x)).$$

In that model, we already encountered an instance of a *saddle-node bifurcation* as two equilibria in the model appeared or disappeared as the parameter that defines the applied external field, H_x, varies. That is, from Sect. 4.10 we know that in the absence of an external field, i.e., $H_x = 0$, there are three equilibrium points: $x_e = \{-1, 0, 1\}$. But as H_x increases, the negative equilibrium at $x_e = -1$ and $x_e = 0$ start to move closer to one another until there is a critical value of H_x beyond which the two equilibria collide and then disappear. The region where this occurs has been marked with a circle in Fig. 4.46. The circle emphasizes the local nature of the bifurcation. That is, the "out-of-the-blue" appearance and disappearance of equilibria occurs locally around a critical value of H_x and it corresponds to a saddle-node bifurcation. The critical value of H_x can be calculated by solving

$$\tanh(c(x + H_e(t) + H_x)) = x$$

for H_x, to get:

$$H_x = -x + \frac{1}{c}\tanh(x),$$

and then, numerically, finding the value of H_x where this last equation has a unique real-valued solution for x. For $c = 3$, the numerical calculation yields, approximately, $H_x = 0.4348$ and $x = -0.8168$. Then a Taylor series expansion around this point yields (after some manipulation) an equation of the form

$$\frac{dx}{dt} = \lambda + ax^2 + \text{higher order terms}, \tag{5.27}$$

where $\lambda \in \mathbb{R}$ is the distinguished bifurcation parameter and a is a real constant. In fact, one can also show that a can be scaled to $a = \pm 1$. Without loss of generality, we consider the case $a = -1$ and later we comment on the opposite case $a = +1$.

Now, one can verify that $f(x, \lambda) = \lambda - x^2$ satisfies the bifurcation condition (5.11). To explore the nature of the bifurcation, we use Fig. 5.16 to visualize the changes in the phase portraits of Eq. (5.27), as the main parameter, λ, is varied.

From left-to-right, when $\lambda < 0$ we can see that $f(x, \lambda) < 0$, so there are no equilibria. Furthermore, observe that $\dot{x} < 0$, which means that the flow always decreases along the x-axis. At $\lambda = 0$, $f(x, 0) = 0$ only at one single point $(0, 0)$, so there is only one equilibrium point $x_e = 0$. Once again, $\dot{x} < 0$, so the flow also decreases

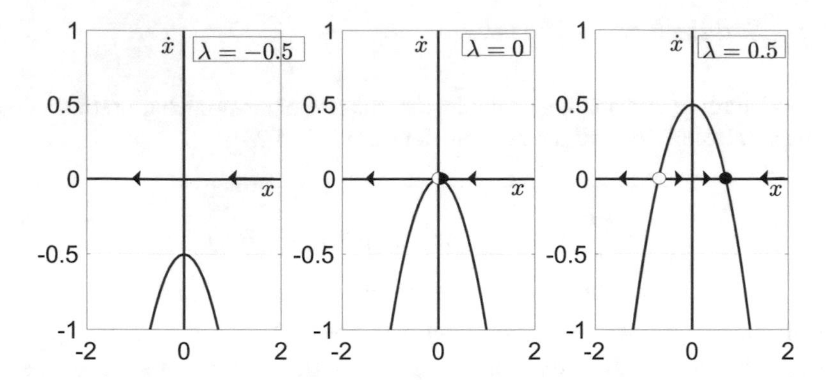

Fig. 5.16 Changes in phase portrait due to passage through a saddle-node bifurcation

along the x-axis. When $\lambda > 0$, there are two points of intersection for the graph of $f(x, \lambda)$, so that $x_e = \pm\sqrt{\lambda}$.

In summary, all equilibrium points are defined by $x_e = \pm\sqrt{\lambda}$, with $\lambda \geq 0$. To determine their stability properties, we compute the derivative of f:

$$\frac{\partial f}{\partial x} = -2x,$$

and then evaluate it at the two equilibrium points to get:

$$\begin{aligned} \tfrac{\partial f}{\partial x}\,|_{x_+} &= -2\sqrt{\lambda} < 0, \quad \text{where} \quad x_+(\lambda) = +\sqrt{\lambda} \\ \tfrac{\partial f}{\partial x}\,|_{x_-} &= +2\sqrt{\lambda} > 0, \quad \text{where} \quad x_-(\lambda) = -\sqrt{\lambda}. \end{aligned}$$

Consequently, we conclude that $x_+(\lambda) = +\sqrt{\lambda}$ is locally asymptotically stable while $x_-(\lambda) = -\sqrt{\lambda}$ is unstable. These stability properties are marked with arrows in Fig. 5.16. But perhaps the best way to summarize and visualize the results of the stability analysis is with the aid of the bifurcation diagram of Fig. 5.17. The diagram shows the state variable x as a function of λ. Solid curves represent branches of stable equilibria, $x = +\sqrt{\lambda}$ in this case, while dashed lines represent unstable equilibria, i.e., $x = -\sqrt{\lambda}$. now, the case of $a = -1$ is similar except that the bifurcation occurs to the left $\lambda = 0$. In a similar manner, one can show that, in that case, the equilibria $x = +\sqrt{\lambda}$ is unstable while $x = -\sqrt{\lambda}$ is stable.

It is worthwhile pointing out that, among all codimension-one bifurcation problems, the saddle-node bifurcation is the **generic** bifurcation in the sense that this is the type of bifurcation that one can expect when there are no additional restrictions imposed on the function $f(x, \lambda)$ beyond its smoothness and the bifurcation conditions (5.11). Roughly speaking, this means that the perturbed problem

$$\frac{dx}{dt} = f(x, \lambda) + \varepsilon\, p(x, \lambda),$$

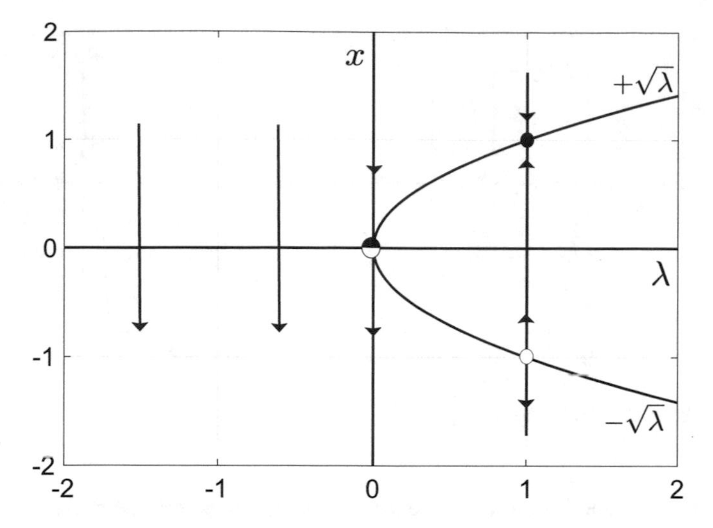

Fig. 5.17 Saddle-node bifurcation diagram. Two new branches of equilibrium points emerge on one side of the parameter λ, while on the other side there are no equilibrium points

where $\varepsilon > 0$ is sufficiently small, and p is the perturbation function, will also exhibit a saddle-node bifurcation, provided that no additional constrains are imposed on f.

5.5.2 Transcritical Bifurcation

We now return to the laser beam model discussed previously in Chap. 4. In the model Eq. (4.18), we encountered a case of a *transcritical bifurcation* as two branches of equilibria exchange stability properties. Recall that the laser beam model has the following form

$$\boxed{\frac{dx}{dt} = \lambda x + ax^2,} \tag{5.28}$$

where, once again, $\lambda \in \mathbb{R}$ is the bifurcation parameter and a is a real constant that can be scaled to $a = \pm 1$. Without loss of generality, we choose $a = -1$. Direct calculations show that $f(x, \lambda) = \lambda x - x^2$ also satisfies the bifurcation condition (5.11). The changes in phase portraits, as λ is varied, are illustrated in Fig. 5.18.

Observe that this time $f(x, \lambda)$ has two roots for all values of λ, i.e., $x_e = 0$, and a nonzero root, $x_e = \lambda$. These roots are equilibrium points, which collide with one another exactly at $\lambda = 0$. Both equilibrium points exchange their stability as λ changes sign.

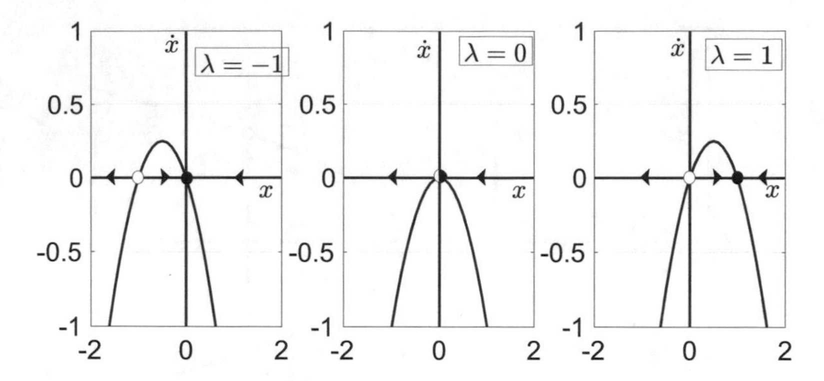

Fig. 5.18 Transcritical bifurcation

To make this description more precise, we compute below the derivative of f, to get:

$$\frac{\partial f}{\partial x} = \lambda - 2x.$$

To determine the stability properties of the equilibrium points, we evaluate the derivative of f at each of them, to get:

$$\frac{\partial f}{\partial x}\,|_{x=0} = \lambda,$$
$$\frac{\partial f}{\partial x}\,|_{x=\lambda} = -\lambda.$$

It follows that $x_e = 0$ is stable whenever $\lambda < 0$ and unstable for $\lambda > 0$. On the contrary, $x_e = \lambda$ is unstable whenever $\lambda < 0$ and unstable for $\lambda > 0$. At $\lambda = 0$ there is only one equilibrium point $x_e = 0$. Since $\dot{x} = -x^2 < 0$ then any solution trajectory with initial condition x_0 decreases along he x-axis.

The bifurcation diagram of Fig. 5.19 summarizes the location and stability properties of both branches of equilibrium points. As in the previous case of a saddle-node bifurcation, solid lines represent branches of stable equilibrium points while dashed lines correspond to unstable equilibria. At $\lambda = 0$ we have a bifurcation point at which two branches of equilibria exchange stability properties. This scenario, i.e., where two branches meet and exchange stability properties, is a distinctive feature of a transcritical bifurcation. As indicated at the beginning of this section, the laser beam model exhibits a transcritical bifurcation. The bifurcation occurs as the laser dynamics transitions from a regular "lamp" to a synchronized laser emission.

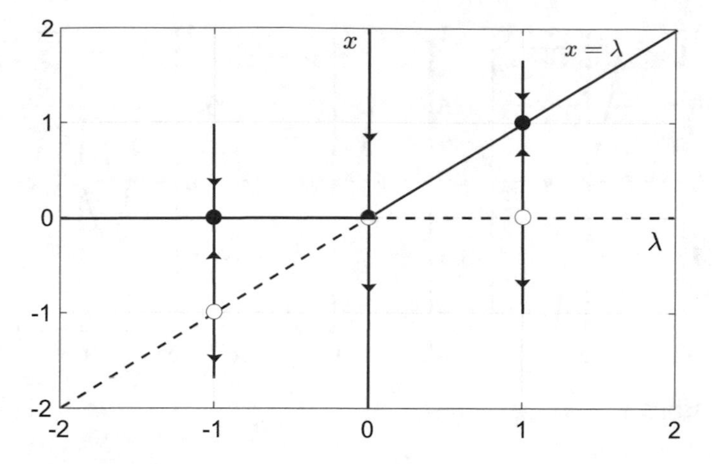

Fig. 5.19 Transcritical bifurcation diagram. Two branches of equilibrium points exchange stability properties at the critical bifurcation point $\lambda = 0$

5.5.3 *Pitchfork Bifurcation*

We now consider the revised Euler beam's model Eq. (5.8), which we rewrite to facilitate the description of thc analysis:

$$\boxed{\frac{dx}{dt} = \lambda x + ax^3,} \tag{5.29}$$

where, once again, $\lambda \in \mathbb{R}$ is the bifurcation parameter and a is a real constant that can be scaled to $a = \pm 1$. Without loss of generality, we choose $a = -1$. Recall that this equation serves as a phenomenological model for the buckling experiment in which an elastic beam is subjected to a compressive force. As usual, we start the analysis by verifying that $f(x, \lambda) = \lambda x - x^3$ satisfies the bifurcation conditions (5.11). The changes in the phase portraits are illustrated in Fig. 5.20.

For $\lambda \leq 0$, the graph of $f(\mathbf{x}, \lambda)$ suggests that there is only one equilibrium point $x_e = 0$. Since $\dot{x} < 0$ then $x_e = 0$ appears to be locally stable. For $\lambda > 0$ two new equilibria have appeared, one positive and one negative. As expected, these two new equilibria correspond to each of the two buckled states. Again, visual inspection suggests the nonzero equilibria to be locally stable but the zero one is now unstable. Next we verify this assertion more precisely. We compute the derivative of f:

$$\frac{\partial f}{\partial x} = \lambda - 3x^2,$$

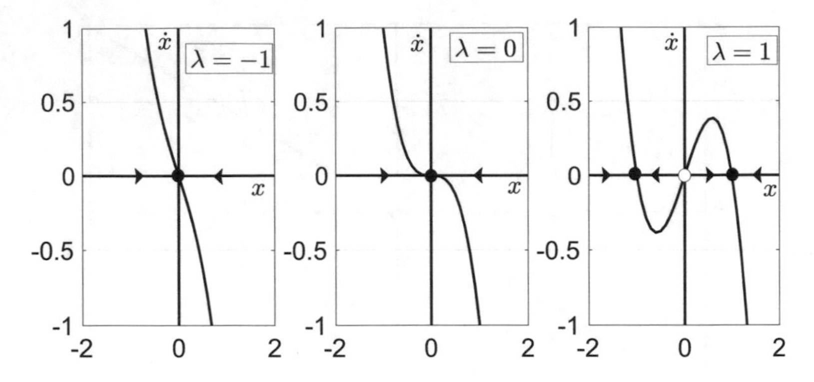

Fig. 5.20 Pitchfork bifurcation

and then evaluate it at the three equilibrium points to get:

$$\begin{aligned} \frac{\partial f}{\partial x}\Big|_{x=0} &= \lambda, \\ \frac{\partial f}{\partial x}\Big|_{x=\pm\sqrt{\lambda}} &= -2\lambda. \end{aligned}$$

Consequently, $x_e = 0$ is stable whenever $\lambda < 0$ and unstable otherwise. The two new equilibria $x_e = \pm\sqrt{\lambda}$ only exist when $\lambda > 0$ and they are always stable. Notice that when $\lambda = 0$, $\dot{x} = -x^3$, so if an initial condition satisfies $x_0 > 0$ then $\dot{x} < 0$ and the associated solution trajectory would decrease towards the zero equilibrium. Similarly, if $x_0 < 0$ then $\dot{x} < 0$ and the associated solution trajectory would increase towards the zero equilibrium.

What does this all mean for the experiment of the buckle beam?

From the experiment standpoint, when the beam buckles due to the compressive force the buckled states are always stable, so we can observe them. Which one of the two states would be observed in an actual experiment depends mainly on the actual conditions of the experiment, i.e., material imperfections of the beam and exact location of the force. Figure 5.21 summarizes the results of the analysis through the corresponding bifurcation diagram.

5.5.4 Nondegeneracy Conditions

In addition to the bifurcation conditions (5.11), it is also possible to determine **non-degeneracy conditions**, which can be used to specify the direction of the bifurcation. These conditions can be derived, for instance, by developing an asymptotic expansion for the locus of equilibrium points in the (x, λ) plane [22] or by consideration of the geometry of the graph of equilibrium points in the (x, λ) plane as well. Here, we choose the latter approach and apply it to the saddle-node bifurcation problem,

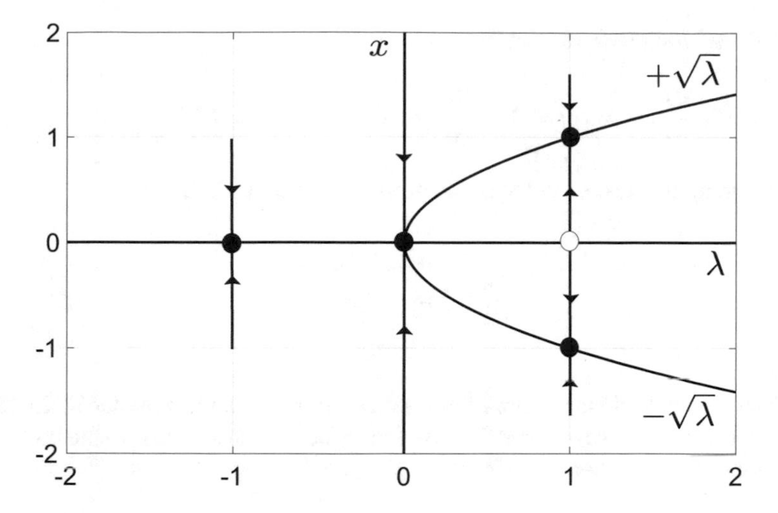

Fig. 5.21 Pitchfork bifurcation diagram. Two stable branches of equilibrium points appear at the critical bifurcation point $\lambda = 0$. At the same point, a trivial branch of equilibrium exchanges stability properties

though a similar process can be applied to the transcritical and pitchfork bifurcations. The calculations follow closely the process used by Wiggins in [6].

Consider again the bifurcation conditions (5.11) and the saddle-node bifurcation problem $\dot{x} = \lambda - x^2$, which arose within the context of the mathematical model of a fluxgate magnetometer. According to the bifurcation diagram of Fig. 5.17, there are two branches of equilibria $x = \pm\sqrt{\lambda}$ that lie locally on one side of $\lambda = 0$. If we interpret these two branches of solutions as a unique curve of equilibria of the form $\lambda(x)$, which is parameterized by x as it passes tangent to the bifurcation point $(0, 0)$, then the Implicit Function Theorem applies to f with the roles of x and λ reversed, that is

$$\frac{\partial f}{\partial \lambda}(0, 0) \neq 0. \tag{5.30}$$

Furthermore, the fact that $\lambda(x)$ lies locally on one side of $\lambda = 0$ is the same as saying that $\lambda(x)$ is either concave up or down as a function of x. That is, no inflection points. Together, these two conditions become

$$\frac{d\lambda}{dx}(0) = 0 \tag{5.31a}$$

$$\frac{d^2\lambda}{dx^2}(0) \neq 0. \tag{5.31b}$$

Now, Eq. (5.30) implies that we can write the solutions to $f = 0$ as:

$$f(x, \lambda(x)) = 0.$$

Differentiating with respect to x yields

$$\frac{df}{dx}(x, \lambda(x)) = 0 = \frac{\partial f}{\partial x}(x, \lambda(x)) + \frac{\partial f}{\partial \lambda}(x, \lambda(x))\frac{d\lambda}{dx}(x). \tag{5.32}$$

Evaluating this last equation at the bifurcation point (0, 0) we get

$$\frac{d\lambda}{dx}(0) = \frac{-\dfrac{\partial f}{\partial x}(0, 0)}{\dfrac{\partial f}{\partial \lambda}(0, 0)}.$$

Observe that the denominator of this last equation cannot be zero due to Eq. (5.30). The numerator is, however, exactly zero due to the second equation in the bifurcation conditions (5.11). It then follows that

$$\frac{d\lambda}{dx}(0) = 0.$$

This last result was expected because it just confirms what we already knew, i.e., that the curve of equilibria $\lambda(x)$ is tangent to the line $\lambda = 0$ at $x = 0$.

Differentiating once again Eq. (5.32) with respect to x yields

$$\begin{aligned}\frac{d^2 f}{dx^2}(x, \lambda(x)) = 0 = \frac{\partial f^2}{\partial x^2}(x, \lambda(x)) + 2\frac{\partial^2 f}{\partial x \partial \lambda}(x, \lambda(x))\frac{d\lambda}{dx}(x) +\\ 2\frac{\partial^2 f}{\partial \lambda^2}(x, \lambda(x))\left(\frac{d\lambda}{dx}(x)\right)^2 + \frac{\partial f}{\partial \lambda}(x, \lambda(x))\frac{d^2\lambda}{dx^2}(x).\end{aligned}$$

Evaluating this last equation at the bifurcation point (0, 0) we get

$$\frac{d^2\lambda}{dx^2}(0) = \frac{-\dfrac{\partial^2 f}{\partial x^2}(0, 0)}{\dfrac{\partial f}{\partial \lambda}(0, 0)}.$$

Again, the denominator of this last equation cannot be zero due to Eq. (5.30). The numerator cannot be zero either due to the concavity condition Eq. (5.31b).

In summary, in addition to the bifurcation conditions (5.11) we have also found two nondegenerate conditions

$$\begin{aligned}\frac{\partial f}{\partial \lambda}(0, 0) &\neq 0\\ \frac{\partial^2 f}{\partial x^2}(0, 0) &\neq 0.\end{aligned}$$

Table 5.2 Classification of codimension one bifurcations in continuous models

Bifurcation	$\frac{\partial f}{\partial \lambda}$	$\frac{\partial^2 f}{\partial x^2}$	$\frac{\partial^2 f}{\partial x \partial \lambda}$	$\frac{\partial^3 f}{\partial x^3}$	Normal form
Saddle-Node (fold)	$\neq 0$	$\neq 0$			$\dot{x} = \lambda \pm x^2$
Transcritical	0	$\neq 0$	$\neq 0$		$\dot{x} = \lambda x \pm x^2$
Pitchfork	0	0	$\neq 0$	$\neq 0$	$\dot{x} = (x^2 \pm \lambda)x$

These two conditions are listed in Table 5.2. Derivatives are evaluated at the bifurcation point, which can be assumed to be at $(x_e, \lambda_c) = (0, 0)$. A similar process can be applied to find out the nondegenerate conditions for transcritical and pitchfork bifurcation.

5.5.5 *Hopf Bifurcation*

All of the bifurcation examples we have studied so far have involved transitions between stationary or equilibrium points. These transitions occur at a codimension one bifurcation when a real-valued eigenvalue passes through zero. But it is also possible to have codimension one transitions when a complex-valued eigenvalues crosses the imaginary axis at nonzero speed. This type of transition is known as a Hopf bifurcation and it involves equilibrium points and oscillatory behavior. As a case study, we consider the Brusselator model Eq. (5.9). To investigate the emergence of periodic solutions, we compute the Jacobian matrix

$$J = \begin{bmatrix} 2XY - (1+\beta) & X^2 \\ -2XY + \beta & -X^2 \end{bmatrix}.$$

Evaluating at the equilibrium point $(X_e, Y_e) = (\alpha, \beta/\alpha)$ we get

$$J\left|_{(X_e,Y_e)}\right. = \begin{bmatrix} \beta - 1 & \alpha^2 \\ -\beta & -\alpha^2 \end{bmatrix}.$$

The characteristic polynomial can be written explicitly as

$$\sigma^2 - (\beta - \alpha^2 - 1)\sigma + \alpha^2 = 0.$$

Recall that when $\mathrm{Tr}(J) < 0$ and $\Delta = \mathrm{Tr}^2(J) - 4\det(J) < 0$ the equilibrium is a spiral sink. This is exactly what happens in the phase portrait of Fig. 5.5 (left). But when $\mathrm{Tr}(J) > 0$, while $\Delta > 0$, the equilibrium changes to an unstable spiral source, while a limit cycle oscillation emerges. This case corresponds to Fig. 5.5 (right).

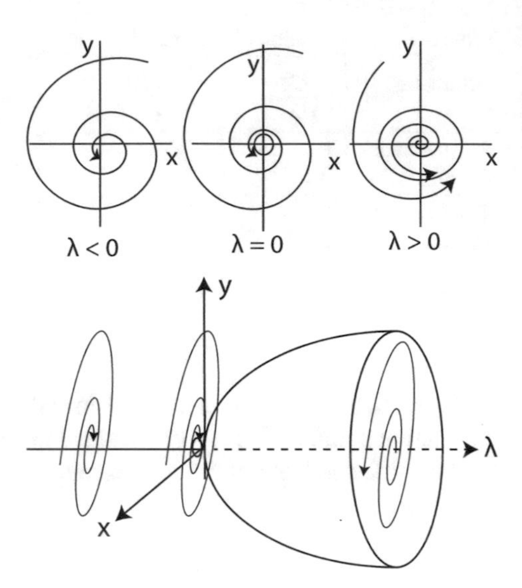

Fig. 5.22 Supercritical Hopf bifurcation

Thus, the condition for a Hopf bifurcation to occur is

$$\mathrm{Tr}(J) = 0, \quad \text{and} \quad \det(J) > 0.$$

Since $det(J) = \alpha^2 > 0$ is always satisfied then the Hopf bifurcation in the Brusselator model occurs along the curve $\beta = \alpha^2 + 1$. This is a one dimensional curve inside a two-parameter space. In other words, a codimension $2 - 1 = 1$ bifurcation. The bifurcation is classified as **supercritical Hopf bifurcation** because the equilibrium looses stability while the limit cycle becomes stable. The opposite scenario, in which a stable equilibrium gives way to an unstable limit cycle, is classified as a **subcritical Hopf bifurcation**. These two scenarios are illustrated in Figs. 5.22 and 5.23. The solid closed curve that appears, at $\lambda = 0$, for the supercritical bifurcation corresponds to stable periodic oscillations. The dashed closed curve that emerges in the subcritical case represents unstable periodic solutions.

The following theorem tells us how to detect the Hopf bifurcation in a general system and how to classify it in terms of supercritical or subcritical.

Theorem 5.8 (Hopf Bifurcation) *Consider the following planar system of ODEs*

$$\begin{aligned} \dot{x} &= f(x, y, \lambda) \\ \dot{y} &= g(x, y, \lambda) \end{aligned} \tag{5.33}$$

where λ is a distinguished bifurcation parameter. Without loss of generality, assume Eq. (5.33) has an equilibrium at $(0, 0)$. Let $F = (f, g)$ and assume the Jacobian matrix, $(dF)_{(0,0)}$, evaluated at the $(0, 0)$ equilibrium to have eigenvalues of the form $\sigma(\lambda) = \alpha(\lambda) \pm \beta(\lambda)$. Suppose also that at a critical value λ_c (which can also be assumed to be zero) the following conditions are satisfied:

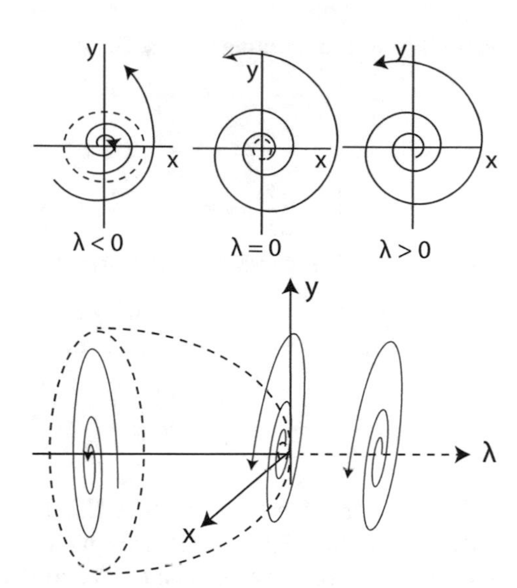

Fig. 5.23 Subcritical Hopf bifurcation

(i) $\alpha(0) = 0$, $\beta(0) = \omega \neq 0$, *where* $\text{sgn}\,\omega = \text{sgn} \left.\frac{\partial g}{\partial x}\right|_{\lambda=0} (0, 0)$

(ii) $\left.\frac{d\alpha(\lambda)}{d\lambda}\right|_{\lambda=0} = d \neq 0$

(iii) $a \neq 0$, *where*

$$a = \frac{1}{16}(f_{xxx} + f_{xyy} + g_{xxy} + g_{yyy}) + \frac{1}{16\omega}\left[f_{xy}(f_{xx} + f_{yy}) - g_{xy}(g_{xx} + g_{yy}) - f_{xx}g_{xx} + f_{yy}g_{yy}\right]$$

all derivatives are evaluated at $\lambda_c = 0$ *and at the equilibrium point* $(0, 0)$.

Then a unique curve of periodic solutions bifurcate from the origin into the region $\lambda > 0$ *if* $ad < 0$ *or* $\lambda < 0$ *if* $ad > 0$. *The origin is a stable fixed point for* $\lambda > 0$ *(resp.* $\lambda < 0$*) and an unstable fixed point for* $\lambda < 0$ *(resp.* $\lambda > 0$*) if* $d < 0$ *(resp.* $d > 0$*) while the periodic solutions are stable (resp. unstable) if the origin is unstable (resp. stable) on the side of* $\lambda = 0$ *where the periodic solutions exist. Alternatively, periodic solutions are stable when* $a < 0$ *and unstable when* $a > 0$. *The amplitude of the periodic orbits grows like* $\sqrt{\lambda}$ *while their periods tend to* $2\pi/|\omega|$ *as* $|\lambda|$ *tends to zero.*

Condition (ii) is known as the *transversality condition*. Geometrically, it means that the eigenvalues cross the imaginary axis with nonzero speed. Condition (iii) is known as the *genericity condition*.

Example 5.9 We continue now the analysis of the Brusselator model. We start by shifting the origin to the $(0, 0)$ through the substitution $x = X - \alpha$ and $y = Y - \beta/\alpha$.

After this shift of coordinates the Brusselator Eq. (5.9) become

$$\boxed{\begin{aligned} \frac{dx}{dt} &= (\beta - 1)x + \alpha^2 y + \frac{\beta}{\alpha}x^2 + x^2 y + 2\alpha xy \\ \frac{dy}{dt} &= -\beta x - \alpha^2 y - \frac{\beta}{\alpha}x^2 - x^2 y - 2\alpha xy. \end{aligned}} \tag{5.34}$$

At the Hopf bifurcation condition $\beta = \alpha^2 + 1$, the eigenvalues of the jacobian matrix $J\big|_{(X_e, Y_e)}$ are $\sigma = \pm\alpha i$. The corresponding eigenvector is:

$$V = \begin{bmatrix} \alpha^2 i \\ \alpha + (-\beta + 1)i \end{bmatrix}$$

We now transform the system of ODEs from $X = (x, y)$ coordinates to $U = (u, v)$ coordinates using the linear transformation $X = PU$, where the columns of the P matrix are the real and imaginary components of the eigenvectors associated with the eigenvalue $\sigma = \pm\alpha i$, explicitly:

$$P = \begin{bmatrix} 0 & \alpha^2 \\ \alpha & -\beta + 1 \end{bmatrix}.$$

Under this transformation into Eq. (5.34) becomes:

$$\begin{bmatrix} \dot{u} \\ \dot{v} \end{bmatrix} = \begin{bmatrix} 0 & -\alpha \\ \alpha & 0 \end{bmatrix} + \begin{bmatrix} (-3 + 3\beta + \alpha^2\beta - 3\alpha)v^2 \\ (2 - \beta)\alpha v^2 + \alpha^3 uv^2 + (1 - \beta)\alpha^2 v^3 + 2\alpha^2 uv \end{bmatrix} \tag{5.35}$$

Let

$$\begin{aligned} f(u, v) &= -\alpha v + (-3 + 3\beta + \alpha^2\beta - 3\alpha)v^2, \\ g(u, v) &= \alpha u + (2 - \beta)\alpha v^2 + \alpha^3 uv^2 + (1 - \beta)\alpha^2 v^3 + 2\alpha^2 uv. \end{aligned}$$

Direct computations of the partial derivatives of f and g yields the following nonzero derivatives:

$$f_{uv} = 1,\ f_{vv} = 2(-3 + 3\beta + \alpha^2\beta - 3\alpha^2),\ g_{vvv} = 6(1 - \beta)\alpha^2,\ g_{uv} = 2\alpha^2,$$

all other derivatives that are required to apply the Hopf theorem are zero. Direct substitution yields:

$$a = -\frac{\alpha^2(2 + \alpha^2)}{8} < 0.$$

Consequently, we conclude that the Hopf bifurcation that yields periodic oscillations in the Brusselator model is supercritical.

Example 5.10 A dimensionless version of the Colpitts oscillator that was introduced earlier on Chap. 4 has the following form:

$$\begin{aligned}
\frac{dx_1}{dt} &= \frac{g}{Q(1-\kappa)}\left[-\left(e^{-x_2}-1\right)+x_3\right] \\
\frac{dx_2}{dt} &= \frac{g}{Q\kappa}x_3 \\
\frac{dx_3}{dt} &= -\frac{Q\kappa(1-\kappa)}{g}(x_1+x_2)-\frac{1}{Q}x_3,
\end{aligned} \tag{5.36}$$

where Q represents the quality factor of the resonant network, see Fig. 4.50, while g is related to the gain in the circuit.

In this example we show that a Hopf bifurcation at $g = 1$ leads to the oscillatory behavior shown in Fig. 5.24.

The oscillations emerge from the trivial equilibrium (0, 0, 0). Thus, linearizing at this equilibrium point leads to the Jacobian matrix

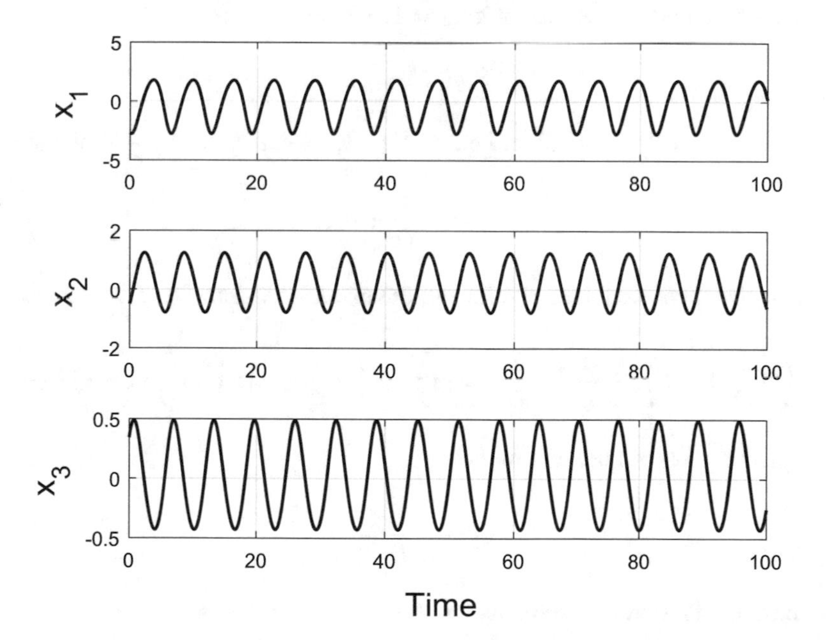

Fig. 5.24 Time series solution of the dimensionless Colpitts oscillator model illustrate periodic oscillations. These oscillations emerge via a Hopf bifurcation. Parameters are: $g = 1.1$, $Q = 1$, and $\kappa = 0.5$

$$J_{(0,0,0)} = \begin{bmatrix} 0 & \frac{g}{Q(1-k)} & \frac{g}{Q(1-k)} \\ 0 & 0 & \frac{g}{Qk} \\ -\frac{Qk(1-k)}{g} & -\frac{Qk(1-k)}{g} & -\frac{1}{Q} \end{bmatrix}.$$

Then the characteristic polynomial det $(J - \lambda I)$ yields the following cubic polynomial in λ

$$\lambda^3 + \frac{1}{Q}\lambda^2 + \lambda + \frac{g}{Q} = 0. \tag{5.37}$$

At $g = 1$, the eigenvalues are:

$$\lambda_1 = -\frac{1}{Q}, \qquad \lambda_{1,2} = 0 \pm i.$$

We cannot yet classify or even identify the bifurcation as a Hopf bifurcation without calculating the nondegenerate conditions indicated in Theorem 5.8. To do those calculations we first find an approximate analytical solution to the eigenvalues of the Jacobian matrix near the point $g = 1$. We set

$$g = 1 + \varepsilon,$$

where ε is a small perturbation, i.e., $\varepsilon \ll 1$. Equivalently, $\varepsilon = g - 1$. We start by setting

$$\lambda_1 = -\frac{1}{Q} + a_1\varepsilon,$$

where a_1 is an undetermined coefficient. Substituting λ_1 into Eq. (5.37) yields

$$\left(-\frac{1}{Q} + a_1\varepsilon\right)^3 + \frac{1}{Q}\left(-\frac{1}{Q} + a_1\varepsilon\right)^2 + \left(-\frac{1}{Q} + a_1\varepsilon\right) + \frac{1}{Q}(1+\varepsilon) = 0.$$

Collecting like powers of ε we get

$$a_1 = -\frac{Q}{Q^2+1}.$$

Thus, to a first-order approximation, the first eigenvalue is

$$\lambda_1 = -\frac{1}{Q} - \frac{Q}{Q^2+1}(g-1) = -\frac{1+gQ^2}{Q(Q^2+1)}.$$

We now perform a similar set of calculations with the remaining two eigenvalues. Since they are complex conjugate, it makes sense to set

$$\lambda_{2,3} = 0 \pm i + a_2\varepsilon.$$

Substituting $\lambda_{2,3}$ into Eq. (5.37) now yields

$$(i + a_2\varepsilon)^3 + \frac{1}{Q}(i + a_2\varepsilon)^2 + (i + a_2\varepsilon) + \frac{1}{Q}(1 + \varepsilon) = 0.$$

Collecting like powers of ε we get

$$a_2 = \frac{Q}{2\left(Q^2+1\right)} + \frac{1}{2\left(Q^2+1\right)i}.$$

Thus, to a first-order approximation, the remaining two eigenvalue can be written as

$$\lambda_{2,3} = \frac{Q}{2\left(Q^2+1\right)}(g-1) \pm \left[1 + \frac{1}{2\left(Q^2+1\right)}(g-1)\right]i.$$

We can now proceed to classify the bifurcation. Observe first that $\lambda_1 < 0$. This suggests that solutions converge to the eigenspace spanned by the eigenvectors associated with $\lambda_{2,3}$. Now, if we let $\mu = g - 1$ then the bifurcation point $g = 1$ corresponds to $\mu = 0$. From the expression above for $\lambda_{2,3}$ we see that $\sigma(\mu) = \alpha(\mu) \pm \beta(\mu)$, where

$$\alpha(\mu) = \Re(\lambda_{2,3}) = \frac{Q}{2\left(Q^2+1\right)}\mu, \qquad \beta(\mu) = \Im(\lambda_{2,3}) = 1 + \frac{1}{2\left(Q^2+1\right)}\mu.$$

Then we can immediately verify the first set of nondegenerate conditions, $\alpha(0) = 0$ and $\beta(0) = \omega = 1$. The second condition can be written as

$$\left.\frac{d\alpha(\mu)}{d\mu}\right|_{\mu=0} = d = \frac{Q}{2\left(Q^2+1\right)} > 0, \quad \forall Q > 0.$$

The third nondegenerate condition yields

$$a = -\frac{1}{16}\frac{Q^5}{(1+4Q^2)(1+Q^2)^2} < 0.$$

It follows from Theorem 5.8 that a supercritical Hopf bifurcation in the Colpitts oscillator model (5.36) occurs at $g = 1$. Figure 5.25 shows a numerical computation of a two-parameter bifurcation diagram.

Example 5.11 We consider again the FitzHugh-Nagumo model of neuron excitability. For completeness purposes, we re-write the model as

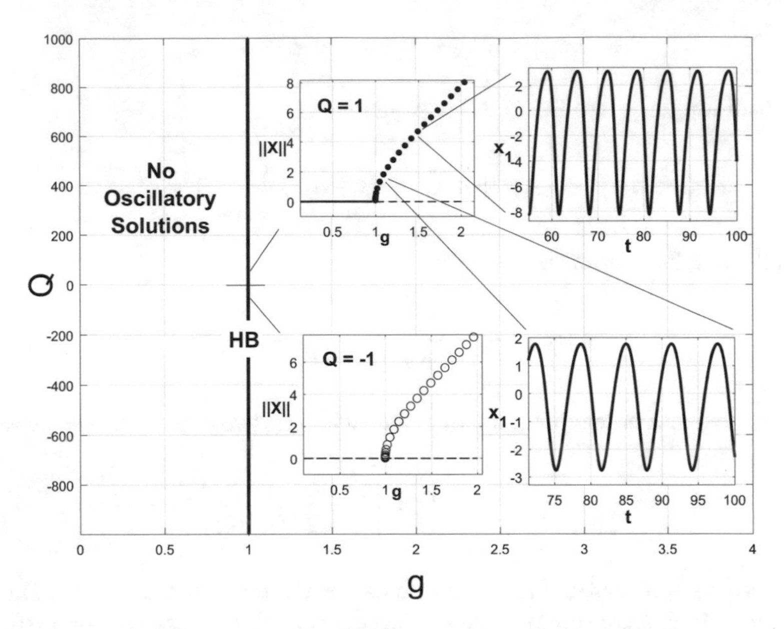

Fig. 5.25 Two-parameter bifurcation diagram of a simplified model of a Colpitts oscillator. The locus of a Hopf bifurcation that leads to limit cycle oscillation is located at $g = 1$, for all values of Q. For values of $Q < 0$ a subcritical Hopf bifurcation at $g = 1$ produces unstable limit cycles. The physically relevant case, $Q > 0$, yields a supercritical Hopf bifurcation at $g = 1$, which produces stable limit cycles. Dashed lines represent unstable equilibrium points; solid lines represent stable equilibria; open circles denote unstable limit cycles; closed circles represent stable limit cycles. The stable limit cycles become more sinusoidal as g is chosen near the Hopf point. This can be seen in the subfigures, which show the time evolution of the oscillations

$$\begin{aligned} \frac{dx}{dt} &= x - \frac{x^3}{3} - y + I \\ \frac{dy}{dt} &= \frac{1}{c}(x + a - by). \end{aligned} \tag{5.38}$$

Recall that x represents the membrane potential of the neuron. Since y describes slow currents, we make the assumption here that $c \gg 1$. In Chap. 4 we showed that if the parameter b lies in the interval $0 < b < 1$ then the model Eq. (5.38) admits one unique equilibrium point.

Let's assume (x_e, y_e) to represent this equilibrium point. The Jacobian evaluated at this equilibrium point is

$$J\,\Big|_{(x_e,y_e)} = \begin{bmatrix} 1 - x^2 & -1 \\ \dfrac{1}{c} & -\dfrac{b}{c} \end{bmatrix}.$$

We know the condition for a Hopf bifurcation is

$$\mathrm{Tr}(J) = 0, \quad \text{and} \quad \det(J) > 0.$$

These conditions are satisfied when

$$(1 - x_e^2) - \frac{b}{c} = 0, \qquad -\frac{b}{c}(1 - x_e^2) + \frac{1}{c} > 0,$$

which implies

$$x_e^2 = 1 - \frac{b}{c}, \qquad x_e^2 > 1 - \frac{1}{b}.$$

Since $0 < b < 1$, the inequality above is always satisfied. It follows that the Hopf bifurcation occurs along the curve

$$x_{HB} = \pm\sqrt{1 - \frac{b}{c}}. \tag{5.39}$$

Equation (5.39) represents the loci of the Hopf bifurcation parameterized by the constants b and c. Observe that for very large values of c, the term b/c vanishes. In this limit case, the loci reduce to $x_{1,2} = \pm 1$, which correspond to the maximum and minimum values of the cubic nullcline $y = x - x^3/3$. In fact, the assumption $c \gg 1$ implies that $b/c < 1$, so that the Hopf bifurcation condition is always satisfied.

We now attempt to find a lower bound on a for the Hopf bifurcation to occur. To do this, we consider again (see Chap. 4) the location of the equilibrium point, which along the x-axis it should satisfy

$$x^3 + px + q = 0, \tag{5.40}$$

where

$$p = -\frac{3(b-1)}{b}, \quad q = \frac{3a}{b} - 3I.$$

The stable equilibrium point, $x_e(a, b)$, should exist when $x_e(a, b) < x_{HB}$. Solving Eq. (5.40) for a and equating with Eq. (5.39), yields the desired expression for the lower bound on a. Assuming the limit case $b/c \ll 1$, the expression for the bound becomes

$$a > -\frac{2}{3}b + 1. \tag{5.41}$$

5.6 Global Bifurcations

The previous section examined generic local bifurcations, which relied on the variation of a single parameter changing the qualitative behavior near an equilibrium. However, mathematical models can change behavior in phase space on a much larger

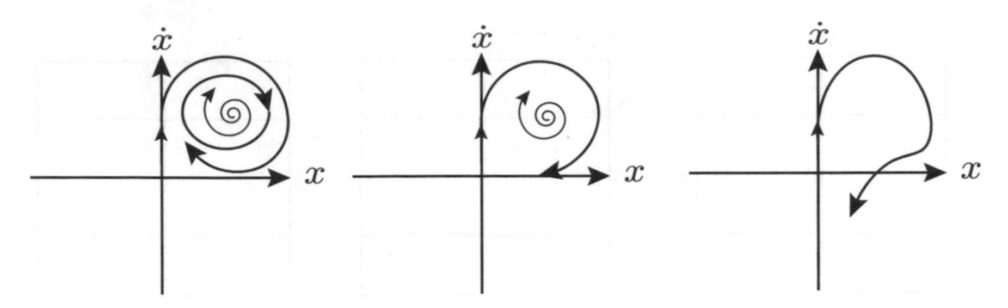

Fig. 5.26 The left phase portrait shows a saddle point at the origin with its unstable manifold approaching a limit cycle. The middle phase portrait shows the unstable manifold connecting to the stable manifold, a homoclinic orbit at some critical parameter value. As the parameter varies more, the right phase portrait shows the limit cycle disappearing

scale, and these *global bifurcations* result in qualitative behavior changes in the topology of the solution trajectories, which are not confined to a local neighborhood. There are a number of these types of bifurcations; however, this section only discusses a few types and gives a model from diabetes for one type of global bifurcation.

Global bifurcations occur for flows from a model when small changes in a parameter result in one invariant set changing into a topologically different invariant set, which is not restricted to a small neighborhood in phase space. One example of this is a *homoclinic bifurcation*, where a saddle point and a limit cycle collide as a parameter varies. Figure 5.26 shows a homoclinic bifurcation, where as a parameter varies, a limit cycle collides with a saddle point. At a critical parameter value, the unstable manifold of the saddle node connects with its stable manifold, creating a *homoclinic orbit*. For smaller values of the parameter, the phase portrait on the left exists with a stable periodic orbit having the unstable manifold of the saddle node converging to the limit cycle. For larger parameter values, the phase portrait on the right shows the loss of the periodic orbit.

A *heteroclinic bifurcation* occurs when a limit cycle collides with two or more saddle points. The classic example of this type of bifurcation is seen in the phase portraits of a pendulum as seen in Fig. 5.27. The governing ODE model for the pendulum is given by:

$$\ddot{\theta} + \delta\dot{\theta} + \frac{g}{l}\sin(\theta) = 0, \tag{5.42}$$

where g is gravity, l is the length of the pendulum, δ is viscous damping, and θ is the angular displacement. There are infinitely many saddle nodes when the pendulum is vertically up with the unstable manifolds of these equilibria connecting to the stable manifolds of a neighboring vertically up equilibria for the undamped pendulum ($\delta = 0$). These heteroclinic orbits surround the periodic solutions of the undamped pendulum. When damping ($\delta > 0$) is included, this neutrally stable behavior changes to the attracting equilibria where the pendulum is attracted to one of its resting positions (down).

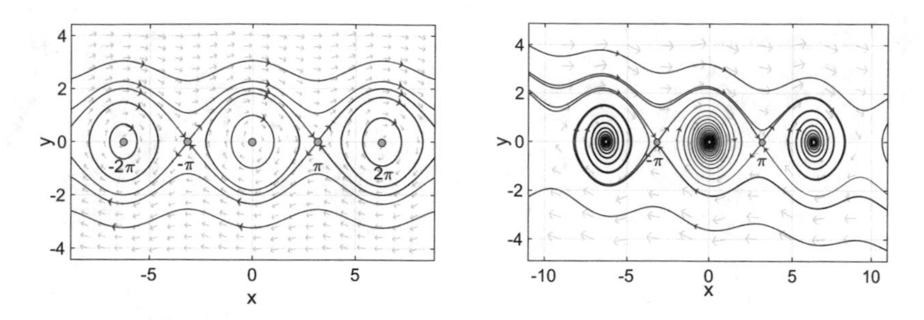

Fig. 5.27 This figure shows phase portraits for the pendulum equation (5.42), exhibiting a heteroclinic bifurcation. The phase portrait on the left is the undamped pendulum, while the one on the right includes damping

One can easily see how these situations become complex and cause a variety of different behaviors. In three or more dimensions, the stable and unstable manifolds can become tangled and can result in what are called *chaotic attractors*. This results in the interesting behavior of models that demonstrate chaos, which is very unpredictable behavior.

5.7 The Role of Symmetry

Symmetry is a geometrical concept that describes the set of transformations that leave an object unchanged. In models of systems with either discrete or continuous behavior, the *objects* are the governing equations, which typically consist of systems of difference equations, ordinary differential equations (ODEs) or partial differential equations, and the transformations are the changes in the underlying variables that leave the equations unchanged.

From a modeling standpoint, symmetry has been recognized for a long time as being an important principle underlying the behavior of many physical systems. Regardless of the type of equations, the set of transformations that leave a model unchanged form an abstract **group**. In this sense, we can say that symmetry is encoded in mathematical models through group theory. Thus, we need to define formally the concept of a group.

Definition 5.4 A group is a set G with an associative operation $G \times G \to G$, i.e., $(ab)c = a(bc)$, $\forall a, b, c \in G$, such that the identity $e \in G$ exists, i.e., $eg = ge = g$, $\forall g \in G$. Inverses $g^{-1} \in G$ also exist $\forall g \in G$, such that $gg^{-1} = e = g^{-1}g$.

Example 5.12 (*The symmetry group of a pentagon*) The symmetries of the pentagon of Fig. 5.28 are described by the dihedral group $\mathbf{D}_5$ of order 10:

$$\mathbf{D}_5 = \{e, \rho, \rho^2, \rho^3, \rho^4, \kappa, \kappa\rho, \kappa\rho^2, \kappa\rho^3, \kappa\rho^4\},$$

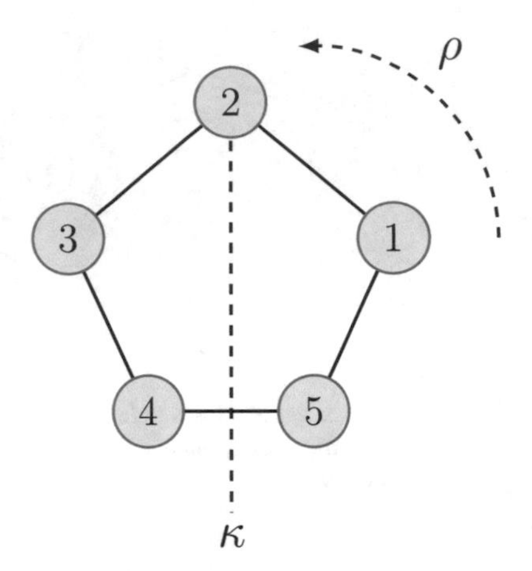

Fig. 5.28 Symmetries of a pentagon are described by the dihedral $\mathbf{D}_5$ group

where ρ represents a rotation by $2\pi/5$ and κ is a reflection across the line that runs between any of the vertices and the middle point of the opposite line. It can be shown that all symmetries of the pentagon result from the combination of rotations and reflections. Thus, we say the group $\mathbf{D}_5$ is generated by ρ and κ and write $\mathbf{D}_5 = \{\rho, \kappa\}$.

In fact, if all the elements of a group can be written as the product of a smaller subset then those elements in the subset are called the **generators** of the group, and the group itself is said to be generated by the subset.

Example 5.13 (*The symmetry group of a directed ring*) The symmetries of an unidirectionally connected ring with N elements, see Fig. 5.29, are described by the group $\mathbf{Z}_N$ of order N, which describes cyclic permutations of N objects $\{1, \dots, N\}$. The elements of the groups can be written as

$$\mathbf{Z}_N = \{e, \rho, \rho^2, \dots, \rho^{n1}\},$$

where $\rho = 2\pi/N$. In this case, all elements of the group are generated by a single element ρ, so we can write $\mathbf{Z}_N = \{\rho\}$.

The order of any group element γ is the minimum integer n such that γ^n. Observe in our previous example that $\rho^N = e$, so ρ has order N. However, consider for instance the case $N = 4$, so that $\mathbf{Z}_4 = \{e, \rho, \rho^2, \rho^3\}$, where $\rho = 2\pi/4$. Notice that $(\rho^2)^2 = e$. Thus ρ^2 has order 2 (Fig. 5.29).

In the example above, we defined the generator ρ of the group $\mathbf{Z}_n$ as a counterclockwise rotation around the ring. Without loss of generality, the same results can be obtained if the rotation is defined clockwise. That is, the groups $\mathbf{Z}_N = \{\rho\}$ and $\mathbf{Z}_N = \{-\rho\}$ are **isomorphic** groups.

Example 5.14 (*The general linear group* $\mathbf{GL}(n)$) The set of all $n \times n$ real and invertible matrices form the general linear group $\mathbf{GL}(n)$. This set is a group under matrix multiplication. Observe that the product of any two invertible $n \times n$ matrices

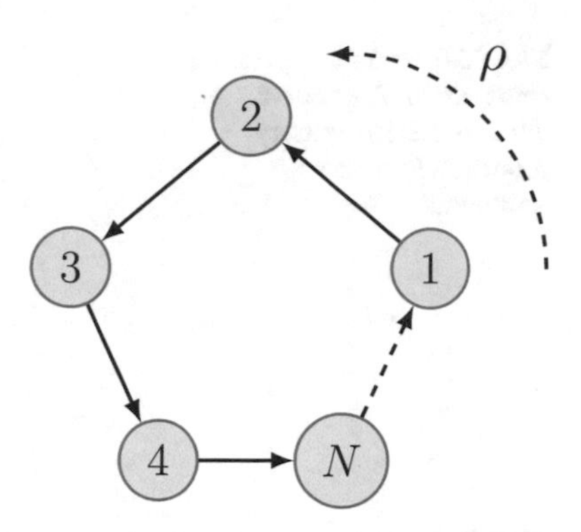

Fig. 5.29 Symmetries of an unidirectionally connected ring with N elements

is an associative operation, which yields an invertible $n \times n$ matrix, so the group is closed under multiplication. There exists an identity element, i.e., the identity matrix I_n, which is also invertible. Finally, every matrix in $\mathbf{GL}(n)$ is invertible. Consequently, the set $\mathbf{GL}(n)$ forms a group.

Definition 5.5 Two groups Γ and Δ are isomorphic, written as $\Gamma_1 \cong \Delta$ if there exists a bijective (one-to-one and onto) map $h : \Gamma \to \Delta$, such that

$$h(\gamma_1\gamma_2) = h(\gamma_1)h(\gamma_2) \in \Delta, \quad \forall \gamma_1, \gamma_2 \in \Gamma.$$

Example 5.15 As another example, consider the dihedral group $\mathbf{D}_3$ of symmetries of a triangle and the group $\mathbf{S}_3$ of all permutations of 3 objects. These two groups are isomorphic to one another, so $\mathbf{D}_3 \cong \mathbf{S}_3$.

Later on, we will describe the set of transformations that leave a mathematical model unchanged through **Lie** groups. By a Lie group we mean a closed subgroup of $\mathbf{GL}(n)$, so the general linear group will be a very important group for a modeling standpoint. But first, we finish this section with on more example of a group.

Example 5.16 (*The symmetry group of a ring pattern*) Consider a ring pattern as is shown in Fig. 5.30. The ring remains unchanged under continuations rotations by an arbitrary angle θ and by reflections (represented by κ) across the plane dissecting through the middle of the ring. These two operations, rotations and reflections on the plane generate the continuous group $\mathbf{O(2)}$ of symmetries of a ring, i.e., $\mathbf{O(2)} = \{\theta, \kappa\}$.

It turns out that, in many cases, one can use the underlying group of symmetries of a physical system to derive mathematical models and, more importantly, predict the behavior of the system without the need for computer simulations or experimental work. This approach works well even if the system does not exhibit exact symmetries. But, first, we need to formalize what we mean by symmetries of a model and how groups are used to describe the symmetries of the corresponding model equations.

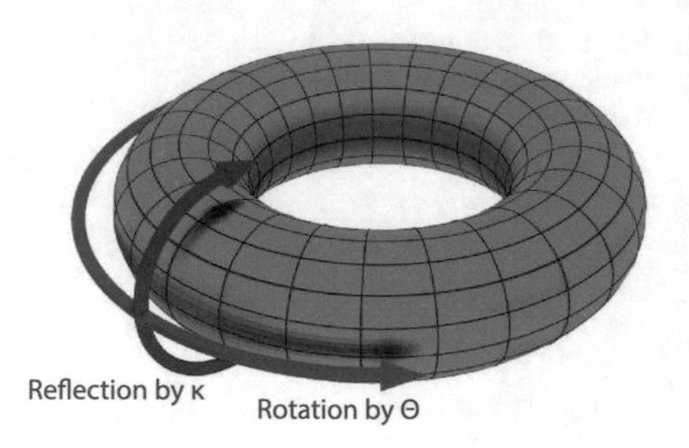

Fig. 5.30 Symmetries of a ring pattern. Blue curve corresponds to arbitrary rotations. Red curve represents reflections

5.7.1 Continuous Models with Symmetry

We now start to formalize the ideas of how symmetry can be incorporated into a modeling approach. First, we define what it means for a model to be symmetric. Then, we study in more detail how symmetry might affect the solution sets of a mathematical model.

Definition 5.6 Consider, for instance, a system with continuous behavior, modeled through the following system of ODEs

$$\frac{dx}{dt} = f(x, \lambda), \tag{5.43}$$

where $x \in \mathbb{R}^n$, $\lambda \in \mathbb{R}^p$ is a vector of parameters and $f : \mathbb{R}^n \times \mathbb{R}^p \to \mathbb{R}$ is a smooth function. Let Γ be a compact **Lie group** acting on $V = \mathbb{R}^n$. The mathematical model (5.43) is said to have Γ-symmetry if

$$f(\gamma x, \lambda) = \gamma f(x, \lambda), \tag{5.44}$$

for all $x \in \mathbb{R}^n$ and for all $\gamma \in \Gamma$. By a Lie group we mean a closed subgroup of $\mathbf{GL}(n)$, the group of all invertible linear transformations of the vector space $\mathbb{R}^n$ into itself. That is, Γ is a matrix group. By compact we mean matrices with bounded entries.

Certain groups of symmetries arise more frequently in mathematical models. Examples include: $\mathbf{D}_n$, the dihedral group of order $2n$, which describes rotation and reflection symmetries of an n-gon; $\mathbf{Z}_n$, the cyclic group of order n, which describes rotational symmetries of an n-gon; $\mathbf{S}_n$, the group of all possible permutations of n objects; $\mathbf{O}(n)$, the orthogonal group of $n \times n$ matrices A that satisfy $AA^T = \mathbf{I}$; $\mathbf{S}(n)$, the special orthogonal group which also satisfies $\det(A) = 1$; $\mathbf{S}^1$, the circle group; and the n-torus $\mathbf{T}^n = \underbrace{\mathbf{S}^1 \times \ldots \times \mathbf{S}^1}_{n \text{ times}}$.

In the next section and chapter we will discuss various examples that involve these groups.

Whenever Eq. (5.44) is satisfied, the function f is said to be **Γ-equivariant**. More importantly, Γ-equivariance implies that if $x(t)$ is a solution of (5.43) then so is $\gamma x(t)$ for all $\gamma \in \Gamma$. To verify this claim, we substitute $\gamma x(t)$ in Eq. (5.43)

$$\frac{d\{\gamma x(t)\}}{dt} = f(\gamma x(t), \lambda),$$

and since γ is simply a constant transformation we get

$$\gamma \frac{dx}{dt} = \gamma f(x(t), \lambda),$$

where the right-hand is obtained through the Γ-equivariant property of f. This shows that $\gamma x(t)$ satisfies the ODE (5.43). And it also shows that the entire model is Γ-symmetric.

In fact, the collection of points $\gamma x(t)$, for all $\gamma \in \Gamma$, forms a set called the **group orbit** of Γ:

$$\Gamma x = \{\gamma x : \gamma \in \Gamma\}.$$

Furthermore, the concept of group orbit applies to any point $x(t)$, not just equilibrium solutions.

Example 5.17 As an example, consider the van der Pol circuit depicted in Fig. 5.31. I_L and I_C are the currents across the inductor L and capacitor C, respectively. I_R is the current across two resistors R_1 and R_2 located inside the rectangle labeled R in which $F(V) = -V/R_1 + V^3/(3R_2^2)$.

The dynamics of the circuit shown in Fig. 5.31, after rescaling, is governed by the following second order scalar ODE

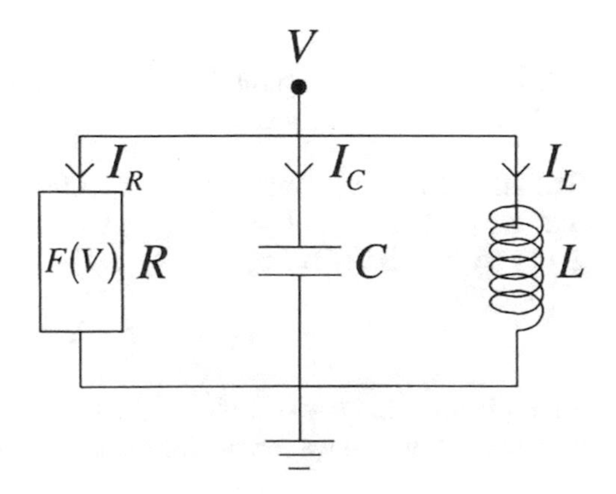

Fig. 5.31 Circuit realization of a Van der Pol oscillator. The governing equations exhibit reflectional symmetry with respect to the state variables

$$\boxed{\frac{d^2V}{dt^2} - \delta(p - V^2)\frac{dV}{dt} - \omega^2 V = 0,} \tag{5.45}$$

where $\delta = 1/(R_2C)$, $p = R_2/R_1$, $\omega = 1/\sqrt{LC}$. After a change of variables, we can rewrite the model equation (5.45) as a first order system of the form

$$\begin{aligned} \frac{dx}{dt} &= \delta\left(px - \frac{x^3}{3}\right) + \omega y \\ \frac{dy}{dt} &= -\omega x, \end{aligned} \tag{5.46}$$

where $x(t) = V(t)$. We can then find two transformations that leave this system unchanged: the identity transformation $\gamma_1 = id$, where $\gamma_1(x, y) \mapsto (x, y)$, and a second transformation, which can be described abstractly as $\gamma_2 = -1$, so that $\gamma_2(x, y) \mapsto (-x, -y)$. The identity transformation is always a symmetry of any system, while the second transformation γ_2 corresponds to a reflection through the origin in the phase space $\mathbb{R}^2$. Furthermore, it can be shown that γ_1 and γ_2 are the only transformations that leave (5.46) unchanged. Together, γ_1 and γ_2 form the group $\mathbf{Z}_2 = \{\gamma_1, \gamma_2\}$ of symmetries of the Van der Pol oscillator (5.46).

5.7.2 *Isotropy Subgroups*

In addition to the symmetries of a model, individual solutions can also exhibit symmetry. For instance, consider equilibrium solutions x_e of Eq. (5.43), which we already know satisfy

$$f(x_e) = 0.$$

The symmetries of equilibrium points of a Γ-equivariant ODE form a subgroup of Γ, which we define next.

Definition 5.7 Let x_e represent an equilibrium or steady-state solution of a Γ-equivariant system of ODEs. The symmetries of x_e form the **isotropy subgroup** Σ of Γ, which is defined by

$$\Sigma_{x_e} = \{\gamma \in \Gamma : \gamma \cdot x_e = x_e\}. \tag{5.47}$$

Example 5.18 Let's revisit the original Euler beam's model (5.7) and the reduced version (5.8), which we rewrite for the purpose of this section:

$$\begin{aligned} &EI\theta''(x) + P\sin\theta(x) = 0, && \text{original model} \\ &\frac{dx}{dt} = \lambda x - x^3, && \text{reduced order model.} \end{aligned}$$

Recall that x is the material coordinate, $\theta(t)$ is the deflection angle between the undeformed rod and the tangent of the deformed rod, E is the elastic modulus, I is moment of inertia, P is the compressive force and L is the length of the beam. Both models possess reflectional symmetry: $\theta \mapsto -\theta$ in the original model or $x \mapsto -x$ in the reduced model. And in both cases the symmetry is described by the group $\mathbf{Z}_2$, which we write in abstract form as $\Gamma = \mathbf{Z}_2 = \{1, -1\}$. Notice that this is the same group of symmetries of the Van der Pol model, except that now the group acts on θ or x instead of (x, y). Nevertheless, in abstract form it is exactly the same group.

In Sect. 5.1 we indicated that the Lyapunov Schmidt Reduction procedure was used to derive the reduced order model. We now show that this can also be accomplished on the grounds of the $\mathbf{Z}_2$–symmetry alone. Consider the schematic diagram in Fig. 5.32 of the experimental apparatus that was introduced earlier on through Fig. 5.3.

We are looking for a mathematical model in the form of a system of ODEs

$$\frac{dx}{dt} = f(x, \lambda), \tag{5.48}$$

for the evolution of the material coordinate $x(t)$ as a function of time. We know equilibria of the model are solutions of $f(x, \lambda) = 0$. By $\mathbf{Z}_2$–symmetry, we know from Fig. 5.32 that if x is a solution so is $-x$, that is:

$$f(x, \lambda) = 0, \quad \overset{Symmetry}{\Longrightarrow} \quad f(-x, \lambda) = 0.$$

To be consistent with the schematic diagram, we can also assume that f is an odd function in x, i.e., $f(-x, \lambda) = -f(x, \lambda)$, which implies that $f(0, \lambda) = 0$, so that $x = 0$ is indeed an equilibrium (the unbuckled state). Since f is odd, we can write

$$f(x, \lambda) = b(x, \lambda)x$$

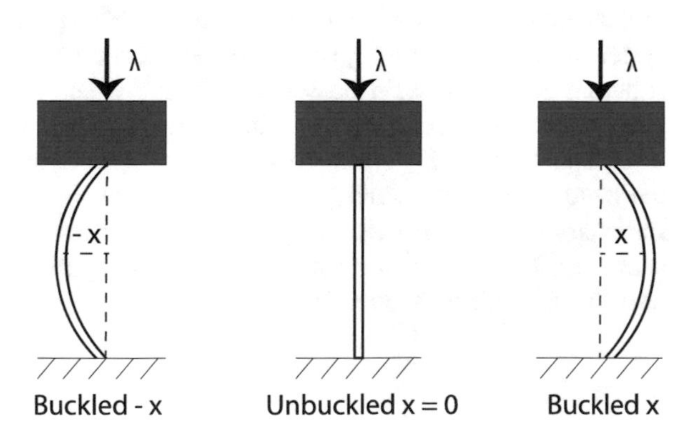

Fig. 5.32 Schematic diagram of Euler Beam's experiment

where b is even in x, i.e., $b(-x,\lambda) = b(x,\lambda)$. Let $b(x,\lambda) = a(x^2,\lambda)$. Then a Taylor expansion of f yields

$$f(x,\lambda) = a(x^2,\lambda)x \stackrel{Taylor}{=} \left(a(0,0) + a_{x^2}(0,0)x^2 + a_\lambda(0,0)\lambda + \ldots\right) x$$

Observe that the first of the two bifurcation conditions (5.11), i.e, $f(0,0) = 0$, is already satisfied by f. The second condition, $f_x(0,0) = 0$ implies $a(0,0) = 0$. Thus, the Taylor expansion for f takes the form

$$f(x,\lambda) = \left(\alpha x^2 + \beta\lambda + \ldots\right) x,$$

where $\alpha = a_{x^2}(0,0)$ and $\beta = a_\lambda(0,0)$. Substituting into Eq. (5.48) we get

$$\frac{dx}{dt} = \beta\lambda x + \alpha x^3.$$

Re-scaling time by $\tau = \beta t$ and setting $\gamma = \alpha/\beta$ yields the reduced order model

$$\frac{dx}{d\tau} = \lambda x + \gamma x^3. \tag{5.49}$$

As expected, Eq. (5.49) remains unchanged under the change of coordinates $x \mapsto -x$. The model equation is said to exhibit reflectional symmetry, which is a direct consequence of the experimental set up. Now, the unperturbed unbuckled state ($\theta = 0$ or $x = 0$) is the trivial solution with Γ-symmetry. Assuming boundary conditions $\theta(0) = \theta(L) = 0$, two nontrivial solutions of the original model are $\pm\sin(\pi x/L)$ and for the reduced order model we already know two nontrivial solutions $\pm\sqrt{\lambda}$. In both cases, the *isotropy subgroup* of these nontrivial solutions is the trivial group **1**.

The underlying mechanism that leads to the emergence of the buckled states in the Beam models is known as **spontaneous symmetry-breaking bifurcation**. The fundamental principle is that changes in a model parameters induce the physical system to transition into a new state of less symmetry, encapsulated by the isotropy subgroup Σ. Of course, in the Beam model this means that the system transitions into a state with no symmetry since Σ is the trivial subgroup. It is still the simplest example of an spontaneous $\mathbf{Z}_2$ symmetric-breaking bifurcation, which serves to illustrate the main ideas. Interestingly, most problems involve transitions into states (either stationary or dynamic ones) where the isotropy subgroup is other than the trivial subgroup. This transition is formally defined as symmetry breaking and we present next the main ideas in more detail.

5.8 Symmetry-Breaking Bifurcations

Equivariant systems of ODEs always posses a trivial solution x_0 whose isotropy subgroup is the entire group of symmetries of the model equations [23–25]. That is, $\Sigma_{x_0} = \Gamma$. But as parameters are varied the system can exhibit a new solution with less symmetry. That is, $\Sigma_x \subset \Gamma$. When this happens, it is then said that the system has undergone a *spontaneous symmetry-breaking bifurcation*. We consider in this chapter two types of symmetry-breaking bifurcations, steady-state and Hopf bifurcations. The former case leads to new equilibrium solutions while the latter to periodic oscillations. We describe next each of these two cases.

5.8.1 Steady-State Bifurcations

To illustrate the concept of spontaneous-symmetry breaking bifurcations to steady-states, we employ two examples, which, from a scientific point of view, e.g., physics and chemistry, they appear to be completely unrelated. One deals with the formation of a crown-like structure over a glass filled with milk. The other, is about the formation of cellular flame patterns over a burner. From a geometric standpoint, these two systems or experiments have a lot more in common, however. Let's take a look in more detail and find out what happens in each experiment when their underlying circular symmetry is broken.

Example 5.19 (*Milk Drop Coronet*) The iconic picture in Fig. 5.33 of the milk drop coronet was captured with the aid of the pioneering work on speed photography by Harold E. Edgerton. The picture illustrates best the phenomenon of steady-state symmetry-breaking bifurcation.

The pool of milk in its unperturbed or trivial state is symmetric under arbitrary rotations and reflections on a plane, which form the orthogonal group $\mathbf{O(2)}$. The perturbation by the droplet breaks, however, the $\mathbf{O(2)}$ symmetry of the trivial solution and it induces a crown-like shape with lesser symmetry. The 24-sided polygon that appears by joining the individual clumps now has $\mathbf{D}_{24}$-symmetry, where $\mathbf{D}_N$ is the dihedral group of symmetries of an N-gon.

Example 5.20 (*Premixed Flame Dynamics*) A mixture of either isobutane and air, or propane and air, are burned on a circular porous plug burner in a low pressure (0.3 to 0.5 atm) combustion chamber, see Fig. 5.34 (left). The process allowed for control of the pressure, flow rate, and fuel to oxidizer ratio to within 0.1%. The simplest cellular pattern generated by the burner is a large single cell with $\mathbf{O(2)}$ symmetry, as is shown in Fig. 5.34 (top-right).

Changes in the experimental parameters (type of fuel, pressure, total flow, and equivalence ratio) lead to spontaneous $\mathbf{O(2)}$ symmetry-breaking bifurcations and, as a result, stationary states with less symmetry appear. In this case, a stationary pattern with $\mathbf{D}_6$ symmetry is shown in the bottom part of the figure.

Fig. 5.33 This iconic picture of the milk drop illustrates the phenomenon of symmetry-breaking bifurcations. An unperturbed pool of milk is invariant under arbitrary rotations and reflections on a plane, which form the orthogonal group $\mathbf{O(2)}$. The crown-like shape that emerges under the perturbation by the droplet is a 24-sided polygon whose symmetryes are described by the dihedral group $\mathbf{D}_{24}$. Source: Harold E. Edgerton, Milk Drop Coronet, 1957. 2010 Massachusetts Institute of Technology

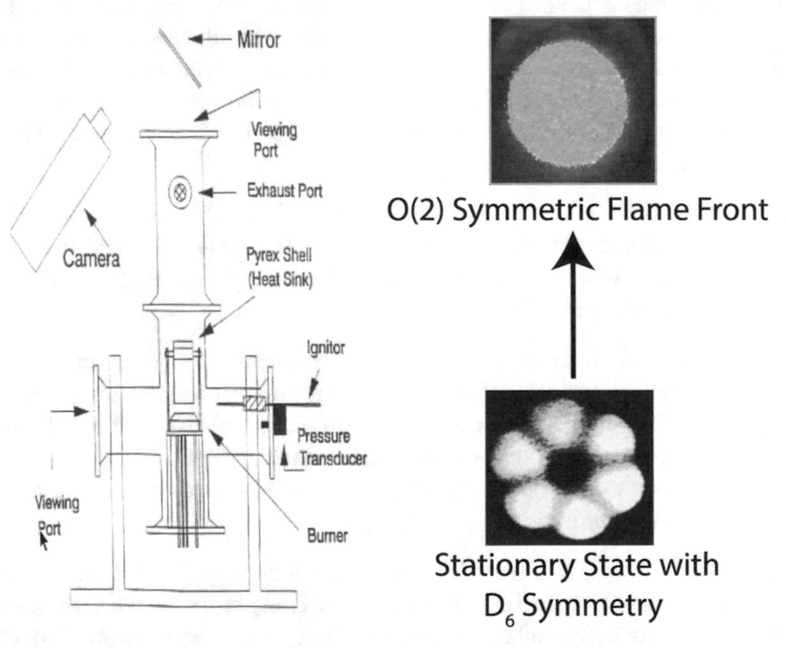

Fig. 5.34 Combustion experiments conducted by M. Gorman, et al., at the University of Houston [26] showcase cellular flame pattern instability. (Left) Experimental apparatus. (Top-right) Simplest pattern that appears is a homogeneous flame front with the same $\mathbf{O(2)}$ symmetry as that of the circular burner. (Bottom-right) spontaneous symmetry-bifurcations lead to a stationary cellular flame pattern with $\mathbf{D}_6$–symmetry

At first glance, the two experiments are completely unrelated. But, as we said before, they have a lot more in common. They are both experiments carried out on a system that has circular symmetry. Thus, it is not surprising that two distinct experiments produce the same type of stationary patterns, one with d_{24} symmetry and the other with $\mathbf{D}_6$ symmetry because both types of patterns belong to the same class of $\mathbf{D}_n$ symmetric solutions that emerge when $\mathbf{O(2)}$ symmetry is broken. It is in this context that we refer to the two problems and their solutions as being **model-independent**.

5.8.2 *Equivariant Branching Lemma*

We have discussed so far symmetries of objects through *abstract* groups. They are called abstract because on their own the group elements do not transform an object, they only serve to describe the geometry of the object. Thus, to each group element we need to associate the corresponding transformations that will act on the actual objects, i.e., model equations. This is done through a critical concept known as the **representation of a group.**

Let Γ be a Lie group and V a vector space. A representation of a finite group Γ is a homomorphism from Γ to the group of general linear matrices $\mathbf{GL}(V)$. That is,

$$\rho : \Gamma \to \mathbf{GL}(V).$$

Thus, a group element $\gamma \in \Gamma$ describes the abstract structure of the group, while $\rho(\gamma) = A \in \mathbf{GL}(V)$ indicates how each group element acts, through the matrix A, on the objects. The dimension of the representation is $\dim(V)$, so it is common practice to refer to V as "the representation of γ".

Example 5.21 (*The Dihedral Group* $\mathbf{D}_3$ *of a Triangle*) Consider for instance a ring structure with three elements, as is shown below in Fig. 5.35. The symmetries of this triangular ring are described by the dihedral group $\mathbf{D}_3 = \{e, \rho, \rho^2, \kappa, \kappa\rho, \kappa\rho^2\}$, where $\rho = 2\pi/3$ and κ is the reflection across the vertical dashed line shown in the figure.

The **natural representation** of the group is a 2D representation given by six matrices $\{A_e, A_\rho, A_{\rho^2}, A_\kappa, A_{\kappa\rho}, A_{\kappa\rho^2}\}$, one for each group element, where

$$A_e = \begin{bmatrix} 1 & 0 \\ 0 & 1 \end{bmatrix}, \quad A_\rho = \begin{bmatrix} -\frac{1}{2} & -\frac{\sqrt{3}}{2} \\ \frac{\sqrt{3}}{2} & -\frac{1}{2} \end{bmatrix}, \quad A_{\rho^2} = \begin{bmatrix} -\frac{1}{2} & \frac{\sqrt{3}}{2} \\ -\frac{\sqrt{3}}{2} & -\frac{1}{2} \end{bmatrix},$$
$$A_\kappa = \begin{bmatrix} -1 & 0 \\ 0 & 1 \end{bmatrix}, \quad A_{\kappa\rho} = \begin{bmatrix} \frac{1}{2} & \frac{\sqrt{3}}{2} \\ \frac{\sqrt{3}}{2} & -\frac{1}{2} \end{bmatrix}, \quad A_{\kappa\rho^2} = \begin{bmatrix} \frac{1}{2} & -\frac{\sqrt{3}}{2} \\ -\frac{\sqrt{3}}{2} & -\frac{1}{2} \end{bmatrix}.$$

It is called the natural representation because it is directly associated with the "natural" geometry of the triangle, so it can be visualized almost immediately. There

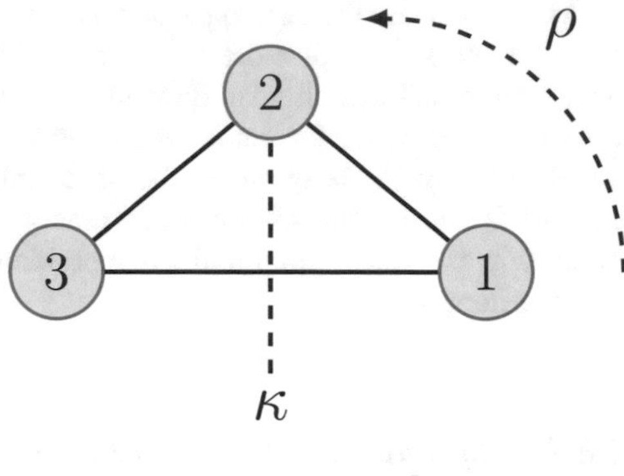

Fig. 5.35 Symmetries of a triangle ring are described by the dihedral $\mathbf{D}_3$ group

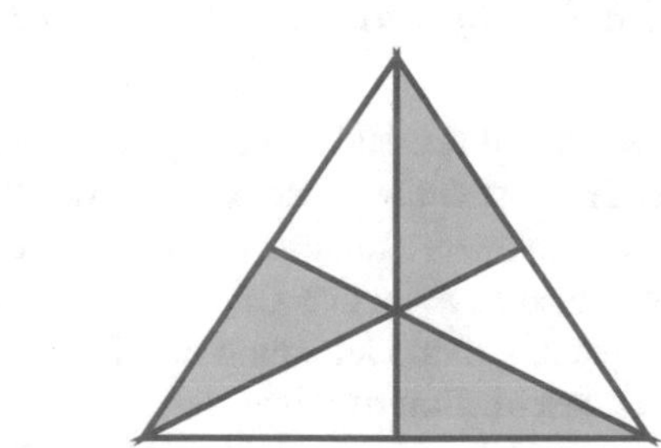

Fig. 5.36 One-dimensional complex representation of $\mathbf{D}_3$

are, however, other representations associated with $\mathbf{D}_3$. For instance, there is a one-dimensional complex representation where e, ρ and ρ^2 act as the identity, while κ, $\kappa\rho$ and $\kappa\rho^2$ act as $z \mapsto -z$, where $z \in \mathbf{C}$. We can try to visualize this representation with the aid of Fig. 5.36.

We can see that the rotations ρ and ρ^2 map white/gray subtriangles onto themselves, so they act as the identity. Adding a reflection by κ to any of the rotations has the same effect as flipping the lines across the origin. This is the same as multiplying by -1 any point in $z \in \mathbf{C}$. Thus, $z \mapsto -z$. This representation is called the **alternating representation**.

Finally, there is one more representation, the identity or trivial representation in which $\rho(\gamma) = \mathbf{1}$, for every $\gamma \in \Gamma$. Every group admits the trivial representation. In our example, we can now think of the "triangular region" as the entire plane, so that any combination of rotations and reflections act as the identity element. In Sect. 8.8 of Chap. 8 we illustrate in more detail the use of these irreducible representations to predict the formation of certain spatial patterns.

It is worthwhile mentioning that there are rigorous techniques for determining the number and type of irreducible representations of a finite group. The techniques belong to an area of mathematics known as **representation theory**. The scope of that work is beyond the scope of this book, so we refer interested readers to various references [8].

Since every linear Lie group Γ is a group of matrices in $\mathbf{GL}(n)$ then it can be shown that every Lie group Γ also admits the natural action on $\mathbb{R}^n$, which is given by matrix multiplication. Now, among all representations of a group, including the natural one, there are two types of representations that are particularly important from the symmetry-breaking bifurcation point of view. One type are called **irreducible representations**. To define them, we need the concept of Γ-invariance.

A subspace $W \subset V$ is said to be Γ-invariant if

$$\rho(\gamma)w \in W, \quad \text{for all } w \in W \text{and } \gamma \in \Gamma.$$

Definition 5.8 A representation or action of Γ on a vector subspace V is said to be irreducible if the only Γ-invariant subspaces of V are $\{0\}$ and V itself.

Example 5.22 (*The special orthogonal group* $\mathbf{SO}(\mathbf{2})$) This group is a subgroup of the orthogonal group $\mathbf{O}(n)$. It consists of 2×2 matrices A that satisfy $AA^T = \mathbf{1}$ and, in addition, $\det(A) = 1$. The standard action of $\mathbf{SO}(\mathbf{2})$ on $\mathbb{R}^2$ is defined by rotation by an angle θ. That is,

$$R_\theta = \begin{bmatrix} \cos\theta & -\sin\theta \\ \sin\theta & \cos\theta \end{bmatrix}.$$

Since the only subspaces of $\mathbb{R}^2$ that remain invariant under arbitrary rotations are the origin and the entire plane then this action is irreducible.

Absolute irreducibility is yet another concept that is used systematically to determine the generic type of bifurcations that can occur in a symmetric system of ODEs. For instance, absolute irreducibility excludes the existence of purely imaginary eigenvalues in the linearization of a model. Consequently, periodic oscillations cannot emerge via Hopf bifurcations for absolutely irreducible spaces. A brief definition follows but more details can be found in [8, 25].

Definition 5.9 A representation of a group Γ on a vector space V is *absolutely irreducible* if the only linear mappings on V that commute with Γ are scalar multiples of the identity.

Example 5.23 (*The Dihedral Group* $\mathbf{D}_3$ *of a Triangle*) All three representations (trivial, alternating and natural) of the dihedral group $\mathbf{D}_3$ are absolutely irreducible.

Example 5.24 (*The special orthogonal group* $\mathbf{SO}(\mathbf{2})$) We just showed above that this group is irreducible. However, it is not absolutely irreducible. To see why, observe that the composition of any two arbitrary rotations matrices R_θ and R_α commute. That is,

$$R_{\theta+\alpha} = R_\theta R_\alpha = R_\alpha R_\theta = R_{\alpha+\theta}.$$

Example 5.25 (*The orthogonal group* $\mathbf{O}(\mathbf{2})$) This group is made up of arbitrary rotations on the plane and reflections. In its standard action or representation, it contains $\mathbf{SO}(\mathbf{2})$ as a subgroup with the rotation matrix R_θ and a subgroup of reflections given by the matrices

$$M_\kappa = \begin{bmatrix} 1 & 0 \\ 0 & -1 \end{bmatrix}.$$

Since we know that **SO(2)** is irreducible then it follows that the standard representation of **O(2)** is also irreducible. Notice that this is the case even though R is an invariant subspace for the reflection M_κ because it is not invariant under **SO(2)**. In other words, reducibility requires invariance under all group elements.

Now, we must determine if the standard representation is also absolutely irreducible. To do this, we need to determine which matrices commute with $\mathbb{R}_\theta$ and which ones with M_κ. Thus, let A be an arbitrary matrix

$$A = \begin{bmatrix} a & b \\ c & d \end{bmatrix}.$$

Direct calculations show that

$$M_\kappa A = \begin{bmatrix} a & b \\ -c & -d \end{bmatrix}, \qquad AM_\kappa = \begin{bmatrix} a & -b \\ c & -d \end{bmatrix}.$$

Thus, $M_\kappa A = AM_\kappa$ implies $b = c = 0$. In addition, commutativity with R_θ requires

$$R_\theta A = \begin{bmatrix} a\cos\theta & -d\sin\theta \\ a\sin\theta & d\cos\theta \end{bmatrix}, \qquad AR_\theta = \begin{bmatrix} a\cos\theta & -a\sin\theta \\ d\sin\theta & d\cos\theta \end{bmatrix}.$$

Then, $R_\theta A = AR_\theta$ requires $a = d$. Consequently, A is a multiple of the identity and it follows that the standard representation of **O(2)** is absolutely irreducible.

It is also a well-known fact that symmetry forces systems of ODEs to have invariant linear subspaces. In particular, the **fixed-point subspace** of a solution is the invariant subspace where the isotropy subgroup acts trivially.

Definition 5.10 Suppose that $\Sigma \subset \Gamma$ is a subgroup. Then the fixed-point subspace is the vector subspace of $\mathbb{R}^n$ where the subgroup Σ acts trivially. Formally:

$$\text{Fix}(\Sigma) = \left\{x \in \mathbb{R}^n : \sigma x = x \quad \forall \sigma \in \Sigma\right\}$$

Claim The fixed-point subspace is a flow invariant subspace [25].

Proof Consider a mathematical model in the form of a continuous system of Γ-equivariant ODEs

$$\frac{d\mathbf{x}}{dt} = f(\mathbf{x}, \lambda).$$

Let $\sigma \in \Sigma$ and $x \in \text{Fix}(\Sigma)$. Then

$$f(x) = f(\sigma x) = \sigma f(x).$$

The first equality follows from the fact that that $x \in \text{Fix}(\Sigma)$, while the second one arises from the Γ-equivariance of f. □

Fixed point subspaces describe the regions of phase space where a particular solution resides. This suggests a model-independent strategy to find solutions of symmetric systems of ODEs. Restrict the equations to $\mathrm{Fix}(\Sigma)$ and then solve for the solutions. Since $\mathrm{Fix}(\Sigma)$ is, in general, lower dimensional that the entire space then it might be significantly easier to solve the restricted equations. A critical observation is the fact that it might be possible, under certain conditions, to predict the type of solutions of a symmetric system of ODEs without having to solve for the solutions. Details are formalized by the *Equivariant Branching Lemma* (EBL).

Theorem 5.9 (Equivariant Branching Lemma [25]) *Let $\Gamma \subseteq \mathbf{O(n)}$ be a compact Lie group acting absolutely irreducibly on $\mathbb{R}^n$. Let*

$$\frac{dx}{dt} = f(x, \lambda), \quad x \in \mathbb{R}^n, \quad \lambda \in \mathbb{R} \tag{5.50}$$

be a Γ-equivariant bifurcation problem so that

$$\begin{aligned} f(0, \lambda) &= 0 \\ (df)_{0,\lambda} &= c(\lambda)I. \end{aligned}$$

Assume $c'(0) \neq 0$ and let $\Sigma \subseteq \Gamma$ satisfy

$$\dim \mathrm{Fix}(\Sigma) = 1.$$

Then there exists a unique branch of solutions to $f(x, \lambda) = 0$ bifurcating from $(0, 0)$, where the symmetry of the solution is Σ.

Example 5.26 Consider again the Euler beam experiment. Recall that the symmetries of the experiment are described by the group $\Gamma = \mathbf{Z}_2$. Furthermore, $\Sigma = \mathbf{1}$ is an isotropy subgroup in which $\mathrm{Fix}(\Sigma) = \mathbb{R}$, so that $\dim \mathrm{Fix}(\Sigma) = 1$. Thus, by the Equivariant Branching Lemma, we can predict the existence of a branch of steady-state solutions with $\Sigma = \mathbf{1}$ symmetry in the idealized model

$$\dot{x} = \lambda x - x^3.$$

This reduced problem satisfies all conditions of the Equivariant Branching Lemma. The steady-state solutions with trivial symmetry are $\pm\sqrt{x}$. They emerge via a pitchfork bifurcation.

Example 5.27 (*Cylindrical Euler Beam*) If the Euler beam were to be cylindrical instead of rectilinear then the group of symmetries of the experiment would become $\Gamma = \mathbf{O(2)}$, the orthogonal group of rotations and reflections on the plane. Without loss of generality, assume the rotations by an arbitrary angle θ is equivalent to shifting $\alpha \mapsto \alpha + \theta$. Similar results can be obtained if the shift is defined in the opposite direction, i.e., $\alpha \mapsto \alpha - \theta$. We can also assume the reflections to be across the x axis, though similar results are obtained if the reflection were to be defined across the y axis.

To see how the $\mathbf{O(2)}$ group acts on the plane, let $\mathbb{R}^2 \cong \mathbf{C}$ and consider a typical point $z = re^{\alpha i} \in \mathbf{C}$. First we look at the rotations. Substituting the shift, $\alpha \mapsto \alpha + \theta$, in z yields:

$$re^{(\alpha+\theta)i} = e^{\theta i}\, re^{\alpha i} = e^{\theta i}\, z.$$

Thus, we have shown that $\theta \in \Gamma$ acts on $\mathbf{C}$ as

$$\theta \cdot z = e^{\theta i} z.$$

Similarly, a reflection across the x-axis by κ is equivalent to $\alpha \mapsto -\alpha$. Substituting in z we get

$$re^{-\alpha i} = \overline{re^{\alpha i}} = \bar{z}.$$

This shows that $\kappa \in \Gamma$ acts on $\mathbf{C}$ as

$$\kappa \cdot z = \bar{z}.$$

Combining the actions of θ and κ, we conclude that $\Gamma = \mathbf{O(2)}$ acts on $\mathbf{C}$ as follows:

$$\begin{aligned} \theta \cdot z &= e^{\theta i} z \\ \kappa \cdot z &= \bar{z}. \end{aligned}$$

These result are illustrated in Fig. 5.37.

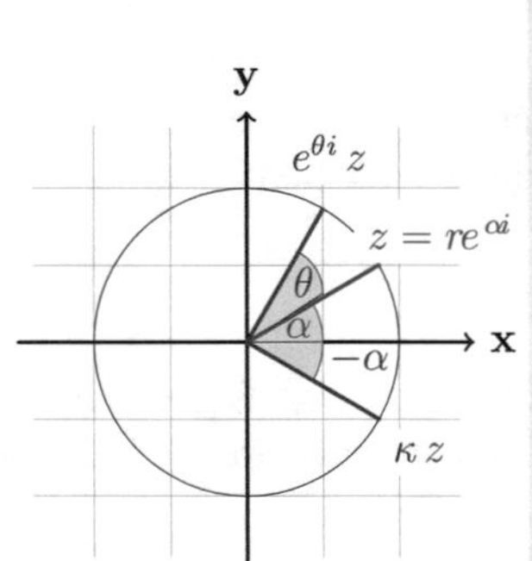

Let $z = re^{\alpha i}$ be any point on the plane. A rotation by an angle θ is equivalent to a shift $\alpha \mapsto \alpha + \theta$. A reflection across the x-axis is equivalent to $\alpha \mapsto -\alpha$. Combined these two operations lead to

$$\begin{aligned} re^{(\alpha+\theta)i} &= e^{\theta i} re^{\alpha i} = e^{\theta i} z \\ re^{-\alpha i} &= \overline{re^{\alpha i}} = \bar{z} \end{aligned}$$

Since z is any point on the plane, this shows that $\Gamma = \{\theta, \kappa\}$ acts on $\mathbf{C}$ as

$$\begin{aligned} \theta \cdot z &= e^{\theta i} z \\ \kappa \cdot z &= \bar{z}. \end{aligned}$$

Fig. 5.37 The action of the orthogonal group $\mathbf{O(2)}$ on the plane

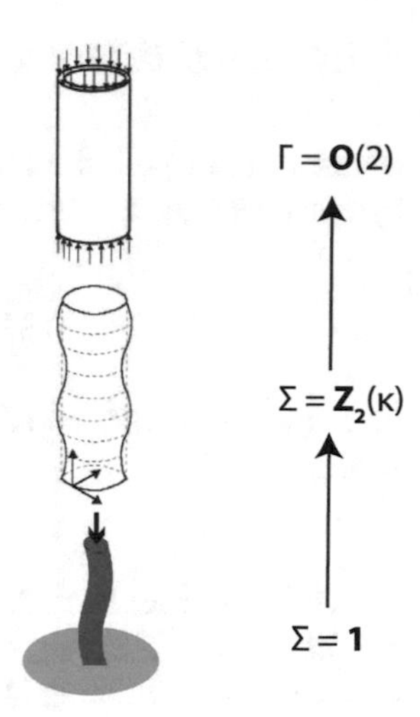

Fig. 5.38 Buckling of a cylindrical beam

Now, the critical observation. The group $\Sigma = \mathbf{Z}_2(\kappa) = \{1, \kappa\}$ is a subgroup of $\mathbf{O(2)}$, which consists of only the reflections on the plane, without rotations. Recall the definition of the fixed point subspace of a subgroup Σ: the vector subspace (of $\mathbb{R}^2$ in this example) in which the subgroup Σ acts trivially. In our example, the only vector subspace where reflections across the x-axis are trivial is precisely the x-axis. Mathematically, this means that

$$\mathrm{Fix}(\Sigma) = \mathbb{R}.$$

Since $\dim(\mathbb{R}) = 1$ then $\dim \mathrm{Fix}(\Sigma) = 1$, and, thus, by the Equivariant Branching Lemma, an $\mathbf{O(2)}$ steady-state bifurcation problem can lead to equilibria with reflectional symmetry. From the modeling standpoint, this means that in an Euler Beam experiment with a cylindrical beam, generically, there will be buckle modes with reflectional symmetry in addition to the buckled state with trivial symmetry. The possible transitions or bifurcations are illustrated in Fig. 5.38. On the left, the figure shows the unperturbed state of the beam. On the right, the symmetry-breaking breaking bifurcations as they are observed through the **lattice of isotropy subgroups**.

We have discussed so far bifurcations that lead to new branches of equilibrium points. However, it is also possible, under certain conditions, for periodic solutions to emerge through Hopf symmetry-breaking bifurcations. Next we discuss those conditions and introduce a revised version of the Equivariant Branching Lemma for periodic solutions.

5.8.3 Hopf Bifurcation with Symmetry

Symmetries of periodic solutions may arise in one of two forms. As purely spatial symmetries or as a combination of space and time symmetries. To get insight, let

$$\frac{dx}{dt} = f(x, \lambda), \qquad x \in \mathbb{R}^n, \lambda \in \mathbb{R}^p \tag{5.51}$$

be a system of differential equations with Γ-symmetry. Let $x(t)$ be a T-periodic solution, so that

$$x(t + T) = x(t).$$

We know from Sect. 5.7.1 that Γ-equivariance of $f(x)$ implies that if $x(t)$ is a solution of Eq. (5.51) so is $\gamma x(t)$ for all $\gamma \in \Gamma$. The two trajectories, $x(t)$ and $\gamma x(t)$ may either have nothing in common. That is

$$\gamma\{x(t)\} \cap \{x(t)\} = \emptyset,$$

as is shown in Fig. 5.39 (left), Or, if the two solutions were to intersect at any point t, say $t = 0$, then the intersection point must the be the same initial point for both solutions. Then by uniqueness of solutions the trajectories must be identical. That is

$$\gamma\{x(t)\} \cap \{x(t)\} = \{x(t)\}.$$

Now, let's assume the two periodic solutions are indeed symmetrically related so that the trajectories are identical. This can happen in two ways, via purely spatial symmetries or through a combination of spatial and temporal symmetries. In the former case, a periodic solution $x(t)$ is fixed at every moment in time by some $\gamma \in \Gamma$, so that γ is a purely spatial symmetry. Formally, we can define the **group of purely spatial symmetries** as

$$K = \{\gamma \in \Gamma : \gamma x(t) = x(t)\} \tag{5.52}$$

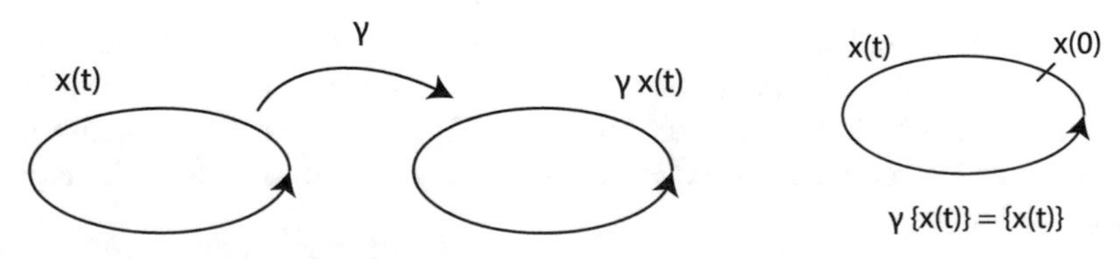

Fig. 5.39 Symmetry of periodic solutions. (Left) $\gamma x(t)$ and the original solution $x(t)$ are disjoint, so they are not symmetrically related. (Right) $\gamma x(t)$ and $x(t)$ intersect at a common point $x(0)$, so by uniqueness of solutions the two trajectories must be the same

Notice that K is defined in a manner that resembles the symmetries of steady-states discussed above.

In the latter case, spatio-temporal symmetries, the solution trajectory is fixed by a combination of the spatial action of $\gamma \in \Gamma$ and a phase-shift $\theta \in \mathbf{S}^1$, where $\mathbf{S}^1$ is the circle group of phase shifts acting on 2π periodic functions. That is,

$$(\gamma, \theta) \cdot x(t) = \gamma x(t + \theta) = x(t), \quad \forall t.$$

As it was the case of steady-state solutions, both types of symmetries can be formally described through an extended version of the isotropy subgroup for periodic oscillations.

Definition 5.11 Let $x(t)$ represent a periodic solution of a Γ-equivariant system of ODEs. The symmetries of $x(t)$ form the isotropy subgroup, which is defined by

$$\Sigma_{x(t)} = \left\{(\gamma, \theta) \in \Gamma \times \mathbf{S}^1 : \gamma x(t + \theta) = x(t)\right\}. \tag{5.53}$$

Observe that the case $\theta = 0$ corresponds to purely spatial symmetries, captured by the subgroup $K \subset \Sigma_x$. But we can also identify one more subgroup of Σ_x,

$$H = \{\gamma \in \Gamma : \gamma\{x(t)\} = \{x(t)\}\}. \tag{5.54}$$

In this case, the subgroup H describes all the symmetries that preserve the trajectory of $x(t)$ without attention to the temporal shift, i.e, H is the subgroup of spatial components of the spatio-temporal symmetries of $x(t)$.

In the next chapter we will make extensive use of these two subgroups H and K when we describe the symmetries of networks of coupled cell systems.

Generically, the existence of Hopf bifurcations in a symmetric system of ODEs is determined by Γ-simple irreducible representations. Formally,

Definition 5.12 A representation W of Γ is Γ-simple if either W is composed of two copies of an absolutely irreducible representation, so that $W = V \oplus V$, or W is non-absolutely irreducible for Γ.

In either of these two cases, it can be shown [8, 25] that Γ-simple representations lead to Jacobian matrices with the following structure

$$(df)_{(0,0)} = \begin{bmatrix} 0 & -I_m \\ I_m & 0 \end{bmatrix}, \tag{5.55}$$

where $m = n/2$.

We can now state the equivalent of the Equivariant Branching Lemma for Hopf bifurcations with symmetry.

Theorem 5.10 (Equivariant Hopf Theorem [25])
Let

$$\frac{dx}{dt} = f(x, \lambda), \quad x \in \mathbb{R}^n, \quad \lambda \in \mathbb{R} \tag{5.56}$$

be a Γ*-equivariant bifurcation problem with* $\Sigma \subset \Gamma \times \mathbf{S}^1$*. Assume the linearization of (5.56) satisfies Eq. (5.55) with eigenvalues* $\sigma(\lambda) \pm w(\lambda)i$*, each of multiplicity* m *and also*

$$\sigma'(0) \neq 0.$$

If the action of Γ *is* Γ*-simple on* $\mathbb{R}^n$ *and* Σ *satisfies*

$$dim\ Fix(\Sigma) = 2,$$

then there exists a unique branch of periodic solutions to $f(x, \lambda) = 0$*, with period near* 2π*, bifurcating from* $(0, 0)$*, where the symmetry of the solution is* Σ*.*

Example 5.28 (*Premixed Flame Dynamics* (continued)) In addition to stationary symmetric cellular flame patterns, it is also possible to observe nonstationary ones within the same experimental setup described earlier. We know the experiment is performed over a circular burner, so it has $\mathbf{O(2)}$-symmetry. We also know from Example 5.25 that the standard action of $\mathbf{O(2)}$ on $\mathbb{R}^2 \equiv \mathbf{C}$ is absolutely irreducible. Thus, we cannot use the Hopf EBL over $\mathbb{R}^2$ because the commuting matrices are multiples of the identity, so they cannot have complex eigenvalues as is required for a Hopf bifurcation. However, if we set $\mathbf{O(2)}$ to act on $\mathbb{R}^4$ through a diagonal action:

$$\begin{aligned} \rho \cdot (z_1, z_2) &= (e^{\rho i} z_1, e^{\rho i} z_2) \\ \kappa \cdot (z_1, z_2) &= (z_2, z_1) \end{aligned}$$

then it is possible for the commuting matrices to have complex eigenvalues, though they will be repeated. To verify this assertion, write an arbitrary 4×4 matrix in block form

$$A = \begin{bmatrix} A_{11} & A_{12} \\ A_{21} & A_{22} \end{bmatrix}.$$

And the rotation matrix in $\mathbf{O(2)}$ as

$$R = \begin{bmatrix} R_\rho & O_{2\times 2} \\ O_{2\times 2} & R_\rho \end{bmatrix}.$$

Then direct computations show that A commutes with R if and only if each of its blocks commutes with R_θ. Since the only matrices that commute with R_ρ are multiple of the identity, this means that A has the form

$$A = \begin{bmatrix} a\mathbf{I}_2 & b\mathbf{I}_2 \\ c\mathbf{I}_2 & d\mathbf{I}_2 \end{bmatrix}.$$

The eigenvalues of A are those of

$$\begin{bmatrix} a & b \\ c & d \end{bmatrix}$$

repeated twice. Hence, setting $a = d = 0$ and $b = -1$ and $c = 0$ we get a pair of purely imaginary eigenvalues $\pm i$, repeated twice. Thus, Hopf symmetry-breaking bifurcation is possible over $\mathbb{R}^4$. Consider the idealization of the flame experiment by a $\mathbf{O(2)}$-equivariant system of ODEs

$$\begin{aligned} \dot{z}_1 &= f(z_1, z_2, \mu) \\ \dot{z}_2 &= g(z_1, z_2, \mu), \end{aligned} \tag{5.57}$$

where μ is a vector of parameters. When the model equations (5.57) are written in Birkhoff Normal Forms [5, 6], it can be shown that the model equations commute not only with Γ but also with the circle group $\mathbf{S}^1$. An intuitive way of explaining this is that $\mathbf{S}^1$-symmetry corresponds to the phase shift that leave periodic solutions invariant. Thus, our flame model is then $\Gamma \times \mathbf{S}^1$ symmetric. It can also be shown that the action of this spatio-temporal group of symmetries acts on $\mathbf{C}^2$ as follows

$$\begin{aligned} \rho \cdot (z_1, z_2) &= (e^{-\rho i} z_1, e^{\rho i} z_2) \\ \kappa \cdot (z_1, z_2) &= (z_2, z_1) \\ \theta \cdot (z_1, z_2) &= (e^{\theta i} z_1, e^{\theta i} z_2), \end{aligned} \tag{5.58}$$

where $\rho, \kappa \in \Gamma = \mathbf{O(2)}$ and $\theta \in \mathbf{S}^1$. We know that the action of $\Gamma = \mathbf{O(2)}$ on $\mathbb{R}^4$ is Γ-simple, since it is made of two copies of absolutely irreducible representations. Then in order to apply the Equivariant Hopf Theorem we need to find isotropy subgroups Σ of Γ with 2D fixed-point subspaces. We can do this task by examining the action of $\Gamma \times \mathbf{S}^1$. This leads us to the following classification of isotropy subgroups and their corresponding solutions.

Standing Waves. Periodic solutions with purely spatial symmetry are called *standing waves* because they must maintain that spatial symmetry at all times, while oscillating periodically. According to the action of Γ, the purely spatial symmetry must be a reflectional symmetry and the corresponding solution must lie on the invariant subspace (z, z). Observe that the axis of reflection is $z = x \in \mathbb{R}$. This axis does not change in time, so the periodic solution can not rotate. In other words, if the solution has a purely spatial symmetry at any given time then it must have that same symmetry at all times, thus rotations are impossible. The corresponding isotropy subgroup is classified as $\mathbf{Z}_2 \times \mathbf{Z}_2^c = \{(0, 0), (\pi, \pi), (\kappa, 0), (\kappa, 0)(\pi, \pi)\}$. A standing wave flame pattern with $\mathbf{Z}_2 \times \mathbf{Z}_2^c$ is shown in Fig. 5.40.

Rotating Waves. Periodic solutions that remain invariant after a combination of a purely spatial rotation by θ and a phase shift by $-\theta$ are known as rotating waves because the combination of these two spatio-temporal symmetries indicate the existence of a periodic solution $x(t)$ that satisfies

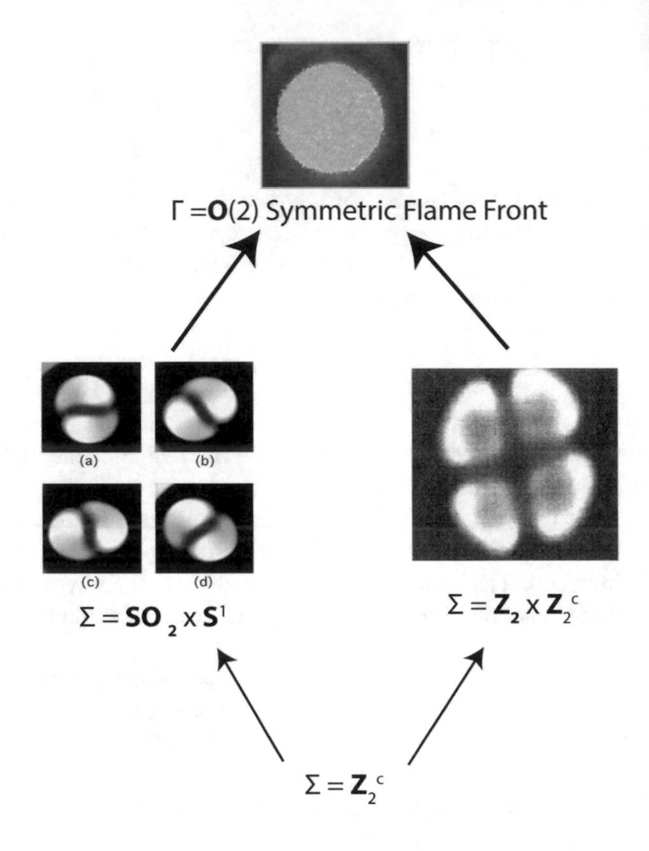

Fig. 5.40 **O(2)** Hopf symmetry breaking bifurcations from a trivial equilibrium (homogeneous flame front) in a combustion system [27] . (Middle-left) Four snapshots from the two-cell state of the flame front that rotates clockwise. States with the opposite geometrical sense (i.e., related by reflections) rotate counter clockwise. (Middle-right) Standing wave patterns

$$R_\theta\, x(t) = x(t+\theta), \quad \text{so that} \quad x(\theta) = R_\theta\, x(0). \tag{5.59}$$

Any periodic solution that satisfies Eq. (5.59) is called a rotating or traveling wave. Their isotropy subgroup is $\mathbf{SO(2)} = \{(\theta, \theta) \in \mathbf{SO(2)} \times \mathbf{S}^1\}$.

Trivial Symmetries. Obviously the trivial spatio-temporal symmetries $(\rho, \theta) = (0, 0)$ leave the entire $\mathbb{R}^4 \equiv \mathbf{C}^2$ phase space unchanged. But $(\rho, \theta) = (\pi, \pi)$ also leave $\mathbf{C}^2$ unchanged. Together, these two spatio-temporal symmetries leave $\mathbf{C}^2$ unchanged and they generate the isotropy subgroup $\mathbf{Z}_2^c$. For this reason, this subgroup appears at the bottom of the lattice in Fig. 5.40.

Trivial Solutions. Our original assumptions indicate that the Γ-equivariant model (5.57) admits a trivial solution of the form $(z_0, z_0) = (0, 0)$. This trivial solution is fixed by the entire symmetry group $\mathbf{O(2)} \times \mathbf{S}^1$. For this reason, this subgroup appears at the top of the lattice in Fig. 5.40.

5.9 Exercises

Exercise 5.1 Forced Damped Pendulum.
The following equation models the dynamics of a forced damped pendulum

$$\ddot{\theta} + b\,\dot{\theta} + k\,\sin\theta = A\cos(\omega t)$$

where $\theta = \theta(t)$ measures the angular position of the pendulum at time t, A is the amplitude of a periodic force applied to the pivot point of the pendulum, and ω is the frequency of the perturbation.

(a) Convert the equation to a first-order system.
(b) Write a Matlab program to integrate and simulate the behavior of the pendulum.
(c) Set $k = 1$, $b = 0.5$, and $\omega = 0.67$. Simulate the dynamics for various values of A in the range $[0, 2]$. Explain how the pendulum changes behavior (upon changing A) using **Time Series** plots and **Phase Portraits** of qualitatively different solutions.
(d) Now fix $A = 1.5$. Then vary ω in the range $[0, 2]$ and repeat part (c).

Exercise 5.2 Laser Dynamics. Milonni and Eberly (1988) show that after certain reasonable approximations, quantum mechanics leads to the following model of a laser

$$\begin{cases} \dfrac{dn}{dt} = GnN - kn \\ \dfrac{dN}{dt} = -GnN - fN + p \end{cases} \tag{5.60}$$

where G is the gain coefficient for stimulated emission, k is the decay rate due to loss of photons by mirror transmission, f is the decay rate for spontaneous emission, and p is the pump strength. All parameters are positive, except p, which can have either sign.

(a) Suppose that N relaxes much more rapidly than n. Then we may make the quasi-static approximation $\dot{N} = 0$. Given this approximation, express $N(t)$ in terms of $n(t)$ and derive a first-order system for n.
(b) Show that $n^* = 0$ becomes unstable for $p > p_c$, where p_c is to be determined.
(c) What type of bifurcation occurs at the laser threshold p_c ?
(d) For what range of parameters is it valid to make the approximation?

Exercise 5.3 Laser Dynamics (continued).
The Maxwell-Bloch equations of a laser describe the dynamics of an electric field E, the mean polarization P of atoms, and the population inversion D, through

$$\begin{aligned} \dot{E} &= \kappa(P - E) \\ \dot{P} &= \gamma_1(ED - P) \\ \dot{D} &= \gamma_2(\lambda + 1 - D - \lambda EP), \end{aligned}$$

where κ is the decay rate in the laser cavity due to beam transmission, γ_1 and γ_2 are decay rates of the atomic polarization and population inversion, respectively, and λ is a pumping energy parameter, which may be positive, negative, or zero; all the other parameters are positive.

(a) Find all the equilibrium points and determine their stability.
(b) Assuming $\dot{P} = 0, \dot{D} = 0$, express P and D in terms of E, and derive a first-order equation for E. Then study and classify the bifurcations of E in terms of λ and draw a bifurcation diagram of E versus λ.
(c) Using the following change of variables, $t = (\sigma/\kappa)\tau$, $E = \alpha x$, $P = \alpha y$, $D = r - z, \gamma_1 = \kappa/\sigma, \gamma_2 = \kappa b/\sigma, \lambda = r - 1$, rewrite Maxwell-Bloch model and then find a set parameter values (of the original laser system) where chaotic behavior is present. Perform computer simulations of Maxwell-Bloch to demonstrate sensitive dependence on initial conditions.

Exercise 5.4 Duffing Oscillator.
Consider a Duffing oscillator of the form

$$\frac{d^2x}{dt^2} + \mu\frac{dx}{dt} + \lambda x - x^3 = 0 \tag{5.61}$$

where μ and λ are constant parameters.

(a) Convert the second order ODE (5.61) to a first-order system of ODE's.
(b) Analytically, calculate *all* equilibrium points and study their stability.
(c) Find the curves in (μ, λ) parameter-space at which the eigenvalues of the linearized Jacobian matrix, J, are purely imaginary, which is equivalent to the condition: $\text{trace}(J) = 0$ and $\det(J) > 0$. This locus of points is called a Hopf bifurcation.
(d) Sketch a diagram in the (μ, λ) plane illustrating the change in stability of each equilibrium point when both μ and λ change. Use pplane or any other equivalent software to sketch phase portraits of different types of behaviors.

Exercise 5.5 Colpitts Oscillator. The following system of equations describes a dimensionless version of a Colpitts oscillator

$$\begin{aligned} \dot{x}_1 &= \frac{g}{Q(1-k)}\left(-e^{-x_2} + 1 + x_3\right) \\ \dot{x}_2 &= \frac{g}{Qk}x_3 \\ \dot{x}_3 &= -\frac{Qk(1-k)}{g}(x_1 + x_2) - \frac{1}{Q}x_3, \end{aligned} \tag{5.62}$$

where g, Q and k are positive parameters.

(a) Compute the Jacobian matrix and evaluate it at the zero equilibrium $(0, 0, 0)$.
(b) Calculate the characteristic polynomial associated with the eigenvalues of the Jacobian matrix.

(c) Calculate the eigenvalues of the Jacobian matrix by finding approximate analytical expressions to the roots of the characteristic polynomials.
(d) Show that Eq. (5.62) undergoes a Hopf bifurcation at $g = 1$. Show, analytically and computationally through computer simulations, that for values of $g > 1$ the trivial equilibrium state $(0, 0, 0)$ loses stability and a stable limit cycle appears around it.

Exercise 5.6 Glycolysis. Selkov (1968) proposed the following model for glycolysis

$$\frac{dx}{dt} = -x + a * y + x^2 y$$
$$\frac{dy}{dt} = b - a * y - x^2 y,$$

where $a > 0$ and $b > 0$ are parameters.

(a) Compute the equilibrium point of this model.
(b) Find the values of a and b where the system undergoes a Hopf bifurcation.

Exercise 5.7 Trimolecular Reactions. Schnackenberg (1979) considered a class of two-species simplest, but chemically plausible, trimolecular reactions which can admit periodic solutions. After using the Law of Mass Action and nondimensionalizing, Schnackenberg reduced the system to

$$\begin{cases} \frac{dx}{dt} = a - x + x^2 y \\ \frac{dy}{dt} = b - x^2 y \end{cases} \tag{5.63}$$

where $a > 0$, $b > 0$ are parameters and $x > 0$, $y > 0$ are dimensionless concentrations.

(a) Show that the system has a unique fixed point, and classify it through the linearization process.
(b) Show that the system undergoes a Hopf bifurcation when $b - a = (a + b)^3$.
(c) Is the Hopf bifurcation subcritical or supercritical? Use a computer to decide.
(d) Plot the stability diagram in (a, b) parameter space. Hint: It is a bit confusing to plot the curve $b - a = (a + b)^3$, since it requires analyzing a cubic. Show that the bifurcation curve can be expressed in parametric form $a = \frac{1}{2}x_E(1 - x_E^2)$, $b = \frac{1}{2}x_E(1 + x_E^2)$, where $x_E > 0$ is the x-coordinate of the fixed point. Then plot the bifurcation curve from these parametric equations.

Exercise 5.8 Autoimmune Diabetes Model. A reduced 3D model for diabetes in NOD mice was introduced by Mahaffy and Edelstein-Keshet [28]. The model has the form:

$$
\begin{aligned}
\frac{dA}{dt} &= (\sigma + \alpha M)\tilde{f}_1(E) - (\beta + \delta_A)A - \varepsilon A^2 \;=\; F_1(A, M, E), \\
\frac{dM}{dt} &= \beta 2^{m_1}\tilde{f}_2(E)A - \tilde{f}_1(p)\alpha M - \delta_M M \;=\; F_2(A, M, E), \\
\frac{dE}{dt} &= \beta 2^{m_2}(1 - \tilde{f}_2(E))A - \delta_E E \;=\; F_3(A, E),
\end{aligned}
\tag{5.64}
$$

where $p \approx (RB/\delta_p)E$, treating B as a slow varying constant, and $\tilde{f}_1(E)$ and $\tilde{f}_2(E)$ are the nonlinear functions:

$$
\tilde{f}_1(p) = \frac{p^n}{k_1^n + p^n} \qquad \text{and} \qquad \tilde{f}_2(p) = \frac{ak_2^m}{k_2^m + p^m}.
$$

The following Table 5.3 shows parameter values that were used in this model:
a. There are three equilibria, (A_e, M_e, E_e), for this model. Find these equilibria and determine their corresponding eigenvalues. Give a brief discussion of the local behavior near each of the equilibria, including a discussion of the dimensions of the stable and unstable manifolds for each equilibrium. Which equilibrium represents the healthy state and which one represents the diseased state? What is the significance of the third equilibrium?
b. Simulate this system with different initial conditions and show the time series solutions of at least two simulations that demonstrate at least two distinct behaviors around the equilibria. (Graphs should appear quite different.) Write a paragraph describing what you expect concerning the global behavior of this system. Also, include a brief discussion connecting the observed behavior in your simulations to the NOD mouse biology.

Exercise 5.9 Henon map.
Consider the map

$$
\begin{bmatrix} x_{n+1} \\ y_{n+1} \end{bmatrix} = \begin{bmatrix} y_n \\ \alpha - \beta x_n - y_n^2 \end{bmatrix},
$$

where $(x_{n+1}, y_{n+1}) \in \mathbb{R}^2$ and α and β are real-valued parameters. Perform the following tasks:

(i) Compute all fixed points and study their stability properties.
(ii) Find an analytical expression for the boundary curve in parameter space (β, α) where a saddle-node bifurcation occurs.
(iii) Find an analytical expression for the boundary curve in parameter space (β, α) where a period-doubling bifurcation occurs.
(iv) Find an analytical expression for the boundary curve in parameter space (β, α) where a Neimark-Sacker bifurcation occurs.
(v) Plot all boundary curves in a single graph.

Table 5.3 Parameters used in simulation for (5.64) for NOD mice with intact β-cells

$\sigma = 0.02$	$\alpha_1 = 20$	$\beta + \delta_A = 1$	$\varepsilon = 1$	$k_1 = 2$	$n = 2$
$\beta 2^{m_1} = 1$	$a = 0.8$	$k_2 = 1$	$m = 3$	$\alpha_2 = 2$	$\delta_M = 0.01$
$\beta 2^{m_2} = 0.1$	$\delta_E = 0.3$	$R = 50$	$B = 1$	$\delta_p = 1$	–

Exercise 5.10 Lozi Map. In 1978, Lozi introduced a 2D map which resembles the Henon map, except that a quadratic term is replaced by a piecewise linear version of it. The equations are:

$$\begin{aligned} x_{n+1} &= 1 + y_n - a|x_n| \\ y_{n+1} &= bx_n, \end{aligned} \tag{5.65}$$

where a and b are real non-vanishing parameters. Assume $|b| \leq 1$ and perform the following tasks.

(a) Find all fixed points and study their stability as a function of a. In particular, study how the number of fixed points changes as a changes in relation to b.
(b) Find all *isolated* period-2 orbits as function of a and b and study their stability properties. Find the range of values in parameter space (b, a) where the period-2 orbit is stable. Sketch the region of parameter space (b, a) where the fixed point and period-2 orbits are stable.
(c) Write a computer program to generate two bifurcation diagrams. In one case, set $b = 0.1$ and vary a in the interval $[0 : 1.8]$ and plot x_n. In the second case, set $a = 1.5$ and now vary b in the interval $[-0.75 : 0.75]$ and plot x_n as a function of b. Make sure that positive and negative initial conditions are used and not to include transient behavior. Compare the diagrams with that of the logistic map. (Turn in the code and TWO plots of the bifurcation diagram).
(d) Set $a = 0.9$ and $b = -1.0$ and set a rectangular grid (100×100) points in (x, y) phase-space: $-50 \leq x, y \leq 50$. Set a few random initial conditions for (x_0, y_0), iterate the code to eliminate transient behavior, and then plot the remaining points. Describe in your own words the emergent pattern.

Exercise 5.11 Linear Systems
Consider the following 2D linear system

$$X_{n+1} = AX_n, \quad \text{where} \quad A = \begin{bmatrix} \frac{3}{4} & -\frac{1}{8} \\ \frac{1}{2} & \frac{1}{4} \end{bmatrix}, \quad X_n = \begin{bmatrix} x_n \\ y_n \end{bmatrix}.$$

(a) Find the general analytical solution of this model.
(b) Find the particular solution that satisfies the initial conditions $X_0 = (400, 600)$.
(c) Compute the limit of X_n as $n \to \infty$.

Exercise 5.12 Nonlinear Discrete System
Consider the following nonlinear system:

$$x_{n+1} = ay_n(1 - y_n)$$
$$y_{n+1} = x_n$$

(a) Find all the nonnegative fixed points as a function of a. For which values of a do the fixed points exist?
(b) Show that a positive fixed point bifurcates from $(0, 0)$.
(c) Find the interval of stability of each fixed point.

Exercise 5.13 Quadratic Map
Consider the quadratic map

$$x_{n+1} = x_n^2 + c, \qquad \text{where } c \text{ is a constant.}$$

(a) Find all the fixed points as a function of c. For which values of c do the fixed points exist?
(b) Plot the graph of $f(x) = x^2 + c$, with $c = -0.5$. Locate the fixed points and then use Cobwebbing with $x_0 = 0.8$ and $x_0 = 1.5$ to determine the stability of each fixed point. Explain the long-term behavior of the model. In particular, can you identify the *basin of attraction* of any stable fixed point?
(c) Plot the graph of $f(x) = x^2 + c$, with $c = -1.0$. Locate the fixed points and then use Cobwebbing with $x_0 = 0.8$ and $x_0 = 1.75$ to determine the stability of each fixed point. Explain the long-term behavior of the model.
(d) Analytically, determine the stability of the fixed points and find the values of c at which these points bifurcate. Classify the bifurcations.
(e) For which values of c is there a stable 2-cycle? Hint: Calculate first the period-2 orbit by solving $f^2(x) = x$ or $f(f(x)) = x$, where $f = x^2 + c$. Factor out the period-1 points or fixed points to simplify the algebra.
(f) Plot a partial bifurcation diagram.

Exercise 5.14 Video Feedback.
Consider the complex-valued map

$$z_{n+1} = z_n^2 + c, \tag{5.66}$$

where $z = x + yi$ and $c = a + bi$. What is most significant about this map is that it possesses virtually every type of period-multiplying cascade imaginable. That is, for (a, b) close to zero a stable fixed point exists that bifurcates into a stable m-cycle for any positive integer m, depending upon the path (a, b) takes in its two dimensional parameter space. Each point of these m-cycles similarly bifurcates into a stable k-cycle, which yields a stable cycle of period mk. As this continues, virtually all period-multiplying cascades can be created.

(a) Find all fixed points of this complex map as functions of c.

(b) Study the stability of each fixed point. In particular, find an analytical expression for the boundary curves, in the (a,b)-plane, where the fixed points are stable. Plot the resulting boundary curves in the (a, b)-plane. Suggestion: Set $f'(z_{fp}) = re^{i\theta}$, where z_{fp} is a fixed point, and then solve for c.
(c) Repeat part (a) and (b) for period-2 points.
(d) Draw the parameter-space of this map with the following algorithm: (i) Set a rectangular grid (200×200) points in (a, b) parameter space: $-2.4 \leq a \leq 1.2$, $-1.5 \leq b \leq 1.5$. (i) For each point (a, b), iterate the map (5.66) starting with $z_0 = 0 + 0i$. This will produce the orbit of c. (iii) If after N iterations (about 50), the orbit remains bounded (within a circle of radius 4) then the point (a, b) is colored black. If the orbit escapes the circle of radius 4 after n iterations, where $1 < n < 50$, then the point is colored black if n is even and white if n is odd.
(e) Generate a bifurcation diagram for the real-valued map $x_{n+1} = x_n^2 + c$, where $-2.4 \leq c \leq 1.2$. In one graph, plot both the parameter-space diagram from part (d) and the bifurcation diagram from (e). Discuss the results. What does the bifurcation diagram tell you about the phase diagram?

Exercise 5.15 Find the irreducible representations of the cyclic group $\mathbf{Z}_3$ over the complex $\mathbf{C}$. Then determine a nonabsolutely irreducible representation.

Exercise 5.16 Show that the symmetry group of a rectangle is *Abelian*. Note: A group Γ is Abelian if all the group elements commute. That is: $\gamma_1\gamma_2 = \gamma_2\gamma_1$ for all $\gamma_1, \gamma_2 \in \Gamma$.

Exercise 5.17 Consider a steady-state bifurcation problem on a square domain with $\Gamma = \mathbf{D}_4$-symmetry. This group is generated by rotations $\rho = R_{2\pi/4}$ and by a reflection κ across the y-axis. Perform the following tasks:

(a) Write down all the elements of the group.
(b) Consider the natural representation and find all the matrices of this representation that correspond to each group element.
(c) Find all the isotropy subgroups Σ of $\mathbf{D}_4$ with one-dimensional fixed point subspaces.
(d) Classify all possible symmetry-breaking steady-state bifurcations according to their isotropy subgroups.

Exercise 5.18 Repeat the previous exercise with $\Gamma = \mathbf{D}_3$ symmetry.

Exercise 5.19 Hopf Bifurcation with $\Gamma = \mathbf{Z}_2$-symmetry. Consider the one-dimensional nontrivial representation in which the group $\mathbf{Z}_2 \times \mathbf{S}^1$ acts as

$$\begin{aligned}(\kappa, 0) \cdot z &= \\ (e, \theta) \cdot z &= e^{\theta i} z.\end{aligned}$$

Study the symmetry-breaking periodic solutions that emerge via Hopf bifurcations. Explain the form of the periodic solutions.

Exercise 5.20 Hopf Bifurcation with $\Gamma = \mathbf{D}_3$-symmetry. Consider the two-dimensional nontrivial representation in which the group $\mathbf{D}_3 \times \mathbf{S}^1$ acts as

$$\begin{aligned} \rho \cdot (z_1, z_2) &= (e^{-\rho i} z_1, e^{\rho i} z_2) \\ \kappa \cdot (z_1, z_2) &= (z_2, z_1) \\ \theta \cdot (z_1, z_2) &= (e^{\theta i} z_1, e^{\theta i} z_2), \end{aligned}$$

where $\rho = 2\pi/3$ and $\theta \in \mathbf{S}^1$. Notice that this is the same action as that of $\mathbf{O(2)} \times \mathbf{S}^1$ restricted to $\rho = 2\pi/3$.

Study the symmetry-breaking periodic solutions that emerge via Hopf bifurcations. Explain the form of the periodic solutions.

Exercise 5.21 Show that the Ricker model (3.13) exhibits a transcritical bifurcation at $\alpha = 1$.

Exercise 5.22 Show that the quadratic model (5.21) exhibits a transcritical bifurcation at $c = 1/4$.

References

1. H. Poincaré, L'Équilibre d'une masse fluide animée d'un mouvement de rotation. Acta Mathematica **7**, 259–380 (1885)
2. M. Braun, *Differential Equations and Their Applications*, vol. 15, 2nd edn. (Springer, New York, 1978)
3. R. Thom, *Structural Stability and Morphogenesis: An Outline of a General Theory of Models*. (Addison-Wesley Co., 1989)
4. E.C. Zeeman, *Catastrophe Theory-Selected Papers 1972-1977*. (Addison-Wesley Co., 1977)
5. J. Guckenheimer, P.J. Holmes, *Nonlinear Oscillations, Dynamical Systems and Bifurcations of Vector Fields*. (Springer, New York, 1993)
6. S. Wiggins, *Introduction to Applied Nonlinear Dynamical Systems*. (Springer, New York, 1990)
7. S.-N. Chow, J.K. Hale, *Methods of Bifurcation Theory*, Grundlehren der mathematischen Wissenschaften, vol. 251. (Springer, New York, 1982)
8. R. Hoyle, *Pattern Formation. An Introduction to Methods* (Cambridge University Press, 2006)
9. J.D. Murray, *Mathematical Biology*, Biomathematics, vol. 19. (Springer, New York, 1989)
10. J. Keller, S.S. Antman, *Bifurcation Theory and Nonlinear Eigenvalue Problems* (W. A. Benjamin, New York, 1969)
11. J. Marsden, T.J.R. Hughes, *Mathematical Foundations of Elasticity*. (Dover Publications, 1983)
12. M. Golubitsky, D.G. Schaeffer, *Singularities and Groups in Bifurcation Theory Vol. I*, vol. 51. (Springer, New York, 1984)
13. G. Nicolis, I. Prigogine, *Self-Organizations in Non-equilibrium Systems*. (Wiley-Interscience, 1977)
14. C. Henry Edwards, *Advanced Calculus of Several Variables* (Dover Publications, 1994)
15. K.T. Alligood, T.D. Sauer, J.A. Yorke, *Chaos: An Introduction to Dynamical Systems*. (Springer, New York, 1996)
16. J.J. Hopfield, Neural networks and physical systems with emergent collective computational abilities. Proc. Natl. Acad. Sci. USA **79**, 2554–2558 (1982)
17. J.J. Hopfield, Neurons with graded response have collective computational properties like those of two-state neurons. Proc. Natl. Acad. Sci. USA **81**, 3088–3092 (1984)

18. R. Haschke, *Bifurcations in discrete-time neural networks: controlling complex network behaviour with inputs*. Ph.D. thesis, Bielefeld University, Germany (2003)
19. J. Argyris, G. Faust, M. Haase, *An Exploration of Chaos*. (North-Holland, Amsterdam, 1994)
20. Y. Kutnetsov, *Elements of Applied Bifurcation Theory* (Springer, 2004)
21. A.N.W. Hone, M.V. Irle, G.W. Thurura, On the neimark-sacker bifurcation in a discrete predator-prey system. J. Biol. Dyn. **4**(6) (2010)
22. P. Glendinning, *Stability, Instability and Chaos: An Introduction to the Theory of Nonlinear Differential Equations*. (Cambridge University Press, Cambridge, UK, 1999)
23. M. Golubitsky, I.N. Stewart, Patterns of oscillations in coupled cell systems, in *Geometry, Mechanics, and Dynamics*. ed. by P. Holmes, A. Weinstein (Springer, New York, 2002), p. 243
24. M. Golubitsky, I. Stewart, *The Symmetry Perspective*. (Birkháuser, Basel, Switzerland, 2000)
25. M. Golubitsky, I.N. Stewart, D.G. Schaeffer, *Singularities and Groups in Bifurcation Theory Vol. II*, vol. 69. (Springer, New York, 1988)
26. M. Gorman, C. Hamill, M. el Hamdi, K. Robbins, Rotating and modulated rotating states of cellular flames. Combust. Sci. Technol. **98**, 25–35 (1994)
27. A. Palacios, G. Gunaratne, M. Gorman, K. Robbins, Cellular pattern formation in circular domains. Chaos **7**(3), 463–475 (1997)
28. J.M. Mahaffy, L. Edelstein-Keshet, Modeling cyclic waves of circulating t cells in autoimmune diabetes. SIAM J. Appl. Math. **67**, 915–937 (2007)

Chapter 6
Network-Based Modeling

Some continuous models are developed and analyzed in Chaps. 4 and 5. The models studied are fairly low dimensional, which allowed complete analysis. This chapter extends our modeling techniques to higher dimensional problems, especially ones with certain symmetries. Systems made up of individual units coupled together, either weakly or tightly, create an important class of models described as *complex networks*. Many studies, including ongoing research, consider individual identical neurons described by ODEs, which are coupled in various ways to each other through excitatory and/or inhibitory connections. These studies of neural networks find patterns of behavior to explain observed phenomena.

Studies of complex networks span a variety of fields. The dynamics of arrays of Josephson junctions [1–4], central pattern generators in biological systems [5–7], coupled laser systems [8, 9], synchronization of chaotic oscillators [10, 11], collective behavior of bubbles in fluidization [12], the flocking of birds [13], and synchronization among interconnected biological and electronic nonlinear oscillators are a few representative examples of these models. There are three primary components to this class of models. First, each unit or compartment forms the *internal dynamics* and is described by an ODE or small set of ODEs. Second, there is a *topology of connections*, where each unit or compartment sends and/or receives information from one or more of the other related components in the model, creating the network. Finally, the *type of coupling*, including the strength of the connection and whether it has a negative or positive feedback, plays an important role in these systems. Much of the interest in these network models arises from the fact that the individual units cannot exhibit the complex behavior of the entire network.

This chapter begins with an analytical tool for determining asymptotic stability of large systems of linear ODEs. Subsequently, a number of network models are developed and analyzed showing variations in the basic model equations governing each individual unit, differing topologies of connections, and a variety of types of coupling between the units. Analysis of these systems show the robustness of certain structures and allow perturbations to the assumption of identical individual units. The

A. Palacios, *Mathematical Modeling*, Mathematical Engineering,
https://doi.org/10.1007/978-3-031-04729-9_6

models introduce how *symmetry* alone can restrict the type of solutions. The work of Golubitsky [14–16] lays down the theoretical foundations for a model-independent analysis to understand and predict the behavior of a dynamical system using the underlying symmetries of the system, while separating the fine details of the model. The examples below illustrate how these complex network models are successful in explaining numerous theoretical and experimental observations, providing valuable analytic modeling tools. Moreover, our examples extend the theoretical studies to practical cases that showcase how such models are used to design novel technologies.

6.1 Routh-Hurwitz Criterion

This chapter examines a number of potentially large systems with symmetries. An important aspect of systems control is knowing when a dynamical system is stable. These systems, which arise in modeling electric circuits, structural design, and other related problems, can have high dimensions, complicating stability analysis. In Chap. 4 local stability is established for a system of ODEs at its equilibria from its linearization and determining if all eigenvalues have negative real part from the characteristic polynomial. (See the Stable Manifold Theorem 4.3.) More specifically, consider a nonlinear autonomous system of the form:

$$\frac{dX}{dt} = F(X, \mu), \tag{6.1}$$

where $X \in \mathbb{R}^N$ is the state variable and $\mu \in \mathbb{R}^p$ is a vector of parameters. Assume X_0 is an equilibrium. By a suitable change of coordinates, it is always possible to shift X_0 to the origin. Thus, without loss of generality, we may assume $X_0 = 0$. The linearization of Eq. (6.1) about X_0 yields the Jacobian matrix

$$(dF)_{(0,0)} = \begin{bmatrix} a_{11} & a_{12} & \dots & a_{1n} \\ a_{21} & a_{22} & \dots & a_{2n} \\ \vdots & \vdots & \ddots & \vdots \\ a_{n1} & a_{n2} & \dots & a_{nn} \end{bmatrix}.$$

The stability of the trivial equilibrium is determined by the signs of the real parts of the eigenvalues of the Jacobian matrix $(dF)_{(0,0)}$. These eigenvalues are obtained by solving the auxiliary equation

$$\det((dF)_{(0,0)} - \lambda I) = 0,$$

which leads to finding the roots of the characteristic polynomial

$$f(\lambda) = a_0\lambda^n + a_1\lambda^{n-1} + a_2\lambda^{n-2} + \cdots + a_n = 0, \qquad a_0 \neq 0. \tag{6.2}$$

However, determining the roots of the characteristic polynomial is a non-trivial problem. One tool from control theory, which allow us to determine whether all roots have negative real parts without solving the characteristic polynomial itself, is the Routh-Hurwitz Criterion [17, 18].

Proposition 6.1 *Consider the characteristic Eq. (6.2) and assume $a_0 > 0$. If any, $a_i < 0$, for $i = 1, \ldots, n$, then there exists some root, λ, with* Re(˘) > 0. *So if (6.2) is the characteristic equation for some system of ODEs, then the zero equilibrium of that system is unstable.*

This Proposition shows that a necessary condition for stability of the equilibrium of a system of ODEs is that its characteristic Eq. (6.2) has all of the $a_i > 0$, for $i = 0, \ldots, n$. The Routh-Hurwitz Criterion gives necessary and sufficient conditions for stability of the equilibrium of a system of ODEs.

Theorem 6.1 (Routh-Hurwitz Criterion) *All roots of the characteristic Eq. (6.2) with $a_0 > 0$ have negative real parts if and only if the following determinant inequalities:*

$$|a_1| > 0, \quad \begin{vmatrix} a_1 & a_3 \\ a_0 & a_2 \end{vmatrix} > 0, \quad \begin{vmatrix} a_1 & a_3 & a_5 \\ a_0 & a_2 & a_4 \\ 0 & a_1 & a_3 \end{vmatrix} > 0, \quad \begin{vmatrix} a_1 & a_3 & a_5 & \cdots & 0 \\ a_0 & a_2 & a_4 & \cdots & 0 \\ 0 & a_1 & a_3 & \cdots & 0 \\ 0 & a_0 & a_2 & \cdots & 0 \\ \vdots & \vdots & \vdots & \ddots & \vdots \\ \cdot & \cdot & \cdot & \cdot & a_n \end{vmatrix} > 0, \tag{6.3}$$

hold.

The proof of this theorem and several alternative forms can be found in Gantmacher [17]. This result provides a valuable tool for determining the stability of equilibrium points in high dimensional ODEs after finding the characteristic equation, which itself may not be a trivial problem.

6.1.1 Spring-Mass System

This section uses the *Routh-Hurwitz Criterion* to prove the stability of a symmetric system of springs and masses. Consider the two mass and three spring system shown in Fig. 6.1, where the masses and the spring are identical. Since this system loses energy through the viscous damping, c, occurring the same on each mass, one expects that the system of ODEs describing this model should be asymptotically stable.

The model for the system shown in Fig. 6.1 uses Newton's Second Law and extends the example shown in Sect. 4.8.1. The two masses are described by their independent positions, x_1 and x_2, where it is assumed the system is at equilibrium for $x_{1e} = 0$ and $x_{2e} = 0$. Applying Newton's Second Law to each of the masses gives the following linear system of second order ODEs:

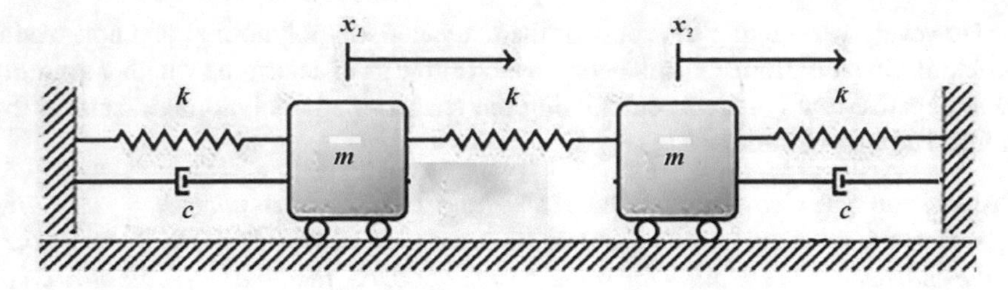

Fig. 6.1 Two mass and three spring system, where the masses, m, and the spring constants, k, are identical. In addition, the masses experience the same viscous damping effects, c

$$\boxed{\begin{aligned} m\ddot{x}_1 &= -kx_1 + k(x_2 - x_1) - c\dot{x}_1 \;=\; -2kx_1 + kx_2 - c\dot{x}_1, \\ m\ddot{x}_2 &= -kx_2 + k(x_1 - x_2) - c\dot{x}_2 \;=\; -2kx_2 + kx_1 - c\dot{x}_2. \end{aligned}} \tag{6.4}$$

To convert Eq. (6.4) into a system of first order ODEs, we let

$$y_1 = x_1, \quad y_2 = \dot{x}_1, \quad y_3 = x_2, \quad \text{and} \quad y_4 = \dot{x}_2.$$

In matrix form this system becomes:

$$\dot{\mathbf{y}} = \begin{pmatrix} \dot{y}_1 \\ \dot{y}_2 \\ \dot{y}_3 \\ \dot{y}_4 \end{pmatrix} = \begin{pmatrix} 0 & 1 & 0 & 0 \\ -\frac{2k}{m} & -\frac{c}{m} & \frac{k}{m} & 0 \\ 0 & 0 & 0 & 1 \\ \frac{k}{m} & 0 & -\frac{2k}{m} & -\frac{c}{m} \end{pmatrix} \begin{pmatrix} y_1 \\ y_2 \\ y_3 \\ y_4 \end{pmatrix} = A\mathbf{y}, \tag{6.5}$$

where y_1 and y_2 are the position and velocity, respectively of the first mass, and y_3 and y_4 are the same for the second mass.

6.1.2 Stability of Spring-Mass System

Since Eq. (6.5) is a system of linear ODEs, the characteristic equation is computed by solving $\det|A - \lambda I| = 0$. As the matrix A has many zeroes, this determinant is readily expanded to give the following characteristic equation:

$$\det|A - \lambda I| = \lambda^4 + \tfrac{2c}{m}\lambda^3 + \tfrac{c^2+4km}{m^2}\lambda^2 + \tfrac{4kc}{m^2}\lambda + \tfrac{3k^2}{m^2} = 0. \tag{6.6}$$

It is obvious that this characteristic equation satisfies the necessary conditions of Proposition 6.1 for the zero equilibrium of Eq. (6.5) to be stable. Thus, it remains to verify the conditions of the Routh-Hurwitz Criterion Theorem 6.1. From the charac-

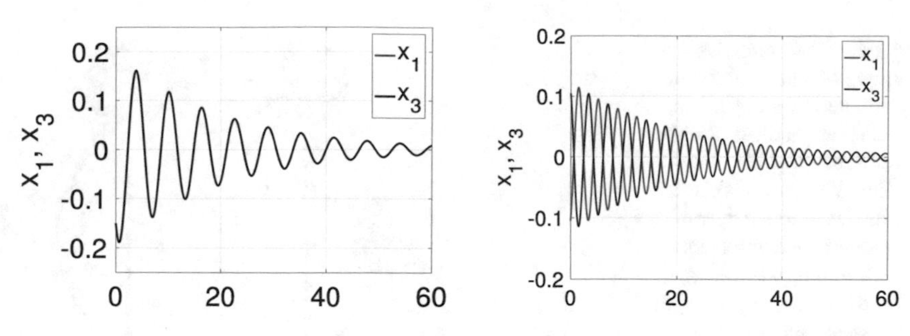

Fig. 6.2 Time series solutions of a two spring-mass system confirm stability of the zero equilibrium, as predicted by the Rouht-Horowitz criterion

teristic equation we have $a_0 = 1 > 0$ and $a_1 = \frac{2c}{m} > 0$. Examining the other determinants gives:

$$\begin{vmatrix} a_1 & a_3 \\ a_0 & a_2 \end{vmatrix} = \begin{vmatrix} \frac{2c}{m} & \frac{4kc}{m^2} \\ 1 & \frac{c^2+4km}{m^2} \end{vmatrix} = \frac{2c^3}{m^3} + \frac{4ck}{m^2} > 0,$$

and

$$\begin{vmatrix} a_1 & a_3 & 0 \\ a_0 & a_2 & a_4 \\ 0 & a_1 & a_3 \end{vmatrix} = \begin{vmatrix} \frac{2c}{m} & \frac{4kc}{m^2} & 0 \\ 1 & \frac{c^2+4km}{m^2} & \frac{3k^2}{m^2} \\ 0 & \frac{2c}{m} & \frac{4kc}{m^2} \end{vmatrix} = \frac{8kc^4}{m^5} + \frac{4c^2k^2}{m^4} > 0,$$

and

$$\begin{vmatrix} a_1 & a_3 & 0 & 0 \\ a_0 & a_2 & a_4 & 0 \\ 0 & a_1 & a_3 & 0 \\ 0 & a_0 & a_2 & a_4 \end{vmatrix} = \begin{vmatrix} \frac{2c}{m} & \frac{4kc}{m^2} & 0 & 0 \\ 1 & \frac{c^2+4km}{m^2} & \frac{3k^2}{m^2} & 0 \\ 0 & \frac{2c}{m} & \frac{4kc}{m^2} & 0 \\ 0 & 1 & \frac{c^2+4km}{m^2} & \frac{3k^2}{m^2} \end{vmatrix} = \frac{24k^3c^4}{m^7} + \frac{12c^2k^4}{m^6} > 0.$$

It follows that the Routh-Hurwitz Criterion Theorem 6.1 holds, so the trivial equilibrium of Eq. (6.5) is asymptotically stable, as we predicted from the physical system. Figure 6.2 illustrates the stability of the zero equilibrium through numerical simulations. Observe that in one case the masses oscillate in-phase with one another, while they approach the zero equilibrium, and in the other case the oscillations are out-of-phase. We will discuss these features (e.g., in-phase and out-of-phase oscillations shortly).

Another way of interpreting the two spring-mass system is as a network of two identical units or cells, which are coupled together to form a larger system. This interpretation leads to the formalism of a *coupled cell system*.

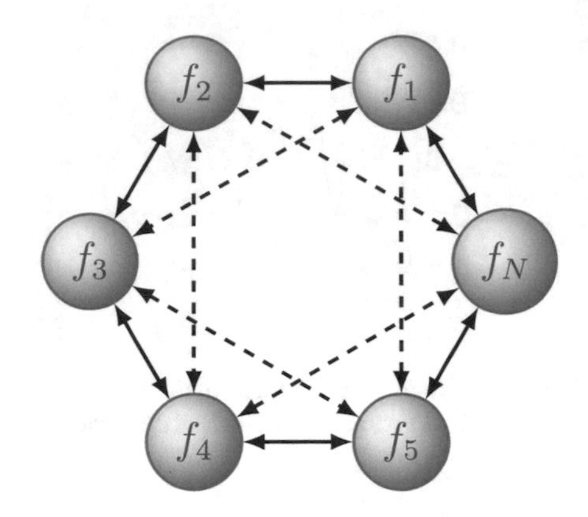

Fig. 6.3 Representative example of a network model represented by a collection of a discrete number of units or cells, coupled in some fashion. Each "cell" or unit has its own internal dynamics governed by a continuous model. In this example, λ_{ij}, defines the coupling strength from node i into node j

6.2 Coupled Cell Systems

A natural mathematical framework for the analysis of network models is that of a *coupled cell system*. By a "cell" we mean an individual component or unit that possesses its own dynamical behavior. Figure 6.3 shows a representative example with N cells and with coupling strengths varying among different nodes.

In what follows, we assume a coupled network with N cells, and consider the internal dynamics of each cell to be governed by a k-dimensional continuous-time system of differential equations of the form

$$\frac{dX_i}{dt} = f_i(X_i, \mu), \tag{6.7}$$

where $X_i = (x_{i1}, \ldots, x_{ik}) \in \mathbb{R}^k$ denotes the state variables of cell i and $\mu = (\mu_1, \ldots, \mu_p)$ is a vector of parameters. Observe that, in this formulation, f depends on i, which implies that the dynamics of each cell can be different. On the other hand, if f is independent of i, then all individual cells would behave identically. The distinction between nonidentical and identical cells lead to two different type of networks. We discuss next these two types of systems in more detail.

Definition 6.1 A *heterogeneous network* is a collection of N distinct cells interconnected in some fashion. We model the network by the following system of coupled differential equations

$$\boxed{\frac{dX_i}{dt} = f_i(X_i, \mu) + \sum_{j \to i} c_{ij} h(X_i, X_j),} \tag{6.8}$$

where h is the coupling function between two cells, the summation is taken over those cells j that are coupled to cell i, and c_{ij} is a matrix of coupling strengths.

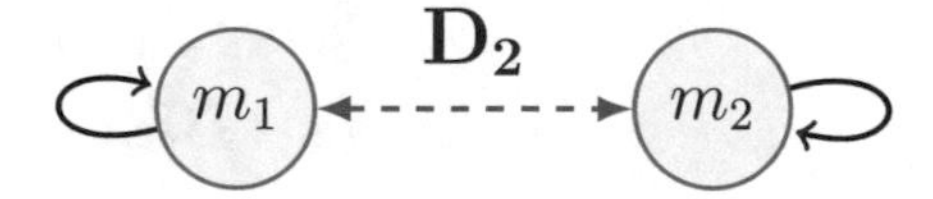

Fig. 6.4 Two spring-mass system visualized as a coupled cell system

Heterogeneous coupled networks of (two-dimensional) biological oscillators, in which the internal dynamics f_i of each individual cell exhibits a limit cycle oscillation, have been extensively studied by Winfree [19, 20]. Winfree's work shows that under certain assumptions, mainly weak coupling, the phase-amplitude dynamics decouples, so that the model equations can be reduced to a network of phase oscillators. The reduction led Kuramoto [21, 22] to show, eventually, that under additional assumptions it was possible to obtain conditions for the existence of analytical solutions.

In contrast, in the studies of diffusion-drive instabilities (see Chap. 8, Sect. 8.2), Alan Turing proposed that under certain conditions, coupled networks with identical cell dynamics can react and diffuse in such a way to destabilize a equilibrium point, thus leading to a heterogenous pattern. These type of networks are defined next.

Definition 6.2 A *homogenous network* is a coupled network in which

$$f_i(X_i) = f(X_i),$$

for all cells in the network.

Most of the networks considered in this chapter are homogeneous networks. Additionally, if we let $X = (X_1, \ldots, X_N)$ denote the state variable of the network, and $F = (f_1, \ldots, f_N)$ as the vector of the internal dynamics of each cell, then we can write (6.8) in the simpler form

$$\frac{dX}{dt} = F(X),$$

where the dependence on the parameters μ has been omitted for brevity. Let's take a look at a couple of concrete examples of network systems to get some insight into the main ideas and concepts.

Example 6.1 (Spring-Mass System) Consider again the two spring-mass system of Fig. 6.1. Since the masses are identical, we can interpret the system as a homogeneous network with two cells, as is shown in Fig. 6.4. Since the cells are identical, it follows that the network has

$$\Gamma = \mathbf{D}_2$$

symmetry, where $\mathbf{D}_2$ is the orthogonal group of symmetries of a rectangle.

If we let $Y_1 = [y_1, y_2]^T$ and $Y_2 = [y_3, y_4]^T$, then we can re-write Eq. (6.5) as a homogeneous coupled cell system of the form

$$\begin{aligned}\dot{Y}_1 &= f(Y_1; \mu) + C(k)(Y_2 - Y_1)\\ \dot{Y}_2 &= f(Y_2; \mu) + C(k)(Y_1 - Y_2),\end{aligned} \tag{6.9}$$

where the internal dynamics of each mass, $j = 1, 2$, and the coupling matrix, C, are given by

$$f(Y_j; \mu) = \begin{bmatrix} 0 & 1 \\ -\dfrac{k}{m} & -\dfrac{c}{m} \end{bmatrix}, \quad C(k) = \begin{bmatrix} 0 & 0 \\ -\dfrac{k}{m} & 0 \end{bmatrix}.$$

with parameters $\mu = [m, c, k]^T$. The linearization of Eq. (6.9) can be expressed as

$$L = \begin{bmatrix} (df)_{(0,0)} - C & C \\ C & (df)_{(0,0)} - C \end{bmatrix}.$$

To study the linearization of Eq. (6.9) about the zero equilibrium $Y_0 = (0, 0, 0, 0)$, we first note that $Y_1 \in \mathbb{R}^2$ and $Y_2 \in \mathbb{R}^2$, so that the phase-space of the coupled cell system is $\mathbb{R}^4$. Then we complexify from $\mathbb{R}^4$ to $\mathbf{C}^4$, and, next, we employ the *isotypic decomposition* of $\mathbf{C}^4$ by $\Gamma = \mathbf{D}_2$, which is given by

$$\mathbf{C}^4 = V_0 \oplus V_1,$$

where the cyclic nature of L leads directly to the subspaces $V_{j's}$:

$$\begin{aligned} V_0 &= [v, v]^T, \\ V_1 &= [v, \zeta v]^T, \end{aligned}$$

where $\zeta = e^{2\pi i/2} = -1$ and $v \in \mathbb{R}^2$. In fact, the subspaces $V_{j's}$ are eigenvectors

$$\begin{aligned} V_0 &= [v, v]^T, \\ V_1 &= [v, -v]^T, \end{aligned}$$

of the linearized matrix L of Eq. (6.11). Direct calculations yield:

$$\begin{aligned} LV_0 &= (df)_{(0,0)} V_0, \\ LV_1 &= ((df)_{(0,0)} - 2C(\lambda))V_1. \end{aligned}$$

Using coordinates along the isotypic components, we find the eigenvalues of the linearization, L, to be those of the matrices:

$$(df)_{(0,0)}, \qquad \text{and} \qquad (df)_{(0,0)} - 2C(k).$$

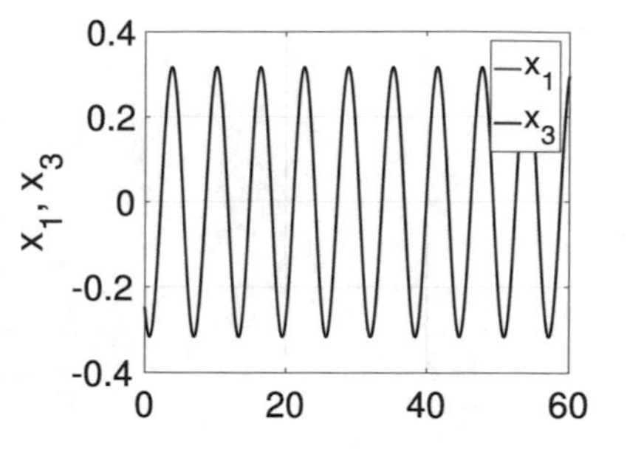

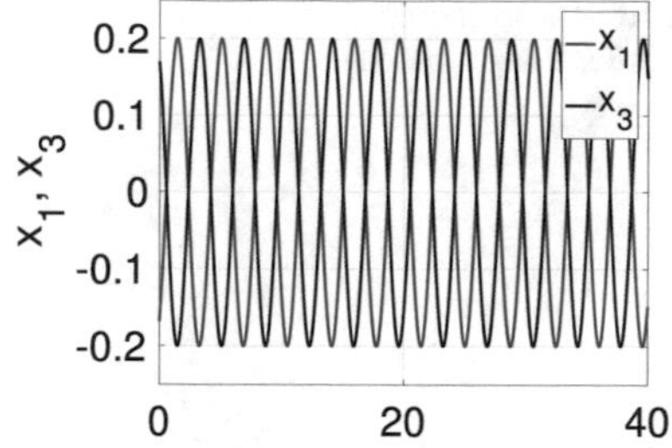

Fig. 6.5 In-phase and out-of-phase oscillations in a two spring mass system, predicted on the basis of symmetry of a coupled network

Consequently, a Hopf bifurcation in the network Eq. (6.9) is possible if the eigenvalues of the matrices $(df)_{(0,0)}$ or $(df)_{(0,0)} - 2C(k)$ are purely imaginary. Direct calculations yield, respectively, the following eigenvalues

$$\sigma_{1,2} = -\frac{c}{m} \pm \frac{1}{2m}\sqrt{c^2 - 4km}, \quad \sigma_{3,4} = -\frac{c}{m} \pm \frac{1}{2m}\sqrt{c^2 - 12km}.$$

Thus, in the absence of damping, i.e., when $c = 0$, we can get purely imaginary eigenvalues:

$$\sigma_{1,2} = \pm\sqrt{\frac{k}{m}}i, \qquad \sigma_{3,4} = \pm\sqrt{\frac{3k}{m}}i.$$

Each pair of eigenvalues is associated with a different mode of oscillation. Since the eigenvalues are simple, it follows that the two modes of oscillations correspond to standard Hopf bifurcation. The first mode is associated with the subspace $V_0 = [v, v]^T$, so the collective pattern corresponds to synchronized oscillations, i.e., same wave form and same phase, with frequency $\omega_1 = \sqrt{k/m}$, as is shown in Fig. 6.5(left). Similarly, the second mode of oscillation is associated with the second subspace, $V_1 = [v, -v]^T$, so the collective pattern corresponds to out-of-phase oscillations, with frequency $\omega_2 = \sqrt{3k/m}$, see Fig. 6.5(right).

These collective patterns of oscillations are model-independent features of the symmetry of the network. That is, they can be observed with completely different sets of oscillators, as long as the symmetries and Hopf bifurcations are preserved. And while the patterns appeared via standard Hopf bifurcations, it is also possible to have patterns that appear via symmetry-breaking bifurcations.

6.3 Self-Oscillating Networks

Many systems are known to oscillate only when they are driven by an external force. A single fluxgate magnetometer discussed in Chap. 4 is one representative example. Indeed, recall that oscillations in the fluxgate model Eq. (4.73) can only

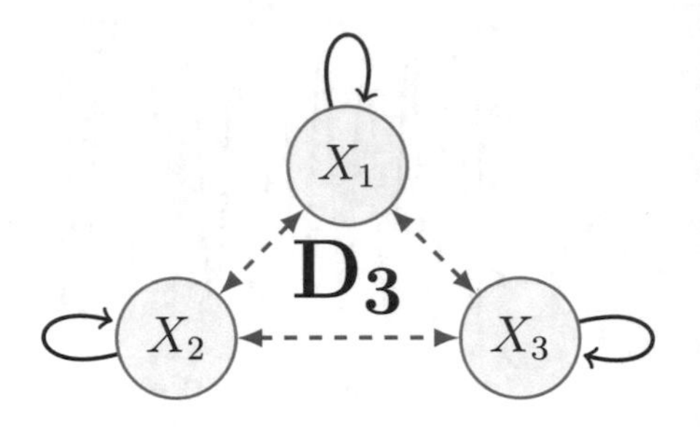

Fig. 6.6 Bidirectionally coupled network with $\mathbf{D}_3$ symmetry

be sustained under the presence of a nonzero external biasing signal $H_e(t)$. That is, only when the model is non-autonomous. Yet, it might be possible, however, for an autonomous system to produce oscillations even if the individual units cannot oscillate. This dichotomy is resolved by considering the effects of coupling. That is, in the absence of coupling (i.e., a coupled cell system with zero coupling matrix $c_{ij} = 0$), the individual cells might not oscillate. But when the coupling is on then the collective network may be able to produce oscillations. This effect was first noticed by Smale [23] in his studies of a two-cell coupled system. In principle, the effect can be applied to create network of fluxgate magnetometer with self-sustained oscillations. But before we study that system in greater detail, let's take a look at a phenomenological model of self-induced oscillations in a (relatively simple) three-cell coupled system.

Example 6.2 (Three Bidirectionally Coupled Oscillators) Consider the three-cell network illustrated in Fig. 6.6, in which each cell is described by a two-state variable $X_j = (x_j, y_j)$, where $j = 1, 2, 3$. And each cell is coupled bidirectionally to its nearest neighbor.

Thus, the internal dynamics of each cell is governed by a two-dimensional system $f(X_j)$. In this particular example, the internal dynamics is given by

$$f(X_j;\lambda) = \begin{bmatrix} -4 & 1 \\ -1 & -4 \end{bmatrix}\begin{bmatrix} x_j \\ y_j \end{bmatrix} + p(x_j^2 + y_j^2)\begin{bmatrix} x_j \\ y_j \end{bmatrix} + q(x_j^2 + y_j^2)\begin{bmatrix} -y_j \\ x_j \end{bmatrix} + 2C(\lambda)\begin{bmatrix} x_j \\ y_j \end{bmatrix}, \tag{6.10}$$

where the coupling matrix $C(\lambda)$ is given by

$$C(\lambda) = \lambda \begin{bmatrix} -4 & 2 \\ -2 & -4 \end{bmatrix}.$$

The homogeneous network equations can then be written as follows

$$\begin{aligned} \dot{X}_1 &= f(X_1;\lambda) + C(\lambda)(X_3 - 2X_1 + X_2) \\ \dot{X}_2 &= f(X_2;\lambda) + C(\lambda)(X_1 - 2X_2 + X_3) \\ \dot{X}_3 &= f(X_3;\lambda) + C(\lambda)(X_2 - 2X_3 + X_1). \end{aligned} \tag{6.11}$$

Figure 6.7 illustrates the effects of coupling on the network's response. When coupling is absent, i.e., $\lambda = 0$, none of the units is capable of oscillating, as is shown

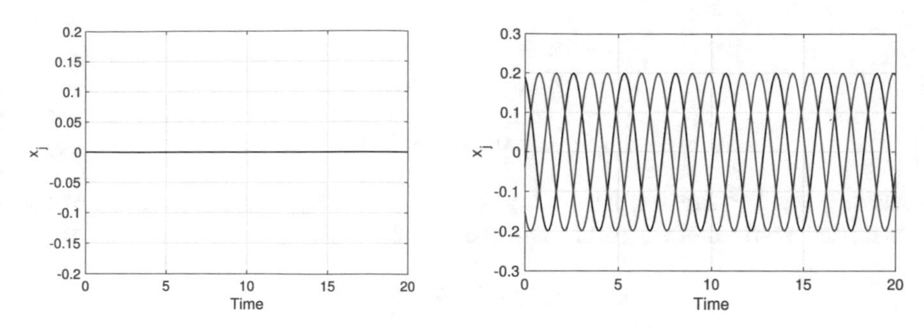

Fig. 6.7 (Left) In the absence of coupling, i.e., $\lambda = 0$, none of the nodes in the network of Fig. 6.6 can oscillate. (right) But with small coupling, $\lambda = 1.05$, the network can, collectively, oscillate in a traveling wave pattern. Parameters are: $p = -5, q = 30$

in Fig. 6.7(left). But with a small amount of coupling strength, e.g., $\lambda = 1.05$, the collective response leads to oscillations, see Fig. 6.7(right).

Bifurcation Analysis

Generically, there are two mechanisms that can lead to oscillatory behavior in the network dynamics: via standard Hopf bifurcations, which can be associated with completely synchronized solutions, i.e., *symmetry-preserving* oscillations, or via *symmetry-breaking* Hopf bifurcations, which can be associated with collective patterns of oscillations. To investigate both types of Hopf bifurcations, we first notice that bidirectional coupling leads to a symmetric network, in which the symmetries are described by the group of symmetries of a triangle, that is

$$\Gamma = \mathbf{D}_3.$$

Next, we compute the linearization of the network equations near the zero equilibrium $X = (X_1, X_2, X_3) = (0, 0, 0)$, which yields

$$L = \begin{bmatrix} (df)_{(0,0)} - 2C(\lambda) & C(\lambda) & C(\lambda) \\ C(\lambda) & (df)_{(0,0)} - 2C(\lambda) & C(\lambda) \\ C(\lambda) & C(\lambda) & (df)_{(0,0)} - 2C(\lambda) \end{bmatrix},$$

where $(df)_{(0,0)}$ is the linearization of the internal dynamics of each cell at the trivial equilibrium $(X, \lambda) = (0, 0)$. Since the internal dynamics of each individual cell is governed by a two-dimensional system, then it follows that $\mathbb{R}^6$ is the phase-space of the entire network. To study the linearization, L, of the network dynamics, we complexify from $\mathbb{R}^6$ to $\mathbf{C}^6$, and then, we employ the well-known *isotypic decomposition* of $\mathbf{C}^6$ by $\Gamma = \mathbf{D}_3$, which is given by

$$\mathbf{C}^6 = V_0 \oplus V_1 \oplus V_2,$$

where, the cyclic nature of L leads directly to the subspaces $V_{j's}$:

$$V_0 = [v, v, v]^T,$$
$$V_1 = [v, \zeta v, \zeta^2 v]^T,$$
$$V_2 = [v, \zeta^2 v, \zeta v]^T,$$

where $\zeta = e^{2\pi i/3}$ and $v \in \mathbb{R}^2$. In fact, the subspaces $V_{j's}$ are eigenvectors of the linearized matrix L of Eq. (6.11). Direct calculations yield:

$$LV_0 = (df)_{(0,0)} V_0,$$
$$LV_1 = ((df)_{(0,0)} - 3C(\lambda))V_1,$$
$$LV_2 = ((df)_{(0,0)} - 3C(\lambda))V_2.$$

Consequently, using coordinates along the isotypic components, we find the eigenvalues of the linearization, L, to be those of the matrices:

$$(df)_{(0,0)} \quad \text{and} \quad (df)_{(0,0)} - 3C(\lambda) \quad \text{(twice)}.$$

It follows that a Hopf bifurcation in the network Eq. (6.11) may occur provided that the eigenvalues of the matrices $(df)_{(0,0)}$ or $(df)_{(0,0)} - 3C(\lambda)$ are purely imaginary. These eigenvalues are: $\sigma_1 = -4 \pm i$ and $\sigma_{2,3} = -4 + 12\lambda \pm 3i$, respectively.

Actually, taking coordinates along the isotypic decomposition allows us to diagonalize the linearization matrix L, so the eigenvalues can be more easily computed. To see this, let e_1, e_2 be the canonical basis of $\mathbb{R}^2$ and define

$$v_{ji} = [e_i, \zeta^j e_i, \zeta^{2j} e_i]^T,$$

for $i = 1, 2$, and $j = 0, 1, 2$. Then a real basis for $\Re^{2N}$ is

$$\left\{v_{0,1}, v_{02}, \Im_{11}, \Im_{12}, \Re_{11}, \Re_{12}\right\},$$

where $\Im_{jk}$ and $\Re_{jk}$ denote the imaginary and real part, respectively, of the vector v_{jk}.

A straightforward computation shows that the transition matrix P which brings the system to block diagonal form of the isotypic decomposition has columns given by the basis vectors. That is,

$$P = \left[v_{0,1}, v_{02}, \Im_{11}, \Im_{12}, \Re_{11}, \Re_{12}\right].$$

Thus, applying the substitution

$$X = PU,$$

to the linear part of Eq. (6.11), we obtain

$$\mathcal{L} = P^{-1}LP = \begin{bmatrix} (df)_{(0,0)} & 0 & 0 \\ 0 & (df)_{(0,0)} - 3C(\lambda) & 0 \\ 0 & 0 & (df)_{(0,0)} - 3C(\lambda) \end{bmatrix}.$$

The same transformation can be applied to the nonlinear part of system (6.11). The important fact here is that the isotypic decomposition of the phase space allows us to write the linear part in diagonal form and then compute its eigenvalues directly from each block.

To uncover the type of collective patterns that may emerge through these Hopf bifurcations, we need to study in more detail the action of the group of symmetries, $\Gamma = \mathbf{D}_3$, on the phase space $\mathbb{R}^6$. $\mathbf{D}_3$ has two generators, a cyclic rotation, γ, by $2\pi/3$, and a reflection, κ. Together, these two generators act on $\mathbb{R}^6$ as follows

$$\gamma \cdot (X_0, X_1, X_2) = (X_1, X_2, X_0)$$
$$\kappa \cdot (X_0, X_1, X_2) = (X_0, X_2, X_1).$$

Consider the first subspace V_0. Observe that both γ and κ act trivially on V_0. Since $v \in \mathbb{R}^2$, it follows that $\mathbf{D}_3$ acts on V_0 by 2 copies of the trivial action on $\mathbb{R}$. Furthermore, the eigenvalues of $(df)_{(0,0)} = -4 \pm i$ are simple, so it follows that this case corresponds to a standard Hopf bifurcation. And since $\mathbf{D}_3$ acts trivially on V_0, the collective pattern is one where all cells oscillates identically, i.e., complete synchronization–same wave form and same phase for each cell.

Now, consider the subspaces V_1 and V_2. In fact, consider the invariant compliment subspace, $V_1 \oplus V_2 = \mathbb{R}^4$, to V_0. It can be shown that this compliment subspace is the sum of two copies of the nontrivial action of $\mathbf{D}_3$ on $\mathbb{R}^2$. And if $[u, v, w] \in \mathbb{R}^6$ then $V_1 \oplus V_2$ is spanned by $[v, w, -v - w]$. This time the corresponding eigenvalues of $(df)_{(0,0)} - 3C(\lambda) = -4 + 12\lambda \pm 3i$ have multiplicity two, and the patterns of collective behavior emerge via $\mathbf{D}_3$ symmetry-breaking Hopf bifurcations. The corresponding theory [15] indicates three possible patterns, which can be described based on their isotropy subgroups, Σ_X, of symmetries, as follows.

(i) $\Sigma_X = \tilde{\mathbf{Z}}_3$.
This pattern is a standard traveling wave in which consecutive cells oscillate out of phase by $2\pi/3$, as is shown in Fig. 6.7(right).

(ii) $\Sigma_X = \mathbf{Z}_2(\kappa)$.
In this case two cells oscillate identically while the third cell oscillates with a different wave form.

(iii) $\Sigma_X = \mathbf{Z}_2(\kappa, \pi)$.
Two cells oscillate with the same waveform but out of phase by π while the third cell oscillates at twice the frequency of the other two cells. In this sense, the third cell is said to be π out of phase with itself.

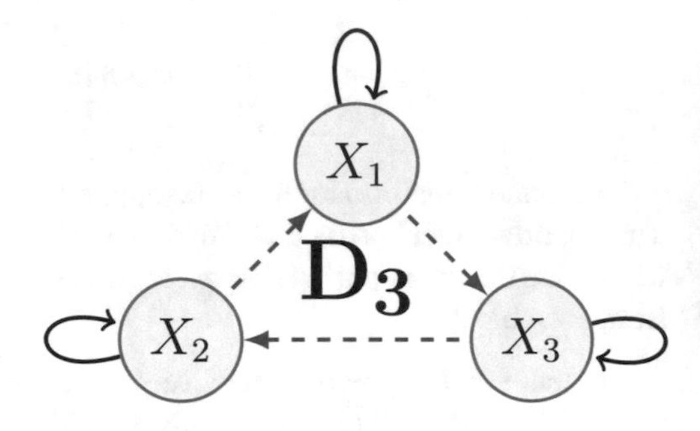

Fig. 6.8 Unidirectionally coupled network with $\mathbf{Z}_3$ symmetry

6.4 Unidirectionally Coupled Colpitts Oscillators

We now consider a three-cell network, similar to that of thew previous example, but with unidirectional coupling, as is shown in Fig. 6.8. In this case, each individual node is capable of producing its own oscillations.

If we let $X_j = \left[x_{j,1}, x_{j,2}, x_{j,3}\right]^T$ then the homogeneous network equations can be written as follows

$$\begin{aligned} \dot{X}_1 &= f(X_1) + C(\lambda)(X_2 - X_1) \\ \dot{X}_2 &= f(X_2) + C(\lambda)(X_3 - X_2) \\ \dot{X}_3 &= f(X_3) + C(\lambda)(X_1 - X_3), \end{aligned} \tag{6.12}$$

where $f(X_j)$ contains the internal dynamics of a Colpitts oscillator (see Exercises in Chap. 4). In fact, we assume a dimensionless, ideal model, where the internal dynamics of each oscillator is of the form:

$$f(X_j) = \begin{bmatrix} \dfrac{g}{Q(1-\kappa)}\left[-\left(e^{-x_{j,2}} - 1\right) + x_{j,3}\right] \\ \dfrac{g}{Q\kappa} x_{j,3} \\ -\dfrac{Q\kappa(1-\kappa)}{g}(x_{j,1} + x_{j,2}) - \dfrac{1}{Q} x_{j,3} \end{bmatrix}$$

and $C(\lambda)$ is the coupling matrix

$$C(\lambda) = \lambda \begin{bmatrix} 1 & 0 & 0 \\ 0 & 0 & 0 \\ 0 & 0 & 0 \end{bmatrix}.$$

Bifurcation Analysis

This time the phase space is $\mathbb{R}^9$ since each cell is governed by a tree-dimensional system of equations. Due to the nature of the unidirectional coupling, the network remains unchanged under cyclic permutations of the nodes. This means that the group

$$\Gamma = \mathbf{Z}_3,$$

of cyclic permutations of $N = 3$ objects, defines the group of symmetries of the network. Next, we compute the linearization of the network equations near the zero equilibrium, $X = (X_1, X_2, X_3) = (0, 0, 0)$, which yields

$$L = \begin{bmatrix} (df)_{(0,0)} - C(\lambda) & C(\lambda) & 0 \\ 0 & (df)_{(0,0)} - C(\lambda) & C(\lambda) \\ C(\lambda) & 0 & (df)_{(0,0)} - C(\lambda) \end{bmatrix},$$

where $(df)_{(0,0)}$ is the linearization of the internal dynamics of each cell at the trivial equilibrium $(X, \lambda) = (0, 0)$. Since the internal dynamics of each individual cell is governed by a three-dimensional system, then it follows that $\mathbb{R}^9$ is the phase-space of the entire network. To study the linearization, L, of the network dynamics, we complexify from $\mathbb{R}^9$ to $\mathbf{C}^9$, and then, we employ the well-known *isotypic decomposition* of $\mathbf{C}^9$ by $\Gamma = \mathbf{Z}_3$, which is given by

$$\mathbf{C}^9 = V_0 \oplus V_1 \oplus V_2,$$

where, the cyclic nature of L leads directly to the subspaces $V_{j's}$:

$$\begin{aligned} V_0 &= [v, v, v]^T, \\ V_1 &= [v, \zeta v, \zeta^2 v]^T, \\ V_2 &= [v, \zeta^2 v, \zeta v]^T, \end{aligned}$$

where $\zeta = e^{2\pi i/3}$ and $v \in \mathbb{R}^3$. As in the previous case of bidirectional coupling, the subspaces $V_{j's}$ are eigenvectors of the linearized matrix L of Eq. (6.12). Direct calculations yield:

$$\begin{aligned} LV_0 &= (df)_{(0,0)} V_0, \\ LV_1 &= ((df)_{(0,0)} - C(\lambda) + \zeta C(\lambda))V_1, \\ LV_2 &= ((df)_{(0,0)} - C(\lambda) + \zeta^2 C(\lambda))V_2. \end{aligned}$$

Consequently, the eigenvalues of L are those of the matrices:

$$(df)_{(0,0)}, \quad (df)_{(0,0)} - C(\lambda) + \zeta C(\lambda), \quad \text{and} \quad (df)_{(0,0)} - C(\lambda) + \zeta^2 C(\lambda).$$

It follows that a Hopf bifurcation in the network Eq. (6.12) may occur provided that the eigenvalues of the matrices $(df)_{(0,0)}$ or $(df)_{(0,0)} - C(\lambda) + \zeta C(\lambda)$ or $(df)_{(0,0)} - C(\lambda) + \zeta^2 C(\lambda)$ are purely imaginary.

We can also compute explicitly the transformation matrix P, which allows us to diagonalize the linearization matrix, L along the isotypic components. To do that, let $\{e_1, e_2, e_3, \}$ be the canonical basis of $\mathbb{R}^3$ and define

$$v_{ji} = [e_i, \zeta^j e_i, \zeta^{2j} e_i]^T,$$

for $i = 1, 2, 3$, and $j = 0, 1, 2$. Then a real basis for $\Re^{3N}$ is

$$\left\{v_{0,1}, v_{02}, v_{03}, \Im_{11}, \Im_{12}, \Im_{13}, \Re_{11}, \Re_{12}, \Re_{13}\right\},$$

where $\Im_{jk}$ and $\Re_{jk}$ denote the imaginary and real part, respectively, of the vector v_{jk}.

A straightforward computation shows that the transition matrix P which brings the system to block diagonal form of the isotypic decomposition has columns given by the basis vectors. That is,

$$P = \left[v_{0,1}, v_{02}, v_{03}, \Im_{11}, \Im_{12}, \Im_{13}, \Re_{11}, \Re_{12}, \Re_{13}\right]$$

Thus, applying the substitution

$$X = PU,$$

to the linear part of Eq. (6.12), we obtain

$$\mathcal{L} = P^{-1}LP = \begin{bmatrix} (df)_{(0,0)} & 0 & 0 \\ 0 & (df)_{(0,0)} - C(\lambda) + \zeta C(\lambda) & 0 \\ 0 & 0 & (df)_{(0,0)} - C(\lambda) + \zeta^2 C(\lambda) \end{bmatrix}.$$

In Sect. 5.5.5 we studied the eigenvalues of $(df)_{(0,0)}$, which correspond to the linearization of the internal dynamics of each (assumed to be identical) colpitts oscillator. It is important to recall them here

$$\begin{aligned} \sigma_1 &= -\frac{1 + gQ^2}{Q\left(Q^2+1\right)}, \\ \sigma_{2,3} &= \frac{Q}{2\left(Q^2+1\right)}(g-1) \pm \left[1 + \frac{1}{2\left(Q^2+1\right)}(g-1)\right] i. \end{aligned}$$

Observe that when $g = 1$ the first eigenvalue, σ_1, is real and negative (since the quality factor Q is always positive), while the remaining eigenvalues $\sigma_{2,3}$ are purely imaginary, i.e., $\sigma_{2,3} = 0 \pm i$. This appears to be a Hopf bifurcation. Indeed, in Sect. 5.5.5 we showed proof that a supercritical Hopf bifurcation occurs at $g = 1$. In the context of coupled colpitts, the block matrix $(df)_{(0,0)}$ corresponds to the trivial representation of the symmetry group $\mathbf{Z}_3$. Since the eigenvalues of $(df)_{(0,0)}$ are simple, it follows that this case corresponds to a standard Hopf bifurcation. In other words, a symmetry-preserving Hopf bifurcation that leads to in-phase oscillations, same wave form and same period. Furthermore, the eigenvalues of $(df)_{(0,0)}$ are invariant under changes of coupling strength. Then, the branch of in-phase oscillations exists for all values of λ.

We now derive expressions to approximate the eigenvalues of the remaining blocks. Let

$$\mathcal{M}_1 = (df)_{(0,0)} - C(\lambda) + \zeta C(\lambda)$$
$$\mathcal{M}_2 = (df)_{(0,0)} - C(\lambda) + \zeta^2 C(\lambda).$$

If we let $\mathcal{A} = -1 + \cos(2\pi/3) \pm \sin(2\pi/3)i$ then the blocks $\mathcal{M}_1$ and $\mathcal{M}_2$ can be written as

$$\mathcal{M}_{1,2} = \begin{bmatrix} \mathcal{A}\lambda & \dfrac{g}{Q(1-\kappa)} & \dfrac{g}{Q(1-\kappa)} \\ 0 & 0 & \dfrac{g}{Qk} \\ \dfrac{-Q\kappa(1-\kappa)}{g} & \dfrac{-Q\kappa(1-\kappa)}{g} & -\dfrac{1}{Q} \end{bmatrix},$$

where $\mathcal{M}_1$ is recovered when the positive sign in $\mathcal{A}$ is used and, likewise, $\mathcal{M}_2$ is obtained when the negative sign in $\mathcal{A}$ is used.

Eigenvalues of both, $\mathcal{M}_{1,2}$, are given by the roots of the characteristic polynomial

$$\sigma^3 + \left(\frac{1}{Q} - \mathcal{A}\lambda\right)\sigma^2 + \left(1 - \frac{\mathcal{A}\lambda}{Q}\right)\sigma + \mathcal{A}\lambda(\kappa - 1) + \frac{g}{Q} = 0. \tag{6.13}$$

Observe that when $\lambda = 0$ (i.e., in the absence of coupling) the characteristic polynomial Eq. (6.13) reduces to that of the single colpitts oscillator of Eq. (5.37). We consider the case of small coupling strength, so we let $\lambda = 0 + \varepsilon$, $|\varepsilon| \ll 1$. Then we seek an approximate value of the eigenvalues and write

$$\sigma_1^{\mathcal{M}} = \sigma_1 + a_1\epsilon$$
$$\sigma_{2,3}^{\mathcal{M}} = \sigma_{2,3} + a_{2,3}\epsilon,$$

where a_1 and $a_{2,3}$ are unknown coefficients that can be found via linear approximations, i.e., asymptotic approximations up to order ε. Indeed, substituting $\sigma_1^{\mathcal{M}}$ into Eq. (6.13) and collecting first-order terms in ε, we find

$$a_1 = \frac{\mathcal{A}(\sigma_1^2 Q + \sigma_1 + Q(1-\kappa))}{3\sigma_1^2 Q + 2\sigma_1 + Q}.$$

This leads to

$$\sigma_1^{\mathcal{M}_1} = \left[\frac{-\frac{\varepsilon}{2}\left(\frac{\beta^2}{Q} - \frac{\beta}{Q} - Qk + Q\right)}{\left(3\frac{\beta^2}{Q} + Q + -2\frac{\beta}{Q}\right)} - \frac{\beta}{Q}\right] + \left[\frac{\frac{\sqrt{3}}{2}\varepsilon\left(\frac{\beta^2}{Q} - \frac{\beta}{Q} - Qk + Q\right)}{\left(3\frac{\beta^2}{Q} + Q - 2\frac{\beta}{Q}\right)}\right]i,$$

where

$$\beta = \frac{Q^2 g - 1}{Q(Q^2 + 1)}.$$

The eigenvalues $\sigma_1^{\mathcal{M}_2}$ are complex conjugate to those of $\sigma_1^{\mathcal{M}_1}$. The eigenvalues $\sigma_{2,3}^{\mathcal{M}}$ are computed in a similar fashion. We leave this task as an exercise. It is important, however, to point out that the eigenvalues $\sigma_{2,3}^{\mathcal{M}_2}$ are also complex conjugate to those of $\sigma_{2,3}^{\mathcal{M}_1}$. Overall, there are three pairs of complex conjugate eigenvalues between $\mathcal{M}_1$ and $\mathcal{M}_2$. At a point of Hopf bifurcation, in which the real part of the eigenvalues is zero, purely imaginary eigenvalues are repeated twice. It follows that the pattern of collective behavior, a traveling wave, emerges via symmetry-breaking Hopf bifurcation.

6.5 Multifrequency Patterns

The process of generating new frequencies from an original oscillatory signal, either up-converting or down-converting the input signal, is of great interest in Physics and Engineering with applications that include: radio frequency communications, sensitive optical detection, music synthesis, acoustic and optical resonators, amplitude modulation, image extraction, and phase-noise measurements [24–30].

In this chapter we describe some innovative methods [31–34], and the corresponding models, to achieve frequency up- and down-conversion. The fundamental idea is to exploit the inherent symmetry of networks to produce collective behavior in which certain oscillators tend to oscillate at different frequencies. This concept is significantly different from other techniques, e.g., master-slave systems, in the sense that the collective behavior arises naturally from the mutual interactions of the individual units, and without any external forcing.

6.5.1 Frequency Up-Conversion

In this section, we explain the methodology for using a network of coupled nonlinear oscillators in a systematic way to achieve frequency up-conversion. Without loss of generality, we will assume the individual oscillators or units to be made up of van der Pol oscillators [35–38], as is depicted in Fig. 4.26. The approach is, however, model-independent in the sense that similar results can be obtained with other type of oscillators, so long as the symmetry requirements are satisfied. We start with a description of the network configuration, followed by a description of the action of the group of global symmetries on the oscillators. The action of the group will lead to the prediction of the multi-frequency patterns. A linear stability analysis is performed to determine the conditions for symmetry-preserving and symmetry-breaking Hopf bifurcations that can lead to different patterns of oscillations. It is shown that a

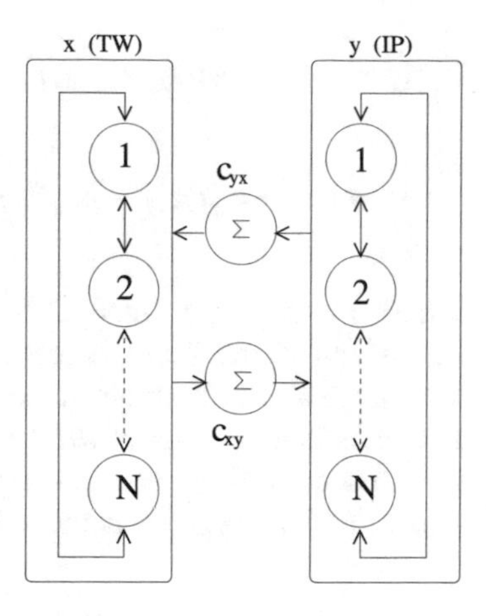

Fig. 6.9 Schematic diagram of a coupled cell system formed by two arrays of van der Pol oscillators. Each arrays contains N oscillators, each coupled to its two nearest neighbors

traveling-wave pattern along one array and a complete synchronization pattern along the opposite array provide the necessary conditions for a multi-frequency pattern in which one array (the in-phase array) oscillates at N times the frequency of the other. Theoretical predictions are validated by numerical simulations for the special case of $N = 3$. A bifurcation analysis shows the stability properties of the ensuing patterns. Robustness of the muti-frequency pattern to noise effects are demonstrated with larger networks of up to $N = 19$ oscillators per array.

6.5.2 Network Configuration

We consider a network of oscillators made up of two arrays of van der Pol oscillators coupled to one another, as is shown schematically in Fig. 6.9. Each array contains N identical oscillators, which are each coupled bidirectionally in a ring array to its two nearest neighbors via diffusive coupling. The internal dynamics of each oscillator cell is governed by Eq. (5.45), which, written in normal form [39], is

$$\dot{z} = (\alpha + \omega i)z - |z|^2 z, \tag{6.14}$$

where $z \in \mathbf{C}$ is now the state variable and α and ω are parameters. Observe that the $\mathbf{Z}_2$-symmetry of the original model Eq. (5.46) is preserved by the normal form. Thus, $\mathbf{Z}_2 = \{I_2, -I_2\}$ is the group of local symmetries of each cell, where I_2 is the 2×2 identity matrix. Then a model for the network dynamics can be written as a system of coupled differential equations of the form:

$$\dot{z_{1j}} = (\alpha_x + \omega_x i)z_{1j} - |z_{1j}|^2 z_{1j} + c_x(z_{1,j-1} - 2z_{1j} + z_{1,j+1}) + c_{yx}\sum_{k=1}^{N}|z_{2k}|^2$$

$$\dot{z_{2j}} = (\alpha_y + \omega_y i)z_{2j} - |z_{2j}|^2 z_{2j} + c_y(z_{2,j-1} - 2z_{2j} + z_{2,j+1}) + c_{xy}\sum_{k=1}^{N}|z_{1k}|^2, \tag{6.15}$$

where $j = 1, \ldots, N \mod (N)$, $x_j \in \mathbf{C}$ and $y_j \in \mathbf{C}$ describe the state of the j-th cell of the X and Y arrays respectively, c_x and c_y represent the coupling strength within the X- and Y-arrays, respectively; and c_{xy} and c_{yx} describe the cross-coupling strengths from the X-array to the Y-array and vice-versa, respectively. Using cartesian coordinates, with $z_{1j} = x_{1j} + x_{2j}$ and $z_{2j} = y_{1j} + y_{2j}$, and letting $X_j = (x_{1j}, x_{2j})$ and $Y_j = (y_{1j}, y_{2j})$, the network Eq. (6.15) can be rewritten as

$$\begin{aligned} \frac{dX_j}{dt} &= F_X(X_j) + C_X(X_{j+1} - 2X_j + X_{j-1}) + c_{yx}G(Y) \\ \frac{dY_j}{dt} &= F_Y(Y_j) + C_Y(Y_{j+1} - 2Y_j + Y_{j-1}) + c_{xy}G(X), \end{aligned} \tag{6.16}$$

where

$$F_X(X_j) = \begin{bmatrix} \alpha_x & -\omega_x \\ \omega_x & \alpha_x \end{bmatrix} \begin{bmatrix} x_{1j} \\ x_{2j} \end{bmatrix} - (x_{1j}^2 + x_{2j})^2 \begin{bmatrix} x_{1j} \\ x_{2j} \end{bmatrix},$$

$$C_X = \begin{bmatrix} c_x & 0 \\ 0 & c_x \end{bmatrix}, \quad G(Y) = \begin{bmatrix} \sum_{k=1}^{N}(y_{1j}^2 + y_{2j}^2) \\ 0 \end{bmatrix},$$

$$F_Y(Y_j) = \begin{bmatrix} \alpha_y & -\omega_y \\ \omega_y & \alpha_y \end{bmatrix} \begin{bmatrix} y_{1j} \\ y_{2j} \end{bmatrix} - (y_{1j}^2 + y_{2j})^2 \begin{bmatrix} y_{1j} \\ y_{2j} \end{bmatrix},$$

$$C_Y = \begin{bmatrix} c_y & 0 \\ 0 & c_y \end{bmatrix}, \quad G(X) = \begin{bmatrix} \sum_{k=1}^{N}(x_{1j}^2 + x_{2j}^2) \\ 0 \end{bmatrix}.$$

6.5.3 *Linear Stability Analysis*

Due to the nature of the connections along each array, the underlying group of global symmetries of each array is $\mathbf{D}_N$, i.e., the group of symmetries of an N-gon. It follows that

$$\Gamma = \mathbf{D}_N \times \mathbf{D}_N,$$

is the group of global symmetries of the network, including the two interconnected arrays. To study the effects of the Γ-symmetry on the network, we represent the state of the X-array as

$$X(t) = (X_1(t), \ldots, X_N(t)),$$

and the Y-array as

$$Y(t) = (Y_1(t), \ldots, Y_N(t)).$$

We now conduct a study of the linearization of the network model Eq. (6.16) near the origin, $(X, Y) = (0, 0)$, which is given by

$$L = \text{diag}(L_X, L_Y), \tag{6.17}$$

where

$$L_X = \begin{bmatrix} J_X & C_X & 0 & 0 & \ldots & 0 & C_X \\ C_X & J_X & C_X & 0 & \ldots & 0 & 0 \\ 0 & C_X & J_X & C_X & 0 & \ldots & 0 \\ \vdots & \vdots & \vdots & \ddots & \ddots & \ddots & \vdots \\ 0 & 0 & \ldots & 0 & C_X & J_X & C_X \\ C_X & 0 & 0 & \ldots & 0 & C_X & J_X \end{bmatrix},$$

with $J_X = (dF_X)_{(0,0)} - 2C_X$, and

$$L_Y = \begin{bmatrix} J_Y & C_Y & 0 & 0 & \ldots & 0 & C_Y \\ C_Y & J_Y & C_Y & 0 & \ldots & 0 & 0 \\ 0 & C_Y & J_Y & C_Y & 0 & \ldots & 0 \\ \vdots & \vdots & \vdots & \ddots & \ddots & \ddots & \vdots \\ 0 & 0 & \ldots & 0 & C_Y & J_Y & C_Y \\ C_Y & 0 & 0 & \ldots & 0 & C_Y & J_Y \end{bmatrix},$$

with $J_Y = (dF_Y)_{(0,0)} - 2C_Y$.

We complexify from $\mathbb{R}^{8N}$ to C^{8N} and employ the well-known isotypic decomposition of C^{8N} by $\mathbf{D}_N \times \mathbf{D}_N$, which is given by

$$\mathbf{C}^{8N} = V_0 \oplus V_0 \oplus V_1 \oplus V_1 \oplus \cdots \oplus V_{N-1} \oplus V_{N-1},$$

where

$$V_j = \mathbf{C}\{v_j\} \text{ with } v_j = (v, \zeta^j v, \zeta^{2j} v, \ldots, \zeta^{(N-1)j} v)^T,$$

$j = 0, \ldots, N-1$, and $\zeta = \exp(2\pi i/N)$, for some $v \in \mathbb{R}^2$. Observe that each isotypic subspace is repeated twice, one time for the X-array and one time for the Y-array. Using coordinates along the isotypic components, we find the eigenvalues of the linearization of Eq. (6.16) to be those of

$$\begin{aligned}L_{Xj} &= (dF_X)_{(0,0)} - 2C_X + (\zeta^j + \zeta^{(N-1)j})C_X \\ &= (dF_X)_{(0,0)} - 2C_X + (\zeta^j + \overline{\zeta}^j)C_X \\ &= (dF_X)_{(0,0)} - 2C_X + 2\cos(2\pi j/N)C_X,\end{aligned}$$

and (through similar calculations) those of

$$L_{Yj} = (dF_Y)_{(0,0)} - 2C_Y + 2\cos(2\pi j/N)C_Y.$$

For the special case of $N = 3$, the spectrum of eigenvalues are those of

$$\begin{aligned}L_X &= \text{diag}(J_X + 2C_X, J_X - C_X, J_X - C_X) \\ L_Y &= \text{diag}(J_Y + 2C_Y, J_Y - C_Y, J_Y - C_Y),\end{aligned}$$

which are the eigenvalues of the blocks

$$\begin{aligned}&(dF_X)_{(0,0)}, \ (dF_X)_{(0,0)} - 3C_X, \quad \text{(twice)} \\ &(dF_Y)_{(0,0)}, \ (dF_Y)_{(0,0)} - 3C_Y, \quad \text{(twice)}.\end{aligned}$$

Observe that these blocks are the same as those found in the linearization of the three-cell system that we studied earlier in Example 6.2. Thus, we can leverage the previous work and conclude that the patterns of oscillations that each array can support are the same as those described in Example 6.2. More specifically, each array can produce in-phase oscillations via $\mathbf{D}_3$ symmetry-preserving bifurcations and three other types of collective patterns with isotropy subgroups, Σ, given by: $\tilde{\mathbf{Z}}_3$, $\mathbf{Z}_2(\kappa)$, and $\mathbf{Z}_2(\kappa, \pi)$. These latter patterns emerge via $\mathbf{D}_3$ symmetry-breaking bifurcations.

We would like to emphasize that this approach allows us to predict the emergence of certain patterns of oscillations, whose spatio-temporal symmetries are described by their respective isotropy subgroups. The approach does not provide, however, a direct link to determining whether the bifurcations that lead to each pattern are super- or subcritical. Also, it does not lead to an identification of which branches (if any) are stable. In fact, in order to determine the criticality of the bifurcations and which branches are stable, one alternative is to perform, either, a Lypunov Schmidt reduction or Center Manifold reduction of the network dynamics into the relevant eigenspaces. These tasks are beyond the scope of the present book. We refer the reader to related works [40], which provide alternative derivations of criticality and stability properties.

6.5.4 The Role of Spatio-Temporal Symmetries

A critical observation is the fact that certain spatio-temporal symmetries, encoded in the isotropy subgroups, may lead to spatio-temporal patterns in which one of the arrays oscillates N times faster than the other [15, 16]. To verify this assertion, consider a spatio-temporal pattern, $P(t)$, generated by the network, and whose evolution, at any given time, t, can be described by

$$P(t) = (X(t), Y(t)).$$

Let us assume that this pattern, $P(t)$, is a periodic solution of period T with the following characteristics. On one side of the network, for instance, the X-array, undergoes a symmetry-breaking Hopf bifurcation that leads to a pattern of oscillations with isotropy subgroup $\tilde{\mathbf{Z}}_N$. That is, the oscillators form a *traveling wave* (TW), i.e., same wave form X_0 shifted (delayed) by a constant time lag $\phi = T/N$:

$$X_k(t) = X_0(t + (k-1)\phi), \ k = 1, \ldots, N.$$

On the opposite side, the Y-array undergoes a symmetry-preserving Hopf bifurcation that leads to a pattern of oscillations with isotropy subgroup D_N. Thus, the oscillators are assumed to be *in-phase* (IP) with identical wave form Y_0, i.e., a synchronous state:

$$Y_k(t) = Y_0(t), \ k = 1, \ldots, N.$$

Together, $\tilde{\mathbf{Z}}_N \times \mathbf{D}_N \times \mathbf{S}^1$ act on $P(t)$ as follows. First, $\tilde{\mathbf{Z}}_N$ cyclically permutes the oscillators of the X-array, while $\mathbf{D}_N$ acts trivially on the oscillators of the Y-array:

$$\begin{aligned}
\tilde{\mathbf{Z}}_N \cdot X_{TW}(t) &= \{X_N(t + (N-1)\phi), X_1(t), \ldots, X_{N-1}(t + (N-2)\phi)\}, \\
\mathbf{D}_N \cdot Y_{IP}(t) &= \{Y_1(t), Y_2(t), \ldots, Y_N(t)\}.
\end{aligned}$$

Then $\mathbf{S}^1$ shifts time by ϕ so that

$$\begin{aligned}
\tilde{\mathbf{Z}}_N \times \mathbf{S}^1 \cdot X_{TW}(t) &= \{X_N(t), X_1(t+\phi), \ldots, X_{N-1}(t + (N-1)\phi)\}, \\
\mathbf{D}_N \times \mathbf{S}^1 \cdot Y_{IP}(t) &= \{Y_N(t+\phi), Y_1(t+\phi), \ldots, Y_{N-1}(t+\phi)\}.
\end{aligned}$$

Since the oscillators are identical, we get

$$\begin{aligned}
\tilde{\mathbf{Z}}_N \times \mathbf{S}^1 \cdot X_{TW}(t) &= X_{TW}(t), \\
\mathbf{D}_N \times \mathbf{S}^1 \cdot Y_{IP}(t) &= Y_{IP}(t+\phi).
\end{aligned}$$

It follows that in order for $Y_{IP}(t)$ to have $\mathbf{D}_N \times \mathbf{S}^1$ symmetry, the in-phase oscillators must oscillate at N times the frequency of the oscillations of the traveling wave. The same conclusion is reached if the roles of the X and Y arrays are interchanged.

Suppose that we change the coupling topology of the network to have the individual arrays unidirectionally coupled. Then,

$$\Gamma = \mathbf{Z}_N \times \mathbf{Z}_N,$$

becomes now the group of global symmetries of the network. Traveling wave solutions, can also arise in $\mathbf{Z}_N$ symmetric systems through $\mathbf{Z}_N$ symmetry-breaking Hopf bifurcation [15]. Likewise, $\mathbf{Z}_N$ symmetry-preserving Hopf bifurcations can also lead to synchronized oscillations. This means that a network with $\Gamma = \mathbf{Z}_N \times \mathbf{Z}_N$ symmetry can support a spatio-temporal pattern with isotropy subgroup $\Sigma = \tilde{\mathbf{Z}}_N \times \mathbf{Z}_N \times \mathbf{S}^1$. Then, a similar set of calculations (not shown for brevity) will lead to the conclusion that

$$\begin{aligned}
\tilde{\mathbf{Z}}_N \times \mathbf{S}^1 \cdot X_{TW}(t) &= X_{TW}(t), \\
\mathbf{Z}_N \times \mathbf{S}^1 \cdot Y_{IP}(t) &= Y_{IP}(t + \phi).
\end{aligned}$$

Consequently, a network with $\mathbf{Z}_N \times \mathbf{Z}_N$-symmetry, i.e., unidirectionally coupled arrays, can also produce the same multi-frequency effect in which the in-phase oscillators are induced to oscillate at N times the frequency of the traveling-wave oscillators. This conclusion serves to highlight the model-independent features of the symmetry-based approach, in which the results and conclusions depend, mainly, on the underlying symmetries of the system as opposed to the internal characteristics of each individual oscillator.

We remark that this approach to manipulate the frequency of oscillations is significantly different from that of sub-harmonic and ultra-harmonic motion generated via a forced system as is described by Hale and Gambill [41] and later by Tiwari and Subramanian [42]. In our case, the multifrequency behavior arises from the mutual interaction of two arrays of oscillators. None of the oscillators is forced and, consequently, the arrays are naturally modeled by an autonomous system instead of the non-autonomous system that is described in the same references [41, 42]. Furthermore, observe that the approach is applicable to arrays of arbitrary size, N.

6.5.5 Numerical Simulations

We now provide evidence of the existence of a multifrequency pattern, such as the one described above, in which the Y-array oscillates at N times the frequency of the oscillators in the X-array, through numerical simulations. Results with larger N are also provided later on. Figure 6.10 shows the results of integrating the model Eq. (6.15) for the particular case of $N = 3$ oscillators per array.

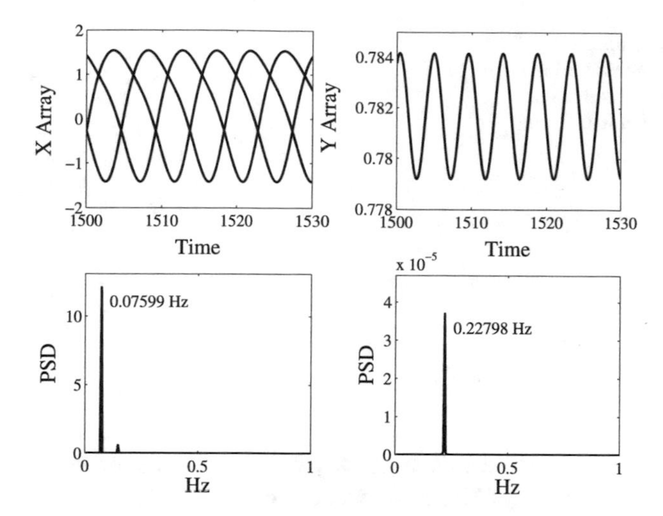

Fig. 6.10 Multifrequency oscillations found in simulations of network equations (6.15) with $N = 3$. (Left) The X-array generates a traveling wave pattern, while the (right) Y-array yields an in-phase pattern that oscillates at three times the frequency of the traveling wave. The bottom panels depict the corresponding power spectral density (PSD) where it can be checked that the in-phase pattern has a frequency three times greater than the traveling wave pattern ($0.22798 \approx 3 \times 0.07599$). Parameters are: $\alpha_x = \alpha_y = 1.0$, $\omega_x = \omega_y = 0.5$, $c_x = -0.4$, $c_y = 0.4$, and $c_{xy} = c_{yx} = 0.12$

Coupling parameters within each array are: $c_x = -0.4$, for the X-array, and $c_y = 0.4$ for the Y-array. Coupling parameters across the arrays are: $(c_{xy}, c_{yx}) = (0.12, 0.12)$. Under these conditions, a stable traveling wave pattern emerges via a $\mathbf{D}_3$ symmetry-Hopf bifurcation along the X array, with period, approximately $T \approx 13.272$ sec (frequency ≈ 0.0753 Hz). The traveling wave pattern has spatio-temporal symmetry described by the isotropy subgroup $\Sigma_X = \tilde{\mathbf{Z}}_3 \times \mathbf{S}^1$. Similarly, a $\mathbf{D}_3$ symmetry-preserving Hopf bifurcation leads to synchronized oscillations along the Y-array. The isotropy subgroup of the synchronized pattern is the entire group, i.e., $\Sigma_Y = \mathbf{D}_3 \times \mathbf{S}^1$. It follows that the spatio-temporal pattern, TW-IP, is described by the isotropy subgroup $\Sigma = \tilde{\mathbf{Z}}_3 \times \mathbf{D}_3 \times \mathbf{S}^1$.

As predicted by theory, the $\tilde{\mathbf{Z}}_3 \times \mathbf{D}_3 \times \mathbf{S}^1$-symmetry leads to a collective pattern in which the in-phase oscillations of the Y-array are three times faster (see power spectra densities in lower panels) than the traveling wave produced by the X-array.

A generalization of the existence of similar multifrequency patterns in larger arrays depends upon network connections that can satisfy the necessary conditions for the network to exhibit a traveling wave pattern in one array and a synchronized solution in the opposite array. The network configuration shown in Fig. 6.9 with N odd, in particular, shows similar multifrequency results. We have tested (through computer simulations) this network and a network with $\mathbf{Z}_N \times \mathbf{Z}_N$-symmetry, with up to $N = 19$ oscillators, and they both can produced the desired patterns of oscillations, including the multi-frequency effects. When N s even, however, other coupling schemes need to be considered so that the network, and the solutions generated

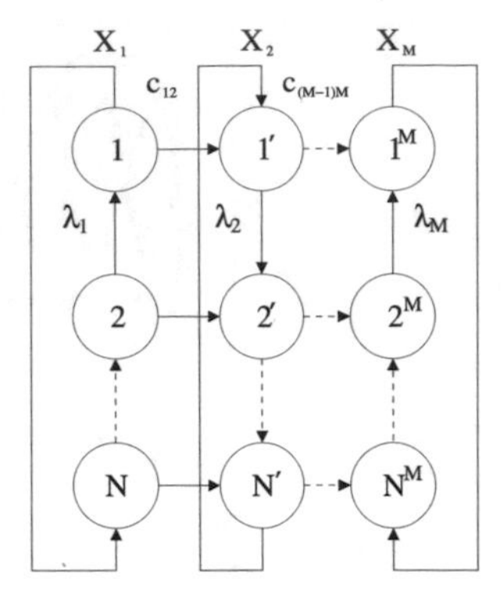

Fig. 6.11 Generalized network configuration of M arrays of coupled cell units. Each array contains N elements

by the connectivity, can meet the necessary conditions for symmetry-breaking and symmetry-preserving Hopf bifurcations to occur.

6.5.6 Frequency Down-Conversion

In this section we show how a high frequency signal can be down-converted by passing it through a cascade of arrays of unidirectionally coupled overdamped bistable elements. As an example, we find that the frequency down-conversion can be by a factor of 1/2, 1/5, or 1/11 for two coupled arrays of three elements, $N = 3$, $M = 2$, where N is the size of each array and M is number of interconnected arrays. A generalization to larger M is also provided. We use analyses tools that emphasize the symmetry of the networks to help us better understand the organization and stability properties of the ensuing behavior while providing the means for determining both invariance and changes in the system without going deep into the analysis of its dynamics. We note that the robustness of a given network guarantees that certain patterns of oscillation persist regardless of the internal dynamics of each individual nonlinear element.

6.5.7 Network Configurations and Symmetries

We start with a special case of the more general setup of cascade arrays depicted in Fig. 6.11.

For the special case of two arrays, the network dynamics is described by the following set of differential equations:

$$\begin{aligned}
\dot{x}_i &= f(x_i, \alpha) + \sum_{j \to i} \lambda_{ij} h(x_i, x_j) \\
\dot{y}_i &= f(y_i, \alpha) + \sum_{j \to i} \lambda_{ij} h(y_i, y_j) + c_{ij} k(y_i, x_j),
\end{aligned} \tag{6.18}$$

where $y_i = (y_{i1}, \ldots, y_{ik}) \in \mathrm{R}^k$ denote the state variables of cell i in the second array, k is an inter-array coupling function, c_{ij} being the corresponding coupling strength. Notice the unidirectional coupling in each array and also between adjacent cells in the two arrays. The opposite directions of the intra-array couplings should also be noted. The unidirectional inter-array coupling yields a network with global symmetry described by the direct product group

$$\Gamma = \mathbf{Z}_N \times \mathbf{Z}_N,$$

in which $\mathbf{Z}_N$ is the group of cyclic permutations of N objects. Each element of the direct product group permutes, simultaneously, each element of the corresponding arrays. For the moment, there is no externally applied signal.

To study the patterns of behavior for the $M = 2$ case of the network in Fig. 6.11, we use

$$X_1(t) \equiv X(t) = (x_1(t), \ldots, x_N(t))$$

to represent the state of one array and

$$X_2(t) \equiv Y(t) = (y_1(t), \ldots, y_N(t))$$

to denote the state of the second array. Thus, at any given time t, a spatio-temporal pattern generated by the network can be described by

$$P(t) = (X(t), Y(t)).$$

To begin the analysis, let us assume that both arrays exhibit a traveling wave (TW) pattern with period T, and isotropy subgroup $\Sigma_X = \Sigma_Y = \tilde{\mathbf{Z}}_3$. That is, the waveforms produced by each array are identical, but out-of-phase by a constant time lag $\phi = T/N$. We also make a second assumption that the X_2 array oscillates at m times the period of the X_1 array, where m is a nonzero integer. Thus, $P(t)$ has the form

$$\begin{aligned}
P(t) = (&x(t), x(t + (N-1)\phi), \ldots, x(t + \phi), \\
&y(t), y(t + m\phi), \ldots, y(t + (N-1)m\phi)),
\end{aligned} \tag{6.19}$$

where the X_1 array exhibits a TW in the opposite direction of the X_2 array, a direct result of the opposite orientation of their coupling schemes. For simplicity, we further assume that $N = 3$, and that the units are coupled as is shown in Fig. 6.11. From Eqs. (6.19), a solution to this network has the form

$$P(t) = \left(x(t), x(t + \frac{2T}{3}), x(t + \frac{T}{3}), y(t), y(t + \frac{mT}{3}), y(t + \frac{2mT}{3}) \right). \quad (6.20)$$

Under the previous assumptions, the pattern $P(t)$ has spatio-temporal symmetry described by the cyclic group $\tilde{\mathbf{Z}}_3 \times \tilde{\mathbf{Z}}_3$ and by the group S^1 of temporal shifts. Together, $\Sigma = \tilde{\mathbf{Z}}_3 \times \tilde{\mathbf{Z}}_3 \times S^1$ acts on $P(t)$ as follows. First, Σ acts as a permutation:

$$\Sigma \cdot (1, 2, 3, 1', 2', 3') \mapsto (3, 1, 2, 3', 2', 1'),$$

so that

$$\Sigma \cdot P(t) = \left(x(t + \frac{T}{3}), x(t), x(t + \frac{2T}{3}), y(t + \frac{2mT}{3}), y(t), y(t + \frac{mT}{3}) \right). \quad (6.21)$$

Then S_1 shifts time by $mT/3$ so that

$$\begin{aligned} \left(\Sigma, \tfrac{mT}{3}\right) \cdot P(t) = \Big(& x(t + \tfrac{m+1}{3}T), x(t + \tfrac{m}{3}T), x(t + \tfrac{m+2}{3}T), \\ & y(t + mT), y(t + \tfrac{m}{3}T), y(t + \tfrac{2m}{3}T) \Big). \end{aligned} \quad (6.22)$$

Since the cells are assumed to be identical, it follows that $\Sigma = \tilde{\mathbf{Z}}_3 \times \tilde{\mathbf{Z}}_3 \times S^1$ is a spatio-temporal symmetry of the network provided that

$$X(t) = X(t + \frac{m+1}{3}T) \quad \text{and} \quad Y(t) = Y(t + mT).$$

But X_1 is T-periodic, which implies that $m = 3k - 1$, where k is a nonzero integer. As k increases (starting at one) we obtain the following values for m : $2, 5, 8, 11, 14, 17, 20, 23, \ldots$. When $m = 2$, for instance, the X_2-array oscillates at $1/2$ the frequency of the X_1-array. Likewise, $m = 5$ suggests that the X_2-array oscillates at $1/5$ the frequency of the X_1-array. The case when $m = 8$ should be excluded, however, since $m = 8 = 2^2 \times 2$.

As N increases, similar frequency down-conversion ratios emerge. A bifurcation analysis shows that the regions of existence of these frequency ratios form an Arnold tongue structure in parameter space (λ_2, c_{xy}). In general we find (noting that N is odd) $\omega_{X_1}/\omega_{X_2} = N - 1, 2N - 1 \ldots Nk - 1$. Table 6.1 summarizes the downconversion ratios.

6.5.8 Simulations

To verify the existence of these oscillations, we define the individual dynamics of each cell to be that of a prototypical bistable system, an overdamped Duffing oscillator with internal dynamics given by

Table 6.1 Down-conversion ratios between the frequency of the X array, ω_X, and Y array, ω_Y, for a network of two coupled arrays interconnected as is shown in Fig. 6.11. k is a positive integer

Number of cells		ω_X/ω_Y		
3	2	5	...	3k-1
5	4	9	...	5k-1
7	6	13	...	7k-1
9	8	17	...	9k-1
⋮	⋮	⋮	⋱	⋮
N	N-1	2N-1	...	Nk-1

$$f(x) = ax - bx^3,$$

and the (unidirectional) intra-array coupling functions by

$$h(x_i, x_{i+1}) = x_i - x_{i+1} \quad \text{and} \quad h(y_i, y_{i-1}) = y_i - y_{i-1},$$

respectively. The inter-array connections are unidirectional, as is shown in Fig. 6.11, hence, the X_1-array dynamics has no dependence on the X_2-array dynamics. Then, the network dynamics are represented by the system

$$\begin{aligned} \tau \dot{x}_i &= ax_i - bx_i^3 + \lambda_1(x_i - x_{i+1}) \\ \tau \dot{y}_i &= ay_i - by_i^3 + \lambda_2(y_i - y_{i-1}) + c_{xy}x_i, \end{aligned} \tag{6.23}$$

where $i = 1, \ldots, N \bmod N$, a and b are positive constants that describe the dynamics of the individual cells, λ_1, and λ_2 define the intra-array coupling strengths for the X_1 and X_2 arrays, respectively, with c_{xy} the inter-array coupling coefficient. τ is a system time constant.

First assume that there is no cross coupling, i.e. $c_{xy} = 0$. Then [43], $\lambda_{1c} = a/2$ is the critical coupling strength beyond which the X_1 elements oscillate. Accordingly, if the coupling strength of the X_2 array is below the critical coupling strength, i.e. $\lambda_2 < \lambda_c$, and the coupling strength of the X_1 array is above, $\lambda_1 > \lambda_c$, then we would obtain the pattern shown in the left panel of Fig. 6.12 for the X_1 elements, but the X_2 array would be quiescent.

Increasing the cross-coupling strength $c_{xy} > 0$ induces the X_2-array to oscillate (above a critical value of c_{xy}) with frequency $\omega_{X_2} = \omega_{X_1}/5$; this is shown in the right panel of Fig. 6.12. Increasing further the cross coupling c_{xy} causes the X_2 array to oscillate at 1/2 the frequency of the X_1-array. Additional frequency down-conversion ratios, $(1/2, 1/5, 1/(3k-1)$, where $k = 1, 2, 3, \ldots$, are also observed as the cross-coupling, c_{xy}, increases further.

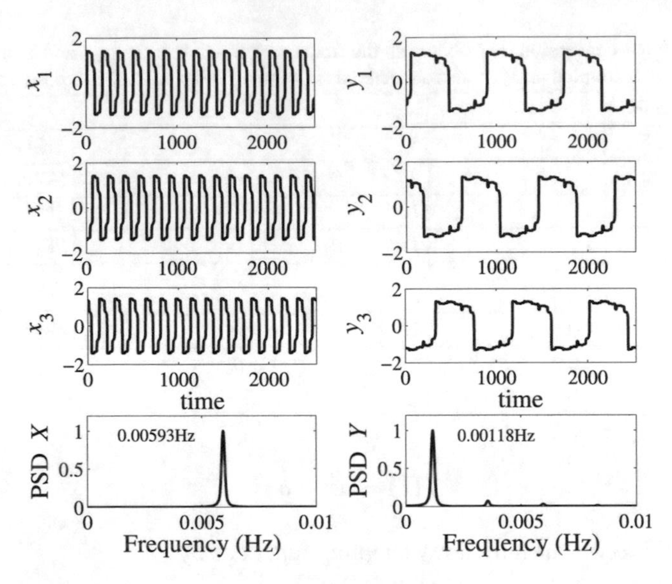

Fig. 6.12 Numerical simulations showing the frequency down-conversion effect. Each element in the X_2-array oscillates at 1/5 the frequency of each element in the X_1-array ($0.001186Hz \approx \frac{1}{5}0.00593Hz$). Parameters are: $c_{xy} = 0.14$, $\lambda_1 = 0.51$, $\lambda_2 = 0.3$, $a = 1$, $b = 1$, and $\tau = 1$

Experiments on frequency up- and down-conversion have been conducted to validate theory. Those works are beyond the scope of the present book but readers interested in more details can find additional information in [31, 33].

6.6 Feedforward Networks

Feedforward networks are a specific type of network characterized by a homogeneous chain of unidirectionally coupled nodes, as is shown in Fig. 6.13. The first node may or may not be self-coupled.

The unidirectional coupling prevents feedback in the system, so that one system may influence another without being itself affected. Observe that the cascade array

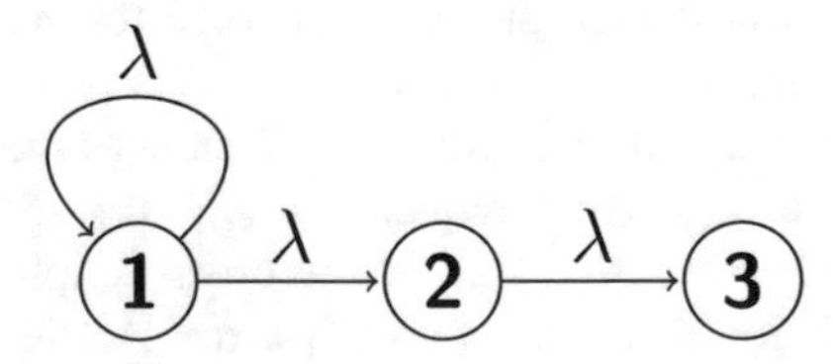

Fig. 6.13 Representative example of a three-cell feedforward network. Arrows indicate coupling, with coupling strength λ. Each cell represents a dynamical system assumed to be operating near a Hopf bifurcation

that was used for frequency-downconversion, see Fig. 6.11, can be thought of as an extension of a feedforward network. We show now that under the right conditions, the feedforward network causes certain bifurcations to exhibit accelerated growth rates.

6.6.1 Hopf Bifurcation

Consider, for example, the three-cell feedforward network shown in Fig. 6.13. Assume the internal dynamics of each cell to be governed by a Hopf bifurcation, which in normal form [44–46], it takes the form

$$\dot{z} = (\mu + \omega i)\, z - (1 + \gamma i)|z|^2 z. \tag{6.24}$$

For $\mu > 0$, the origin loses stability and tends to oscillate at frequency ω with some positive amplitude. For $\mu < 0$, the origin is a stable equilibrium and the cell will tend to zero. Equation (6.24) is the *normal form* of a Hopf bifurcation, and therefore this is the simplest possible system which displays a Hopf bifurcation. The network is then modeled by the following system of equations

$$\begin{aligned} \dot{z}_1 &= (\mu + i\omega)\, z_1 - (1 + i\,\gamma)\ |z_1|^2\, z_1 - \lambda z_1 \\ \dot{z}_2 &= (\mu + i\omega)\, z_2 - (1 + i\,\gamma)\ |z_2|^2\, z_2 - \lambda z_1 \\ \dot{z}_3 &= (\mu + i\omega)\, z_3 - (1 + i\,\gamma)\ |z_3|^2\, z_3 - \lambda z_2. \end{aligned} \tag{6.25}$$

The authors in [46–48] have found that coupling causes the amplitudes of oscillation that arise from the onset of the Hopf bifurcation to grow at a larger rate. If μ is the bifurcation parameter, and $\mu = 0$ is the onset of a supercritical Hopf bifurcation, then the third cell undergoes oscillations of amplitude approximately equal to $\mu^{1/6}$, rather than the *expected* amplitude of $\mu^{1/2}$. This phenomenon showcases an accelerated growth rate that has the potential for the design and fabrication of advanced filters in signal processing [49, 50]). An example of a time series, obtained from simulations of Eq. (6.25), which exhibits this growth phenomenon is shown in Fig. 6.14.

As the feedforward network grows in size, the authors report that the growth rate of oscillations in the final cell are determined by taking successive cube roots [49]. Thus, in a five-cell feedforward network, the growth rate should be proportional to the 54th root of the bifurcation parameter, which has also been proved in [51].

This phenomenon of such large-amplitude oscillations in the third cell can be understood as a type of nonlinear resonance as well as being the result of the combination of the unidirectional coupling and the higher-degree nonlinearities [52]. Due to the network topology of the feedforward network, the third (or last) oscillator is being periodically forced by the previous one. This line of inquiry led to

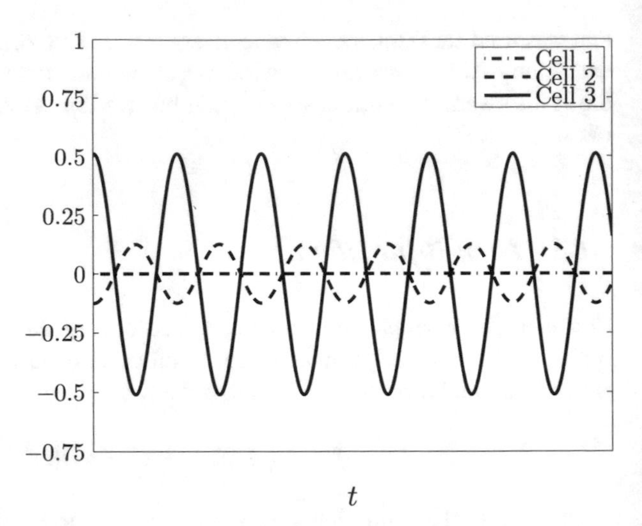

Fig. 6.14 Amplitude of oscillations produced by the onset of a Hopf bifurcations in the feedforward network shown in Fig. 6.13. Parameters are: $\mu = (1/2)^6$, $\omega = 1$, $\gamma = 0$, $\lambda = 1$. Observe the signal amplification effect on the third cell

investigations of periodically forced Hopf bifurcations, and even periodically forced feedforward networks of Hopf bifurcations [46, 49, 53]. Related articles, proving anomalous growth rates can occur for equilibria in (unusual) regular networks, for bifurcations at simple eigenvalues are in [54, 55].

6.6.2 Analysis

Consider the case when $\gamma = 0$. The model Eq. (6.25) becomes

$$\begin{aligned} \dot{x} &= (\mu + i\omega)\, x - |x|^2\, x - \lambda x \\ \dot{y} &= (\mu + i\omega)\, y - |y|^2\, y - \lambda x \\ \dot{z} &= (\mu + i\omega)\, z - |z|^2\, z - \lambda y, \end{aligned} \tag{6.26}$$

where ω is the rotational velocity in the complex plane and μ is the bifurcation parameter. Note that we can rewrite the first equation in (6.26) as

$$\dot{x} = (\tilde{\mu} + i\omega)\, x - |x|^2\, x,$$

where $\tilde{\mu} = \mu - \lambda$. Then the bifurcation parameter in the first cell is different than the bifurcation parameters of the second and third cells. We will use this fact later on in the analysis. We introduce the two time scales $\xi = \omega t$ and $\eta = \varepsilon t$. Applying the chain rule we can conclude that

$$\frac{d}{dt} = \omega \frac{\partial}{\partial \xi} + \varepsilon \frac{\partial}{\partial \eta}. \tag{6.27}$$

We assume that each solution can be expressed as a Taylor series expansion of the form:

$$\begin{aligned} x(\xi,\eta) &= x_0(\xi,\eta) + \varepsilon x_1(\xi,\eta) + \varepsilon^2 x_2(\xi,\eta) + \cdots \\ y(\xi,\eta) &= y_0(\xi,\eta) + \varepsilon y_1(\xi,\eta) + \varepsilon^2 y_2(\xi,\eta) + \cdots \\ z(\xi,\eta) &= z_0(\xi,\eta) + \varepsilon z_1(\xi,\eta) + \varepsilon^2 z_2(\xi,\eta) + \cdots \end{aligned} \tag{6.28}$$

We suppress the nonlinearities, exponential growths, and couplings in (6.26) with ε. This gives us

$$\begin{aligned} \dot{x} &= (\varepsilon\mu + i\omega)\,x - \varepsilon\,|x|^2\,x - \varepsilon\lambda x \\ \dot{y} &= (\varepsilon\mu + i\omega)\,y - \varepsilon\,|y|^2\,y - \varepsilon\lambda x \\ \dot{z} &= (\varepsilon\mu + i\omega)\,z - \varepsilon\,|z|^2\,z - \varepsilon\lambda y. \end{aligned} \tag{6.29}$$

Substituting (6.27) and (6.28) into (6.29) and collecting like powers of ε, we get the terms of order ε^0:

$$\frac{\partial x_0}{\partial \xi} = i x_0, \quad \frac{\partial y_0}{\partial \xi} = i y_0, \quad \frac{\partial z_0}{\partial \xi} = i z_0, \tag{6.30}$$

with solutions

$$\begin{aligned} x_0\,(\xi,\eta) &= A(\eta)\exp\left(i\,\phi_1(\eta)\right)\exp\left(i\xi\right) \\ y_0\,(\xi,\eta) &= B(\eta)\exp\left(i\,\phi_2(\eta)\right)\exp\left(i\xi\right) \\ z_0\,(\xi,\eta) &= C(\eta)\exp\left(i\,\phi_3(\eta)\right)\exp\left(i\xi\right), \end{aligned} \tag{6.31}$$

respectively. This essentially serves to express an approximation to x, y, and z in polar coordinates. From now on, we will implicitly assume the dependence of A, B, C, and ϕ_i on η. For equations of order ε, we get

$$\begin{aligned} \omega\frac{\partial x_1}{\partial \xi} &= i\omega x_1 - \frac{\partial x_0}{\partial \eta} + \mu x_0 - |x_0|^2\,x_0 - \lambda x_0 \\ \omega\frac{\partial y_1}{\partial \xi} &= i\omega y_1 - \frac{\partial y_0}{\partial \eta} + \mu y_0 - |y_0|^2\,y_0 - \lambda x_0 \\ \omega\frac{\partial z_1}{\partial \xi} &= i\omega z_1 - \frac{\partial z_0}{\partial \eta} + \mu z_0 - |z_0|^2\,z_0 - \lambda y_0. \end{aligned} \tag{6.32}$$

We can see that to eliminate secular terms, we must set the sum of all the terms involving $x_{i,0}$ or its derivatives equal to zero. Thus we get the set of equations

$$\begin{aligned} \frac{\partial x_0}{\partial \eta} &= \mu x_0 - |x_0|^2\,x_0 - \lambda x_0 \\ \frac{\partial y_0}{\partial \eta} &= \mu y_0 - |y_0|^2\,y_0 - \lambda x_0 \\ \frac{\partial z_0}{\partial \eta} &= \mu z_0 - |z_0|^2\,z_0 - \lambda y_0. \end{aligned} \tag{6.33}$$

By substituting (6.31) into (6.33), and subsequently dividing through by $\exp(i\ (\xi + \phi_i))$, we get

$$\begin{aligned} A' + iA\phi_1' &= \mu A - A^3 - \lambda A \\ B' + iB\phi_2' &= \mu B - B^3 - \lambda A \exp\left(i(\phi_1 - \phi_2)\right) \\ C' + iC\phi_3' &= \mu C - C^3 - \lambda B \exp\left(i(\phi_2 - \phi_3)\right). \end{aligned} \tag{6.34}$$

Now let us separate (6.34) into its real and imaginary parts, and introduce the variables $\psi_1 = \phi_1 - \phi_2$ and $\psi_2 = \phi_2 - \phi_3$. Noting that the first equation of Eq. (6.34) implies that $\phi_1' = 0$, we rewrite (6.34) as

$$\begin{aligned} A' &= \mu A - A^3 - \lambda A \\ B' &= \mu B - B^3 - \lambda A \cos(\psi_1) \\ C' &= \mu C - C^3 - \lambda B \cos(\psi_2) \\ \psi_1' &= \frac{\lambda A}{B} \sin(\psi_1) \\ \psi_2' &= \frac{\lambda B}{C} \sin(\psi_2) - \frac{\lambda A}{B} \sin(\psi_1)\,. \end{aligned} \tag{6.35}$$

Here we have implicitly assumed that $B, C \neq 0$. The Jacobian of this system of equations is difficult to analyze, and its eigenvalues even more so. Let us then make an observation about the equations for ψ_1' and ψ_2'. These have fixed points, regardless of the values of A and B, only when $\psi_1 = n_1\pi$ and $\psi_2 = n_2\pi$ for $n_i \in \mathbb{Z}$. It follows then that to obtain a fixed point, we must have $\cos(\psi_1) = \pm 1$ and $\cos(\psi_2) = \pm 1$. Making this substitution yields the following system:

$$\begin{aligned} A' &= \tilde{\mu}\, A - A^3 \\ B' &= \mu\, B - B^3 \pm \lambda A \\ C' &= \mu\, C - C^3 \pm \lambda B. \end{aligned} \tag{6.36}$$

Now let us note that the Jacobian $\mathbf{J}$ of this simplified system is a triangular matrix, with the property that the added uncertain plus or minus signs do not appear in the diagonal entries; *i.e.*, these do not influence the eigenvalues, and by extension, the stability properties of any fixed points.

$$\mathbf{J} = \begin{pmatrix} \tilde{\mu} - 3\,A^2 & 0 & 0 \\ \pm\lambda & \mu - 3\,B^2 & 0 \\ 0 & \pm\lambda & \mu - 3\,C^2 \end{pmatrix}. \tag{6.37}$$

Since A, B, and C represent the amplitudes of oscillation in each cell, it will suffice to study the values of $|A|$, $|B|$, and $|C|$, respectively. This does not influence the stability—note that the powers of A, B, and C that appear in the eigenvalues are

Table 6.2 Fixed point solutions of System (6.36) and their associated stability properties

$\|A\|$	$\|B\|$	$\|C\|$	Eigenvalues			Stability
0	0	0	$\tilde{\mu}$	μ	μ	Saddle
0	0	$\sqrt{\mu}$	$\tilde{\mu}$	μ	-2μ	Saddle
0	$\sqrt{\mu}$	$\phi(\mu) \sim \lambda^{1/3}\mu^{1/6}$	$\tilde{\mu}$	-2μ	$\mu - 3\phi^2(\mu)$	Stable

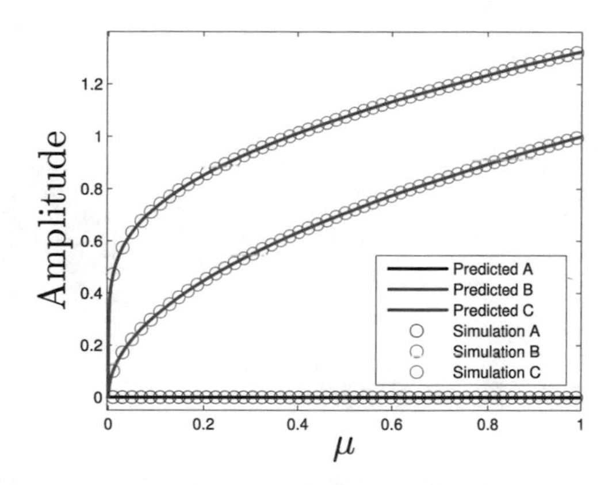

Fig. 6.15 Bifurcation diagrams for Eq. (6.26) showing the close match between the predicted (solid lines) amplitudes and the exact amplitudes extracted from the numerical simulations (circles). The bifurcation parameter μ varies in the interval $0 < \mu < \lambda$

even. A negative amplitude on the oscillation corresponds to shifting the phase by π. Table 6.2 summarizes the stability properties of the amplitudes of A, B, and C for $0 < \mu < \lambda$.

We can now substitute the solutions for A, B and C into Eq. (6.31) to get an order ε^0 approximation. Once the secular terms have been eliminated, Eq. (6.32) becomes

$$\frac{\partial x_1}{\partial \xi} = ix_1, \quad \frac{\partial y_1}{\partial \xi} = iy_1, \quad \frac{\partial z_1}{\partial \xi} = iz_1, \tag{6.38}$$

which can be readily solved directly to yield $x_1 = x_{10}e^{i\xi}$, $y_1 = y_{10}e^{i\xi}$, and $z_1 = z_{10}e^{i\xi}$, where x_{10}, y_{10}, and z_{10} represent arbitrary initial conditions. These are order ε solutions. Together, order ε^0 and order ε solutions are ensembled into Eq. (6.28) to yield analytical approximations to the solutions for the three-cell feedforward network, up to order ε. Figure 6.15 shows a computational bifurcation diagram of the original model Eq. (6.26). Circles represent the branches of bifurcations produced numerically. The analytical approximations predicted by the asymptotic analysis are shown in solid lines. The diagrams show that a good approximation between the exact solutions and the predicted asymptotic ones for the restricted parameter range $0 < \mu < \lambda$.

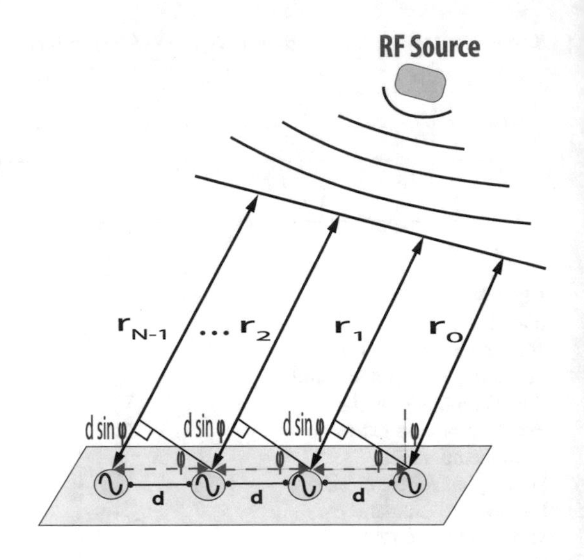

Fig. 6.16 An array of nonlinear oscillators drive individual radiation elements to form a radiation beam pattern

6.7 Beam Steering

Beam steering is about manipulating the direction of a radiating far-field intensity pattern. Applications usually include: optics, acoustics, and, antenna and radar systems. In antennas and radar systems, for instance, beam steering can be achieved either by switching the antenna elements or by controlling the phase differences between oscillating components, which typically consist of arrays of nonlinear oscillators. The oscillators are usually arranged in an array, as is shown, schematically, in Fig. 6.16.

Common point sources for beam steering in active antennas and radar systems consist of multiple nonlinear oscillators, e.g., van der Pol oscillators, each driving a separate radiation patch element. This lead to a mathematical model, which using the coupled cell formalism, can be expressed as

$$\frac{dz_j}{dt} = f_j(z_j, \mu) + \kappa e^{i\Phi}(z_{j+1} - 2z_j - z_{j-1}) + f_e(t), \tag{6.39}$$

where f is the internal dynamics of each van der Pol oscillator, written in normal form:

$$f_j(z_j, \mu) = (\alpha + w_j i)z_j - |z_j|^2 z_j,$$

where z_j is a complex-valued state variable for each oscillator j, with $j = 1, \ldots, N$, with boundary conditions $z_0 = z_{N+1} = 0$, α is the main excitation bifurcation parameter, which determines the amplitude of the ensuing oscillations, ω_i is related to the natural frequency of each oscillator, κ is the coupling strength among nearest neigh-

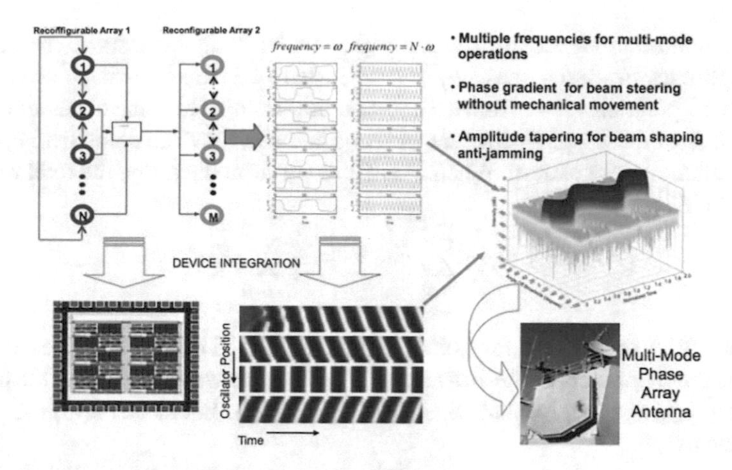

Fig. 6.17 Reconfigurable antenna made up of thousands of van der Pol circuits coupled in a chain. Small frequency perturbations of the end-points of the chain allows for beam shaping and steering without the need to mechanically rotate the antenna device

bors, Φ is a coupling phase parameter, and $f_e(t)$ is an external incoming signal, usually of the form

$$f_e(t) = a(t)e^{i(\Omega t+\varphi)},$$

where $a(t)$ is a complex amplitude factor that allows for slow changes (relative to the oscillating period) in the magnitude or phase, φ, of the incoming signal, with frequency Ω. Equation (6.39) serves as a model for a reconfigurable antenna [56], see Fig. 6.17.

Indeed, Fig. 6.17 illustrates that while applying a small perturbation to the end-points of the chain of oscillators, the beam can be shaped and steered [39] without the need to mechanically rotate the antenna device. Let us explore in more detail how this can be done.

6.7.1 Array Factor

It has been shown [39] that the far-field intensity pattern of an antenna or radar system consists of alternating light and dark bands, also known as *interference fringes*. These interference fringes arise from the phase differences incurred by the different path lengths between the sources. When a constant phase shift between neighboring sources is introduced, then the positions of the interference fringes, and, consequently, the radiating pattern, will change. Thus, controlling the phase shift between nonlinear oscillators is of critical importance for achieving beam steering. Common methods for controlling the phase differences of the nonlinear oscillators include: phase shifters [57], injection current [58, 59], and frequency de-tuning [60, 61].

The radiation produced by each patch element has both an (assumed to be identical) amplitude, E_0, and a phase, $\xi_j = \vec{k} \cdot \vec{r}_j$, where $\vec{k}$ is the free-space wave vector, and $\vec{r}_j$ is the position vector from the jth radiation element to some observation point P. When N of these signals interact with one another, they can collectively produce a total radiation field pattern. When P is far away from the array, the total radiating electric field is

$$E(P) = \frac{1}{N} \sum_{j=0}^{N-1} E_0 e^{i\xi_j} = \frac{1}{N} \sum_{j=0}^{N-1} E_0 e^{i\vec{k}\cdot\vec{r}_j}. \tag{6.40}$$

Figure 6.16 shows that for points in the far-field, i.e., for distances from the elements much grater than the array size, $(N-1)d$, the wave vector and the position vector are, approximately, parallel, so that $\vec{k} \cdot \vec{r}_j \approx kr_j$. Equation (6.40) can then be re-written as

$$E(P) = \left(\frac{1}{N} \sum_{j=0}^{N-1} e^{ik(r_j - r_0)} \right) E_0 e^{ikr_0}. \tag{6.41}$$

Since $r_j - r_0 = jd \sin\varphi$, where φ is the angle of incidence or transmission of the radiation wave, we can, once again, re-write Eq. (6.41) as

$$E(P) = \left(\frac{1}{N} \sum_{j=0}^{N-1} e^{ijkd \sin\varphi} \right) E_0 e^{ikr_0}. \tag{6.42}$$

Equation (6.42) is also known as the *array pattern multiplication property*. It indicates that the total radiation pattern of an antenna array is the product of the electric field produced by a single patch element, $E_0 e^{ikr_0}$, multiplied by an *Array Factor*, $A(\Psi)$, which is the term in parenthesis written as

$$A(\Psi) = \frac{1}{N} \sum_{j=0}^{N-1} e^{ij\Psi}, \tag{6.43}$$

where $\Psi = kd \sin\varphi$. Direct calculations show that the array factor given by Eq. (6.43) can also be expressed as

$$A(\Psi) = \frac{\sin\left(\frac{N\Psi}{2}\right)}{N \sin\left(\frac{\Psi}{2}\right)} e^{i(N-1)\Psi/2}. \tag{6.44}$$

Figure 6.18 shows the array factor for an array antenna with $N = 8$ elements, plotted in rectangular and polar coordinates. Observe that $|A(\Psi)|$ is symmetric with respect to $\Psi = 0$, and it always attains a maximum at $\Psi = 0$, which corresponds to an angle of incidence of $\varphi = 0$. This angle is also known as the *broadside direction*

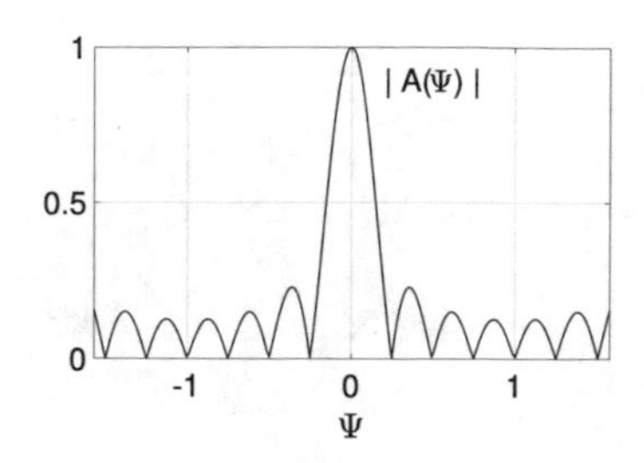

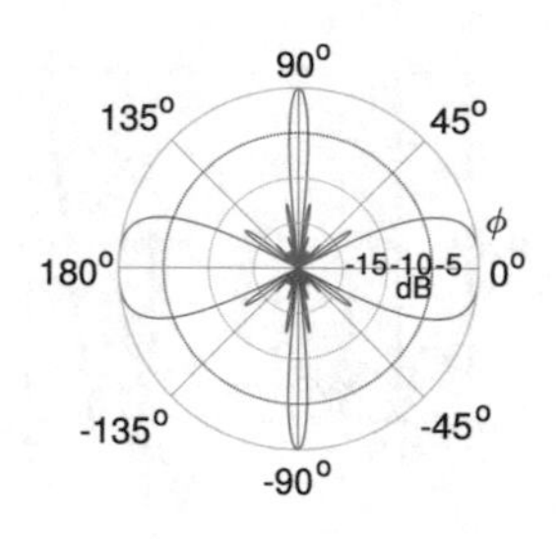

Fig. 6.18 Array factor for an array antenna with $N = 8$ identical elements, plotted in (top) rectangular coordinates and (bottom) polar coordinates

as it is normal to the plane of the array. Notice also that the width of the main lobe decreases as N increases.

Consider now the model Eq. (6.39). When $f_e(t) = 0$, the antenna operates in *transmission* mode, while when $f_e(t) \neq 0$ then the antenna functions as a receiver. In the former case, the emphasis is on the radiating patterns that emanate from the sources. The latter case concerns the response of the nonlinear beamformer to incident signals and noise. In both cases, transmission and receiving, one seeks solutions to the model Eq. (6.39) with a spatially uniform phase gradient across the array, i.e., in the form

$$z_j(t) = A_j e^{i\phi_j(t)}. \tag{6.45}$$

Previous calculations of the total radiating electric field assume the radiating source elements to be in-phase. Let us assume now a uniform phase gradient, introduced by the individual phase of the oscillators, across the array, such that

$$\xi_j = kr_j + \phi_j.$$

Substituting ξ_j into Eq. (6.40), while assuming a constant phase difference among neighboring oscillators, i.e., $\phi_{j+1} - \phi_j = \theta$, so that $\phi_j - \phi_1 = j\theta$, then a similar set of calculations yield

$$E(P) = \left(\frac{1}{N} \sum_{j=0}^{N-1} e^{ij(kd \sin\varphi + \theta)} \right) E_0 e^{i(kr_0 + \phi_1)}. \tag{6.46}$$

It follows that the array factor becomes

$$A(\Psi + \theta) = \frac{\sin\left(\dfrac{N(\Psi + \theta)}{2}\right)}{N \sin\left(\dfrac{\Psi + \theta}{2}\right)} e^{i(N-1)(\Psi+\theta)/2}. \tag{6.47}$$

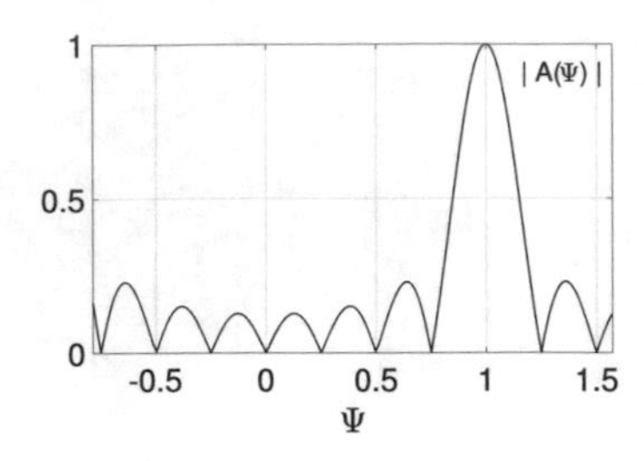

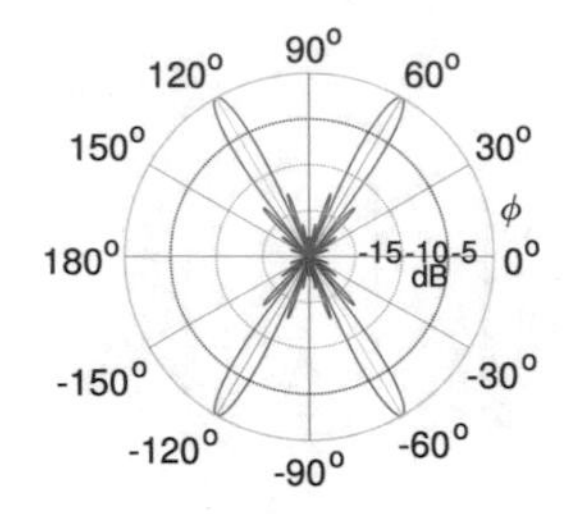

Fig. 6.19 Array factor for an array antenna with $N = 8$ identical elements, steered by an angle of -60^0, plotted in (top) rectangular coordinates and (bottom) polar coordinates

Consequently, the constant phase difference among the oscillators leads to a shift or steering of the beam pattern, from the angle Ψ to $\Psi + \theta$. Figure 6.19 illustrates the effect of steering the beam pattern shown in Fig. 6.18 by a constant phase difference of $\theta = -60^0$.

Now, assume the phase of each individual oscillator to be defined as

$$\phi_j = \omega t + (j - 1)\theta.$$

Substituting the desired solution (6.45), i.e., $z_j(t) = A_j e^{i\omega t + (j-1)\theta}$ into the model Eq. (6.39), and assuming identical amplitudes of oscillations, where, $E_0 = A_j$, yields the following conditions on the natural frequencies:

$$\begin{aligned} \omega_1 &= \omega + \kappa \sin(\theta + \Phi) \\ \omega_j &= \omega + (j - 1)\dot{\theta} \\ \omega_N &= \omega + (N - 1)\dot{\theta} - \kappa \sin(\theta + \Phi). \end{aligned} \tag{6.48}$$

In the *static* approach to beam steering, phase differences must be controlled so that the array produces a stationary far-field pattern at a fixed location. This case implies that $\dot{\theta} = 0$. It follows that the natural frequencies of only the two end elements, ω_1 and ω_N, need to be manipulated. By contrast, in the *dynamic* approach, one seeks a far-field intensity pattern that moves continuously. In this case, $\dot{\theta} \neq 0$, which implies that to achieve a continuously scanning beam then the natural frequency of every individual oscillator must be adjusted in a time-dependent manner [60, 62, 63].

In addition, we could re-arrange the oscillators into a cascade network, so we can exploit the symmetry of the interconnected oscillators to manipulate the collective frequency of oscillation over a broad range of frequencies. These two schemes, beam steering and multi-frequency oscillations, can lead to a multi-purpose antenna device. Next, we explore the idea of using a feedforward network to achieve, in addition to beam steering, signal amplification.

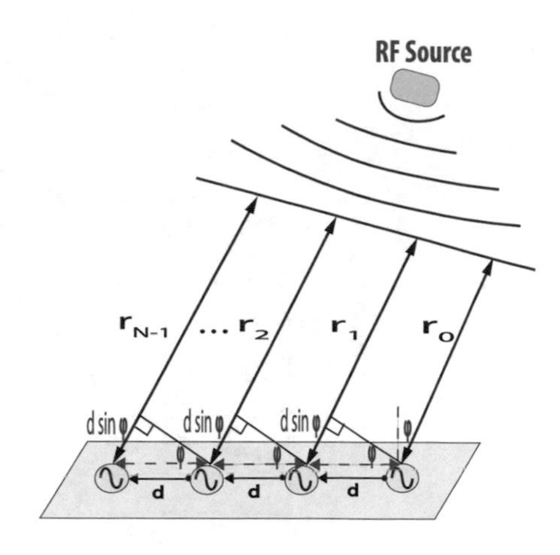

Fig. 6.20 Common point sources for beam steering in active antennas and radar systems consist of multiple nonlinear oscillators, e.g., van der Pol oscillators, each driving a separate radiation patch element. The oscillators in this array are coupled in a feedforward configuration

6.7.2 *Signal Amplification*

Figure 6.20 illustrates the concept of an array of nonlinear van der Pol oscillators connected in a feedforward fashion. The first oscillator (cell) is coupled to itself, and each successive oscillator is coupled to the next one.

When the first oscillator in the feedforward network shown in Fig. 6.20 is coupled to itself, the model equations, written in normal form, can be expressed as follows.

$$\begin{aligned} \dot{z}_1 &= (\alpha + \omega_1 i)\, z_1 - |z_1|^2 z_1 - \kappa e^{i\Phi} z_1 + F e^{i\omega t}, \\ \dot{z}_j &= \left(\alpha + \omega_j i\right) z_j - |z_j|^2 z_j - \kappa e^{i\Phi} z_{j-1} + F e^{i(\omega t + (j-1)\Delta\varphi)}, \end{aligned} \tag{6.49}$$

where $j = 1, \ldots, N$, α is the main excitation bifurcation parameter, which determines the amplitude of the ensuing oscillations, κ is the coupling strength, Φ is a coupling phase parameter, $f_e(t) = Fe^{i(\omega t + \Delta\varphi)}$ is the external incoming signal with constant amplitude F, frequency ω, and $\Delta\varphi$ is a constant phase difference introduced by the directionality of the signal. Without self coupling on the first cell, the term $\kappa e^{i\Phi} z_1$ can be removed. We showed above that in the transmission problem, where the external signal is absent, both type of arrays, with and without self coupling, phase-locking and synchronization exist and, under certain conditions, they are locally asymptotically stable solutions of the array. Furthermore, the third cell, and subsequent cells, oscillate (via a Hopf bifurcation) with a larger growth of $\alpha^{1/6}$, as opposed to the standard $\alpha^{1/2}$, which is characteristic of Hopf bifurcations. Figure 6.21 shows that the signal amplification persists under the presence of a small incident signal of amplitude $F = 0.1$ and frequency $\omega = 1$.

We assume $\Phi = 0$, so there is no coupling of the phases, and $\Delta\varphi = 0$, which corresponds to a broadside angle of incidence for the external signal. To see the amplification best, it is important to set the parameters in the model Eq. (6.49)

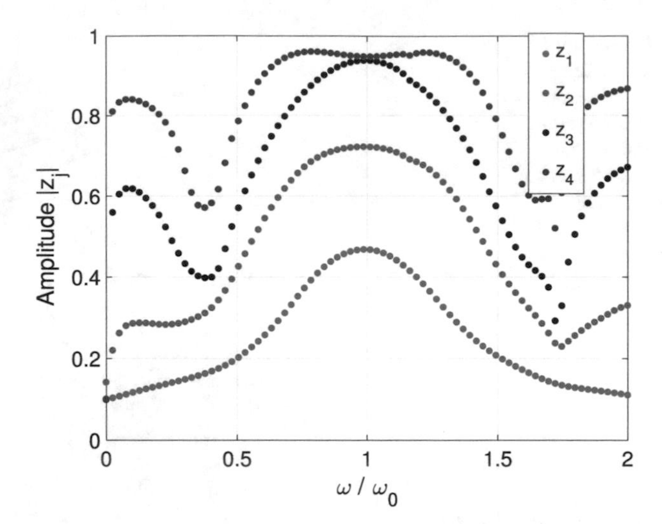

Fig. 6.21 Signal amplification in a feedforward network of van der Pol oscillators. Parameters are: $\alpha = 0.01$, $\omega_j = \omega_0 = 1.0$, $\kappa = 1.0$, and $\Phi = 0$, $F = 0.1$, $\Delta\varphi = 0.0$

near the onset of the Hopf bifurcation, so we assume $\alpha = 0.01$. Figure 6.21 shows a resonant effect that leads to signal amplification, with optimal response when the oscillators are synchronized with the external signal and the frequency of the external signal is close to the natural frequency of the oscillators.

When the frequency detuning is large, the individual oscillators undergo a secondary Hopf bifurcation that leads to quasi-periodic oscillations, and it decreases the amplification effect. This can be observed in Fig. 6.21 in the regions where $\omega/\omega_0 < 0.5$ and in the interval $\omega/\omega_0 > 1.5$.

6.8 Coupled Fluxgate System

Consider, for instance, the fluxgate magnetometer that was described earlier on, in Chap. 4, Sect. 4.10. Recall that such magnetometer contains a ferromagnetic core wound by two coils, one is the excitation coil to induce oscillations, and one is the pick-up coil, which is designed to detect the presence of external fields and to record the oscillations. Recall also that a one-dimensional ODE, Eq. (4.73), serves as a model for the single-core fluxgate magnetometer. We could then create a network of N fluxgate magnetometers, by connecting the flux output from one fluxgate into the next one in a ring fashion, as is shown schematically in Fig. 6.22.

Mathematically speaking, a model for the sensor network would take the form

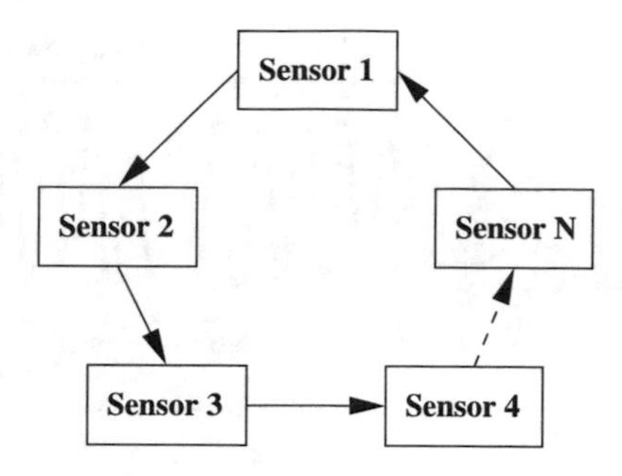

Fig. 6.22 (Left) Schematic of a netws ork-based sensor architecture. Individual sensors are coupled unidirectionally in a ring fashion. It will be shown later that a network-based system may offer certain advantages, e.g., performance enhancements, that cannot be achieved by individual sensor units

$$\tau_i \frac{dx_i}{dt} = -x_i + \tanh(c_i(x_i + \lambda_{i+1}x_{i+1} + H_e(t) + H_x)), \tag{6.50}$$

where $x_i(t)$ represents the magnetic flux at the output (i.e. in the secondary coil) of each individual i^{th} unit, with $i = 1, \ldots N \bmod N$; λ_{i+1} describes the coupling strength connecting the flux output from unit $i+1$ to unit i; and τ_i is the time constant of each individual fluxgate. If we make the underlying assumption of the fluxgates to be identical, then the revised model can be cast as

$$\tau \frac{dx_i}{dt} = -x_i + \tanh(c(x_i + \lambda x_{i+1} + H_e(t) + H_x)). \tag{6.51}$$

In both cases, non-identical and identical fluxgates, the model equations represent a network model because they contain a discrete number of units, N in this case, coupled together; and each unit is governed by a continuous model.

A fundamental question that arises almost immediately is: why would we want to create a network of fluxgate magnetometers?

Well, there are several reasons for building such a network. One of then has to do with the fact that, collectively, a network may exhibit oscillatory behavior even if none of the units can oscillate on their own. Figure 6.23 illustrates this point.

The fundamental idea is based on the fact that when certain systems are interconnected in some fashion, the symmetry of the resulting topology of connections, i.e., which units are coupled with each other, and the nonlinear characteristics of each individual unit, might be exploited to induce the interconnected network to generate a collective pattern of oscillation via an appropriate coupling function, see Fig. 6.23.

From a mathematical standpoint, the choice of coupling function can be any type of function, leading to a wide range of network solutions. From an engineering standpoint, the coupling function is restricted, however, by the type of system or

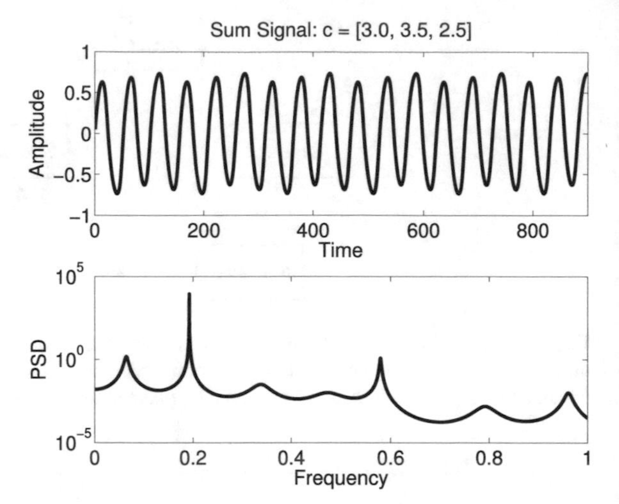

Fig. 6.23 Coupling-induced oscillations can appear (under certain coupling strength) in a network of three non-identical fluxgate magnetometers coupled unidirectionally as is shown in Fig. 6.22

technology being used. This is, in other words, a *model-dependent* feature of the system. For instance, fluxgate magnetometers coupled through magnetic flux are restricted to unidirectional coupling since directing magnetic flux in both directions can be extremely complicated to achieve. Mechanical gyroscopes can be coupled, however, bidirectionally through a series of mass-spring systems.

The purpose of this approach is to develop a robust and programmable dc sensing device that can be used to investigate the theoretical limit [64, 65] of magnetic-field sensitivity. In fact, in this section we describe the overall endeavor, which lead to the model, design and, ultimately, to the fabrication of the most sensitive fluxgate-type of magnetometer on the planet.

6.8.1 *Network Model*

We now consider a network of N fluxgate magnetometers coupled unidirectionally in a ring fashion, as is shown schematically in Fig. 6.22. We assume the individual fluxgates to be identical as well as the common coupling strength λ. Thus the global symmetries of the network are described by the group $\mathbf{Z}_N$ of cyclic permutations of N objects.

The mathematical model that governs the collective behavior of the network of fluxgate sensors of Fig. 6.22 is now a system of Ordinary Differential Equations (ODEs) in the following form:

$$\tau_i \frac{dx_i}{dt} = -x_i + \tanh(c(x_i + \lambda x_{i+1} + \varepsilon)), \tag{6.52}$$

where $x_i(t)$ represents the (suitably normalized) magnetic flux at the output (i.e. in the secondary coil) of each individual i^{th} unit, with $i = 1, \ldots N \bmod N$, and $\varepsilon \ll U_0$ is the externally applied dc magnetic flux (denoted as H_x in the model of a single fluxgate, see Eq. (4.73)), c is again the temperature-dependent nonlinearity parameter (recall that each element is bistable for $c > 1$), and U_0 is the energy barrier height of any of the elements, absent the coupling. Notice that the (uni-directional) coupling term, having strength λ, which is assumed to be equal for all three elements, is *inside* the nonlinearity. This is a direct result of the mean-field nature of the description in the fluxgate magnetometer, the coupling is through the induction in the primary or "pick up" coil. More importantly, observe the absence of the biasing signal $H_e(t)$. This is not a mistake.

Since the limit of magnetic sensitivity depends, mainly, on the ability of the network to produce stable self-biasing oscillations, our goal is to study the behavior of a coupled fluxgate system system in response to changes in parameters. In particular, the existence and stability of periodic solutions in response to changes in the coupling strength, the topology of connections, and the temperature-related parameters. But first, we need to examine the equilibrium points of the model Eq. (6.52). Let $X_e = (x_1, \ldots, x_N)$ denote such equilibrium point. Notice that X_e is actually a collection of N values, i.e., $X_e \in R^N$. Thus, to find an equilibrium point analytically, we would need to solve the following algebraic nonlinear system of N equations

$$-x_i + \tanh(c(x_i + \lambda x_{i+1} + \varepsilon)) = 0, \qquad i = 1, \ldots, N,$$

for the N unknowns $(x_1, \ldots, x_N)$. This is not an easy task, even when N is small. To get insight, we examine next, from a geometric and numerical standpoint, the special case of $N = 3$. However, we will be primarily interested in the N odd case, since it has been shown that when N is even there are no oscillations nor solutions connecting equilibrium points, i.e., heteroclinic cycles [66, 67].

6.8.2 Geometric Description of Solutions by Group Orbits

We start with a brief description of the solution sets for Eq. (6.52) as the coupling strength λ varies. The bifurcation methods generalize to arbitrary N, but the detailed specifics for finding and visualizing the basins of attraction are limited to 3D, so we focus on $N = 3$. Without loss of generality, the external field ε is set to zero and after re-scaling time the time constant can be set to $\tau_i = 1$. The specific system of equations used for our subsequent figures satisfies:

$$\boxed{\begin{aligned} \dot{x_1} &= -x_1 + \tanh(c(x_1 + \lambda x_2)) \\ \dot{x_2} &= -x_2 + \tanh(c(x_2 + \lambda x_3)) \\ \dot{x_3} &= -x_3 + \tanh(c(x_3 + \lambda x_1)), \end{aligned}} \tag{6.53}$$

where $c = 3$. Observe now that since the activation function tanh is odd, Eq. (6.53) remains unchanged under the transformation $x_i \mapsto \pm x_i$. Under unidirectional coupling with positive feedback the network equations are also unchanged under the cyclic transformation $x_i \mapsto x_{i+1}$. It follows that the symmetries of the coupled bistable system (6.53) are captured by the 24-elements group

$$\Gamma \simeq \mathbf{Z}_2^3 \otimes \mathbf{Z}_3,$$

which is generated by

$$\begin{aligned}(x_1, x_2, x_3) &\mapsto (\pm x_1, \pm x_2, \pm x_3)\\ (x_1, x_2, x_3) &\mapsto (x_2, x_3, x_1).\end{aligned}$$

The various type of solutions of Eq. (6.53) can be observed in the bifurcation diagram shown in Fig. 6.24, which was computed with the aid of the continuation software package AUTO [68]. When the coupling parameter is sufficiently large, and negative, then all solutions other than the unstable trivial solution and its 1D symmetric stable manifold, emanating along the line $x_1 = x_2 = x_3$, are attracted to a stable asymmetric periodic orbit with 3-fold symmetry. The oscillations occur for $\lambda < \lambda_c$, where λ_c is a critical coupling strength to be determined later on. At the other end, when the coupling parameter is sufficiently large, and positive, or at least $\lambda > \lambda_c$, then all solutions other than the unstable trivial solution and its 2D stable manifold are attracted to one of two stable symmetric equilibria. The same result ensues if N is even, or if the coupling is bidirectional.

For values of λ slightly less than λ_c, there is a small interval $\lambda_{HB} \leq \lambda \leq \lambda_c$ where global oscillations and synchronous equilibria of the form $(x_1, \ldots, x_N) = (x*, \ldots, x*)$ can coexist. In this interval, complex transitions that involve multiple equilibrium points, periodic solutions, and heteroclinic connections are observed. A close-up view of the interval of bistability of large amplitude oscillations and stable synchronous equilibria is also included in Fig. 6.24. The four branches of unstable equilibria that appear via saddle-node bifurcations (labeled LP) correspond to nonsynchronous equilibria.

To unravel those transitions we start with a large negative value of λ and replot the bifurcation diagram using x_1 in Fig 6.25. As λ increases, there is a pitchfork bifurcation, producing two new symmetric equilibria moving away from the origin along its 1D stable manifold. As λ further increases, the 1D stable manifold expands into two conical regions symmetric about the origin, which morph into two 3-sided pyramidal shaped regions surrounding two symmetric stable equilibria with additional increases in λ. The 3-fold stable limit cycle has its period increase until it spends longer and longer times near six points, which appear as stable equilibria through a saddle node bifurcation. A second saddle node bifurcation produces six other asymmetric equilibria, which arise and generate separatrices. These divide our space into eight basins of attraction. Two are the small symmetric pyramidal shaped regions centered on the line $x_1 = x_2 = x_3$, while the other six attracting regions surround these symmetric regions.

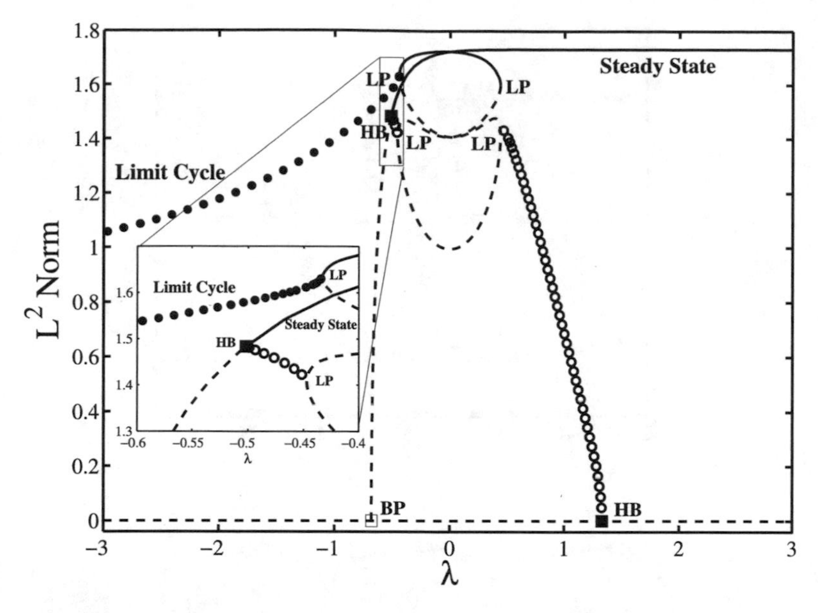

Fig. 6.24 Bifurcation diagram for a system of three identical bistable elements coupled unidirectionally and without delay. Solid (dotted) lines indicate stable (unstable) equilibrium points. Filled-in (empty) circles represent stable (unstable) periodic oscillations. (Insert) Close-up view of the region of bistability between large-amplitude oscillations and synchronous equilibria.

As λ increases to zero, the uncoupled state, these eight basins of attraction shrink or expand until they become simply the eight octants in 3-space with stable equilibria near $x_i \approx \pm 1$ for $i = 1, 2, 3$. As λ becomes more positive, the two symmetric basins of attraction increase in size, while the six asymmetric basins shrink in size. Another saddle node bifurcation occurs with the loss of six stable equilibria in the six asymmetric basins of attraction, and the separatrices between these equilibria and the remaining symmetric equilibria vanish. The remaining 2D stable manifold (separatrix) with 3-fold symmetry divides our 3-space into two basins of attraction containing our only two remaining stable equilibria. Further increases in λ only result in a flattening of this separatrix between the symmetric stable equilibria.

Recall from Chap. 5, Sect. 5.7, the definition of the group orbit of any point $x(t)$:

$$\Gamma x = \{\gamma x : \gamma \in \Gamma\}.$$

Thus, collectively, there are 27 equilibrium points, which can be arranged into one of four group orbits generated by the symmetry group $\Gamma \simeq \mathbf{Z}_2^3 \otimes \mathbf{Z}_3$, see Fig. 6.26. Thus, Fig. 6.26 can be interpreted as a color-coded evolution of four distinct group orbits as a function of λ, which yields: a straight-line in the middle connecting the two symmetric equilibria $(\pm x, \pm x, \pm x)$ and the origin; six gray-to-black corner segments connecting six asymmetric stable nodes of the form $(\pm x, \pm x, \mp x)$; twelve blue curves for the group orbit of twelve asymmetric unstable nodes with representative

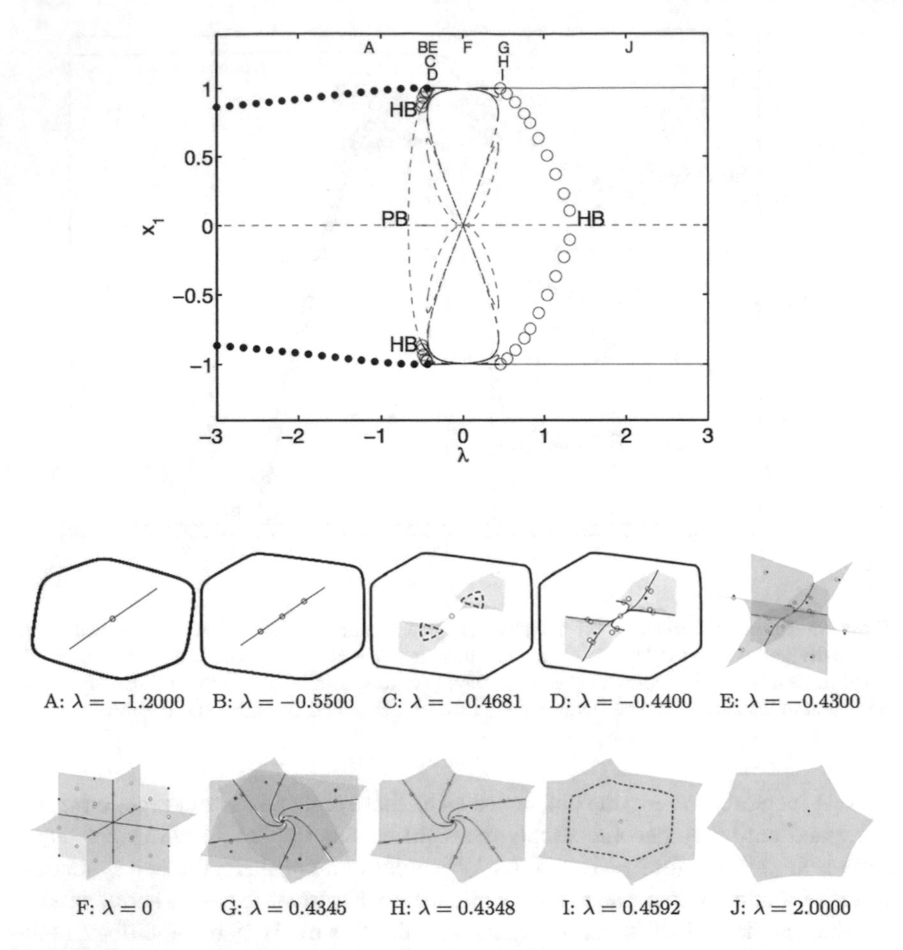

Fig. 6.25 Bifurcation diagram for system (6.53) computed in the equilibrium continuation program AUTO [68]. The diagram provides the values of x_1 at equilibria and the maximum and minimum values of x_1 at periodic orbits. Solid (dashed) lines indicate stable (unstable) equilibrium points. For unstable equilibria, black (dashed) indicates a 3D unstable manifold, red (dashed) has a 1D stable manifold, and blue (dashed) presents a 2D stable manifold. Filled-in (empty)circles represent stable (unstable) periodic oscillations. Parameters are $c = 3$ and $\epsilon = 0$. Notation: HB denotes Hopf bifurcation points and PB is a pitchfork bifurcation point

$(0, \pm x, \pm x)$; six red curves which connect the remaining six asymmetric unstable saddle nodes of the type $(0, 0, \pm x)$. All other equilibria can be readily obtained by applying directly the 24-elements of the group Γ to the representative elements listed above. We observe in Fig. 6.26 that if one begins at any one of the asymmetric equilibria and increases and decreases λ between the saddle node bifurcation values, then one can continuously reach all the remaining 23 asymmetric equilibria.

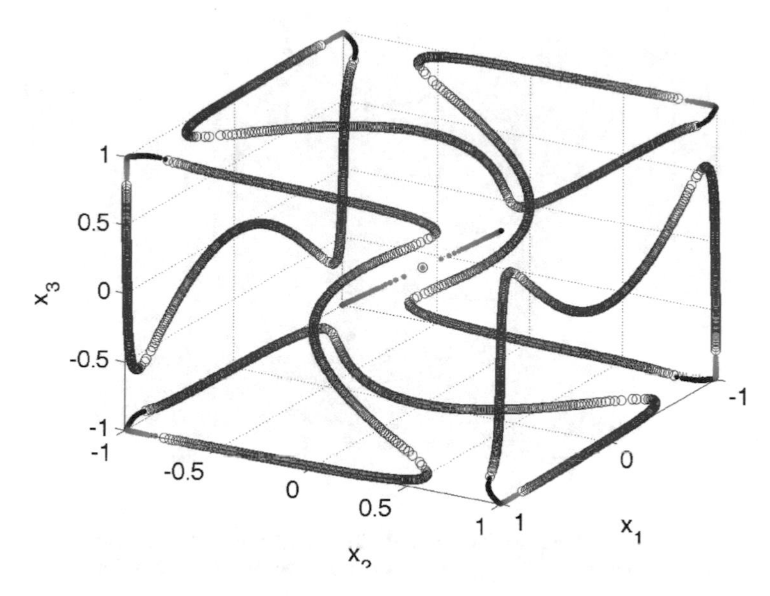

Fig. 6.26 There are 27 equilibria of Eq. (6.53) shown in this diagram at various values of λ. Three are the origin and the two symmetric equilibria. There are 24 asymmetric equilibria, which over the range of λ connect in a long chain. The gray-black equilibria are stable. The blue equilibria have 2D stable manifolds, and the red equilibria have 1D stable manifolds. The darker the shade of blue or red the higher the value of λ. Changes in color occur at saddle node bifurcations. A gray-black straight line for the graph of the group orbit of two symmetric equilibria, including the origin and three additional curves, red, blue and gray, for the 24 asymmetric equilibria

6.8.3 *Onset of Large Amplitude Oscillations*

We now investigate the global bifurcation that leads to the onset of stable infinite-period oscillations, and seek an analytic expression for the critical point λ_c. It is well-known that a generic feature of symmetric nonlinear systems is the existence of *heteroclinic cycles*, defined as a collection of solution trajectories that connect sequences of equilibria and/or periodic solutions [69]. Heteroclinic cycles are highly degenerate. Certain symmetries, however, can facilitate the existence of cyclic trajectories that can "travel" through invariant subspaces while connecting, via saddle-sink connections, one solution to another. In Eq. (6.52), in particular, we find six near-invariant planar regions (with $\lambda < 0$):

$$\begin{aligned} \delta_i &= \{x_i : \lambda x_i < 1, \quad x_{(i+2 \bmod 3)} = -1\}, \quad i = 1, 2, 3, \\ \delta_i &= \{x_i : \lambda x_i > -1, \, x_{(i+2 \bmod 3)} = 1\}, \qquad i = 4, 5, 6. \end{aligned}$$

Then the solution trajectories on the cycle lie on flow-invariant lines, see Figure 6.27, defined by the intersection of the invariant planes. A typical trajec-

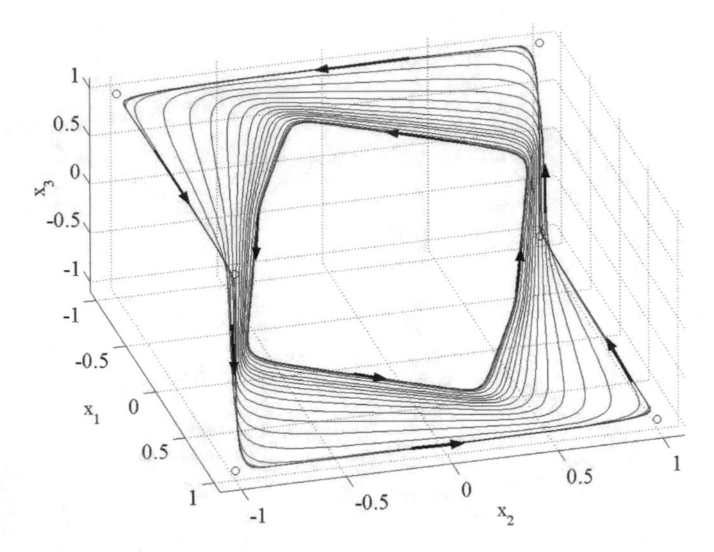

Fig. 6.27 Stable limit cycle solutions with amplitude $O(1)$ appear in system (6.52) for large negative values of λ. Fixed parameters are: $c = 3$, $\epsilon = 0$. Arrows indicate the direction of the flow

tory on the cycle connects six saddle points located near the points: $(1, -1, -1)$, $(1, 1, -1)$, $(-1, 1, -1)$, $(-1, 1, 1)$, $(-1, -1, 1)$, and $(1, -1, 1)$.

The saddle points exist only for $\lambda > \lambda_c$ and are annihilated when the periodic solutions appear. This suggests that we could determine the exact location of the heteroclinic cycle by finding the regions of parameter space where the saddle points exist, but leads to the complicated task of finding roots of polynomials of high order. On the other hand, we can use the fact that, at the birth of the cycle, solutions are confined to invariant lines. The flow on these lines cannot be obstructed by other equilibrium points, unless they are part of the cycles. This leads to the following conditions for existence of a cyclic solution:

$$-x + \tanh(c(x - \lambda + \epsilon)) > 0 \tag{6.54}$$

$$-x + \tanh(c(x + \lambda + \epsilon)) < 0. \tag{6.55}$$

When $\epsilon = 0$, the lhs of (6.54) and (6.55) each have a local minimum and a local maximum for $x \in (-1, 1)$. When $\epsilon > 0$, both extrema are shifted vertically. Thus, (6.54) is satisfied for $\epsilon = 0$ as well as $\epsilon > 0$. Hence, we only have to worry about condition (6.55). To find the critical point λ_c, we then compute the local maximum of (6.55), set it to zero, and solve for λ. We get:

$$\lambda_c = -\epsilon + \frac{1}{c}\ln(\sqrt{c} + \sqrt{c-1}) - \tanh(\ln(\sqrt{c} + \sqrt{c-1})). \tag{6.56}$$

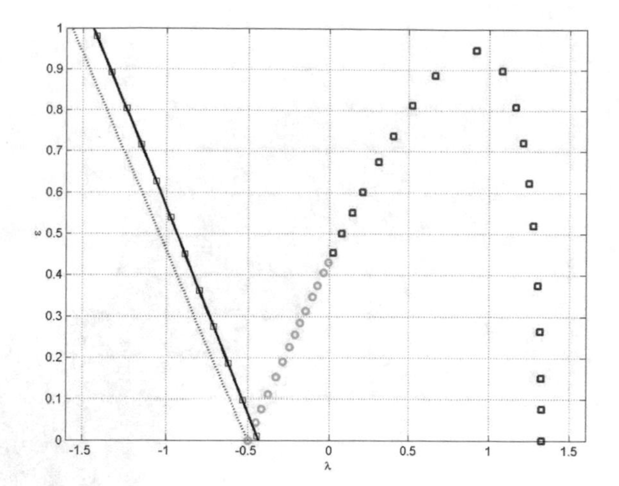

Fig. 6.28 Two-parameter continuation of Hopf bifurcation points (dash line, empty circles, and squares) and heteroclinic connections (black line obtained numerically via AUTO, superimposed squares obtained analytically). Periodic solutions are globally stable only for parameter values (λ, ϵ) below the black line, and unstable everywhere else. The temperature-related parameter is $c = 3$

An alternative derivation shows that this critical value of coupling strength can be expressed as

$$\lambda_c = -\varepsilon - x_{inf} + c^{-1} \tanh^{-1} x_{inf},$$

with $x_{inf} = \sqrt{(c-1)/c}$ is the inflection point of the energy function

$$U(x) = \frac{x^2}{2} - T \ln \left(\cosh \left(\frac{x+h}{T} \right) \right).$$

To verify this result, we conducted, numerically, a two-parameter continuation analysis using AUTO with $c = 3$, see Fig. 6.28. The dark diagonal line represents the loci of the heteroclinic cycle obtained numerically by AUTO, which shows very good agreement with the analytic loci determined by (6.56) (superimposed square points). The other curves represent the loci of HB points, which in all cases lead to unstable periodic solutions.

6.8.4 *Frequency Response*

The oscillation frequency ω, as a function of the system parameters, can be calculated from its period T. Near the onset λ_c of oscillations, T is essentially the time required for a solution to travel along the invariant lines of the heteroclinic cycle. By symmetry, the time spent on each branch is approximately the same. Hence,

$$T \approx 6 \int_{-1}^{1} dt, \quad \text{where} \quad dt \approx dx/(-x + \tanh(c(x - \lambda + \epsilon))),$$

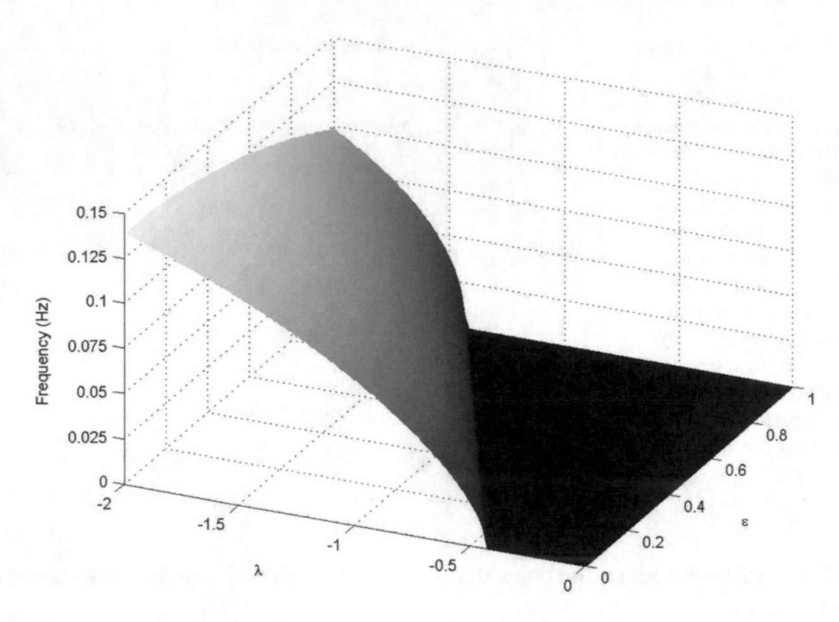

Fig. 6.29 Frequency response vs. system parameters λ and ϵ, for coupled system (6.52) with $N = 3$ and $c = 3$

and the integral must be evaluated numerically. In Fig. 6.29, we examine the relation between frequency and system parameters λ and ϵ with $c = 3$. The zero-frequency line in the $(\lambda,\ \epsilon)$ plane is in very good agreement with our expression (6.56) for the critical coupling strength.

Then a numerical approximation (for the special case of $N = 3$) for the frequency dependence on the system parameters can be obtained:

$$\omega = 0.115\sqrt{-\lambda - 0.85\epsilon - 0.4345}. \tag{6.57}$$

6.8.5 *Sensitivity Response*

Residence Times Detection. A new sensing technique, the *Residence Times Detection*, consists of measuring the "residence times" of the oscillations of the sensor device about the two stable states of the potential energy function $U(x)$. In the absence of noise and of external signals, the potential energy function is symmetric; hence, the two residence times are identical, i.e., $T_+ = T_-$. In the presence of a target signal, however, the hysteresis loop is skewed and the crossing-times are no longer equal. Then either the difference $|T_+ - T_-|$ or the ratio T_+/T_- of residence times can be used to quantify the signal, see Fig. 6.30. In the presence of noise, the residence times must be replaced by their ensemble averages.

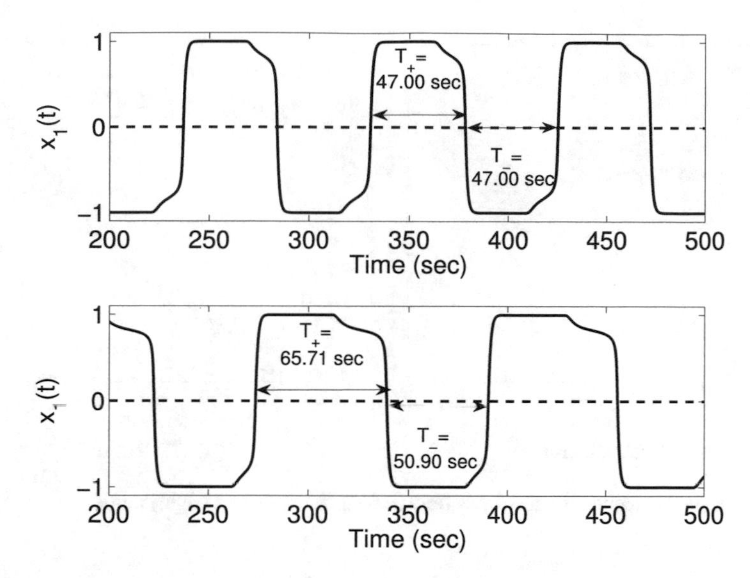

Fig. 6.30 Residence Time Detection. (Top) Without an external field, i.e., $\varepsilon = 0$, the wave form has top-to-bottom symmetry. (Bottom) With an external field, $\varepsilon = 0.01$ in this case, the wave forms develops an asymmetry. Then the difference or ratio of crossing times can used to quantify the external signal

Advantages of this procedure are:

- The RTD procedure can be implemented on-chip without the computationally demanding power spectral of the system output;
- Large-period oscillations yield large differences/ratios of residence times, i.e., better sensitivity;
- RTD can be optimized to require very low onboard power.

Numerical simulations show that, near the onset point λ_c, the period of the summed waveform becomes very large, which causes the waveform to yield larger values of the RT difference/ratio when an external signal is present, i.e., higher sensitivity. For illustrative purposes, Figure 6.31 compares the theoretical sensitivity of a single fluxgate with that of a network of $N = 3$ sensors. The slope of a RTD curve is proportional to the level of sensitivity. It follows that a network of three fluxgates with RTD ratio readout can be, approximately, 200 times more sensitivity than a single fluxgate.

6.9 Heteroclinic Connections

In simple terms, a *heteroclinic cycle* is a collection of solution trajectories that connects sequences of equilibria, periodic solutions, and/or chaotic sets [69–74]. As time evolves, a typical nearby trajectory stays for increasingly longer periods

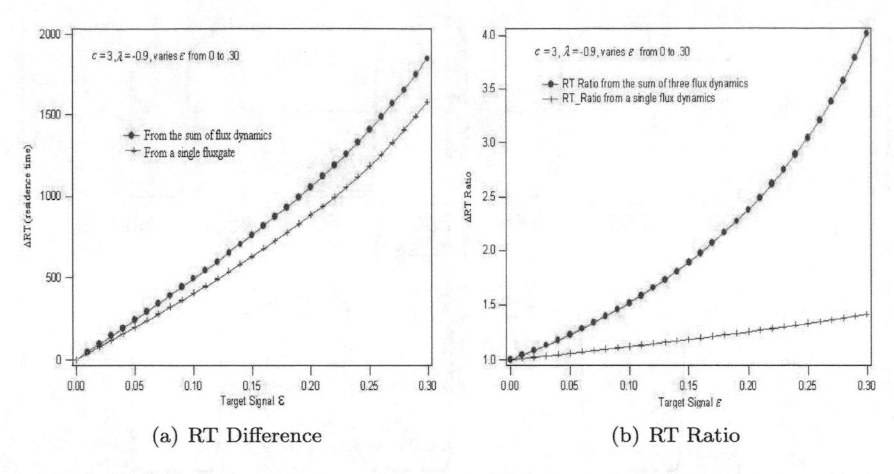

(a) RT Difference (b) RT Ratio

Fig. 6.31 Signal detection via residence time (left) differences and (right) ratios

near each solution before it makes a rapid excursion to the next solution. For a more precise description of heteroclinic cycles and their stability, see Melbourne et al. [74], Krupa and Melbourne [73], the monograph by Field [71], and the survey article by Krupa [69]. The existence of structurally stable heteroclinic cycles is considered a highly degenerate feature of both types of systems, continuous and discrete. In other words, typically they do not exist. In continuous systems, where the governing equations normally consist of systems of differential equations, it is well-known that the presence of symmetry can, however, lead to structurally stable, asymptotically stable, cycles [41, 75]. Let's consider an example.

Example 6.3 (The Guckenheimer-Holmes Cycle) Figure 6.32 illustrates a cycle involving three steady-states of a system of ODE's proposed by Guckenheimer and Holmes [41]. Observe that as time evolves a nearby trajectory stays longer on each equilibrium.

The group of symmetries, Γ, in this example, has 24 elements and is generated by the following symmetries

$$\begin{aligned}(x, y, z) &\mapsto (\pm x, \pm y, \pm z)\\ (x, y, z) &\mapsto (y, z, x)\end{aligned}$$

Note that, in fact, this is a homoclinic cycle since the three equilibria are on the group orbit given by the cyclic generator of order 3. The actual system of ODE's can be written in the following form

$$\begin{aligned}\dot{x}_1 &= \mu x_1 - (ax_1^2 + bx_2^2 + cx_3^2)x_1\\ \dot{x}_2 &= \mu x_2 - (ax_2^2 + bx_3^2 + cx_1^2)x_2\\ \dot{x}_3 &= \mu x_3 - (ax_3^2 + bx_1^2 + cx_2^2)x_3.\end{aligned} \tag{6.58}$$

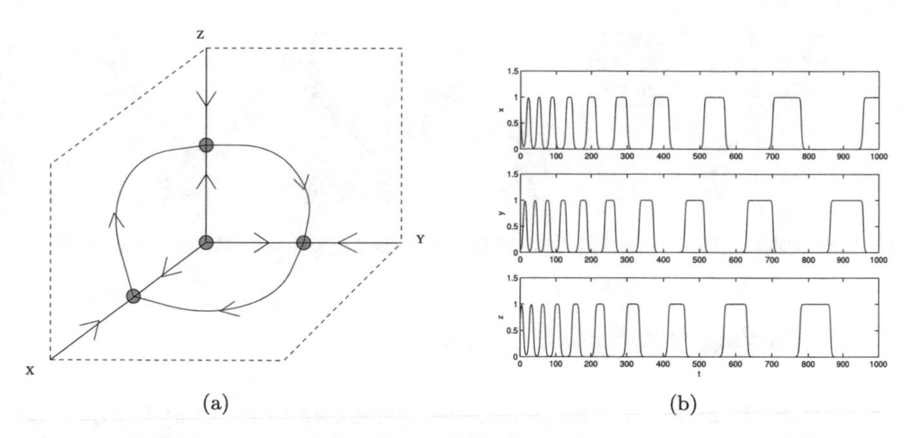

Fig. 6.32 Heteroclinic cycle found between three equilibrium points of the Guckenheimer and Holmes system. **a** Saddle-sink connections in phase-space, **b** Time series evolution of a typical nearby trajectory. Parameters are: $\mu = 1.0$, $a = 1.0$, $b = 0.55$, $c = 1.5$

In related work that describes cycling chaos, Dellnitz et al. (1995) point out that the Guckenheimer-Holmes system (6.58) can be interpreted as a coupled cell system (with three cells) in which the internal dynamics of each cell is governed by a pitchfork bifurcation of the form $f(x_i) = \mu x_i - a x_i^3, i = 1, 2, 3$. The network can be expressed as

$$\frac{dx_i}{dt} = f(x_i) + h(x_j, x_i)x_i,$$

where $h(x_j, x_i) = -(b x_{i+1}^2 + c x_{i+2})$, with all indices evaluated mod 3. As μ varies from negative to positive thropugh zero, a bifurcation from the trivial equlibrium $x_i = 0$ to nontrivial equlibria $x_i = \pm\sqrt{\mu}$ occurs. Guckenheimer and Holmes (1988) show that when the strength of the remaining terms in the system of ODE's (which can be interpreted as coupling terms) is large, an asymptotically stable hetroclinic cycle connecting these bifurcated equilibria exists. The connection between the equilibria in cell one to the equilibria in cell two occurs through a saddle-sink connection in the x_1x_2–plane (which is forced by the internal symmetry of the cells to be an invariant plane for the dynamics). As Dellnitz et al. (1995) further indicate, the global permutation symmetry of the three-cell system guarantees connections in both the x_2x_3–plane and the x_3x_1–plane, leading to a heteroclinic connection between three equlibrium solutions.

Although heteroclinic cycles are said to be non-generic features of nonlinear systems (either because they typically do not exist or because it is very difficult to produce them), one can systematically use the *lattice of isotropy subgroups* to find a heteroclinic cycle. Next we show how this can be done.

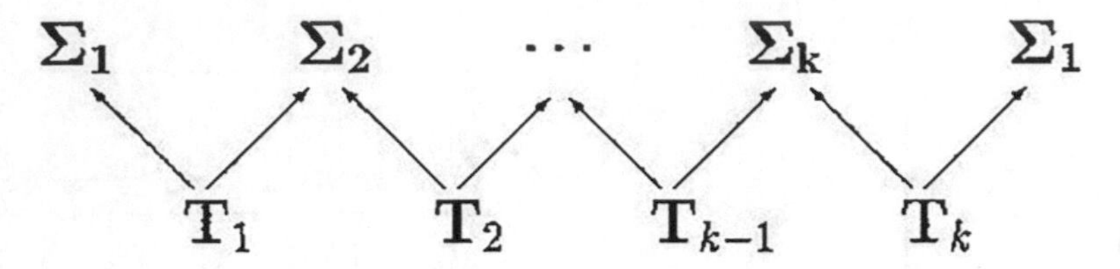

Fig. 6.33 Pattern inside lattice of subgroups that suggests the existence of heteroclinic cycles

6.9.1 Finding Heteroclinic Cycles

For systems whose symmetries are described by the continuous group $\mathbf{O(2)}$, i.e. the group of rotations and reflections on the plane, Armbruster et al. [76] show that heteroclinic cycles between steady-states can occur stably, and Melbourne et al. [74] provide a method for finding cycles that involve steady-states as well as periodic solutions. Let $\Gamma \subset \mathbf{O}(N)$ be a Lie subgroup (where $\mathbf{O}(N)$ denotes the orthogonal group of order N) and let $g : \mathbb{R}^N \to \mathbb{R}^N$ be Γ-equivariant, that is,

$$g(\gamma X) = \gamma g(X),$$

for all $\gamma \in \Gamma$. Consider the system

$$\frac{dX}{dt} = g(X).$$

Note that $N = kn$ in an n-cell system with k state variables in each cell. Equivariance of g implies that whenever $X(t)$ is a solution, so is $\gamma X(t)$. Using fixed-point subspaces, Melbourne et al. (1989) suggest a method for constructing heteroclinic cycles connecting equilibria. Suppose that $\Sigma \subset \Gamma$ is a subgroup. Then the fixed-point subspace

$$\mathrm{Fix}(\Sigma) = \{X \in \mathbb{R}^N : \sigma X = X \quad \forall \sigma \in \Sigma\}$$

is a flow invariant subspace. The idea is to find a sequence of maximal subgroups $\Sigma_j \subset \Gamma$ such that $\dim \mathrm{Fix}(\Sigma_j) = 1$ and submaximal subgroups $T_j \subset \Sigma_j \cap \Sigma_{j+1}$ such that $\dim \mathrm{Fix}(T_j) = 2$, as is shown schematically in Figure 6.33. In addition, the equilibrium in $\mathrm{Fix}(\Sigma_j)$ must be a saddle in $\mathrm{Fix}(T_j)$ whereas the equlibrium in $\mathrm{Fix}(\Sigma_{j+1})$ must be a sink in $\mathrm{Fix}(T_j)$.

Such configurations of subgroups have the possibility of leading to heteroclinic cycles if saddle-sink connections between equilibria in $\mathrm{Fix}(\Sigma_j)$ and $\mathrm{Fix}(\Sigma_{j+1})$ exist in $\mathrm{Fix}(T_j)$. It should be emphasized that more complicated heteroclinic cycles can exist. Generally, all that is needed to be known is that the equilibria in $\mathrm{Fix}(\Sigma_j)$ is a saddle and the equlibria in $\mathrm{Fix}(\Sigma_{j+1})$ is a sink in the fixed-point subspace $\mathrm{Fix}(T_j)$ (see Krupa and Melbourne (1995)) though the connections can not, in general, be proved. Since saddle-sink connections are robust in a plane, these heteroclinic cycles are stable to perturbations of g so long as Γ-equivariance is preserved by the perturbation. For a detailed discussion of asymptotic stability and nearly asymptotic stability of heteroclinic cycles, which are also very important topics, see Krupa and Melbourne (1995).

Near points of Hopf bifurcation, this method for constructing heteroclinic connections can be generalized to include time periodic solutions as well as equilibria. Melbourne, Chossat, and Golubitsky (1989) do this by augmenting the symmetry group of the differential equations with S^1—the symmetry group of Poincare-Birkhoff normal form at points of Hopf bifurcation—and using phase-amplitude equations in the analysis. In these cases the heteroclinic cycle exists only in the normal form equations since some of the invariant fixed-point subspaces disappear when symmetry is broken. However, when that cycle is asymptotically stable, then the cycling like behavior remains even when the equations are not in normal form. This is proved by using asymptotic stability to construct a flow invariant neighborhood about the cycle and then invoking normal hyperbolicity to preserve the flow invariant neighborhood when normal symmetry is broken. Indeed, as is shown by Melbourne (1989), normal form symmetry can be used to produce stable cycling behavior even in systems without any spatial symmetry. More generally, it also follows that if an asymptotically stable cycle can be produced in a truncated normal form equation (say truncated at third or fifth order), then cycling like behavior persists in equations with higher order terms—even when those terms break symmetry—and the cycling like behavior is robust.

6.9.2 A Cycle in a Coupled-Cell System

Buono, Golubitsky, and Palacios [70] proved the existence of heteroclinic cycles involving steady-state and time periodic solutions in differential equations with $\mathbf{D}_n$ symmetry. In their approach, they studied various mode interactions—in particular, the six-dimensional steady-state/Hopf mode interaction where $\mathbf{D}_n$ acts by its standard representation on the critical eigenspaces. The exact cycles they discussed are found in the normal form equations which have $\mathbf{D}_n \times S^1$ symmetry when $n = 6$ and $n = 5$—though much of their discussion is relevant for a general $\mathbf{D}_n$ system.

Consider for instance a system of differential equations with the symmetries of a hexagon, which are described by the dihedral group $\mathbf{D}_6$. Reflectional symmetries of a hexagon come in two (nonconjugate) types: those whose line of reflection connects opposite vertices of the hexagon (κ) and those whose line of symmetry connects midpoints of opposite sides ($\gamma\kappa$). It is known that $\mathbf{D}_6$ symmetry-breaking steady-state bifurcations produce two nontrivial equilibria—one with each type of reflectional symmetry—and $\mathbf{D}_6$ symmetry-breaking Hopf bifurcations produce two standing waves—one with each type of reflectional symmetry. In normal form the

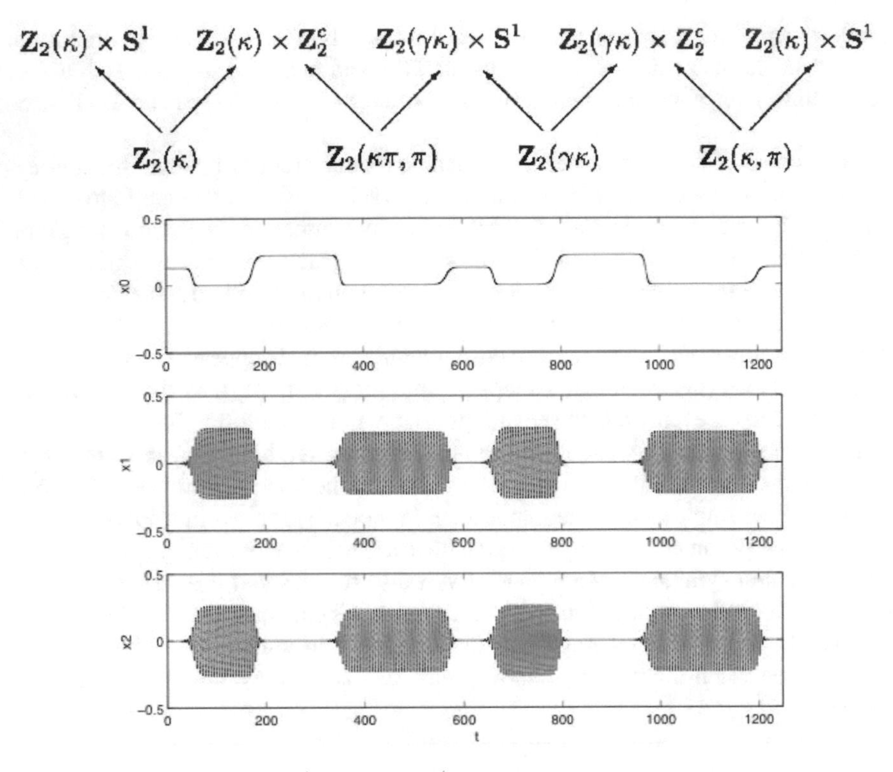

Fig. 6.34 Subgroups in $\mathbf{D}_6 \times S^1$ lattice that permit the existence of heteroclinic cycles

symmetry groups of these four solutions are $Z_2(\kappa) \times S^1$, $Z_2(\gamma\kappa) \times S^1$, $Z_2(\kappa) \times Z_2^c$, and $Z_2(\gamma\kappa) \times Z_2^c$ where $Z_2^c = Z_2(\pi, \pi)$. Using the ideas described by Melbourne et al. (1989), the lattice shown in Figure 6.34(top) suggests that robust, asymptotically stable heteroclinic cycles can appear in unfoldings of $\mathbf{D}_6$ normal form symmetry-breaking steady-state/Hopf mode interactions.

The cycle would connect the first steady-state with the first standing wave with the second steady-state with the second standing wave and back to the first steady-state. A general system of ODE's with $\mathbf{D}_6 \times S^1$-symmetry has the form

$$\frac{dz}{dt} = g(z, \lambda, \mu) = (C(z), Q(z)) \in C \times C^2,$$

where

$$C(z) = C^1 z_0 + C^3 \bar{z}_0 z_1 \bar{z}_2 + C^5 \bar{z}_0^5 + C^7 \bar{z}_0 (\bar{z}_1 z_2)^2 + C^9 \bar{z}_0^3 \bar{z}_1 z_2 + C^{11} z_0 (z_1 \bar{z}_2)^3$$

$$Q(z) = Q^1 \begin{bmatrix} z_1 \\ z_2 \end{bmatrix} + Q^2 \delta \begin{bmatrix} z_1 \\ -z_2 \end{bmatrix} + Q^3 \begin{bmatrix} z_0^2 z_2 \\ \bar{z}_0^2 z_1 \end{bmatrix} + Q^4 \delta \begin{bmatrix} z_0^2 z_2 \\ -\bar{z}_0^2 z_1 \end{bmatrix} +$$

$$Q^5 \begin{bmatrix} \bar{z}_0^4 z_2 \\ z_0^4 z_1 \end{bmatrix} + Q^6 \delta \begin{bmatrix} \bar{z}_0^4 z_2 \\ -z_0^4 z_1 \end{bmatrix} + Q^7 \begin{bmatrix} \bar{z}_0^2 \bar{z}_1 z_2^2 \\ z_0^2 z_1^2 \bar{z}_2 \end{bmatrix} +$$

$$Q^8 \delta \begin{bmatrix} \bar{z}_0^2 \bar{z}_1 z_2^2 \\ -z_0^2 z_1^2 \bar{z}_2 \end{bmatrix} + Q^9 \begin{bmatrix} (\bar{z}_1 z_2)^2 z_2 \\ (z_1 \bar{z}_2)^2 z_1 \end{bmatrix} + Q^{10} \delta \begin{bmatrix} (\bar{z}_1 z_2)^2 z_2 \\ -(z_1 \bar{z}_2)^2 z_1 \end{bmatrix},$$

where $\delta = |z_2|^2 - |z_1|^2$, $C^j = c^j + i\delta c^{j+1}$, c^j are real-valued $\mathbf{D}_6 \times S^1$-invariant functions and $Q^j = p^j + q^j i$ are complex-valued $\mathbf{D}_6 \times S^1$-invariant functions depending on two parameters λ and μ. Numerical integration of this $\mathbf{D}_6 \times S^1$-equivariant system (in normal form) yields the cycle shown in Fig. 6.34(bottom).

6.10 Exercises

Exercise 6.1 For each of the following problems, write the equations as a first-order system and then apply the Routh-Horowitz criterion to investigate the stability of the zero equilibrium solution.

(a) $x''' + 6x'' + 3x' + 2x = 0$.
(b) $x^{IV} + 2x''' + 4x'' + 7x' + 3x = 0$.

Exercise 6.2 Apply the Routh-Horowitz criterion to determine the region of parameter space, (μ, ν), where the zero equilibrium solution of the system

$$x^{IV} + x''' + \mu x'' + \nu x' + x = 0,$$

is asymptotically stable.

Exercise 6.3 Apply the Routh-Horowitz criterion to investigate for which values of the parameters μ and ν the zero equilibrium solution of the following system is asymptotically stable:

$$\frac{dx}{dt} = \mu x + \nu y$$
$$\frac{dy}{dt} = x - z$$
$$\frac{dz}{dt} = -x + y.$$

Exercise 6.4 Consider the two spring-mass system model Eq. (6.4).

(a) Apply the transformation: $y_1 = x_1 - x_2$, $y_2 = y_1 + y_2$ and rewrite Eq. (6.4) in terms of y_1 and y_2. Show that, in these new coordinates, the dynamics of spring-

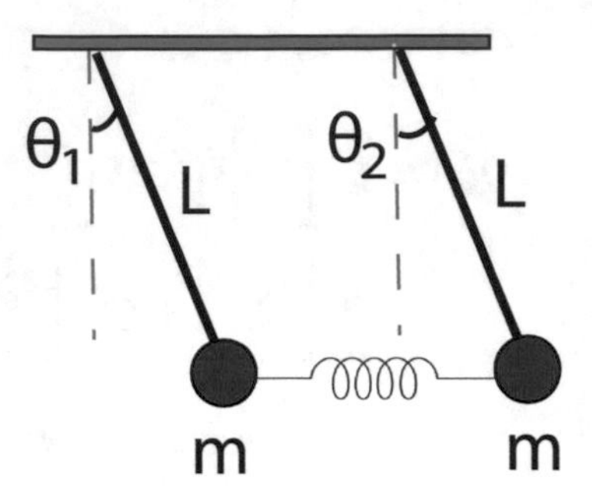

Fig. 6.35 Two coupled pendulums

mass system decouples from one another. Observe that $y_1 = 0$ corresponds to synchronized oscillations of the masses, while $y_2 = 0$ represents anti-phase oscillations.

(b) Solve the decoupled system explicitly.
(c) Find a general solution for $x_1(t)$ and $x_2(t)$.
(d) Consider the initial conditions: $x_1(0) = x_2(0) = 0$ and $\dot{x}_1(0) = \dot{x}_1(0) = 0$. Find a particular solution for $x_1(t)$ and $x_2(t)$.

Exercise 6.5 A mathematical model of a self-sustained electro-mechanical transducer consisting of an electrical part and a mechanical part coupled through Laplace force and Lenz electromotive voltage [77], can be written by coupling a van der Pol oscillator to a Duffing oscillator, leading to:

$$\begin{aligned} \frac{d^2x}{dt^2} + \omega_1^2 x + \mu_1(1 - x^2)\frac{dx}{dt} + f\frac{d^2y}{dt^2} &= 0 \\ \frac{d^2y}{dt^2} + \omega_1^2 y + \mu_2\frac{dy}{dt} + cy^3 - dx &= 0, \end{aligned} \tag{6.59}$$

where ω_j, μ_j, $(j = 1, 2)$, c, d, and f are parameters.

(a) Write Eq. (6.59) as a first-order system in the new variables $[x_1, x_2, y_1, y_2]^T$.
(b) Show that for $c \geq 0$ the first-order system has only one equilibrium point, $(0, 0, 0, 0)$, while for $c < 0$, two new nontrivial equilibrium points appear besides the trivial equilibrium point.
(c) Assume $c \geq 0$, so that Eq. (6.59) admits only the trivial equilibrium. Apply the Routh-Horowitz criterion to determine the conditions (on the parameters) for the trivial equilibrium to be stable.
(d) Determine the parameter conditions under which the system Eq. (6.59) undergoes a Hopf bifurcation.

Exercise 6.6 Consider the following set of two identical pendulums coupled to one another through a spring, as is shown in Fig. 6.35

Assuming the pendulums have length L and mass m, a mathematical model of the pendulums is given by

$$\boxed{\begin{aligned} m\frac{d^2x_1}{dt^2} + \frac{mg}{L}x_1 &= k(x_2 - x_1) \\ m\frac{d^2x_2}{dt^2} + \frac{mg}{L}x_2 &= k(x_1 - x_2), \end{aligned}} \tag{6.60}$$

where $x_1 \approx L\theta_1$, $x_2 \approx L\theta_2$, represent small displacements of the angles formed by the pendulums from a vertical position, g is gravity and k is the spring constant connecting both pendulums.

In this exercise you will need to show, on the basis of symmetry alone, that the coupled pendulums exhibit two modes of oscillations: a synchronized solution in which both pendulums oscillate with the same amplitude and same phase, and an anti-phase solution in which the oscillations are out-of-phase by half a period.

(a) Start by recognizing that the group of symmetries of the coupled pendulums Eq. (6.60) is $\Gamma = \mathbf{D}_2$.

(b) Let $X_1 = [x_1, \dot{x}_1]^T$ and $X_2 = [x_2, \dot{x}_2]^T$. Rewrite the model Eq.(6.60) in the following form:

$$\begin{aligned} \dot{X}_1 &= f(X_1; \lambda) + C(k)(X_2 - X_1) \\ \dot{X}_2 &= f(X_2; \lambda) + C(k)(X_1 - X_2), \end{aligned} \tag{6.61}$$

where λ is a parameter and $C(k)$ is a 2×2 matrix, both to be determined.

(c) Compute the linearization L of the network dynamics as is described by Eq. (6.61).

(d) Since $X = (X_1, X_2) \in \mathbb{R}^4$, it follows that $\mathbb{R}^4$ is the phase-space of the coupled system. Then, apply the isotypic decomposition of $\mathbf{C}^4$ by $\Gamma = \mathbf{D}_2$, which is given by

$$\mathbf{C}^4 = V_0 \oplus V_1,$$

and show that $V_0 = [v, v]^T$ and $V_1 = [v, -v]^T$, where $v \in \mathbb{R}^2$, are eigenvectors of L.

(d) Employ coordinates along the isotypic components to diagonalize L. Then argue that Hopf bifurcations in L allow us to predict the existence of synchronized and anti-phase oscillations. Notice that this argument is a model-independent feature of the symmetry of the network system. In other words, same predictions are valid with other type of oscillators, as long as the Hopf bifurcation conditions are satisfied.

Exercise 6.7 In this exercise you will study the stability properties of the two branches of oscillations, which are predicted to exist on the basis of symmetry, in the coupled pendulums model Eq. (6.60). That is, apply the Routh-Horowitz criterion to determine the conditions (on the parameters) for the synchronized and anti-phase solutions of the coupled pendulums Eq. (6.60) to be stable.

Exercise 6.8 Once again, consider the coupled pendulum Eq. (6.60).

(a) Apply the transformation: $y_1 = x_1 - x_2$, $y_2 = y_1 + y_2$, rewrite Eq. (6.60) in terms of these new coordinates and show that the dynamics of y_1 and y_2 decouple from one another. Observe that in these coordinates, $y_1 = 0$ corresponds to synchronized oscillations of the pendulums, while $y_2 = 0$ represents anti-phase oscillations.

(b) Write solutions for the decoupled system as:

$$y_1 = A \cos \omega_1 t, \qquad y_2 = B \cos \omega_2 t,$$

where A and B are constants, and ω_1 and ω_2 are the *normal frequencies*, which need to be found.

(c) Find a general solution for $x_1(t)$ and $x_2(t)$.

(d) Consider the initial conditions: $x_1(0) = x_2(0) = 0$ and $\dot{x}_1(0) = \dot{x}_1(0) = 0$. Find a particular solution for $x_1(t)$ and $x_2(t)$.

Exercise 6.9 Consider the mathematical model of a fluxgate magnetometer: $\dot{x} = -x + c \tanh(c\,(x + H_e(t) + H_x))$. Compute all equlibrium points and study the bifurcations that lead to the creation or annihilation of equilibria.

Exercise 6.10 Consider the three-cell network of Fig. 6.6 whose dynamics is described by Eq. (6.11). Conduct computer simulations of the network equations to show the existence of a collective pattern of oscillations with $\Sigma_X = \mathbf{Z}_2(\kappa, \pi)$ symmetry. Hint: One of the cells must oscillate at twice the frequency of the other two. Try a slightly larger value of λ than that of the traveling wave pattern. For instance, try $\lambda = 1.1$.

Exercise 6.11 Consider the model of a three-cell feedforward network as it appears in Eq. (6.25). Repeat the analysis carried out in Section 6.6 but this time with $\gamma \neq 0$. Show that the branch of periodic oscillations along the third cell exhibits $\mu^{1/6}$ growth rate.

Exercise 6.12 Perform a computational bifurcation analysis of a network three fluxgate magnetometers, similar to Eq. (6.52), but with non-identical elements. That is, c is no longer constant as it varies from one fluxgate to the next one, so it must be replaced by c_i.

Exercise 6.13 A mathematical model for a *drive-free* gyroscope system is given by

$$\boxed{\begin{aligned} m\ddot{x}_j + c\dot{x}_j + \kappa x_j + \mu x_j^3 &= \lambda x_{j+1} + 2m\Omega_z \dot{y}_j \\ m\ddot{y}_j + c\dot{y}_j + \kappa y_j + \mu y_j^3 &= \qquad\quad - 2m\Omega_z \dot{x}_j, \end{aligned}} \tag{6.62}$$

where λ is the coupling strength and $j = 1, 2, 3 \mod 3$. The system is said to be drive-free because it exploits the concept of coupling-induced oscillations to generate oscillations. Study the bifurcations of the network Eq. (6.62). As a point of reference, consider the same parameter values as those shown in Table 9.1, which were used for the network of gyroscopes studied in Sect. 9.6.

Exercise 6.14 Study the bifurcations of an array of N identical vibratory gyroscopes arranged in a ring configuration, coupled bidirectionally along the driving- and sensing-modes, with equations of motion given by

$$\begin{aligned} m\ddot{x}_j + c\dot{x}_j + \kappa x_j + \mu x_j^3 &= \varepsilon \sin \omega_d t + 2m\Omega_z \dot{y}_j + \lambda(x_{j+1} - 2x_j + x_{j-1}) \\ m\ddot{y}_j + c\dot{y}_j + \kappa y_j + \mu y_j^3 &= \quad - 2m\Omega_z \dot{x}_j + \lambda(y_{j+1} - 2y_j + y_{j-1}). \end{aligned} \tag{6.63}$$

Consider the same parameter values as those shown in Table 9.1, which are used for the stochastic network of gyroscopes studied in Sect. 9.6.

Exercise 6.15 Apply Kirchhoff's law to derive the model equations for a network of crystal oscillators coupled unidirectionally, as is shown in Fig. 9.12.

Exercise 6.16 Apply Kirchhoff's law to derive the model equations for a network of crystal oscillators coupled bidirectionally.

Exercise 6.17 Compute phase drift as a function of the number of oscillators using the synchronized solution produced by a network of crystal oscillators coupled unidirectionally. Discuss the scaling law distribution.

Exercise 6.18 Consider a coupled system made up of two unit masses constrained to move on a straight line while restrained by two springs. One of the masses, labeled x, is restrained by an anchor spring. The second mass, represented by y, is connected to the first mass by a second spring. The entire system is driven by a periodic force $F(t) = f \cos wt$, which is applied to the second mass, y. f represents the amplitude of the applied force and w its frequency.

A mathematical model for the governing equations is given by

$$\begin{aligned} \frac{d^2x}{dt^2} + 2x - y + x^3 &= 0, \\ \frac{d^2y}{dt^2} + y - x \quad &= f \cos wt \end{aligned} \tag{6.64}$$

We seek a periodic response of this two-cell couple system and, since there is no damping, we can set

$$x = A \cos wt, \qquad y = B \sin wt.$$

Substitute x and y, as given above, into Eq. (6.64). Set the coefficients of $\cos wt$ to zero and derive a set of two equations for the amplitudes A and B involving w and f. Fix f and solve for A as a function of w. Also, solve for B as a function of w. Plot both functions.

Exercise 6.19 Consider the following model of a pair of coupled van der Pol Oscillators

$$\begin{aligned} \frac{d^2x}{dt^2} + x - \varepsilon(1 - x^2)\frac{dx}{dt} &= \varepsilon\alpha(y - x) \\ \frac{d^2y}{dt^2} + (1 + \varepsilon\Delta)y - \varepsilon(1 - y^2)\frac{dy}{dt} &= \varepsilon\alpha(x - y), \end{aligned} \tag{6.65}$$

where ε is a small parameter, Δ is also a parameter representing the difference in uncoupled frequencies, also known as the detuning parameter, and α is a coupling constant.

Set the two variable expansion: $\xi = (1 + \kappa_1\varepsilon)t$ and $\eta = \varepsilon t$. Expand x and y up to order $O(\varepsilon)$, i.e., $x = x_0 + \varepsilon$ and $y = y_0 + \varepsilon y_1$, respectively. Apply the chain rule to compute first—and second-order derivates of x with respect to ξ and η. Substitute x and y into the original equations and collect order one and order $O(\varepsilon)$ terms. Let

$$\begin{aligned} x_0(\xi, \eta) &= A(\eta)\cos\xi + B(\eta)\sin(\xi), \\ y_0(\xi, \eta) &= C(\eta)\cos\xi + D(\eta)\sin(\xi), \end{aligned}$$

where A, B, C and D are unknown amplitudes. Derive a system of four ordinary differential equations for these amplitudes and solve them numerically. Discuss the qualitative behavior of these equations.

Exercise 6.20 Ermentrout and Kopell (1990) illustrate the notion of "oscillator death" with the following model:

$$\begin{cases} \dfrac{d\theta_1}{dt} = \omega_1 + \sin\theta_1\cos\theta_2 \\ \dfrac{d\theta_2}{dt} = \omega_2 + \sin\theta_2\cos\theta_1 \end{cases} \tag{6.66}$$

where θ_1 and θ_2 are state variables that represent phase dynamics on a circle, and $\omega_1, \omega_2 \geq 0$ are parameters.

(a) Equation (6.66) can be interpreted as a network of two coupled oscillators, in which the phase dynamics is independent of the amplitude dynamics. Apply the transformation $\phi_1 = \theta_1 - \theta_2$, $\phi_2 = \theta_1 + \theta_2$ and show that the phase dynamics in these new coordinates decouples from one another.
(b) Use the decoupled phase dynamics to find the curves in (ω_1, ω_2) parameter space along which bifurcations occur, and classify the various bifurcations.
(c) Plot the stability diagram in (ω_1, ω_2) parameter space.

References

1. D.G. Aronson, M. Golubitsky, M. Krupa, Coupled arrays of Josephson junctions and bifurcation of maps with s_n symmetry. Nonlinearity **4**, 861–902 (1991)
2. E. Doedel, D. Aronson, H. Othmer, The dynamics of coupled current-biased Josephson junctions: Part I. IEEE Trans. Circuits Syst. **35**(7), 0700–0810–0700–0817 (1988)
3. E. Doedel, D. Aronson, H. Othmer, The dynamics of coupled current-biased Josephson junctions: Part II. Int. J. Bifurc. Chaos **1**(1), 51–66 (1991)
4. P. Hadley, M.R. Beasley, K. Wiesenfeld, Phase locking of Josephson-junction series arrays. Phys. Rev. B **38**, 8712–8719 (1988)
5. N. Kopell, G.B. Ermentrout, Coupled oscillators and the design of central pattern generators. Math. Biosci **89**, 14–23 (1988)
6. N. Kopell, G.B. Ermentrout, Phase transitions and other phenomena in chains of oscillators. SIAM J. Appl. Math. **50**, 1014–1052 (1988)
7. A.H. Cohen, S. Rossignol, S. Grillner (eds.), *Systems of Coupled Oscillators as Models of Central Pattern Generators*, New York (Wiley, 1988)
8. W. Rappel, Dynamics of a globally coupled laser model. Phys. Rev. E **49**, 2750–2755 (1994)
9. K. Wiesenfeld, C. Bracikowski, G. James, R. Rajarshi, Observation of antiphase states in a multimode laser. Phys. Rev. Lett. **65**(14), 1749–1752 (1990)
10. L. Pecora, T.L. Caroll, Synchronization in chaotic systems. Phys. Rev. Lett. **64**, 821–824 (1990)
11. C.W. Wu, L.O. Chua, A unified framework for synchronization and control of dynamical systems. Int. J. Bifurc. Chaos **4**(4), 979–998 (1994)
12. J.S. Halow, E.J. Boyle, C.S. Daw, C.E.A. Finney, *PC-Based, Near Real-Time, 3-Dimensional Simulation of Fluidized Beds* (Fluidization IX Durango, Colorado, 1998)
13. J. Toner, T. Yuhai, Flocks herds and schools: a quantitative theory of flocking. Phys. Rev. E. **58**, 4828–4858 (1998)
14. M. Golubitsky, I. Stewart, Symmetry and pattern formation in coupled cell networks, in *Patternformation in Continuous and Coupled Systems*, In IMA Volumes in Mathematics and its Applications, vol. 115 (Springer, New York, 1999), pp. 65–82
15. M. Golubitsky, I.N. Stewart, D.G. Schaeffer, *Singularities and Groups in Bifurcation Theory Vol. II*, vol. 69 (Springer, New York, 1988)
16. M. Golubitsky, I. Stewart, *The Symmetry Perspective* (Birkháuser Verlag, Basel, Switzerland, 2000)
17. F.R. Gantmacher, *The Theory of Matrices*, vol. II (Chelsea Publishing, New York, 1960)
18. Online Source. Routh–hurwitz stability criterion. https://en.wikipedia.org/wiki/Routh-Hurwitz_stability_criterion
19. A.T. Winfree, *Geometry of Biological Time* (Springer, 2001)
20. A.T. Winfree, *When Time Breaks Down: The Three-Dimensional Dynamics of Electrochemical Waves and Cardiac Arrhythmias* (Princeton University Press, 1987)
21. Y. Kuramoto, Self-entrainment of population of coupled nonlinear oscillators, in *Proceedings of the International Symposium on Mathematical Problems in Theoretical Physics*, ed. by H. Araki, vol. 39, p. 420 (1975)
22. Y. Kuramoto, D. Battogtokh, *Coexistence of Coherence and Incoherence in Nonlocally Coupled Phase Oscillators: A Soluble Sase*, pp. 1–9, Oct.2002. arXiv:cond-mat/0210694v1 [cond-mat.stat-mech]
23. S. Smale, *The Hopf Bifurcation and its Applications*, volume 19, chapter A mathematical model of two cells via Turing's equation (Springer, New York, NY, 1976), pp. 354–367
24. K. Otsuka, Pattern recognition with a bidirectionally coupled nonlinear optical-element system. Opt. Lett. **14**, 925–927 (1989)
25. C. Poynton, *Digital Video and HDTV: Algorithms and Interfaces* (Morgan Kaufmann Publishers, 2003)
26. A. Pikovsky, M. Rosenbleum, J. Kurths, *Synchronization: A Universal Concept in Nonlinear Sciences* (University Press, Cambridege, UK, 2001)

27. E. Mosekilde, Y. Maistrenko, D. Postnov, *Chaotic Synchronization: Applications to Living Systems* (World Scientific, 2002)
28. A. Balanov, N. Janson, D. Postnov, O. Sosnovtseva, *Synchronization: from Simple to Complex* (Springer, 2009)
29. A. Ishida, Y. Inuishi, Time and field variations of acoustic frequency spectrum in amplifying CDS revealed by brillouin scattering measurements. Phys. Lett. A **27**(7), 442–443 (1968)
30. M.A. Cohen, S. Grossberg, Absolute stability of global pattern formation and parallel memory storage by competitive neural networks. IEEE Trans. Syst. Man Cybern. **13**, 815–826 (1983)
31. V. In, A. Kho, J. Neff, A. Palacios, P. Longhini, B. Meadows, Experimental observation of multifrequency patterns in arrays of coupled nonlinear oscillators. Phys. Rev. Lett. **91**(24), 244101–1–244101–4 (2003)
32. A. Palacios, R. Carretero, P. Longhini, N. Renz, V. In, A. Kho, J. Neff, B. Meadows, A. Bulsara, Multifrequency synthesis using two coupled nonlinear oscillator arrays. Phys. Rev. E **72**, 026211 (2005)
33. P. Longhini, A. Palacios, V. In, J. Neff, A. Kho, A. Bulsara, Exploiting dynamical symmetry in coupled nonlinear elements for efficient frequency down-conversion. Phys. Rev. E **76**, 026201 (2007)
34. V. In, P. Longhini, A. Kho, N. Liu, S. Naik, A. Palacios, J. Neff, Frequency down-conversion using cascading arrays of coupled nonlinear oscillators. Physica D **240**, 701–708 (2011)
35. B. Van der Pol, On "relaxation-oscillations". London Edinburgh Dublin Philos. Mag. J. Sci. Ser. **7**(2), 978–992 (1926)
36. B. van der Pol, Forced oscillations in a circuit with non-linear resistance (reception with reactive triode). London Edinburgh Dublin Philos. Mag. J. Sci. Ser. **7**(3), 65–80 (1927)
37. B. van der Pol, J. van der Mark, Frequency demultiplication. Nature **120**, 363–364 (1927)
38. P. Holmes, D.R. Rand, Bifurcation of the forced van der Pol Oscillator. Quart. Appl. Math. **35**, 495–509 (1978)
39. T. Heath, K. Wiesenfeld, R.A. York, Manipulated synchronization: beam steering in phased arrays. Int. J. Bif. Chaos **10**, 2619–2627 (2000)
40. A.S. Landsman, I.B. Schwartz, Predictions of ultraharmonic oscillations in coupled arrays of limit cycle oscillators. Phys. Rev. E **74**, 036204 (2006)
41. J. Guckenheimer, P. Holmes, Structurally stable heteroclinic cycles. Math. Proc. Camb. Phil. Soc **103**, 189–192 (1988)
42. R. Tiwari, R. Subramanian, Subharmonic and superharmonic synchronization in weakly nonlinear systems. J. Sound Vib. **47**, 501–508 (1976)
43. V. In, A. Palacios, A. Bulsara, P. Longhini, A. Kho, J. Neff, S. Baglio, B. Ando, Complex behavior in driven unidirectionally coupled overdamped duffing elements. Phys. Rev. E **73**, 066121 (2006)
44. S. Wiggins, *Introduction to Applied Nonlinear Dynamical Systems* (Springer, New York, 1990)
45. A. Nayfeh, *The Method of Normal Forms* (Wiley-VCH, 2011)
46. Martin Golubitsky, Claire Postlethwaite, Feed-forward networks, center manifolds, and forcing. Disc. Contin. Dyn. Syst. **32**(8), 2913–2935 (2012). (August)
47. Toby Elmhirst, Martin Golubitsky, Nilpotent hopf bifurcations in coupled cell systems. J. Appl. Dyn. Syst. **5**(2), 205–251 (2006)
48. M. Golubitsky, M. Nicol, I. Stewart, Some curious phenomena in coupled cell networks. J. Nonlinear Sci. **14**(2), 207–236 (2004)
49. M. Golubitsky, L.J. Shiau, C. Postlethwaite, Y. Zhang, The feed-forward chain as a filter-amplifier motif, in *Coherent Behavior in Neuronal Networks*, ed. by K. Josìc et al. (Springer Science+Business Media, LLC, 2009)
50. N.J. McCullen, T. Mullin, M. Golubitsky, Sensitive signal detection using a feed-forward oscillator network. Phys. Rev. Lett. **98** (2007)
51. B. Rink, J. Sanders, Coupled cell networks: semigroups, lie algebras, and normal forms. Trans. Am. Math. Soc. **367**, 3509–3548 (2015)
52. T. Levasseur, A. Palacios, Asymptotic analysis of bifurcations in feed-forward networks. Int. J. Bif. Chaos, In Print (2000)

53. Yanyan Zhang, Martin Golubitsky, Periodically forced hopf bifurcations. SIAM J. Appl. Dyn. Syst. **10**(4), 1272–1306 (2011)
54. I. Stewart, M. Golubitsky, Synchrony-breaking bifurcations at a simple real eigenvalue for regular networks 1: 1-dimensional cells. SIAM J. Appl. Dyn. Syst. **10**, 1404–1442 (2011)
55. I. Stewart, M. Golubitsky, Synchrony-breaking bifurcations at a simple real eigenvalue for regular networks 2: higher-dimensional cells. SIAM J. Appl. Dyn. Syst. **13**, 129–156 (2014)
56. B.K. Meadows et al., Nonlinear antenna technology, in *Proceedings of the IEEE*, vol. 90 (IEEE, 2002), pp. 882–897
57. R. Hansen, *Phased Array Antennas* (Wiley, 2009)
58. K.D. Stephan, IEEE Trans. Microwave Theory Tech. **MTT-34**, 1017 (1986)
59. K.D. Stephan, W.A. Morgan, IEEE Trans. Antennas Propagat **AP-35**, 771 (1987)
60. R.A. York, T. Itoh, Injection- and phase-locking techniques for beam control. IEEE Trans. Microwave Theory Tech. **46**, 1920–1929 (1998)
61. R.A. York, Nonlinear analysis of phase relationships in quasi-optical oscillator arrays. IEEE Trans. Microwave Theory Tech. **41**(10), 1799–1809 (1993)
62. B. Meadows, T. Heath, J. Neff, E. Brown, D. Fogliatti, M. Gabbay, V. In, P. Hasler, S. Deweerth, W. Ditto, Nonlinear antenna technology. Proc. IEEE **90**(5), 882–897 (2002)
63. M. Gabbay, M.L. Larsen, L.S. Tsimring, Phased array beamforming using nonlinear oscillators. *Proceedings of the SPIE, Advanced Signal Processing Algorithms, Architectures, and Implementations XIV*, vol. 5559, pp. 146–155 (2004)
64. M. Karlsson, J. Robinson, L. Gammaitoni, A. Bulsara, The optimal achievable accuracy of the advanced dynamic fluxgate magnetometer (ADFM), in *Proceedings of MARELEC*, Stockholm, Sweden (2001)
65. R. Koch, J. Deak, G. Grinstein, Fundamental limits to magnetic-field sensitivity of flux-gate magnetic-field sensors. Appl. Phys. Lett. **75**(24), 3862–3864 (1999)
66. V. In, A. Bulsara, A. Palacios, P. Longhini, A. Kho, J. Neff, Coupling induced oscillations in overdamped bistable systems. Phys. Rev. E **68**, 045102–1–0415102–4 (2003)
67. A. Bulsara, V. In, A. Kho, P. Longhini, A. Palacios, W. Rappel, J. Acebron, S. Baglio, B. Ando, Emergent oscillations in unidirectionally coupled overdamped bistable systems. Phys. Rev. E **70**, 036103 (2004)
68. E. Doedel, X. Wang, *Auto94: Software for Continuation and Bifurcation Problems in Ordinary Differential Equations*. Applied Mathematics Report (California Institute of Technology, 1994)
69. M. Krupa, Robust heteroclinic cycles. J. Nonlin. Sci. **7**(2), 129–176 (1997)
70. P.L. Buono, M. Golubitsky, A. Palacios, Heteroclinic cycles in rings of coupled cells. Physica D **143**, 74–108 (2000)
71. M.J. Field. *Lectures on Bifurcations, Dynamics and Symmetry*, volume 356 of Pitman Research Notes (Addison-Wesley Longman Ltd., Harlow, 1996)
72. M. Krupa, Bifurcations of relative equilibrias. SIAM J. Math. Anal. **21**, 1453–1486 (1990)
73. M. Krupa, I. Melbourne, Asymptotic stability of heteroclinic cycles in systems with symmetry. Ergod. Th. & Dynam. Sys. **15**, 121–147 (1995)
74. I. Melbourne, P. Chossat, M. Golubitsky, Heteroclinic cycles involving periodic solutions in mode interactions with o(2) symmetry. Proc. Roy. Soc. Edinburgh **113A**, 315–345 (1989)
75. M.J. Field, Equivariant dynamical systems. Trans. Am. Math. Soc. **259**(1), 185–205 (1980)
76. D. Armbruster, J. Guckenheimer, P. Holmes, Heteroclinic cycles and modulated traveling waves in systems with o(2) symmetry. Physica D **29**, 257–282 (1988)
77. J.C. Chedjou, P. Woafo, S. Domngang, Shilnikov chaos and dynamics of a self-sustained electromechanical transducer. ASME J. Vib. Acoust **123**, 170–174 (2001)

Chapter 7
Delay Models

Applications in science and engineering frequently have inherent time delays in the dynamics of the systems that are being modeled. The time delays in the differential equations often significantly affect the behavior and analysis of the model. This chapter introduces some mathematical models, which include time delays, and techniques are shown for how to analyze these models.

7.1 Structure and Behavior of Delayed Systems

A fundamental principle in mathematical modeling is that the behavior of a system arises from its own structure [1]. The majority of structures consists of the nonlinear interactions between variables, i.e., *coupling*, and *feedback loops*. For instance, electronic circuits typically involve structure in the form multiple interconnected components. Simplified assumptions of instantaneous coupling could be made to get first insight into the behavior of the system. However, in practice, we must account for the fact that even high-speed, high-precision, circuit components can introduce a delay in the coupling signal that travels between multiple components. Positive feedback in a system can be considered as a self-reinforcing mechanism, which can lead to behavior in the form of exponential growth. That is, the larger the quantity of the state of the system, the greater its net increase, further augmenting the quantity. On the other hand, negative feedback loops seek balance as the structure counteract any disturbances that perturb the state of the system away from an equilibrium state. The structure and behavior of a system with a negative feedback loop is better illustrated with Fig. 7.1.

Every negative feedback loop includes a process to compare the desired and actual state of the system. If a discrepancy is found then a corrective action can be taken. For instance, a delay between the time a discrepancy is observed and the corrective action is taken can lead to *delay-induced* oscillations. This result might sound as counter-

A. Palacios, *Mathematical Modeling*, Mathematical Engineering,
https://doi.org/10.1007/978-3-031-04729-9_7

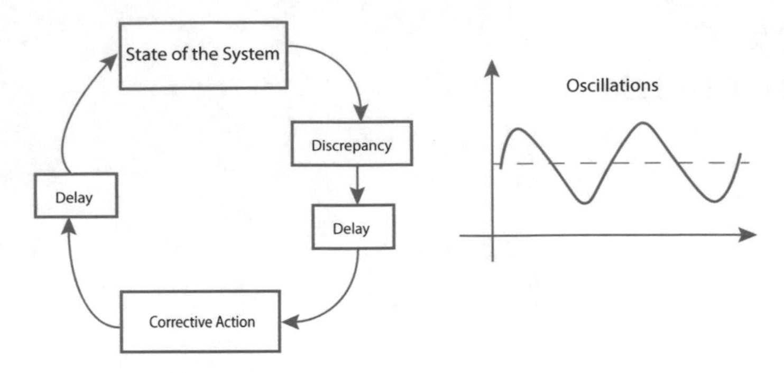

Fig. 7.1 Classic negative feedback loop in a system with time delay

intuitive behavior. But it's actually easy to explain. An oscillatory system typically overshoots an equilibrium state, it reverses course, and then undershoots. The cycle is then repeated many times. Now, overshooting and undershooting can arise from the presence of delay, thus causing continues corrections leading to oscillatory behavior. In military applications, Minorsky [2, 3] studied the negative feedback system in which water is constantly pumped in order to stabilize ships against rolling waves. It was found that the delay in the mechanical response of the pumps could actually lead to enhancing the oscillatory effects, via delay-induced oscillations, of the waves [4]. Nevertheless, engineers often consider negative feedback loops into their designs to help stabilize their systems[5].

7.2 System Dynamics with Negative Feedback

Let us consider a negative feedback loop modeled through the following delay differential equations (DDEs)

$$\dot{x}(t) = g(x(t)) + f(x(t - r)), \tag{7.1}$$

where $x \in \mathbb{R}^n$ is the state of the system at time t, g is just the input-output process of the system, without delay, and the negative sign in the function f indicates negative feedback.

In general, delay differential equations (DDEs) such as Eq. (7.1) are more complicated to study and the ensuing behavior can be significantly harder to analyze. Thus, modeling efforts should carefully consider the importance of any time delays for a particular application.

7.2.1 Equilibrium Points

A significant difference in the analysis of DDEs is that a solution requires knowing the history or initial data of the unknown function, $x(t)$, for $t \in [-r, 0]$, as opposed to just the initial conditions at time $t = 0$ for systems without delay. Consequently, since the history is specified along the interval, $t \in [-r, 0]$, this problem becomes infinite dimensional. For this reason, it is quite often for DDEs to be set in the Banach space of continuous functions with the sup norm, $\mathcal{C}([-r, 0], \mathbb{R})$.

In spite of some of the differences on initial conditions, equilibrium points of DDEs are found in a similar manner to ODEs. However, instead of an equilibrium point, the equilibrium is a constant function, that satisfies the algebraic equation:

$$g(x_e) + f(x_e) = 0 \quad \text{or} \quad f(x_e) = -g(x_e).$$

One standard negative feedback model with delay has $f(x)$ satisfying:

$$f(x) > 0, \quad f'(x) < 0, \quad \text{and} \quad \lim_{x\to\infty} f(x) = 0.$$

In the case of linear (Malthusian) decay, in which $g(x) = a\,x$ with $a < 0$, then it is easy to see that this negative feedback model has a unique positive equilibrium.

7.2.2 Linearization

Let us assume the system to have an equilibrium point, x_e, that satisfies the algebraic equation $g(x_e) + f(x_e) = 0$. To study its stability properties, we apply a small perturbation, $y(t)$, so that $x(t) = x_e + y(t)$.

Treating $x = x(t)$ and $x_r = x(t - r)$ as two separate variables, we perform a Taylor expansion of Eq. (7.1) about the equilibrium and get:

$$\dot{y}(t) = g(x_e) + f(x_e) + \frac{\partial g}{\partial x}(x_e)y(t) + \frac{\partial f}{\partial x_r}(x_e)y(t - r) + \cdots .$$

Since x_e is an equilibrium point, i.e., $g(x_e) + f(x_e) = 0$, the linearization (up to order one) yields

$$\dot{y}(t) = \frac{\partial fg}{\partial x}(x_e)y(t) + \frac{\partial f}{\partial x_r}(x_e)y(t - r). \tag{7.2}$$

If we assume that $g'(x_e) = a$ and $f'(x_e) = b$, then Eq. (7.2) can be rewritten as:

$$\dot{y}(t) = a\,y(t) + b\,y(t - r). \tag{7.3}$$

This is a scalar linear one-delay differential equation.

As with linear ODEs, stability analysis begins by trying a solution of the form, $y(t) = ce^{\lambda t}$. When this solution is inserted into Eq. (7.3), we obtain:

$$c\lambda e^{\lambda t} = ace^{\lambda t} + bce^{\lambda(t-r)}.$$

Dividing by $c\,e^{\lambda t}$ gives the *characteristic equation*

$$\lambda - a = be^{-\lambda r}, \tag{7.4}$$

where λ is the eigenvalue. This is known as an exponential polynomial, and this type of characteristic equation has infinity many solutions for most sets of parameters a, b, and r. This characteristic equation is rarely solved exactly. The linear stability is determined by finding the sign of the real part of the complex eigenvalues with unstable equilibria having eigenvalues with positive real parts.

The *boundary of stability* is a subset of solutions to the characteristic equation with $\lambda = i\omega$ or

$$i\omega - a = be^{-i\omega r} = b\left(\cos(\omega r) - i\,\sin(\omega r)\right),$$

or for $\lambda = 0$, the *real root crossing* satisfies:

$$a = -b.$$

From the characteristic equation with $\lambda = i\omega$, the real and imaginary parts give the parametric equations:

$$\begin{aligned} a(\omega) &= -b(\omega)\cos(\omega r), \\ \omega &= -b(\omega)\sin(\omega r). \end{aligned}$$

Solving these equations for $a(\omega)$ and $b(\omega)$ gives

$$\begin{aligned} a(\omega) &= \omega\cot(\omega r), \\ b(\omega) &= -\frac{\omega}{\sin(\omega r),} \end{aligned} \tag{7.5}$$

which are clearly singular at any $\frac{n\pi}{r}$, $n = 0, 1, \ldots$ This creates distinct curves $\omega \in \left(\frac{(n-1)\pi}{r}, \frac{n\pi}{r}\right)$ for $n \geq 1$.

Figure 7.2 shows the image of the purely imaginary roots (or $\lambda = 0$) for the characteristic Eq. (7.4) in the ab-parameter space. The real root crossing solid green line satisfies $\lambda = 0$ with $a = -b$. This separates the figure into a half plane above the line $a = -b$ or $a > -b$, where there is always a positive real root. Equation (7.5) produces the parametric curves for the image of the purely imaginary roots. A transverse crossing of any of these curves results in either the addition or subtraction of two complex eigenvalues with positive real parts. This set of curves creates a

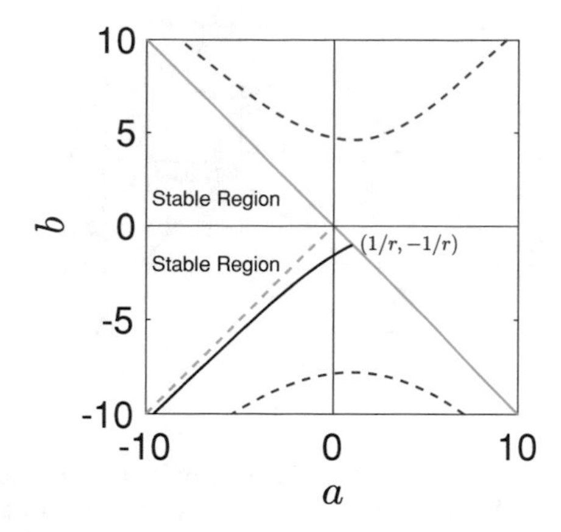

Fig. 7.2 Diagram of the parametric curves in the ab-parameter space for all eigenvalues on the imaginary axis (neutral stability)

D-partitioning[6] of the complex plane into distinct regions with a distinct integer number of eigenvalues with real positive parts.

The analysis above shows that the region where $a < 0$ and $|b| < |a|$ the equilibrium point x_e is stable independent of the delay. Additionally, as $r \to 0$, the DDE approaches the ODE with a stability region $a + b < 0$. In the limit, as $\omega \to 0$, Eq. (7.5) shows that the stability region approaches the point at $\left(\frac{1}{r}, -\frac{1}{r}\right)$. Figure 7.2 also shows that the imaginary root crossings are distinct, non-intersecting curves, leaving this *stability boundary* generated by the parametric equations with $\omega \in \left(0, \frac{\pi}{r}\right)$, which is the red curve just below the negative a-axis.

Example 7.1 Consider the negative feedback problem given by:

$$\dot{x}(t) = -x(t) + \frac{10}{1 + x^5(t-2)}. \tag{7.6}$$

In this example, $g(x) = -x$ and $f(x) = 10/(1 + x^5)$, which is positive and monotonically decreasing. Equilibrium points are found by solving $g(x_e) + f(x_e) = 0$, which in this case is equivalent to solving:

$$\frac{10}{1 + x_e^5} = x_e.$$

This last equations yields, numerically, a unique solution: $x_e = 1.4305$.

Next, we determine linearization of Eq. (7.6). Direct computations show $a = g'(x_e) = 1$, and

$$b = f'(x_e) = -\frac{50x_e^4}{(1 + x_e^5)^2} = -4.2847.$$

It follows from Eq. (7.3) that the linearized DDE at the equilibrium is:

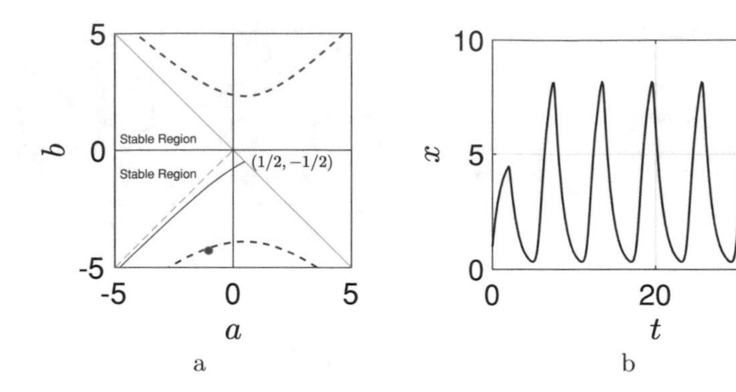

Fig. 7.3 (Left) *D-partitioning* of the linearized DDE with the blue circle indicating the location in ab-parameter space of this specific example. (Right) Simulation of Eq. (7.6) with initial data being the constant function 1

$$\dot{y}(t) = -y(t) - 4.2847\, y(t-2).$$

Figure 7.3a shows the *D-partitioning* of the linear DDE (7.3) with the delay $r = 2$. The linearized form of our specific negative feedback model given by Eq. (7.6) is marked by a blue circle in the figure.

Applying the methods shown in El'sgol'ts and Norkin[6] we find that this linear DDE has 4 eigenvalues with positive real part. Using a nonlinear solver, the leading four pairs of eigenvalues are:

$$\lambda_1 = 0.4156 \pm 1.2160\, i, \qquad \lambda_2 = 0.01305 \pm 4.0496\, i,$$
$$\lambda_3 = -0.2567 \pm 7.1206\, i, \qquad \lambda_4 = -0.4363 \pm 10.2377\, i.$$

As indicated earlier, there are infinitely many eigenvalues satisfying the linear DDE (7.3), but we only list the leading four eigenvalues with positive real part. The nature of the eigenvalues can be seen by connecting a path from the *stable region* to the point in ab-parameter space, $(-1, -4.2847)$. In the stable region there are no eigenvalues with positive real part. To reach the point $(-1, -4.2847)$, the connecting path must cross both the solid red curve and the dashed red curve. The transverse crossing of any of these curves implies two complex eigenvalues gain positive real parts. The frequency (imaginary part) is close to the value of ω, where it crosses the curves in this D-partitioning figure. Figure 7.3b shows a time-series simulation of the negative feedback model (7.6), which depicts sustained oscillations with a period of approximately 5. These oscillations are driven by the leading pair of eigenvalues, which can be seen to have a frequency of 1.2160, so have a period $2\pi/1.2160 \approx 5.167$.

Now consider Eq. (7.6) with the delay being r instead of 2. We know that as $r \to 0$, this model become an ODE, which is asymptotically stable. Since this model is unstable with $r = 2$, it follows that there must be some critical delay, r_c, where this model loses its stability. This value is where this model undergoes a *Hopf bifurcation*. The critical value is found by solving Eq. (7.4) with $a = -1$, $b = -4.2847$, and $\omega \in (0, \pi/r)$, where $r \in (0, 2)$. With a nonlinear solver we find that the Hopf bifurcation occurs at:

$$r_c = 0.4336 \quad \text{and} \quad \omega_c = 4.1664.$$

It follows that the model (7.6) with delay, r, is stable for $r \in (0, 0.4336)$ and unstable for $r > 0.4336$, which is typical for this form of negative control model. It also demonstrates how delays can significantly impact the behavior of negative feedback models.

7.3 Stability Properties

In this section we summarize some stability results (without proofs) that can be very useful in the analysis of differential equations with delay. The appropriate references are included for readers interested in finding rigorous proofs. Let us start with a linear model.

The stability properties of an equilibrium point of a linear delay differential model, of the form given by Eq. (7.3), can be inferred by the following theorem [7].

Theorem 7.1 (Stability of Linear Model) *Consider the linear delay system*

$$\dot{y}(t) = a\, y(t) + b\, y(t - r), \tag{7.7}$$

where a and b are scalars with equilibrium solution $x = 0$. The following applies to this equilibrium.

(i) If $a + b > 0$, then $x = 0$ is unstable.
(ii) If $a + b < 0$ and $b \geq a$, then $x = 0$ is asymptotically stable.
(iii) If $a + b < 0$ and $b < a$, then there exists $r^* > 0$ such that $x = 0$ is asymptotically stable for $0 < r < r^*$ and unstable for $r > r^*$.

Furthermore, in case (iii), there exist a pair of purely imaginary roots at

$$r = r^* = (b^2 - a^2)^{-1/2} \cos^{-1}(-a/b).$$

We already saw that the stability analysis of a linear delay model leads to a characteristic polynomial of the form $\lambda = a + be^{-\lambda r}$, where λ is the eigenvalue that determines the stability properties of the equilibrium. In the case of nonlinear models, often times we encounter a slightly different characteristic polynomial of the form

$$p(\lambda) + q(\lambda)e^{-\lambda r} = 0, \tag{7.8}$$

where p and q are polynomials with real coefficients, and r is our usual delay. The roots of this characteristic polynomial, and consequently, the eigenvalues of the stability analysis, are characterized by the following theorem [7, 8].

Theorem 7.2 (Absolute Stability) *Let p and q be polynomials with real coefficients. Suppose:*

(i) $p(\lambda) \neq 0, \ \Re(\lambda) \geq 0.$

(ii) $|q(iy)| < |p(iy)|, \ 0 \leq y < \infty.$

(iii) $$\lim_{|\lambda| \to \infty, \ \Re(\lambda) \geq 0} \left| \frac{p(\lambda)}{q(\lambda)} \right| = 0.$$

Then $\Re(\lambda) < 0$ for every root λ and all $r \geq 0$.

Note: The concept of absolute stability refers to the fact that the conclusion of the theorem holds for every value of the delay.

Corollary 1 *Let p be a polynomial with real coefficients, and have leading coefficient one. Let* $q = c$ *be a constant. If*

(i) All roots of p are real and negative and $|p(0)| > |c|$, *or*
(ii) $p(\lambda) = \lambda^2 + a\lambda + b$, $a, b > 0$, *and either*

- $b > |c|$, *and* $a^2 \geq 2b$, *or*
- $a\sqrt{4b - a^2} > 2|c|$ *and* $a^2 < 2b$,

then $\Re(\lambda) < 0$ *for every root* λ *and all* $r \geq 0$.

It is also often the case to encounter multiple delays in a mathematical model. For the particular case of two delays, r_1 and r_2, we have the following result [9].

Theorem 7.3 (Stability with Multiple Delays) *Consider a linear delay differential equation with two delays*

$$\frac{dx}{dt} = -ax(t - r_1) - bx(t - r_2), \tag{7.9}$$

where $a, b, r_1, r_2 \in [0, \infty)$. *The characteristic equation for the model Eq. (7.9) is*

$$\lambda + ae^{-\lambda r_1} + be^{-\lambda r_2} = 0, \tag{7.10}$$

where λ *is a complex number.*

(i) Let $b = 0$ *and* $a, r_1 \in (0, \infty)$. *A necessary and sufficient condition for all roots of* $\lambda + ae^{-\lambda r_1} = 0$ *to have negative real parts is* $0 < ar_1 < \pi/2$.
(ii) Let $a, b, r_1, r_2 \in (0, \infty)$. *A sufficient condition for all roots of Eq. (7.10) to have negative real parts is* $ar_1 + br_2 < 1$, *and a necessary condition for the same is* $ar_1 + br_2 < \pi/2$.

7.4 Epidemic Model

A mathematical model, proposed by Cook [10], for the spread of an epidemic is in the form of a delay differential equation through

$$\frac{dx}{dt} = B\,x(t-7)(1-x(t)) - C\,x(t), \tag{7.11}$$

where $x(t)$ describes the fraction of a population which is infected at time t, b and c are positive parameters.

We will investigate the existence and stability of the equilibrium points of the model and conduct computer simulations of the model equation to validate theoretical results.

An equilibrium point, x_e, is found by solving

$$B\,x_e(1-x_e) = C\,x_e.$$

Geometrically, there are two equilibrium points, which correspond to the intersection of the parabola $B\,x_e(1-x_e)$ with the straight line $y = Cx$. Direct calculations show that these two intersection points are

$$x_{e_1} = 0 \quad \text{and} \quad x_{e_2} = \frac{B-C}{B}.$$

Since x describes the fraction of a population that has been infected, the nontrivial equilibrium point x_{e_2} makes sense only when $B > C$.

To study the stability of the equilibrium points, we first note that in the epidemic model the state variables $x(t)$ and $x(t-r)$ do not appear separated into two distinct functions, as it was the case of negative feedback systems. Thus, we cannot apply Eq. (7.3), not just yet. Instead, we consider first, a more general form of a delay system

$$\dot{x} = f(x(t), x(t-r)). \tag{7.12}$$

Let us assume the system to have an equilibrium that satisfies the algebraic equation $f(x_e) = 0$. To study its stability properties, we apply a small perturbation, $y(t)$, so that $x = x_e + y(t)$. Treating $x = x(t)$ and $x_r = x(t-r)$ as two separate variables, we perform a Taylor expansion about the equilibrium

$$\dot{y}(t) = f(x_e) + \frac{\partial f}{\partial x}(x_e)y(t) + \frac{\partial f}{\partial x_r}(x_e)y(t-r) + \cdots.$$

Since x_e is an equilibrium point, i.e., $f(x_e) = 0$, the linearization (up to order one) yields

$$\dot{y}(t) = \frac{\partial f}{\partial x}(x_e)y(t) + \frac{\partial f}{\partial x_r}(x_e)y(t-r). \tag{7.13}$$

In the original model Eq. (7.11),

$$f(x(t), x(t-r)) = B\,x(t-7)(1-x(t)) - C\,x(t).$$

Applying Eq. (7.13) yields the linearization:

$$\frac{dy}{dt} = -(Bx_e + C)y(t) + B(1-x_e)y(t-7).$$

Observe that this equation is now in the same form as the linear model Eq. (7.3), with $a = -(Bx_e + C)$, $b = B(1 - x_e)$, and the delay $r = 7$.

For the trivial equilibrium, $x_{e_1} = 0$, we get

$$\frac{dy}{dt} = -Cy(t) + By(t-7).$$

Applying Theorem 7.1 we have $a + b = B - C > 0$ (which is required for the nontrivial equilibrium to exist). Consequently, $x_{e_1} = 0$ is unstable.

For the nontrivial equilibrium, $x_{e_2} = (B - C)/B$, we get

$$\frac{dy}{dt} = -By(t) + Cy(t-7).$$

Applying Theorem 7.1 again, we get $a + b = -(B - C) < 0$, Thus, the nontrivial equilibrium is asymptotically stable.

Figure 7.4 illustrates numerical solutions of the epidemic model, with and without delay. Parameters values are: $B = 2$ and $C = 1$, and initial history of 0.8. These parameter values lead to equilibrium points: $x_{e_1} = 0$ and $x_{e_2} = 0.5$. The *MatLab script* for numerically solving this model satisfies.

```
sol2 = dde23(@ddefun, [tau], history, tspan,[], B,C);
```

The complete MATLAB code can be found in the Appendix.

The stability analysis indicates that, for this choice of parameters, the nontrivial equilibrium is asymptotically stable. The simulations confirm this result. Furthermore, in the case of no delay, the solution, $x(t)$ decays exponentially towards the nontrivial equilibrium. In the case of delay, $r = 7$, the solution also converges towards the nontrivial equilibrium but it decays in gradual steps before the equilibrium is reached.

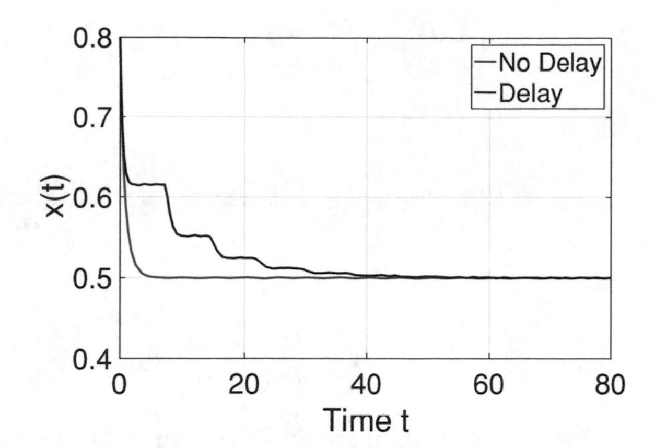

Fig. 7.4 Computer simulation of epidemic model, with and without delay. Parameters are: $B = (2$ and $C = 1$. MATLAB code in Appendix)

7.5 Lotka-Volterra Model

In Chap. 4 we introduced the Lotka-Volterra model to study predator-prey interactions, see Eq. (4.26). In particular, we used the model to analyze the interactions between populations of sharks and tuna. In this section, we revisit the Lotka-Volterra model with the perspective of studying the effects of delay. Thus, consider the following model

$$\begin{aligned} \frac{dx}{dt} &= x(t)\left[\mu_1 - ax(t) - by(t-r)\right] \\ \frac{dy}{dt} &= y(t)\left[\mu_2 - cx(t-r) - dy(t)\right]. \end{aligned} \tag{7.14}$$

Equilibrium solutions, (x_e, y_e), are solutions of the following system of equations:

$$\begin{aligned} x(\mu_1 - ax - by) &= 0 \\ y(\mu_2 - cx - dy) &= 0. \end{aligned} \tag{7.15}$$

Direct computation yields four equilibrium points:

$$(0,0), \quad \left(\frac{\mu_1}{a}, 0\right), \quad \left(0, \frac{\mu_2}{d}\right), \quad \left(\frac{\mu_1 d - \mu_2 b}{(ad-bc)}, \frac{\mu_2 a - \mu_1 c}{(ad-bc)}\right).$$

To study the stability of the equilibrium points, and for convenience, we rewrite the model Eq. (7.14) as

$$\begin{aligned} \frac{dx}{dt} &= f_1(x(t), y(t), x(t-r), y(t-r)) \\ \frac{dy}{dt} &= f_2(x(t), y(t), x(t-r), y(t-r)). \end{aligned} \tag{7.16}$$

The linearization of Eq. (7.16) is given by:

$$\begin{pmatrix} \dot{z}_1(t) \\ \dot{z}_2(t) \end{pmatrix} = \begin{pmatrix} a_{11} & a_{12} \\ a_{21} & a_{22} \end{pmatrix} \begin{pmatrix} z_1(t) \\ z_2(t) \end{pmatrix} + \begin{pmatrix} b_{11} & b_{12} \\ b_{21} & b_{22} \end{pmatrix} \begin{pmatrix} z_1(t-r) \\ z_2(t-r) \end{pmatrix}.$$

where

$$a_{11} = \frac{\partial f_1(x_e, y_e)}{\partial x}, \; a_{12} = \frac{\partial f_1(x_e, y_e)}{\partial y}, \; a_{21} = \frac{\partial f_2(x_e, y_e)}{\partial x}, \; a_{22} = \frac{\partial f_2(x_e, y_e)}{\partial y},$$

and

$$b_{11} = \frac{\partial f_1(x_e, y_e)}{\partial x(t-r)}, \; b_{12} = \frac{\partial f_1(x_e, y_e)}{\partial y(t-r)}, \; b_{21} = \frac{\partial f_2(x_e, y_e)}{\partial x(t-r)}, \; b_{22} = \frac{\partial f_2(x_e, y_e)}{\partial y(t-r)}.$$

Applying these results to Eq. (7.14), we arrive at the linearization about an equilibrium (x_e, y_e):

$$\begin{pmatrix} \dot{z}_1(t) \\ \dot{z}_2(t) \end{pmatrix} = \begin{pmatrix} \mu_1 - 2ax_e - by_e & 0 \\ 0 & \mu_2 - cx_e - 2dy_e \end{pmatrix} \begin{pmatrix} z_1(t) \\ z_2(t) \end{pmatrix} + \begin{pmatrix} 0 & -bx_e \\ -cy_e & 0 \end{pmatrix} \begin{pmatrix} z_1(t-r) \\ z_2(t-r) \end{pmatrix}.$$

We attempt to find solutions of the form $\mathbf{z}(t) = \c{}e^{\lambda t}$, with $\mathbf{z} = (z_1, z_2)^T$. It is not difficult to see that the resulting characteristic equation has the form:

$$\det \begin{bmatrix} \mu_1 - 2ax_e - by_e - \lambda & -bx_e e^{-\lambda r} \\ -cy_e e^{-\lambda r} & \mu_2 - cx_e - 2dy_e - \lambda \end{bmatrix} = 0.$$

This determinant can be simplified by making use of the equilibrium conditions Eq. (7.15), which lead to

$$\det \begin{bmatrix} -ax_e - \lambda & -bx_e e^{-\lambda r} \\ -cy_e e^{-\lambda r} & -dy_e - \lambda \end{bmatrix} = 0.$$

The determinant produces a characteristic polynomial:

$$p(\lambda) + q(\lambda) e^{-\lambda(2r)} = 0,$$

where $p(\lambda) = \lambda^2 + (ax_e + dy_e)\lambda + adx_e y_e$ and $q(\lambda) = -bcx_e y_e$.

We can now apply Theorem 7.2 to investigate the stability properties of equilibrium points. In what follows, we focus on the positive equilibrium for specific cases of parameter values. We will employ $2r$ for the role of r in the characteristic while using Theorem 7.2.

Case I: Stable Positive Equilibrium

Let us consider the case where $\mu_1 = \mu_2 = 2$, $a = d = 2$ and $b = c = 1$ and study the stability of the fourth equilibrium point, which becomes $(x_e, y_e) = (2/3, 2/3)$. In this case, the characteristic polynomial has the form

$$\left(\lambda + \frac{4}{3}\right)^2 - \frac{4}{9}e^{-\lambda(2r)} = 0.$$

If we let $p(\lambda) = (\lambda + 4/3)^2$ and $q(\lambda) = -4/9$, then we find that $p(\lambda)$ has a leading coefficient one, while $q(\lambda) = -4/9$ is negative and constant. In addition, the roots of p are $-4/3$ (twice), so they are real and negative, and $|p(0)| = 16/9 > |q| = 4/9$. Thus, by Corollary 1 of the absolute stability Theorem 7.2 we can conclude that $\Re(\lambda) < 0$ for every root λ and all $r \geq 0$. This implies that the equilibrium point $(2/3, 2/3)$ is asymptotically stable.

Case II: Unstable Positive Equilibrium

Let us consider the case where $\mu_1 = \mu_2 = 2$, $a = d = 1$ and $b = c = 2$. The fourth equilibrium point is again $(2/3, 2/3)$, and the characteristic polynomial is

$$\left(\lambda + \frac{2}{3}\right)^2 - \frac{16}{9}e^{-\lambda(2r)} = 0.$$

If we let $p(\lambda) = (\lambda + 2/3)^2$ and $q(\lambda) = -16/9$, then we find that $p(\lambda)$ has a leading coefficient one, while $q(\lambda) = -16/9$ is negative and constant. In addition, the roots of p are $-2/3$ (twice), so they are real and negative. However, $|p(0)| = 4/9 \ngtr |q(\lambda)| = 16/9$. Consequently, one of the criteria of Corollary 1 of the absolute stability Theorem 7.2 does not hold. Then, we can conclude that the equilibrium point $(2/3, 2/3)$ is unstable.

Case III: Delay-Induced Oscillations

Assume now: $\mu_1 = 1, \mu_2 = -1, a = 1, b = 1, c = -2, d = 1$. This choice of parameters yields a positive equilibrium point $(x_e, y_e) = (2/3, 1/3)$. The characteristic polynomial is

$$\left(\lambda + \frac{2}{3}\right)\left(\lambda + \frac{1}{3}\right) + \frac{4}{9}e^{-\lambda(2r)} = 0. \tag{7.17}$$

If we let $p(\lambda) = (\lambda + 2/3)(\lambda + 1/3)$ and $q(\lambda) = 4/9$, then we find that $p(\lambda)$ has a leading coefficient one, and all roots are negative. The polynomial q is constant, but $|p(0)| = 2/9 \ngtr |q(\lambda)| = 4/9$. Hence, one of the criteria of Corollary 1 of the absolute stability Theorem 7.2 does not hold. Then, we can conclude that the equilibrium point $(2/3, 1/3)$ is unstable.

We are interested in stability changes of the positive equilibrium that may lead to small amplitude oscillations via a Hopf bifurcation. Those changes can only occur when $\lambda = \omega i$. Substituting into Eq. (7.17) we get

$$-\omega^2 + \frac{2}{9} + \omega i = -\frac{4}{9}e^{-\omega(2r)i}.$$

Separating real and imaginary parts we get

$$\begin{aligned} -\omega^2 + \frac{2}{9} &= -\frac{4}{9}\cos\left(\omega(2r)\right) \\ \omega &= \frac{4}{9}\sin\left(\omega(2r)\right). \end{aligned} \tag{7.18}$$

These last two equations lead to the following polynomial in ω:

$$w^4 + \frac{5}{9}\omega^2 - \frac{12}{81} = 0. \tag{7.19}$$

Solving this polynomial (first for ω^2 then for ω), we find only one positive root $\omega_0 = 0.4041$. We can then use Eq. (7.18) to write an expression for the critical value of the delay, r_c, that leads to small amplitude oscillations via Hopf bifurcations:

$$r_c = \frac{1}{2\omega_0}\left[\tan^{-1}\left(\frac{\omega_0}{\omega_0^2 + \frac{2}{9}}\right) + \pi j\right], \quad j = 1, 2, \ldots. \tag{7.20}$$

Observe that due to the periodicity of $e^{i\omega(2r)}$, there are infinitely many solutions of the critical value r_c. The parameters chosen for this example lead to a critical value of the delay: $r_c = 2.13$.

Figure 7.5 illustrates the solutions of the Lotka-Volterra model, with and without delay. Two cases are shown, one where the delay is less than the critical value, i.e., $r < r_c$, and one where the delay is greater than the critical delay value.

7.6 Logistic Growth Model with Multiple Delays

We will now analyze a mathematical model with two delays. As an example, we consider the logistic growth model (studied in Chap. 4).

$$\frac{dx}{dt} = \mu\, x(t)\left[1 - a_1 x(t - \tau_1) - a_2 x(t - \tau_2)\right]. \tag{7.21}$$

Equilibrium points are found by solving:

$$x\,(1 - a_1 x - a_2 x) = 0,$$

which leads to two equilibrium points, $x_{e_1} = 0$, representing extinction of the population, and $x_{e_1} = 1$. The linearization of the model Eq. (7.21) about any of these equilibrium points yield

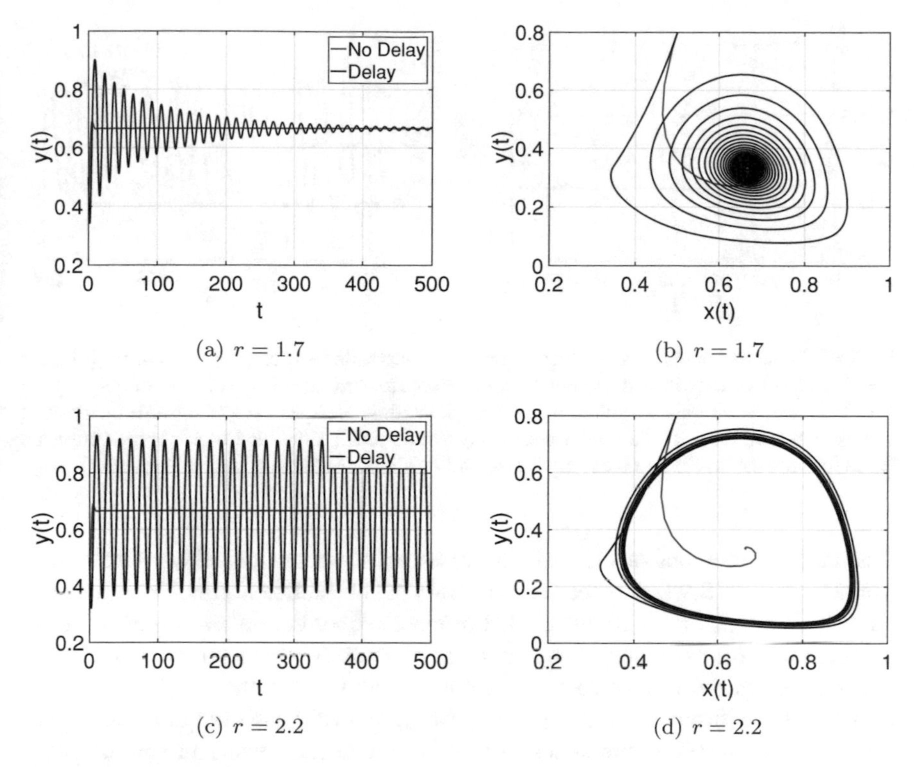

Fig. 7.5 Time series solutions and phase portraits of solutions of the Lotka-Volterra model for various values of a delay. At about $r_c = 2.125$, the model undergoes a Hopf bifurcation that leads to small amplitude oscillations. Parameters are: $\mu_1 = 1, \mu_2 = -1, a = 1, b = 1, c = -2, d = 1$

$$\frac{dy}{dt} = -\mu a_1 y(t - \tau_1) - \mu a_2 y(t - \tau_2). \tag{7.22}$$

We seek a solution of the form $x(t) = ce^{\lambda t}$, which leads to the following characteristic equation:

$$\lambda + \mu a_1 e^{-\lambda \tau_1} + \mu a_2 e^{-\lambda \tau_2} = 0. \tag{7.23}$$

Observe that the characteristic polynomial is already in the same form as in Eq. (7.10) in Theorem 7.3, with $a = \mu a_1$ and $b = \mu a_2$. Applying this theorem we can conclude that if $a_2 = 0$, then a necessary and sufficient condition for all roots of $\lambda + \mu a_1 e^{-\lambda \tau_1} = 0$ to have negative real parts is $0 < \tau_1 < \pi//(2\mu a_1)$.

Furthermore, a sufficient condition for all roots of Eq. (7.23) to have negative real parts is $a_1\tau_1 + a_2\tau_2 < 1/\mu$, and a necessary condition for the same is $a_1\tau_1 + a_2\tau_2 < \pi/(2\mu)$.

Let us consider a specific case where: $\mu = 0.15$, $a_1 = 1/4$, and $b = 3/4$. Then a necessary condition for all roots of Eq. (7.23) to have negative real parts is $\tau_1 + 3\tau_2 < 41.88$, while a sufficient condition for the same is $\tau_1 + 3\tau_2 < 26.7$. Figure 7.6

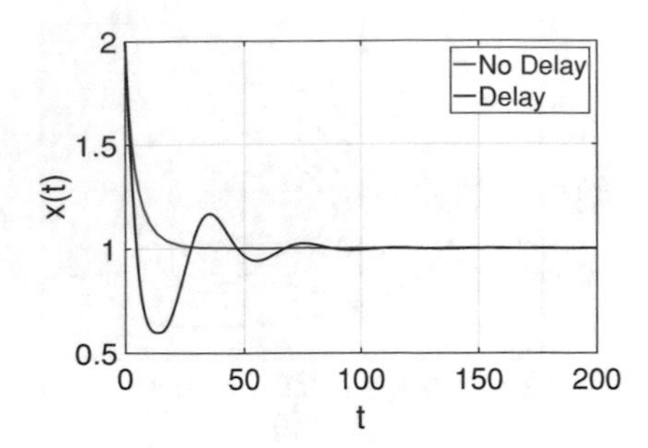

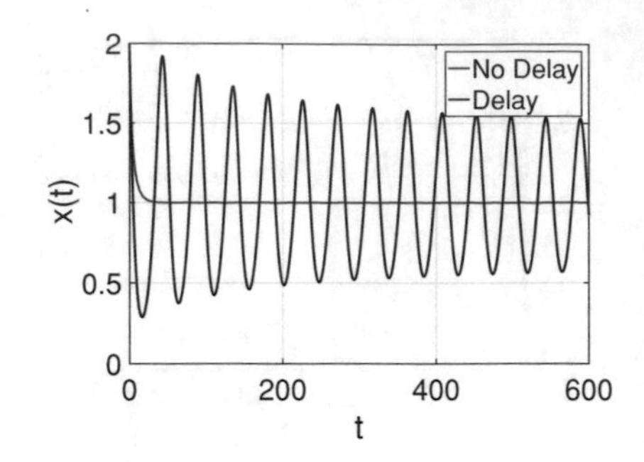

Fig. 7.6 Computer simulations of a logistic growth model with two delays. (Left) For $\tau_1 = 15$ and $\tau_2 = 5$, both solutions (with and without delay) converge towards a positive equilibrium. (Right) For $\tau_1 = 15$ and $\tau_2 = 10$, only the solution of the model without delay converges towards the positive equilibrium, while that of the model with delay shows small amplitude oscillations. Parameter values for both cases are: $\mu = 0.15$, $a_1 = 1/4$, and $b = 3/4$. (MATLAB code in Appendix)

illustrates the solutions of the logistic model with two sets of delays. One where, $\tau_1 = 15$ and $\tau_2 = 5$, which meets the necessary but not sufficient conditions. And one where, $\tau_1 = 15$ and $\tau_2 = 10$, which do not meet neither the necessary nor sufficient conditions for the existence of negative real roots of the characteristic polynomial. Observe that in the former case, the solution (with and without delay) converge towards the positive equilibrium $x_{e_2} = 1$. But in the later case, only the solution of the model without delay approaches the positive equilibrium, while the solution of the model with delay tends to oscillate. The source of these small amplitude oscillations is a delay-induced instability via Hopf bifurcations. The analysis of this later case is left as an exercise.

7.7 Nyquist Stability Criterion

Nyquist criterion is a graphical-based test for determining the stability of a feedback control system. The test is based on the complex analysis work known as Cauchy's principle of argument. In this section, we review its application to study the roots of the characteristic polynomial

$$p(\lambda) + q(\lambda)e^{-\lambda\tau} = 0, \tag{7.24}$$

which determines the stability of most time-delay differential equations.

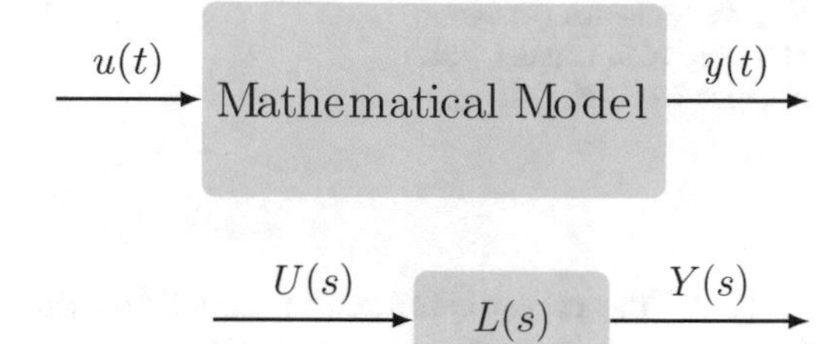

Fig. 7.7 Schematic diagram of an input-output model

Fig. 7.8 Schematic diagram of a transfer function

7.7.1 *Transfer Function*

The *transfer function* of a mathematical model is a function that describes the relationship between the input, $u(t)$, and the output, $y(t)$, of the system being modeled. Consider the following schematic diagram of an input-output process (Fig. 7.7):

For a mathematical model described by a continuous system of differential equations, the transfer function $L(s)$ is defined as the ratio of the Laplace transform of the model's output, $y(t)$, with respect to that of its input $x(t)$, assuming zero initial conditions. That is

$$L(s) = \frac{\mathcal{L}(y(t))}{\mathcal{L}(u(t))} = \frac{Y(s)}{U(s)}.$$

This relation is described, schematically, through Fig. 7.8.

The fundamental idea is that a transfer function is an equivalent way to mathematically describe the dynamics of a system. It contains the same information, including stability properties, as the original ODE does. The only difference is that the information contained in the transfer function is in the s domain as opposed to the time domain.

Suppose, for instance, that the model is described by a differential equation of the form

$$a_n y^{(n)}(t) + a_{n\;1} y^{(n-1)}(t) + \cdots + a_0 y(t) =$$
$$b_m u^{(m)}(t) + b_{m-1} u^{(m-1)}(t) + \cdots + b_0 u(t).$$

Applying the Laplace's transform to both sides of this equation we get

$$a_n s^n Y(s) + a_{n-1} s^{(n-1)} Y(s) + \cdots + a_1 Y(s) =$$
$$b_m s^m U(s) + b_{m-1} s^{(m-1)} U(s) + \cdots + b_1 U(s).$$

Solving for $Y(s)$ yields the desired transfer function

$$L(s) = \frac{Y(s)}{U(s)} = \frac{b_m s^m + b_{m-1} s^{(m-1)} + \cdots + b_1 s + b_0}{a_n s^n + a_{n-1} s^{(n-1)} + \cdots + a_1 s + a_0}. \quad (7.25)$$

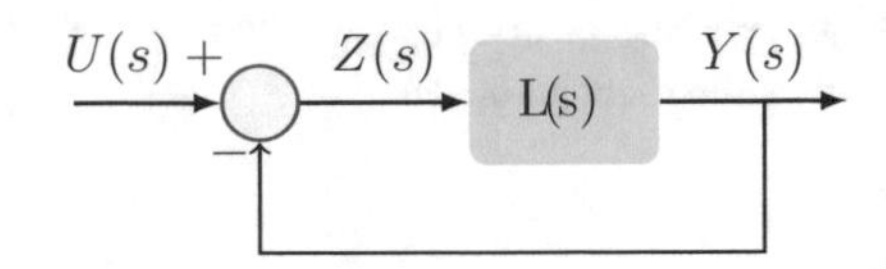

Fig. 7.9 Schematic diagram of a closed-loop model with negative feedback

In the the case of differential equations with delay, we can think of the input function as being a function of the delay, i.e., $u = u(t - \tau)$. Then, $\mathcal{L}(u(t - \tau)) = e^{-s\tau}U(s)$, and Eq. (7.25) becomes

$$L(s) = \frac{b_m s^m + b_{m-1}s^{(m-1)} + \cdots + b_1 s + b_0}{a_n s^n + a_{n-1}s^{(n-1)} + \cdots + a_1 s + a_0}\, e^{-s\tau}. \tag{7.26}$$

If we let $y(t) = ce^{\lambda t}$ and $u = y(t - \tau) = ce^{\lambda(t-\tau)}$, then direct computations show that Eq. (7.26) can be written as

$$L(s) = \frac{q(s)}{p(s)}\, e^{-s\tau}, \tag{7.27}$$

where p and q are the polynomials that make up the characteristic polynomial Eq. (7.24).

We wish to determine whether or not an equilibrium of a delay differential equation, with characteristic polynomial given by Eq. (7.24) is stable. To answer this question, consider now a closed-loop system with negative feedback, as is shown in Fig. 7.9.

Combining $Y(s) = L(s)Z(s)$ and $Z(s) = U(s) - Y(s)$ yields the transfer function

$$T(s) = \frac{Y(s)}{U(s)} = \frac{L(s)}{1 + L(s)}.$$

The roots of the numerator of $T(s)$ are called the zeros of $T(s)$, while the roots of the denominator are the poles of $T(s)$. Then, a critical observation is that the poles of $T(s)$ are also the roots of $1 + L(s) = 0$, which is exactly the same as Eq. (7.24). This means that the stability properties of a delay differential equation with characteristic polynomial as in Eq. (7.24) can be inferred by asking whether the denominator of the transfer function, $T(S)$, has any zeros in the right-half of the s-plane. Nyquist was able to answer this question by applying Cauchy's principle of argument as follows.

7.7.2 *Cauchy's Principle of Argument*

Let $F(s)$ be an analytic function in a closed region of the complex plane s, shown in Fig. 7.10 except at a finite number of points (namely, the poles of $F(s)$). It is also assumed that $F(s)$ is analytic at every point on the contour. Then, as s travels around

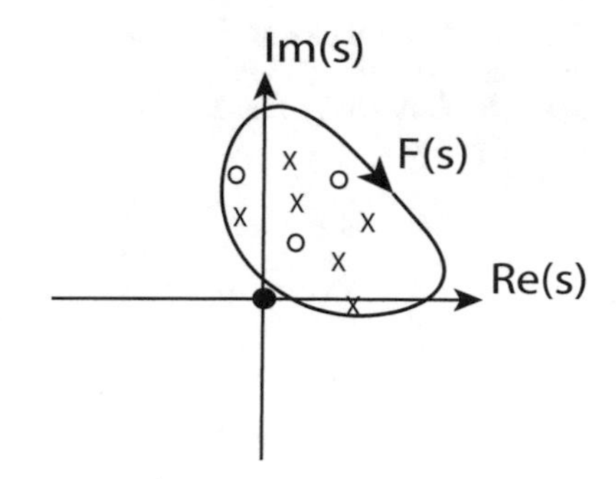

Fig. 7.10 Cauchy's Principle of Argument. A mapping $s \mapsto F(s)$ encircles $Z = 3$ zeros and $P = 6$ poles, so that $N = -3$

the contour in the s-plane in the clockwise direction, the function $F(s)$ encircles the origin in the $(\Re F(s), \Im F(s))$-plane in the same direction N times (see Fig. 7.10), with N given by

$$N = Z - P, \tag{7.28}$$

where Z and P stand for the number of zeros and poles (including their multiplicities) of the function $F(s)$ inside the contour.

This result can also be represented as

$$\arg\{F(s)\} = (Z - P)2\pi = 2\pi N.$$

Now, finding the zeros of $1 + L(s)$ is equivalent to solving $L(s) = -1 + 0\,i$, since the origin of $1 + L(s)$ corresponds to the point $-1 + 0\,i$ of the complex plane. Thus, to solve $L(s) = -1$, one can apply Cauchy's principle, while performing, first, a mapping

$$s \mapsto L(s),$$

in which the path in s encircles the entire right half plane and then count the number of encirclements of the point $-1 + 0\,i$ by $L(s)$ in the clockwise direction. This should yield the total number, N, of encirclements. But what we want is the number of zeros, Z. Thus, we first find the number of poles, P, of $L(s)$ and then apply Cauchy's Eq. (7.28) to solve for Z:

$$Z = N + P.$$

7.7.3 Examples

Example 7.2 Consider the simplest system with a time delay

$$\frac{dx}{dt} = -x(t - \tau). \tag{7.29}$$

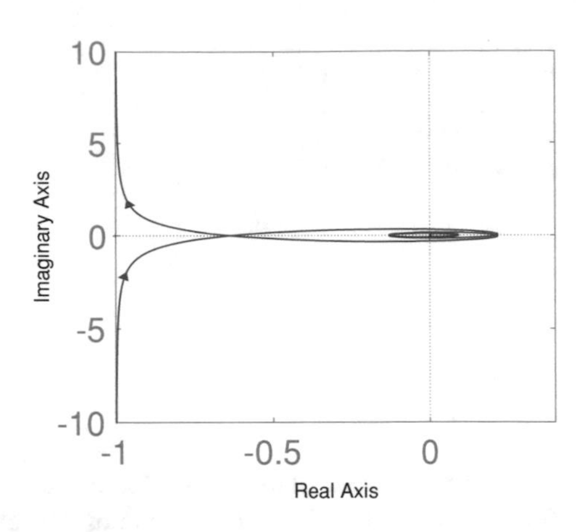

Fig. 7.11 Nyquist plot for the transfer function (7.30). MATLAB code can be found in the Appendix

The characteristic equation associated with this example is

$$\lambda + e^{-\lambda\tau} = 0.$$

Thus, $p(\lambda) = \lambda$ and $q(\lambda) = 1$. Then the transfer function $L(s)$ becomes

$$L(s) = \frac{1}{s} e^{-\lambda\tau} = 0. \tag{7.30}$$

The Nyquist plot produced by the mapping $s \mapsto L(s)$ is shown in Fig. 7.11. The *MatLab script* for computing the transfer function is

```
L1 = tf(1,[1 0],'InputDelay',tau);
```

where the first input "1" represents the numerator of Eq. (7.30) and "" is the denominator. The complete MATLAB code is in the Appendix.

There are no encirclements of the point $-1 + 0\,i$, so $N = 0$. The only pole is at $s = 0$, so $P = 0$ as well. Then the number of zeros of $1 + L(s)$ in the right half of the s-plane is $Z = 0$. Hence, the trivial equilibrium $x = 0$ is asymptotically stable for all values of τ.

Example 7.3 Let us consider Lotka-Volterra predator-prey model Eq. (7.14) with the following parameters: $\mu_1 = \mu_2 = 2, a = d = 2$ and $b = c = 1$. This combination of parameter values is the same as in Case II analyzed earlier, in which we found the nontrivial equilibrium $(x_e, y_e) = (2/3, 2/3)$ to be asymptotically stable. We now explore its stability properties through Nyquist criterion. Recall the characteristic polynomial was found to be

$$\left(\lambda + \frac{4}{3}\right)^2 - \frac{4}{9}e^{-\lambda(2r)} = 0,$$

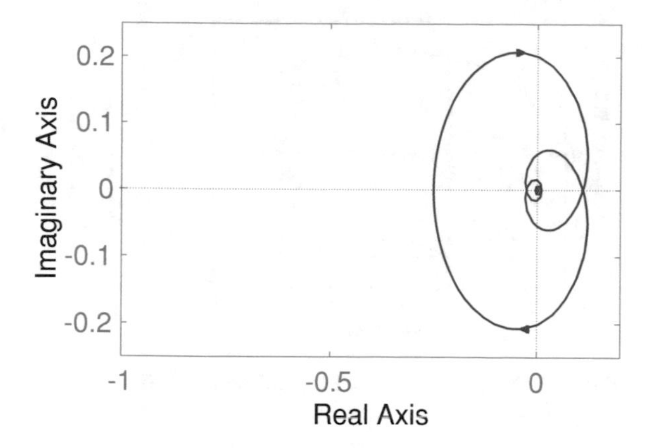

Fig. 7.12 Nyquist plot for the transfer function (7.31). MATLAB code can be found in the Appendix

which shows $p(\lambda) = \lambda^2 + 8/3\lambda + 16/9$ and $q(\lambda) = -4/9$. The corresponding transfer function, $L(s)$, is

$$L(s) = \frac{-\frac{4}{9}}{s^2 + \frac{8}{3}s + \frac{16}{9}} e^{-\lambda(2\tau)} = 0. \tag{7.31}$$

The Nyquist plot produced by the mapping $s \mapsto L(s)$ is shown in Fig. 7.12. The *MatLab script* for computing the transfer function is

```
L1 = tf(-4/9,[1 8/3 16/9],'InputDelay',2*tau);
```

where the first input "-4/9" represents the numerator of Eq. (7.31) and "[1 8/3 16/9]" is the denominator. The complete MATLAB code is in the Appendix.

There are no encirclements of the point $-1 + 0i$, so $N = 0$. There are two (repeated) poles at $s = -4/3$, so $P = 0$ as well. Then the number of zeros of $1 + L(s)$ in the right half of the s-plane is $Z = 0$. Hence, the trivial equilibrium $x = 0$ is asymptotically stable, just as it was determined earlier.

Example 7.4 Let us consider now the case of delay-induced oscillations in the Lotka-Volterra predator-prey model Eq. (7.14) with the following parameters: $\mu_1 = 1, \mu_2 = -1, a = 1, b = 1, c = -2, d = 1$. This combination of parameter values is the same as in Case III analyzed earlier, in which we found the nontrivial equilibrium $(x_e, y_e) = (2/3, 1/3)$ to be asymptotically stable for $0 < \tau < 2.13$, and undergoes, at $\tau = 2.13$, a Hopf bifurcation that leads to delay-induced small amplitude oscillations

We now explore the stability properties through Nyquist criterion. Recall the characteristic polynomial was found to be

$$\left(\lambda + \frac{2}{3}\right)\left(\lambda + \frac{1}{3}\right) + \frac{4}{9}e^{-\lambda(2r)} = 0.$$

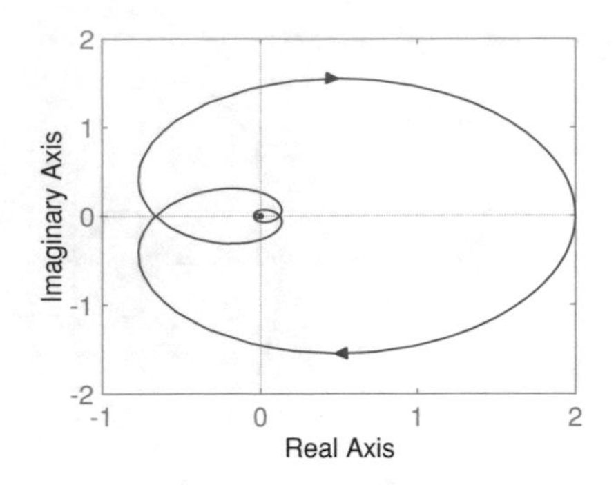

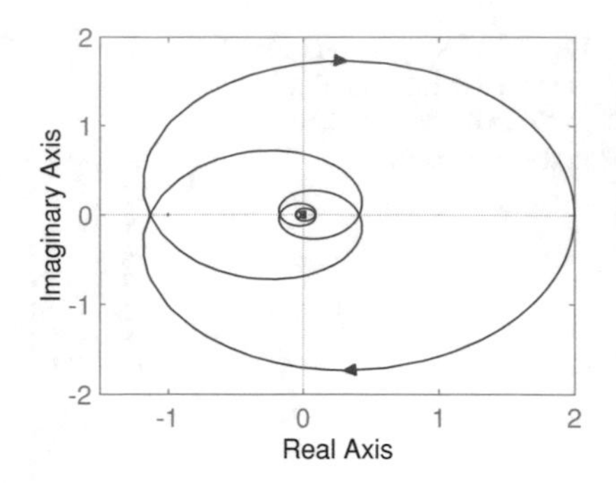

Fig. 7.13 Nyquist plot for the transfer function (7.32). MATLAB code can be found in the Appendix

which shows $p(\lambda) = \lambda^2 + \lambda + 2/9$ and $q(\lambda) = 4/9$. The corresponding transfer function, $L(s)$, is

$$L(s) = \frac{\frac{4}{9}}{s^2 + s + \frac{2}{9}} e^{-\lambda(2\tau)} = 0. \tag{7.32}$$

The Nyquist plot produced by the mapping $s \mapsto L(s)$ is shown in Fig. 7.11. The *MatLab script* for computing the transfer function is

```
L1 = tf(4/9,[1 1 2/9],'InputDelay',2*tau);
```

where the first input "4/9" represents the numerator of Eq. (7.32) and "[1 1 2/9]" is the denominator. The complete MATLAB code is in the Appendix. When $0 < \tau < 2.13$, there are no encirclements of the point $-1 + 0\,i$, so $N = 0$. Figure 7.13(left) shows a specific instance of the Nyquist plot for $\tau = 1$. There are two poles, one at $s = -1/3$, and one at $s = -2/3$, so $P = 0$ as well. We conclude that when $0 < \tau < 2.13$ the number of zeros of $1 + L(s)$ in the right half of the s-plane is $Z = 0$. Hence, the trivial equilibrium $x = 0$ is asymptotically stable. When $\tau \geq 2.13$, the Nyquist plot shows one encirclement of the $-1 + 0\,i$, so $N = 1$. Figure 7.13(right) shows a specific instance of the Nyquist plot for $\tau = 2.3$. The number of poles is still the same, i.e., $P = 0$. Hence there are now $Z = 1$ zero of the characteristic polynomial in the right half of the s-plane. This means that the equilibrium $(x_e, y_e) = (2/3, 1/3)$ is unstable, just as it was previously determined.

7.8 Delay in the Coupled Fluxgate Magnetometer

While the mathematical models and related devices governed by bistable potential functions may assume instantaneous coupling, in practice we must account for the fact that even high-speed, high-precision, circuit components can introduce a delay in the coupling signal. Thus, in this section we investigate the behavior of a ring of overdamped bistable systems with delayed nearest-neighbor connections. We concentrate on the CCFM system as the "test-bed" to study the effects of time delay in generic formulations of coupled bistable systems. In this system, we have already shown that, without delay, large-amplitude oscillations and nontrivial synchronous equilibria can coexist near the onset of the oscillations. Our study shows that delay-induced Hopf bifurcation occurs from the synchronous equilibria but, generically, the small amplitude oscillations that appear are unstable. Thus, delay has the effect of decreasing the size of the basin of attraction of nontrivial synchronous equilibria, which in turn, makes the basin of attraction of the stable large-amplitude oscillations larger. Collectively, this is a positive effect because the sensor device depends mainly on large amplitude oscillations, so a small delay can make it easier to induce the device to oscillate on its own.

7.8.1 Model Equations with Multiple Delays

As a "test bed", we use the model equations of a CCFM device with N fluxgates [11]. The results are, however, generic and applicable to all rings of overdamped bistable units unidirectionally coupled. For $N = 3$ the model equations are

$$
\begin{aligned}
\dot{x}_1(t) &= -x_1(t) + \tanh\left(c(x_1(t) + \lambda x_2(t-\tau_1) + \varepsilon)\right), \\
\dot{x}_2(t) &= -x_2(t) + \tanh\left(c(x_2(t) + \lambda x_3(t-\tau_2) + \varepsilon)\right), \\
\dot{x}_3(t) &= -x_3(t) + \tanh\left(c(x_3(t) + \lambda x_1(t-\tau_3) + \varepsilon)\right),
\end{aligned}
\tag{7.33}
$$

where τ_1, τ_2, and τ_3, denote the corresponding delays in the connectivity scheme. Recall that we are primarily interested in the case where $\lambda < 0$, which is a negative feedback system.

The surface shown in Fig. 7.14 depicts the boundary between the basin of attraction of the synchronous equilibria and the large amplitude periodic oscillations. Initial conditions inside the pyramid-like shape are attracted to equilibrium points, while those outside are attracted to the large-amplitude oscillations. Observe that the size of the basin of attraction of the stable (nontrivial synchronous) equilibrium point gets larger as initial conditions move away from the origin.

7.8.2 Conversion to Single Delay

We now return our attention to Eq. (7.33). We make the following change of variables to create a single delayed term with $\tau = \tau_1 + \tau_2 + \tau_3$

$$
y_1(t) = x_1(t), \qquad y_2(t) = x_2(t-\tau_1), \qquad y_3(t) = x_3(t-(\tau_1+\tau_2)).
$$

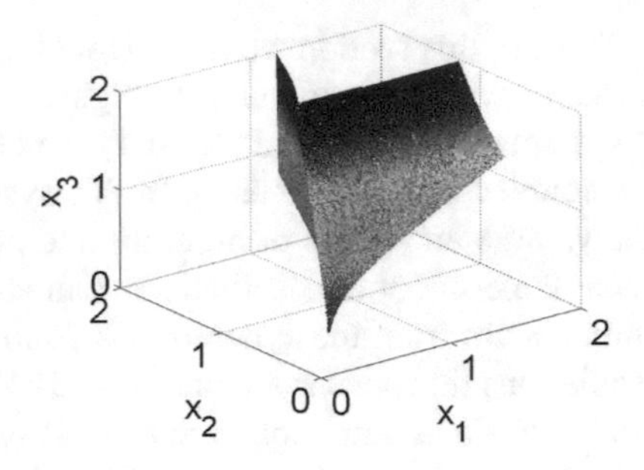

Fig. 7.14 Three dimensional surface defining the boundary between the basins of attraction of equilibrium points and that of periodic oscillations for a coupled-core fluxgate magnetometer, see Eq. (7.33), with $\tau = 0$. Points inside the pyramid-like shape are attracted to synchronous equilibria while those outside are attracted to a global branch of large-amplitude periodic oscillations. Parameters are: $c = 3$, $\lambda = -0.44$, $N = 3$, $\varepsilon = 0.0$

The resulting system of equations is given by

$$\begin{aligned}
\dot{y}_1(t) &= -y_1(t) + \tanh\left(c(y_1(t) + \lambda y_2(t) + \varepsilon)\right), \\
\dot{y}_2(t) &= -y_2(t) + \tanh\left(c(y_2(t) + \lambda y_3(t) + \varepsilon)\right), \\
\dot{y}_3(t) &= -y_3(t) + \tanh\left(c(y_3(t) + \lambda y_1(t-\tau) + \varepsilon)\right),
\end{aligned}$$

where time is shifted for the second and third equations. For convenience (and greater generality), we consider the system given by:

$$\begin{aligned}
\dot{y}_1(t) &= -y_1(t) + f_1(y_1(t), y_2(t)), \\
\dot{y}_2(t) &= -y_2(t) + f_2(y_2(t), y_3(t)), \\
\dot{y}_3(t) &= -y_3(t) + f_3(y_1(t-\tau), y_3(t)),
\end{aligned} \tag{7.34}$$

where $f_1(y_1(t), y_2(t)) = \tanh\left(c(y_1(t) + \lambda y_2(t) + \varepsilon)\right)$, etc.

7.8.3 *Stability Properties of Synchronous Equilibria*

We now wish to investigate the stability properties of the *synchronous* equilibria of the transformed system (7.34), which we denote by $(\bar{y}_1, \bar{y}_2, \bar{y}_3)$. The linearization of the above system is given by:

$$\begin{pmatrix} \dot{y}_1(t) \\ \dot{y}_2(t) \\ \dot{y}_3(t) \end{pmatrix} = \begin{pmatrix} -1+a_{11} & a_{12} & 0 \\ 0 & -1+a_{22} & a_{23} \\ 0 & 0 & -1+a_{33} \end{pmatrix} \begin{pmatrix} y_1(t) \\ y_2(t) \\ y_3(t) \end{pmatrix} + \begin{pmatrix} 0 & 0 & 0 \\ 0 & 0 & 0 \\ a_{31} & 0 & 0 \end{pmatrix} \begin{pmatrix} y_1(t-\tau) \\ y_2(t-\tau) \\ y_3(t-\tau) \end{pmatrix},$$

where

$$a_{11} = \frac{\partial f_1(\bar{y}_1, \bar{y}_2)}{\partial y_1}, \quad a_{12} = \frac{\partial f_1(\bar{y}_1, \bar{y}_2)}{\partial y_2}, \quad a_{22} = \frac{\partial f_2(\bar{y}_2, \bar{y}_3)}{\partial y_2},$$

$$a_{23} = \frac{\partial f_2(\bar{y}_2, \bar{y}_3)}{\partial y_3}, \quad a_{31} = \frac{\partial f_3(\bar{y}_1, \bar{y}_3)}{\partial y_1}, \quad a_{33} = \frac{\partial f_3(\bar{y}_1, \bar{y}_3)}{\partial y_3}.$$

Because we are primarily interested in the negative feedback system, $\partial f_i/\partial x_j < 0$ for $i \neq j$. As usual, we attempt to find solutions of the form $\mathbf{y}(t) = \text{¸}e^{\sigma t}$, with $\mathbf{y} = (y_1, y_2, y_3)^T$. It is not difficult to see that the resulting characteristic equation has the form:

$$\det \begin{bmatrix} -1+a_{11}-\sigma & a_{12} & 0 \\ 0 & -1+a_{22}-\sigma & a_{23} \\ a_{31}e^{-\sigma\tau} & 0 & -1+a_{33}-\sigma \end{bmatrix} = 0,$$

which is easily solved by expanding the first column to give:

$$(\sigma+1-a_{11})(\sigma+1-a_{22})(\sigma+1-a_{33}) - a_{12}a_{23}a_{31}e^{-\sigma\tau} = 0.$$

We are particularly interested in nontrivial *synchronous* equilibria of the form $(\bar{y}_1, \bar{y}_2, \bar{y}_3) = (\bar{y}, \bar{y}, \bar{y})$ so that $a_{11} = a_{22} = a_{33}$ and $a_{12} = a_{23} = a_{31}$. Note from Fig. 6.24 that there are exactly two nontrivial synchronous equilibria $(\bar{y}, \bar{y}, \bar{y})$ and $(-\bar{y}, -\bar{y}, -\bar{y})$. In both cases the characteristic polynomial reduces to:

$$(\sigma - A)^3 = B^3 e^{-\sigma\tau}, \tag{7.35}$$

where $A = a_{11} - 1$ and $B = a_{12}$. Direct calculations show that $0 < a_{11} < 1$ and $-1 < a_{12} < 0$, so that $-1 < A < 0$ and $-1 < B < 0$. We are interested in stability changes of the synchronous equilibria that may lead to small amplitude oscillations via a Hopf bifurcation. Those changes can only occur when $\sigma = \omega i$, and since the left-hand side of (7.35) is a polynomial function, monotonically increasing in magnitude and angle, then by the Argument principle of complex analysis, Eq. (7.35) has a solution whenever $B < A$. To visualize this result, we can also solve (7.35) graphically, first substituting $\sigma = \omega i$ to obtain:

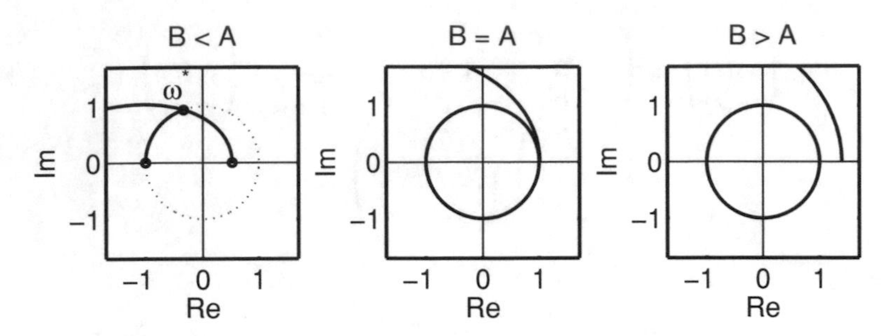

Fig. 7.15 Graphical representation of the solution set of Eq. (7.36)

$$\left(\frac{A^3 - 3A\omega^2}{B^3}\right) + \left(\frac{\omega^3 - 3A^2\omega}{B^3}\right) i = -e^{-\omega\tau i}. \tag{7.36}$$

The left-hand side of (7.36) represents a complex-valued curve $\mathcal{C}_3(\omega)$ parametrized by ω while the right-hand side describes the unit circle in the complex plane, parametrized also by ω and by the delay τ. For $\tau > 0$, as ω increases (starting at zero) the right-hand term traces the unit circle S^1 clockwise starting at the point $(-1, 0)$, as is shown in Fig. 7.15(left).

Simultaneously, the left-hand curve $\mathcal{C}_3(\omega)$ traverses the complex plane counter-clockwise starting at the point $(A^3/B^3, 0)$. When $B < A$ this starting point is in the interval $(0 < A^3/B^3 < 1, 0)$, and since the magnitude and angle of points traversed along $\mathcal{C}_3$ are monotonically increasing, then there is a critical value ω^* at which both the circle S^1 and $\mathcal{C}_3$ intersect. At $\omega = \omega^*$, the point of intersection on S^1 corresponds to a critical angle θ_c measured from the starting point $(-1, 0)$. The critical delay τ_c producing the Hopf bifurcation satisfies $\tau_c = \theta_c/\omega^*$. This critical delay corresponds to the solution of (7.36) at the Hopf bifurcation. By the periodicity of $e^{i\omega\tau}$, there are infinitly many solutions of (7.36), but other solutions produce larger values of τ, which are unstable. Analytically, ω^* is the solution of

$$\omega^6 + 3A^2\omega^4 + 3A^4\omega^2 + A^6 - B^6 = 0.$$

When $B = A$, see Fig. 7.15(middle), the point $(A^3/B^3, 0)$ has moved to $(1, 0)$ but the right-hand side term is still at $(-1, 0)$, so as soon as ω increases the point $(A^3/B^3, 0)$ separates away from the circle due to the monotonic nature of $\mathcal{C}_3$, thus there is no solution. Similarly, when $B > A$, Fig. 7.15(right), the starting point $(A^3/B^3, 0)$ is already separated from the unit circle and so there is no solution either.

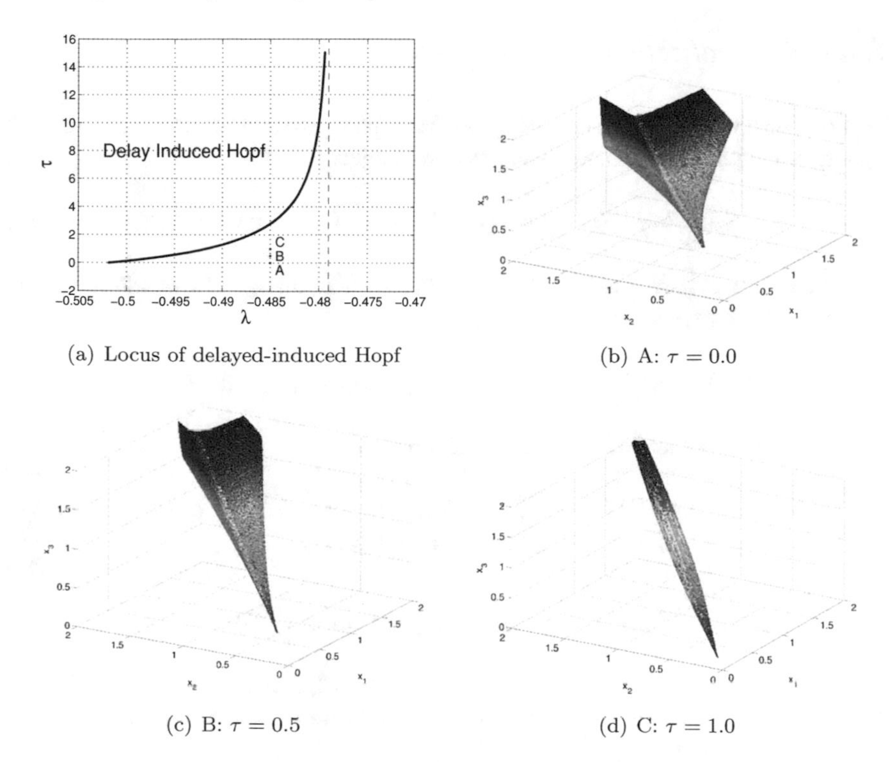

Fig. 7.16 **a** Locus of Hopf bifurcation induced by delayed coupling in a ring with $N = 3$ over-damped bistable systems governed by Eq. (7.37). **b–d** Basins of attraction of synchronous equilibria for delay values labeled A, B, and C in (**a**), respectively. For small τ, both synchronous equilibria and large-amplitude oscillations have reasonably large basins of attraction. As τ increases towards the locus of the Hopf bifurcation, however, the equilibria lose stability, and consequently, their basins of attraction shrink accordingly

7.8.4 Locus of Delay-Induced Oscillations

The locus of the delayed-induced Hopf bifurcation points ω^*, in parameter space (λ, τ), is shown in Fig. 7.16a. The rightmost point along this two-parameter boundary curve corresponds to the condition $A = B$. The leftmost point is the Hopf bifurcation point without delay, i.e., $\tau = 0$. Substituting $\tau = 0$ in (7.36) we can solve for ω, which then yields the condition $B = 2A$ for the Hopf bifurcation without delay. Computational work conducted with the aid of DDE-BIFTOOL [12], a software tool for the bifurcation analysis of delay differential equations, confirms that the delay-induced oscillations exist in the region just above the locus curve. The oscillations are, however, unstable. Figures 7.16b–d illustrate the contraction that occurs in the basin of attraction of synchronous equilibria due to delay-induced instability.

7.8.5 Generalization to Larger Arrays

We now consider the general case of N (odd) fluxgates governed by an N-dimensional system of coupled overdamped bistable units subject to delay:

$$\dot{x}_i(t) = -x_i(t) + \tanh\left(c(x_i(t) + \lambda x_{i+1}(t - \tau_i) + \varepsilon)\right), \tag{7.37}$$

where $i = 1, 2, \ldots, N \mod N$. We perform a similar change of variables to create a single delay term with $\tau = \tau_1 + \cdots + \tau_N$:

$$y_1(t) = x_1(t), \quad y_2(t) = x_2(t - \tau_1), \ldots, y_N(t) = x_N(t - (\tau_1 + \cdots + \tau_{N_1})).$$

The resulting system of equations is given by

$$\begin{aligned}
\dot{y}_1(t) &= -y_1(t) + f_1(y_1(t), y_2(t)), \\
\dot{y}_2(t) &= -y_2(t) + f_2(y_2(t), y_3(t)), \\
&\vdots \\
\dot{y}_N(t) &= -y_N(t) + f_N(y_1(t - \tau), y_N(t)),
\end{aligned} \tag{7.38}$$

where $f_1(y_1(t), y_2(t)) = \tanh\left(c(y_1(t) + \lambda y_2(t) + \varepsilon)\right)$, etc. The linearization of (7.38) with respect to the synchronous equilibrium solution $(y_1, \ldots, y_N) = (\bar{y}, \ldots, \bar{y})$, has the form

$$L = \begin{bmatrix} A & B\ 0\ \ldots 0 \\ 0 & A\ B\ \ldots 0 \\ \vdots & \\ Be^{-z\tau} & 0\ 0\ \ldots A \end{bmatrix},$$

where $A = \partial f_i(\bar{y}, \bar{y})/\partial y_i - 1$ and $B = \partial f_i(\bar{y}, \bar{y})/\partial y_j$, $j \neq i$. Again we are most interested in the negative feedback system where $\partial f_i/\partial x_j < 0$ for $i \neq j$ and N is odd. The characteristic polynomial becomes $(\sigma - A)^N = B^N e^{-\sigma\tau}$, where σ represents again the eigenvalues of the linearized matrix L. Substituting the Hopf bifurcation condition $\sigma = \omega i$, we get

$$\frac{(A - \omega i)^N}{B^N} = -e^{-\omega\tau i}. \tag{7.39}$$

The left-hand side of (7.39) defines again a curve $C_N(\omega)$ that traverses the complex plane counter-clockwise as ω increases from zero. This curve is similar to that of the $N = 3$ case, with $-1 < A < 0$ and $-1 < B < 0$, except that now the starting point is $(A^N/B^N, 0)$. When $B < A$, this starting point is in the interval $(0 < A^N/B^N < 1, 0)$ and so there is a critical value ω^* at which C_N intersects the unit circle at θ_c, i.e., the right-hand side of (7.39), so that the critical delay producing the Hopf bifurcation is once again $\tau_c = \theta_c/w^*$. The critical value ω^* can be found analytically by noticing

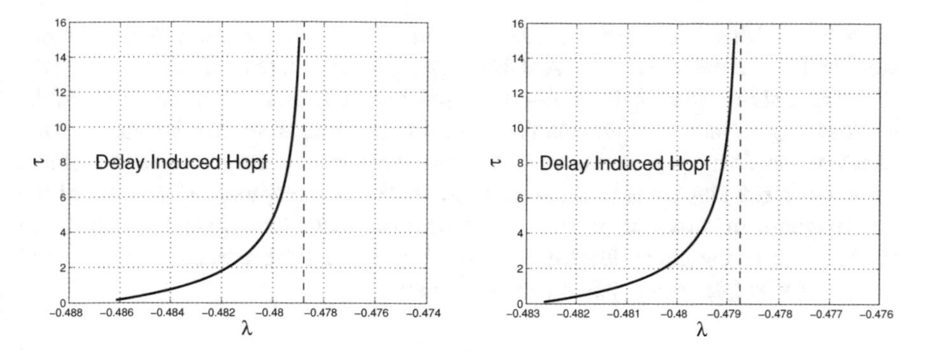

Fig. 7.17 Locus of Hopf bifurcation induced by delayed coupling in a ring with (left) $N = 5$ and (right) $N = 7$ overdamped bistable systems governed by Eq. (7.37)

that both sides of (7.39) are complex-valued expressions, which are equal only when their magnitudes and angles are identical, this produces the polynomial

$$(w^2 + A^2)^N = B^{2N},$$

whose solution ω^* yields the desired eigenvalue for the delayed-induced Hopf bifurcation point. Once again, the solution vanishes when the point $(A^N/B^N, 0)$ is at the opposite end point $(1, 0)$, which yields the condition $B = A$. Delayed-induced oscillations terminate at a regular Hopf bifurcation from the synchronous equilibrium with $\tau = 0$. To find this bifurcation point we note that when $\tau = 0$ the linearized matrix L becomes cyclic, with the eigenvectors space spanned by

$$V_j = \left\{\left[v, \zeta^j v, \zeta^{2j} v, \ldots, \zeta^{(N-1)j} v\right] : v \in R\right\},$$

where $\zeta = e^{2\pi/N}$. Direct calculations yield $L \cdot V_j = (A + \zeta^j B)V_j$. Hence the eigenvalues of $L|V_j$ are those of $A + \zeta^j B = A + \cos(2\pi j/N) + Bi \sin(2\pi j/N)$. It follows that a Hopf bifurcation occurs when

$$B = -\frac{1}{\cos(2\pi/N)} A. \tag{7.40}$$

Notice that this expression yields the previously found condition $B = 2A$ for the special case when $N = 3$. Figure 7.17 shows the locus, in parameter space (λ, τ), of the delayed-induced Hopf bifurcation point for $N = 5$ and $N = 7$, respectively.

AUTO and numerical studies suggest that the delay-induced Hopf bifurcation is unstable leading to almost all solutions beyond the Hopf point approaching the large amplitude periodic orbit, making it a global attractor. However, we have not performed rigorous center manifold analysis [13] to prove instability of the Hopf bifurcation.

In each of these two cases, i.e., $N = 5$ and $N = 7$, small amplitude oscillations are found above the loci curves. A stability analysis conducted with the aid of DDE-BIFTOOL shows the small-amplitude, delay-induced, oscillations to be unstable. But this is good news for the coupled-core fluxgate magnetometer because the net effect of the delay is essentially to increase the basin of attraction of the global branch of large-amplitude oscillations between the two magnetization states of the ferromagnetic materials. In other words, delayed coupling tends to enhance the basin of attraction of the global branch of oscillations so that it becomes more robust to induce a CCFM-based device to oscillate on its own.

7.9 Exercises

Exercise 7.1 (*Logistic Growth Model with Delay*) A modified version of the logistic growth model, with a delay term added into the growth rate, was introduced around 1948 by G.E. Hutchinson [14]. The modified model is

$$\frac{dN}{d\tilde{t}} = rN\left[1 - \frac{N(\tilde{t} - \tau)}{k}\right],$$

where τ is the delay.

(a) Write a dimensionless version of the model by introducing the following change of variables: $x = N/k, t = \tilde{t}/\tau$. Note that this change of variables will lead to a dimensionless model with a delay of exactly 1. Specifically:

$$\frac{dx}{dt} = r\tau x[1 - x(t - 1)].$$

Let $\mu = r\tau$, so that the time delayed is absorbed into one single parameter μ.

(b) Compute the equilibrium points of the latest dimensionless model.
(c) Study the stability properties of the equilibrium points.
(d) Find a critical value of μ that can lead to oscillations via Hopf bifurcation.
(e) Perform computer simulations to illustrate your results.

Exercise 7.2 (*Periodic Breathing*) A one-dimensional model with one-single delay has been proposed to describe the dynamics of CO_2 concentration, $p(t)$, in the lungs

$$V_L \frac{dp}{dt} = M - pV(p(t - \tau)) - T(p),$$

where V_L is lung volume, M is the constant rate of CO_2 production due to metabolism, $V(p(t - \tau))$ is ventilation, processed through brain stems, which monitors CO_2 levels in the blood. The delay τ represents the delay between the measurement of CO_2 levels and the time at which ventilation takes place. $T(p)$ is a transfer function.

(a) Assume p_e to be an equilibrium point. Perform a stability analysis about x_e. Note: you don't need to know the explicit form of $V(p_e)$, or $V'(p_e)$, just that they represent the mean ventilation and the chemoreflex gain, respectively. Thus, assume $V(p_e)$ is known and $T'(p_e) = \beta Q$, where β is the solubility of CO_2 in the blood, and Q is the cardiac output.

(b) Let

$$\mu_1 = \frac{(V(p_e) + T'(p_e))\tau}{V_L}, \qquad \mu_2 = \frac{p_e V'(p_e)\tau}{V_L}.$$

Determine the locus of a Hopf bifurcation in parameter space (μ_1, μ_2).

Exercise 7.3 (*Specific Model of Periodic Breathing*) Glass and Mackey [15] proposed a specific model with one-single delay for periodic breathing

$$\frac{dC}{dt} = \lambda - \alpha V_m C(t) \frac{(C(t-\tau))^n}{\theta^n + (C(t-\tau))^n},$$

where $C(t)$ is the concentration of CO_2 in the lungs at time t, $\dot{V}(t) = (C(t))^n/(\theta^n + C(t))^n)$ is the rate of ventialtion, V_m, θ, n and α are constants.

(a) Compute all equilibrium points.
(b) Linearize the model equation and find the characteristic polynomial associated with the equilibrium points.
(c) Study the stability of the model and determine a critical value of the delay, τ_c that may lead to small amplitude oscillations via Hopf bifurcations.

Exercise 7.4 (*Logistic Growth Model with Two Delays*) Consider again the logistic growth model with two delays, shown in Eq. (7.21), and rewritten in here for completeness purposes:

$$\frac{dx}{dt} = \mu x(t)\left[1 - a_1 x(t-\tau_1) - a_2 x(t-\tau_2)\right].$$

Determine the critical value of the delay $\tau = \tau_1 + \tau_2$ that leads to small amplitude oscillations via a Hopf bifurcation. Perform computer simulations to validate your results.

Exercise 7.5 Consider the following system

$$\begin{aligned}\dot{u}(t) &= a_1 u(t) + b_1 v(t-\tau_2)\\ \dot{v}(t) &= a_2 v(t) + b_2 u(t-\tau_1).\end{aligned}$$

Let $y(t) = [u(t), v(t)]^T$ represent a solution of the above equation. Perform a stability analysis to show that

$$\lim_{t\to\infty} y(t) = 0,$$

if $\Re(a_i) < 0$, $i = 1, 2$, and $|b_1 b_2| < \Re(a_1)\,\Re(a_2)$.

Exercise 7.6 Consider the simple delay equation

$$\dot{x}(t) = a - x\left(t - \frac{\pi}{2}\right).$$

Set $\tau = \pi/2$ and find an analytical exact solution on the interval $t \in [-\pi/2, 0]$. Study the behavior of the solution.

Exercise 7.7 Consider the following form of the logistic model

$$\dot{x}(t) = \alpha x(t-\tau)\left(1 - x(t-\tau)\right).$$

(b) Calculate the equilibrium points of this model.
(c) Study the stability properties of the equilibrium points.
(d) Find a critical value of μ that can lead to oscillations via Hopf bifurcation.
(e) Perform computer simulations to illustrate your results for different values of the parameter α.

Exercise 7.8 (*Red Blood Cells Model*) A mathematical model for the size of a population $x(t)$ of mature red blood cells was proposed by Mackey-Glass [15], through

$$\dot{x}(t) = \alpha \frac{x(t-\tau)}{1 + x^k(t-\tau)} - \beta x(t),$$

where τ is the maturation time of red blood cells, α, β, and k are positive constants. Let $\alpha = 0.2$, $\beta = 0.1$, and $m = 10$. Perform various computer simulations with delays $\tau = 7$, $\tau = 7.75$, $\tau = 9.696$, and $\tau = 12$. Describe the results.

Exercise 7.9 (*Ważewszka-Czyżewska and Lasota Model*) Ważewszka-Czyżewska and Lasota [16] proposed a model for the growth of blood cells, given by

$$\frac{dx}{dt} - \mu x(t) + \rho e^{-\gamma x(t-\tau)},$$

where $x(t)$ represents the number of cells at any time t, μ is the natural death of the red blood cells, ρ and γ are positive constants related to the recruitment term for the red blood cells and τ is the time required for producing red blood cells.

(a) Conduct some numerical simulations with the following parameter values: $\rho = 2$, $\gamma = 0.1$, $\mu = 0.5$ and $\tau = 5$. Describe the results.
(b) Change the values of $\rho, \gamma \in (0, \infty)$ and $\mu \in (0, 1)$, and perform additional numerical simulations. Describe the observed changes in the dynamics.

Exercise 7.10 Consider the following linear equation with one discrete delay

$$\dot{x}(t) = -ax(t) + bx(t-\tau).$$

Compute the equilibrium points and study their stability.

Exercise 7.11 Consider now a system of two equations with one discrete delay

$$\begin{aligned}\dot{x}(t) &= \alpha_1 x(t) + \beta_1 y(t) + \alpha_2 x(t-\tau) + \beta_2 y(t-\tau)\\ \dot{y}(t) &= \gamma_1 x(t) + \delta_1 y(t) + \gamma_2 x(t-\tau) + \delta_2 y(t-\tau).\end{aligned}$$

Compute all equilibrium points and study their stability.

Exercise 7.12 (*Neural Network Model with two Delays*) Consider the following neural network model with two delays

$$\begin{aligned}\dot{u}_1(t) &= -u_1(t) + a_{12} f(u_2(t-\tau_2))\\ \dot{u}_2(t) &= -u_2(t) + a_{21} f(u_1(t-\tau_1)).\end{aligned}$$

(a) Assume $f(0) = 0$, so that $(u_1, u_2) = (0, 0)$ is an equilibrium point. Compute the linearization of the original model equations about the trivial equilibrium, and write the characteristic polynomial associated with it.
(b) Determine the conditions that lead to a Hopf bifurcation, i.e., the critical value of the delay $\tau = \tau_1 + \tau_2$. Also, find the conditions that separate the first quadrant of the (τ_1, τ_2) plane into two parts, one being a stable region, another being unstable, and the boundary between them corresponding to the loci of Hopf bifurcations.

Exercise 7.13 (*Delayed-Food Limited Model*) Gopalswamy et al. [17] introduced the following delayed food-limited model

$$\frac{dx}{dt} = rx(t)\left[\frac{k - x(t-\tau)}{k + rcx(t-\tau)}\right].$$

(a) Write a description of the model.
(b) Let $r = 0.15$, $k = 100$, $c = 1$, and $\tau = 8$. Calculate the equilibrium points and determine their stability.
(c) Perform computer simulations and discuss the results. Conduct additional computer simulations with $\tau = 12.8$ and discuss the results.

Exercise 7.14 Consider Lotka-Volterra predator-prey model Eq. (7.14) with the following parameters: $\mu_1 = \mu_2 = 2, a = d = 1$ and $b = c = 2$. Apply Nyquist criterion to determine the stability of the nontrivial equilibrium point $(2/3, 2/3)$.

Exercise 7.15 (*Delayed-Protein Degradation*) A mathematical model for describing the production of proteins at any time is

$$\frac{dP}{dt} = \alpha - \beta P(t) - \gamma P(t-\tau),$$

where $P(t)$ is the concentration of proteins at time t, α is the rate of production of proteins, β is the rate of nondelayed protein degradation, and γ is the rate of delayed protein degradation. The time delay τ is due to the fact that protein degradation occurs after a time τ after initiation.

(a) Calculate the equilibrium points.
(b) Write down the characteristic polynomial associated with the linearization about an equilibrium point.
(c) Compute the linearization of the model and study the stability of the equilibrium points by applying the appropriate theorems. In particular, determine the critical value of the delay, τ_c, that may lead to delay-induced oscillations.
(d) Apply Nyquist criterion to determine the stability of each equilibrium point. Approximate the critical value of the delay, τ_c, that may lead to delay-induced oscillations.
(e) Let $\alpha = 40, \beta = 0.3, \gamma = 0.1. \tau = 20$ and initial history 150. Perform computer simulations and discuss the results. Repeat the simulations with: $\alpha = 100, \beta = 1.1, \gamma = 1. \tau = 10$ and initial history 20. Discuss the results.

Exercise 7.16 Consider the following delay differential equation:

$$\frac{dx}{dt} = -x(t)^3 - x(t-\tau).$$

(a) Calculate the equilibrium points.
(b) Write down the characteristic polynomial associated with the linearization about an equilibrium point.
(c) Apply Nyquist criterion to determine the stability of each equilibrium point.

Exercise 7.17 (*Fluxgate Magnetometer*) Consider the following model of a single fluxgate magnetometer subject to a delay

$$\frac{dx}{dt} = -x(t) + \tanh\left[cx(t-\tau)\right]$$

(a) Calculate the equilibrium points.
(b) Write down the characteristic polynomial associated with the linearization about an equilibrium point.
(c) Apply Nyquist criterion to determine the stability of each equilibrium point.

Exercise 7.18 (*Gene Regulation*) A negative feedback model for gene regulation is

$$\begin{aligned}\frac{dx_1}{dt} &= \frac{g_m}{1 + \left(\dfrac{x_2(t-\tau)}{k}\right)^n} - \alpha_1 x_1(t)\\ \frac{dx_2}{dt} &= x_1(t) - \alpha_2 x_2(t).\end{aligned}$$

where $x_1(t)$ denotes intracellular mRNA and $x_2(t)$ represents the protein product of the gene. The delay τ represents the time for mRNA to leave the nucleus, undergo protein synthesis in the ribosome, whereupon the protein reenters the nucleus and suppresses its own mRNA production. g_m, α_1, α_2, and k are parameters.

(a) Let $g_m = 1, k = 0.5, n = 3, \alpha_1 = \alpha_2 = 1$, and $\tau = 0.5$. Perform computer simulations and describe the results.
(b) Change the delay to $\tau = 3.5$ and perform additional computer simulations. Describe the observed changes.

Exercise 7.19 (*Ikeda Model for Absorbing Model*) A nonlinear model for an absorbing medium with two level atoms in a ring cavity was proposed by Ikeda [18, 19] in 1979. Using Maxwell-Block equations, Ikeda derived the following model

$$\frac{dx}{dt} = -x(t) + \mu \sin\left[x(t-\tau) - x_0\right],$$

where μ and x_0 are constants.

(a) Calculate the equilibrium points.
(b) Write down the characteristic polynomial associated with the linearization about an equilibrium point.
(c) Study the stability of the equilibrium points by applying the appropriate theorems.
(d) Apply Nyquist criterion to determine the stability of each equilibrium point.
(e) Let $\mu = 20$, $x_0 = \pi/4$, and $\tau = 5$. Numerically solve the model and discuss the results.

Exercise 7.20 (*Allee Effect with Delay*) Recall the Alleė effect exercise from Chap. 3. A continuous version with delay is

$$\frac{dx}{dt} = x(t)[a + bx(t-\tau) - cx^2(t-\tau)],$$

where $x(t)$ is the population density at time t, and the term in bracket represents the per capita growth rate, a, b, and c are all positive real constants.

(a) Calculate the equilibrium points.
(b) Write down the characteristic polynomial associated with the linearization about an equilibrium point.
(c) Let $a = 1$, $b = 1$, and $c = 0.5$. Study the stability of the equilibrium points by applying the appropriate theorems. Determine the critical value of the delay, τ_c, that may lead to delay-induced oscillations.
(d) Repeat part (c) by applying the Nyquist criterion.

Exercise 7.21 (*Another Neural Network Model with two Delays*) The following system of delay differential equations serves as a model for a simple two-neuron network

$$\frac{du_1(t)}{dt} = -u_1(t) + a_1 \tanh\left[u_2(t - \tau_2)\right]$$
$$\frac{du_2(t)}{dt} = -u_2(t) + a_2 \tanh\left[u_1(t - \tau_1)\right].$$

(a) Compute the linearization of the model equations about the trivial equilibrium $(u_1, u_2) = (0, 0)$, and write the characteristic polynomial associated with it.
(b) If $a_1 a_2 < -1$, then there exists a critical value of the delay parameter, $\tau_c > 0$, such that when $0 < \tau = \tau_1 + \tau_2 < \tau_c$, the zero equilibrium solution is asymptotically stable and τ_c corresponds to a Hopf bifurcation point. Determine the critical value, τ_c, by analyzing the characteristic polynomial.
(c) Repeat part (b) using Nyquist criterion.
(d) Let $a_1 = -2$ and $a_2 = 1$. Perform computer simulations of the dynamics for $\tau < \tau_c$ and $\tau > \tau_c$ and verify the results obtained from the analysis.

References

1. Business Dynamics, *Systems Thinking and Modeling for a Complex World* (Irwin McGraw Hill, New York, 2000)
2. N. Minorsky, Self-excited oscillations in dynamical systems possessing retarded actions. J. Appl. Mech. **9**, A65–A71 (1942)
3. N. Minorsky, Experiments with activated tanks. Trans. ASME **69**, 735–747 (1947)
4. K.L. Cooke, Z. Grossman, Discrete delay, distributed delay and stability switches. J. Math. Anal. App. **86**, 592–627 (1982)
5. K. Ogata, *Modern Control Engineering* (5th edn., Prentice Hall, Boston, MA, 2010)
6. L.E. El'sgol'ts, S.B. Norkin, *Introduction to the Theory of Differential Equations with Deviating Arguments* (Academic Press, New York, NY, 1977)
7. H. Smith, *AnIntroduction to Delay Differential Equations with Applications to the Life Sciences*. Texts in Applied Mathematics (Springer, 2010)
8. F. Brauer, Absolute stability in delay equations. J. Diff. Eqns. **69**, 185–191 (1987)
9. S. Banerjee, *Mathematical Modeling* (CRC Press, 2014)
10. *Time Lags in Biological Models* (Springer, Heidelberg, 1978)
11. D. Lyons, Effects of delay on coupled sensor systems. Master's thesis, San Diego State University (2011)
12. K. Engelborghs, T. Luzyanina, G. Samaey, D. Roose, K. Verheyden, *DDE-BIFTOOL v.2.03 MATLAB Package for Bifurcation Analysis of Delay Differential Equations*, Belgium (2007)
13. S.A. Campbell, Calculating centre manifolds for delay differential equations using calculating centre manifolds for delay differential equations using maple, in *Delay Differential Equations: Recent Advances and New Directions*, New York, ed. by T. Kalmár-Nagy B. Balachandran, D. Gilsinn. (Springer, 2008)
14. G.E. Hutchinson, Circular causal systems in ecology. Ann. N.Y. Acad. Sci. **50**, 221–246 (1948)
15. L. Glass, M.C. Mackey, *From Clocks to Chaos* (Princenton University Press, 1988)
16. M. Wazėwszka-Czyzėwska, A. Lasuta, Mathematical problems of the dynamics of the red blood cell system. Ann. Polish Math. Soc. III Appl. Math. **31**, 23–40 (1976)
17. K. Gopalswamy, M.R.S. Kulenovic, G. Ladas, Time lags in a food-limited population model. Appl. Anal. **31**, 225–237 (1988)

18. K. Ikeda, Multiple valued stationary state and its instability of the transmitted light by aring cavity system. Opt. Commun. **30**(2), 257–261 (1979)
19. K. Ikeda, H. Daido, O. Akimoto, Optical turbulence: chaotic behavior of transmitted light from a ring cavity. Phys. Rev. Lett. **45**(9), 709–712 (1980)

Chapter 8
Spatial-Temporal Models

It is now time to study mathematical models that depend simultaneously on space and time. These *spatio-temporal models* serve to investigate phenomena such as reaction-diffusion processes; population dynamics that include diffusion of species; vibrations of a membrane; and pattern formation such as Turing patterns, which are driven by diffusion instabilities. We also use this chapter to introduce *Agent-based Models*. These type of models have gained much popularity in recent years. They are used to describe emergent behavior based on a set of rules that govern how multiple "agents" or units interact with one another. For instance, bubbles in fluidization processes can be treated as agents and the collective interactions can be described through a set of rules that determines the mutual interaction between bubbles. In all the examples, we develop quantitative and qualitative methods to study the conditions for the existence and stability of spatio-temporal solutions.

8.1 Reaction-Diffusion Models

Let $c(x, t)$ be the concentration of a species, e.g., cells, amount of chemicals, number of animals, or heat along a one-dimensional region or interval $I : x_0 < x < x_1$. Later on we will consider the case of higher dimensions. Let $J(x, t)$ represent the amount of material being transported, also known as the *flux*. According to Fick's first law of diffusion, the flux varies from regions of high concentration to regions of low concentration. The rate of variation is, in fact, proportional to the concentration gradient, i.e., spatial derivative. Mathematically, this means that

$$J \propto -\frac{\partial c}{\partial x}.$$

A. Palacios, *Mathematical Modeling*, Mathematical Engineering,
https://doi.org/10.1007/978-3-031-04729-9_8

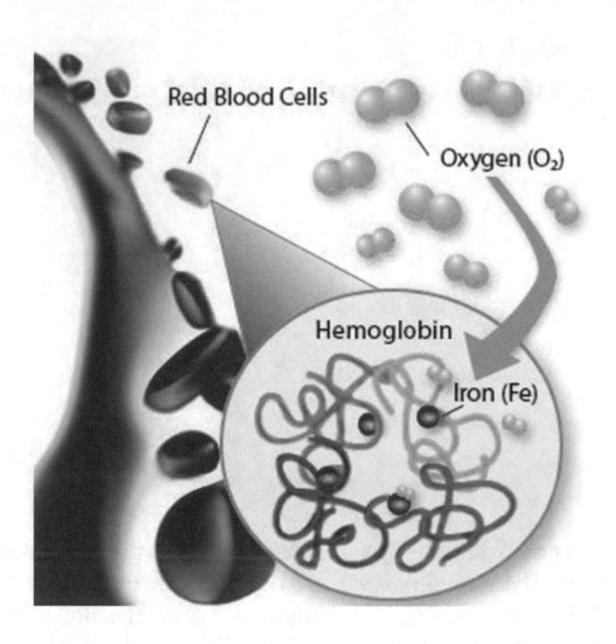

Fig. 8.1 Hemoglobin molecules in red blood cells. *Source* MedicineNet

In practice, the rate of variation of the flux depends on material properties. Thus, if we let $D = D(x, t)$ be the diffusion coefficient or diffusivity of a given material, i.e., a measure of how efficiently the particles in the material disperse from areas of high-density to regions of low-density, then the flux can be expressed as

$$J = -D\frac{\partial c}{\partial x}.$$

For references purposes, hemoglobin molecules in red blood cells have a diffusion coefficient of approximately, $D = 10^{-7}\,\text{cm}^2/\text{s}$, see Fig. 8.1. And oxygen in blood diffuses at a rate of $D = 10^{-5}\,\text{cm}^2/\text{s}$.

Assuming, for right now, the absence of external sources of material production, a *conservation equation* over the interval $I : x_0 < x < x_1$ can be established by balancing changes in material with changes in the flux, as follows:

$$\frac{\partial}{\partial t}\int_{x_0}^{x_1} c(x, t)dt = J(x_0, t) - J(x_1, t) = -\int_{x_0}^{x_1} \frac{\partial J}{\partial x}dx.$$

This equation indicates that the rate of change of the amount of material (left term) on the region I must be equal to the change in flux across I or rate of flow across the boundaries. Combining left- and right-hand sides we arrive at an *Integral Conservation Law*

$$\int_{x_0}^{x_1} \left(\frac{\partial c}{\partial t} + \frac{\partial J}{\partial x}\right) dx = 0. \tag{8.1}$$

This integral must be satisfied for arbitrary regions, yielding a *Conservation Law*

$$\frac{\partial c}{\partial t} + \frac{\partial J}{\partial x} = 0. \tag{8.2}$$

Applying Fick's first law of diffusion to the flux J we arrive at the *Diffusion Equation*

$$\frac{\partial c}{\partial t} = \frac{\partial}{\partial x}\left(D\frac{\partial c}{\partial x}\right). \tag{8.3}$$

In practice, it is quite common for the diffusion coefficient D to remain constant across the material. This particular case yields another common form a diffusion model

$$\frac{\partial c}{\partial t} = D\frac{\partial^2 c}{\partial x^2}. \tag{8.4}$$

Now, consider a 3D region V and let S be an arbitrary surface enclosing the volume of space V. This time we will assume, however, an external source of material represented by $f = f(c, x, t)$. Balancing again the changes in material with the flux and the external source of material, we obtain an equivalent conservation equation

$$\frac{\partial}{\partial t}\int_V c(x,t)dv = -\int_S J \cdot ds + \int_V f dv.$$

Assuming c to be a continuous function of x and t and applying the Divergence theorem, we get

$$\int_V \left[\frac{\partial c}{\partial t} + \nabla \cdot J - f(c,x,t)\right] dv = 0,$$

where

$$\nabla \cdot J = \frac{\partial}{\partial x}J_x + \frac{\partial}{\partial y}J_y + \frac{\partial}{\partial z}J_z.$$

Again, since the volume V is arbitrary, then the integrand in the above equation must be zero. We then arrive at a conservation equation for $c(x, t)$

$$\frac{\partial c}{\partial t} + \nabla \cdot J = f(c,x,t). \tag{8.5}$$

In 3D, the gradient of the concentration can be written as ∇c. Then, applying Fick's first law of diffusion so that $J = -D\nabla c$, yields the first version of a *Reaction-Diffusion Model*

$$\frac{\partial c}{\partial t} = f + \nabla \cdot (D \nabla c), \tag{8.6}$$

in which $D = D(x, c)$. A space-dependent diffusion coefficient is quite common in biomedical applications. For instance, in diffusion of genetically engineered organisms in heterogenous environments. It is also found in studies of the effects of white and grey matter in the growth and spread of brain tumors.

For a constant diffusion rate D we get the common form of a reaction-diffusion model

$$\frac{\partial c}{\partial t} = f + D \nabla^2 c, \tag{8.7}$$

where $\nabla^2 c$ is the Lapacian of the concentration, i.e.,

$$\nabla^2 c = \frac{\partial^2 c}{\partial x^2} + \frac{\partial^2 c}{\partial y^2} + \frac{\partial^2 c}{\partial z^2}.$$

8.1.1 Logistic Growth with Diffusion

In this section we revisit the logistic growth model with the addition of diffusion. We show, computationally and analytically, that diffusion leads to solutions in the form of traveling wave patterns.

Example 8.1 (*Logistic Population Growth*) Assuming a constant diffusion rate of a "population" $u(x, t)$, we can recast the Logistic Growth model of earlier chapters in the form of a reaction-diffusion model

$$\frac{\partial c}{\partial t} = rc\left(1 - \frac{c}{K}\right) + D\,\nabla^2 c. \tag{8.8}$$

Observe that the reaction term is the standard logistic growth model with r representing the linear reproduction rate and k the carrying capacity of the environment. In this case, it is more appropriate to interpret c as the population density instead of concentration.

We know from Chap. 4 that a solution to the logistic model (8.8) without diffusion, i.e., $D = 0$, is given by

$$c(t) = \frac{c_0 K e^{rt}}{K + c_0(e^{rt} - 1)},$$

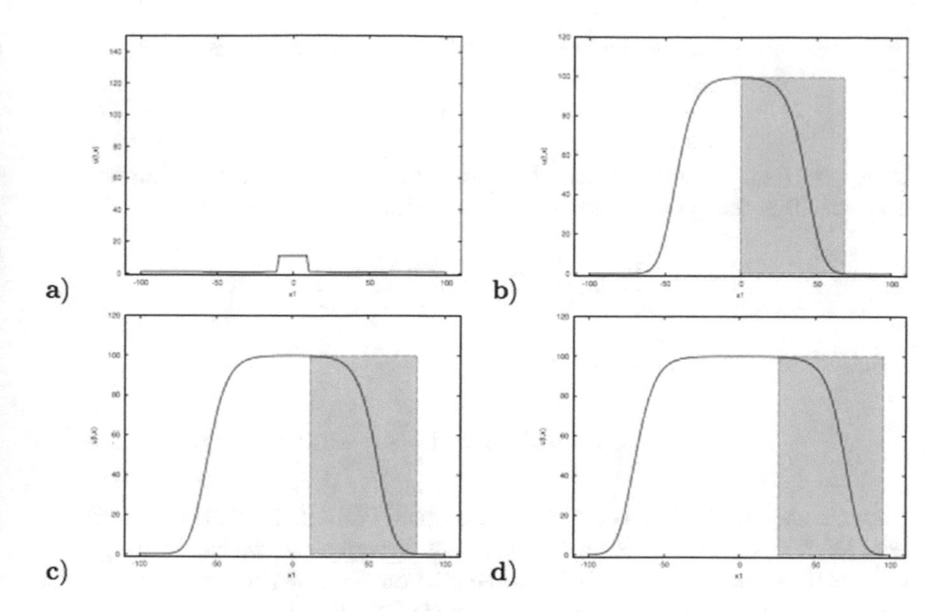

Fig. 8.2 Traveling wave solutions in the logistic growth model with diffusion. Parameters are: $D = 10$, $r = 1.0$. *Source* Kandler and Unger [1]

where $c(t = 0) = c_0$. When diffusion is added, numerical simulations of the model (8.8) show [1], see Fig. 8.2, the existence of traveling wave solutions.

The simulations reveal that for an initial small population, especially smaller than the carrying capacity, the population can grow until the carrying capacity of the system is reached. This is reminiscent of the behavior without diffusion. But then, once the carrying capacity is reached, the population saturates spatially and starts to decrease towards a lower density while forming a spatial wave form. As time evolves, the wave travel spatially. The decay and speed of the wave are both proportional to the diffusivity coefficient D.

We now seek an analytic solution to the traveling wave patterns. But, first, it is convenient to re-write the model equation in dimensionless form by letting

$$u = \frac{c}{K}, \qquad \tau = rt, \qquad \tilde{x} = \sqrt{\frac{r}{D}}x.$$

Substituting into (8.8) and simplifying (with relabeling $\tilde{x}$ as x), yields a dimensionless version of the Logistic growth model with diffusion

$$\boxed{\frac{\partial u}{\partial t} = u(1 - u) + D\,\nabla^2 u.} \tag{8.9}$$

The traveling wave solution we are looking for can be written as

$$u(x,t) = U(x - st) = U(z),$$

where $s > 0$ is the assumed speed of the wave and $z = x - st$ is a moving frame of reference. This change of variables leads to the following relations

$$\frac{\partial u}{\partial t} = -s\frac{dU}{dz}, \qquad \frac{\partial^2 u}{\partial x^2} = \frac{d^2U}{dz^2}.$$

Substituting into (8.9) we arrive at

$$\frac{d^2U}{dz^2} + s\frac{dU}{dz} + U(1-U) = 0. \tag{8.10}$$

In this way, we have thus reduced the original PDE model (8.9) into a second-order ODE model. The ODE is nonlinear but, nevertheless, we can investigate its behavior using the techniques we learned earlier on. For instance, we can convert the second-order model into a first-order system of ODEs by letting $V = U'$, this yields

$$\begin{aligned} U' &= V \\ V' &= -sV - U(1-U). \end{aligned} \tag{8.11}$$

Two equilibrium points are found along the U axis: $(U, V) = (0, 0)$ and $(U, V) = (1, 0)$. The Jacobian matrix is

$$J = \begin{bmatrix} 0 & 1 \\ -1+2U & -s \end{bmatrix}.$$

At $(U, V) = (0, 0)$, the eigenvalues for J are

$$-\frac{s}{2} \pm \frac{1}{2}\sqrt{s^2 - 4},$$

while at $(U, V) = (1, 0)$, the eigenvalues are

$$-\frac{s}{2} \pm \frac{1}{2}\sqrt{s^2 + 4}.$$

Consequently, it follows that $(0, 0)$ is a stable spiral sink for $s < 2$ or a stable node for $s \geq 2$. The other equilibrium point, $(1, 0)$, is always an unstable saddle for all values of $s > 0$. Figure 8.3 shows the corresponding phase portraits for $0 < s < 2$ and $s \geq 2$.

A stable spiral at $(0, 0)$ implies that $U(t)$ oscillates periodically as is shown in Fig. 8.4(left). But from a biological standpoint, this solution makes no sense because

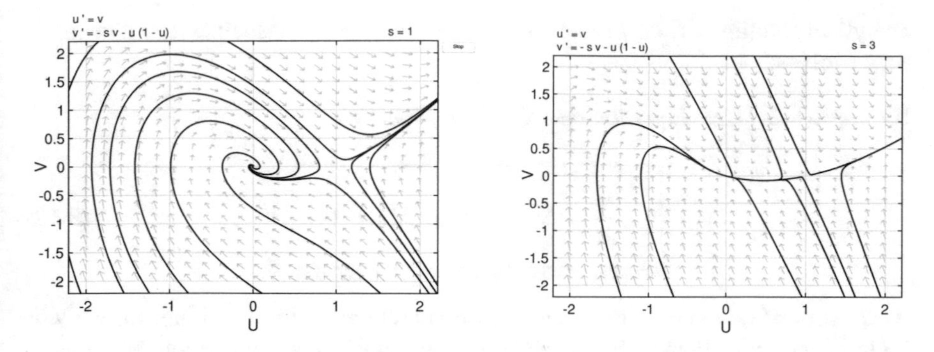

Fig. 8.3 Phase space solutions of Eq. (8.11) with (left) $s = 1$ and (right) $s = 3$

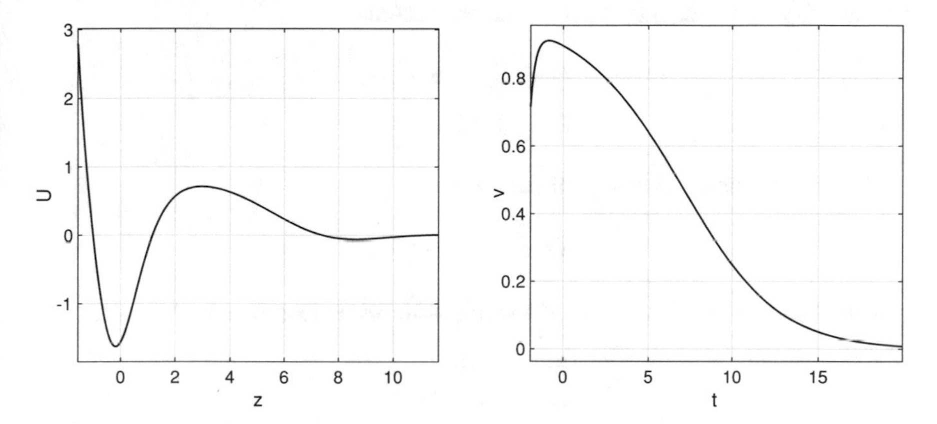

Fig. 8.4 Time series solutions of Eq. (8.11) with (left) $s = 1$ and (right) $s = 3$

as $U(z)$ oscillates it acquires negative values which are inconsistent with population sizes. Consequently, this solution has to be discarded.

Now, for $s \geq 2$, $(0, 0)$ is also stable but of a node type, i.e., there are no oscillations into negative values of U. Instead, the velocity field satisfies $V = du/dz < 0$, which shows the decay in the traveling wave profile as $z \to \infty$. This is now shown in Fig. 8.4(right). Consequently, we conclude that $s \geq 2$ is a necessary condition for the existence of a traveling wave solution. Finding an exact analytical solution to this problem is beyond the scope of the present book.

8.1.2 Heat Equation on Circular Domain

A second example is a model that governs the evolution of heat over a circular domain. The model is the Heat equation. In this problem, we apply some basic techniques from the field of Partial Differential Equations (PDEs) to obtain complete analytical solutions. Then we use computer simulations to visualize the behavior as time evolves.

Example 8.2 (*Heat Equation on a Circular Domain*)

$$\boxed{\frac{\partial u}{\partial t} = \nabla^2 u, \quad \mathbf{x} \in \Omega,\ t > 0,} \tag{8.12}$$

where $u = u(\mathbf{x}, t)$ and Ω represents some subregion within $\mathbb{R}^3$. Later on we will focus on a circular domain but for right now we consider any subregion. We assume zero boundary conditions along the boundary $\partial\Omega$ and an initial (valid at $t = 0$) spatial profile $f(x)$. Together, these two conditions become

$$\begin{aligned} u(\mathbf{x}, t) &= 0, \qquad \mathbf{x} \in \partial\Omega, \\ u(\mathbf{x}, 0) &= f(\mathbf{x}), \qquad \mathbf{x} \in \Omega. \end{aligned}$$

We seek a solution to $u(\mathbf{x}, t)$ of the form

$$u(\mathbf{x}, t) = T(t)X(\mathbf{x}).$$

Direct substitution into Eq. (8.12) yields, after some rearrangements

$$\frac{T'}{T} = \frac{\nabla^2 X}{X} = -\lambda.$$

The first two terms arise from separating the two variables X and T. The last term arises from the fact that two functions of two independent variables, X and T in this case, can only be the same, at every value of the respective variables, if they are equal and constant. Without loss of generality, the constant is assumed to be $-\lambda$ but the same results can be obtained with λ.

The zero boundary condition leads to

$$X(\mathbf{x}) = 0, \quad \mathbf{x} \in \partial\Omega.$$

To solve for $X(x)$ we must solve the following Sturm–Liouville problem

$$\begin{aligned} \nabla^2 X &= -\lambda X, \qquad \mathbf{x} \in \Omega \\ X(\mathbf{x}) &= 0, \qquad \mathbf{x} \in \partial\Omega. \end{aligned} \tag{8.13}$$

So far we have assumed any subregion $\Omega \in \mathbb{R}^3$, while the separation of variables and Sturm–Liouville problem are valid for any general subregion. We now wish to consider a 2D circular domain

$$\Omega = \{(x, y) : \; x^2 + y^2 \leq 1\}.$$

Figure 8.5 shows a polar grid over the circular domain.
Letting $X(\mathbf{x}) = v(x, y)$, the Sturm–Liouville problem can be written as

$$\begin{aligned} \frac{\partial^2 v}{\partial x^2} + \frac{\partial^2 v}{\partial y^2} &= -\lambda v, \quad (x, y) \in \Omega \\ v(x, y) &= 0, \qquad x^2 + y^2 = 1. \end{aligned}$$

For convenience and consistency with the circular domain of the PDE, it is better to work on polar coordinates: $x = r\cos\theta$, $y = r\sin\theta$, where $0 \leq r \leq 1$ and $-\pi \leq \theta < \pi$. Relabeling $v = v(r, \theta)$, direct substitution and simplification yields the Sturm–Liouville problem in polar coordinates

$$\begin{aligned} \frac{1}{r}\frac{\partial}{\partial r}\left(r\frac{\partial v}{\partial r}\right) + \frac{1}{r^2}\frac{\partial^2 v}{\partial \theta^2} &= -\lambda v, \\ v(1, \theta) = 0, \qquad -\pi \leq \theta < \pi \end{aligned} \tag{8.14}$$

We now apply separation of variables again on the spatial components (r, θ) by setting: $v(r, \theta) = R(r)\,\Psi(\theta)$, which yields (after separation and simplification):

$$r\frac{d}{dr}\left(r\frac{dR}{dr}\right)\frac{1}{R(r)} + \lambda r^2 = -\frac{d^2\Psi}{d\theta^2}\frac{1}{\Psi(\theta)} = \mu,$$

and the boundary condition becomes $v(1, \theta) = R(1)\Psi(\theta) = 0$. Since $\Psi(\theta) \neq 0$ is required for nontrivial solutions $v(r, \theta)$ to exist, then $R(1) = 0$. In addition, continuity of $v(r, \theta)$ and its derivative along the azimuthal direction θ require

$$v(r, -\pi) = v(r, \pi), \qquad v_\theta(r, -\pi) = v(r, \pi).$$

These two smoothness conditions translate into

$$\Psi(-\pi) = \Psi(\pi), \qquad \frac{d\Psi}{d\theta}(-\pi) = \frac{d\Psi}{d\theta}(\pi).$$

And since we are assuming $v(x, y)$ to be bounded on Ω, then $R(r)$ must also be bounded in $0 \leq r \leq 1$. To summarize, to solve the Sturm–Liouville problem in polar coordinates we must solve

$$\frac{d^2\Psi}{d\theta^2} + \mu\Psi(\theta) = 0, \tag{8.15a}$$

$$r\frac{d}{dr}\left(r\frac{dR}{dr}\right)\frac{1}{R(r)} + \lambda r^2 = \mu, \tag{8.15b}$$

with boundary conditions

$$\Psi(-\pi) = \Psi(\pi), \quad \Psi_\theta(-\pi) = \Psi_\theta(\pi).$$

We start by solving Eq. (8.15a) for $\Psi(\theta)$. We must consider three cases.

Case I. If $\mu < 0$ then the roots of the characteristic polynomial associated with Eq. (8.15a) are real-valued, $\pm\sqrt{-\mu}$. This yields a possible solution

$$\Psi(\theta) = C_1 e^{\sqrt{-\mu}\theta} + C_2 e^{-\sqrt{-\mu}\theta},$$

where C_1 and C_2 are arbitrary constants. Applying the boundary conditions $\Psi(-\pi) = \Psi(\pi)$ and $\Psi_\theta(-\pi) = \Psi_\theta(\pi)$ yields $C_1 = C_2 = 0$. Thus, only the trivial solution exists.

Case II. If $\mu = 0$ then a solution to Ψ is of the form $\Psi(\theta) = C_1\theta + C_2$. Boundary conditions imply $C_1 = 0$, so $\Psi(\theta) = C$, where C is just an arbitrary constant.

Case III. The case of $\mu > 0$ leads to a solution is of the form

$$\Psi_m(\theta) = a_m \cos(m\theta) + b_m \sin(m\theta),$$

where $\mu = m^2$, $m = 0, 1, 2, \ldots$, a_m and b_m are constant coefficients. Details of the derivation of this solution re left as an exercise.

Next, we solve Eq. (8.15b), but first rewrite it as

$$r^2 \frac{d^2 R_m}{dr^2} + r \frac{dR_m}{dr} + (\lambda r^2 - m^2) R_m = 0, \qquad R_m(1) = 0, \qquad |R_m(0)| < \infty.$$

This last equation is not exactly a Sturm–Liouville problem due to the boundary condition at the origin $r = 0$, i.e., due to $|R_m(0)| < \infty$. Instead, it can be treated as a "singular" version of a Sturm–Liouville problem. In particular, it can be shown that the boundary condition $X(\mathbf{x}) = 0$ for $\mathbf{x} \in \partial\Omega$ associated with the equation $\nabla^2 X = -\lambda X$, restricts $\lambda > 0$. Then, we can re-scale by $s = \sqrt{\lambda} r$, to arrive at the **Bessel's differential equation of order** m

$$\boxed{s^2 \frac{d^2 R_m}{ds^2} + s \frac{dR_m}{ds} + (s^2 - m^2) R_m = 0.} \tag{8.16}$$

The general solution of Eq. (8.16) is

$$R_m(s) = c_{m1} J_m(s) + c_{m2} Y_m(s),$$

where $J_m(s)$ and $Y_m(s)$ are known, respectively, as the Bessel functions of the first and second kinds [2]. Now, since $Y_m(s)$ is unbounded at $s = 0$ then $c_{m2} = 0$ in order for the boundedness condition $|R_m(0)| < \infty$ to be satisfied. Thus, the solution for R_m simplifies to

$$R_m(s) = c_m J_m(s).$$

Additionally, the condition $R(1) = 0$ implies

$$J_m(\sqrt{\lambda}) = 0,$$

which, in turn, yields the eigenvalues: $\lambda_{mn} = j_{mn}^2$, where j_{mn} represents the nth zero of $J_m(s)$. Substituting the solutions for R_m and Ψ_m in $v(r, \theta) = R(r)\Psi(\theta)$, we get

$$v_{mn}(r, \theta) = J_m(j_{mn}r)\,(a_m \cos(m\theta) + b_m \sin(m\theta)) .$$

As a final step, we must put together the solution for $u(x, y, t)$ by considering the differential equation for $T(t)$:

$$T' = -\lambda T.$$

The solution is straightforward:

$$T_n(t) = T_n(0)e^{-\lambda_n t}.$$

Then the solution for $u = u(x, y, t) = T(t)X(x)$ can be expressed as

$$u(x, y, t) = e^{-\lambda_{mn}t} v_{mn}(x, y) = e^{-j_{mn}^2 t} v_{mn}(x, y).$$

The linear superposition of all individual *modes*, i.e., solutions for each combination of m and n, yields the final solution:

$$u(x, y, t) = \sum_{m=1}^{m=\infty} \sum_{n=1}^{n=\infty} e^{-\lambda_{mn}t} J_m(j_{mn}r)\,(a_m \cos(m\theta) + b_m \sin(m\theta)) . \tag{8.17}$$

Figure 8.6 shows a few snapshots of the solution of the 3D Heat equation, reconstructed using Eq. (8.17), at various values of time.

8.1.3 Vibrating Membrane on a Rectangular Domain

In this third example, we consider a model of the vibrations of a membrane on a rectangular domain.

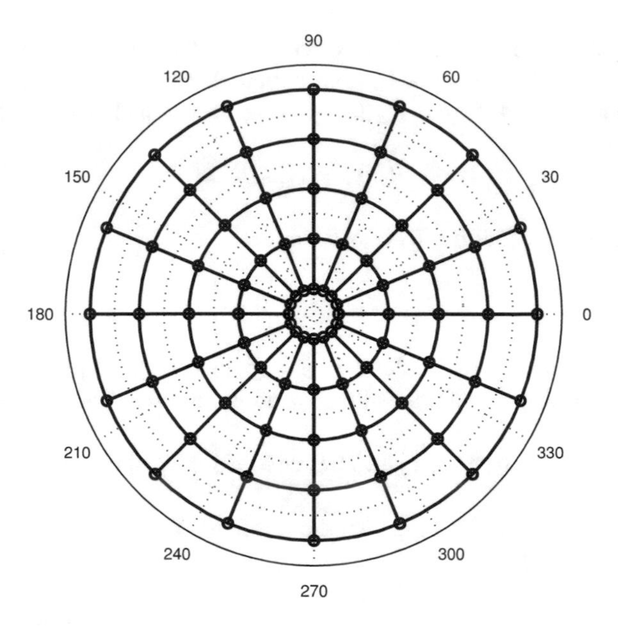

Fig. 8.5 Polar grid for a 2D circular domain

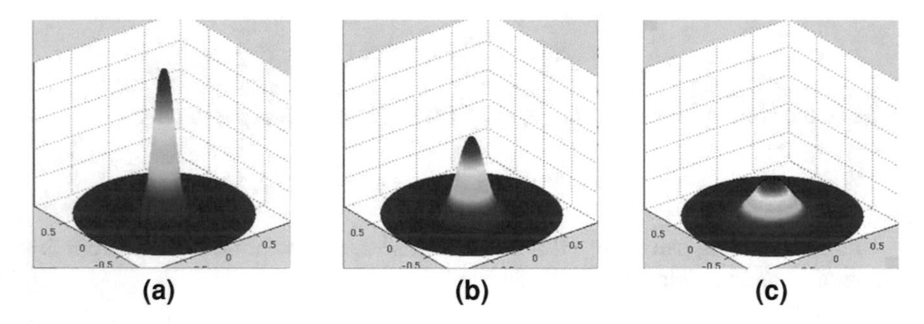

Fig. 8.6 Simulations of the solutions of the 3D Heat equation over a polar grid

Example 8.3 (*Vibrating Membrane*) We now consider the wave equation for modeling and simulating the vibrations of a membrane over a rectangular domain through

$$\boxed{\frac{\partial^2 u}{\partial t^2} = \nabla^2 u, \quad \mathbf{x} \in \Omega,\ t > 0,} \tag{8.18}$$

We assume the following boundary conditions:

$$\begin{aligned} u(\mathbf{x}, t) &= 0, && \mathbf{x} \in \partial\Omega, \\ u(\mathbf{x}, 0) &= f(\mathbf{x}), && \mathbf{x} \in \Omega, \\ u_t(\mathbf{x}, 0) &= g(\mathbf{x}), && \mathbf{x} \in \Omega. \end{aligned}$$

We seek a solution to $u(\mathbf{x}, t)$ of the form

$$u(\mathbf{x}, t) = T(t)X(\mathbf{x}).$$

Direct substitution into Eq. (8.12) yields, after some rearrangements

$$\frac{T''}{T} = \frac{\nabla^2 X}{X} = -\lambda.$$

The first two terms arise from separating the two variables X and T. The last term arises from the fact that two functions of two independent variables, X and T in this case, can only be the same, at every value of the respective variables, if they are equal and constant. Without loss of generality, the constant is assumed to be $-\lambda$ but the same results can be obtained with λ.

The zero boundary condition leads to

$$X(\mathbf{x}) = 0, \quad \mathbf{x} \in \partial\Omega.$$

To solve for $X(x)$ we must solve the following Sturm–Liouville problem

$$\begin{aligned} \nabla^2 X &= -\lambda X, && \mathbf{x} \in \Omega \\ X(\mathbf{x}) &= 0, && \mathbf{x} \in \partial\Omega. \end{aligned} \tag{8.19}$$

Observe that this is almost the same Sturm–Liouville problem as that of Eq. (8.13), which was derived from the Heat equation. The only difference is in the set of boundary conditions.

We now wish to consider a 2D rectangular domain

$$\Omega = \{(x, y) : \quad 0 \le x < x_0, 0 \le y < y_0\}.$$

Letting $X(\mathbf{x}) = v(x, y)$, the Sturm–Liouville problem can be written as

$$\begin{aligned} \frac{\partial^2 v}{\partial x^2} + \frac{\partial^2 v}{\partial y^2} &= -\lambda v, && (x, y) \in \Omega \\ v(0, y) = v(x_0, y) &= 0, && 0 \le y \le y_0 \\ v(x, 0) = v(x, y_0) &= 0, && 0 \le x \le x_0. \end{aligned} \tag{8.20}$$

Let $v(x, y) = X(x)Y(y)$. Substituting into Eq. (8.19) and separating variables yields

$$\frac{Y''}{Y} + \lambda = -\frac{X''}{X} = \mu.$$

The following step is left as an exercise. Solving for $X(x)$ and $Y(y)$ leads to

$$\begin{aligned} X_m(x) &= a_m \sin\left(\frac{m\pi x}{x_0}\right), \qquad \mu_m = \left(\frac{m\pi}{x_0}\right)^2, \quad m = 1, 2, 3, \ldots \\ Y_n(y) &= b_n \sin\left(\frac{n\pi y}{y_0}\right), \qquad \nu_n = \left(\frac{n\pi}{x_0}\right)^2, \quad n = 1, 2, 3, \ldots \end{aligned}$$

Combining these two solutions we can write an expression for $v(x, y) = X(x)\,Y(y)$:

$$v_{mn}(x, y) = c_{mn} \sin\left(\frac{m\pi x}{x_0}\right) \sin\left(\frac{b\pi y}{y_0}\right), \qquad m, n = 1, 2, 3, \ldots,$$

and the eigenvalues are

$$\lambda_{mn} = \mu_m + \nu_n = \pi^2 \left(\frac{m^2}{x_0^2} + \frac{n^2}{y_0^2}\right).$$

Next we must solve the temporal part of the differential equation

$$T'' + \lambda T = 0.$$

Since we already know that $\lambda > 0$ then the solution can be found immediately as

$$T_n(t) = \alpha_n \cos\left(\sqrt{\lambda} t\right) + \beta_n \sin\left(\sqrt{\lambda} t\right).$$

A linear superposition of all individual modes yields the final solution

$$u(x, y, t) = \sum_{m=1}^{m=\infty} \sum_{n=1}^{n=\infty} \left[\alpha_{nm} \cos\left(\sqrt{\lambda} t\right) + \beta_{nm} \sin\left(\sqrt{\lambda} t\right)\right] \sin\left(\frac{m\pi x}{x_0}\right) \sin\left(\frac{n\pi y}{y_0}\right). \quad (8.21)$$

The actual values of α_{mn} and β_{mn} are found using the boundary conditions. This task is left as an exercise. Figure 8.7 shows a few snapshots of the solution of the wave equation, reconstructed using Eq. (8.21), at various values of time.

8.1.4 Generalization to Higher Dimensions

The generalization of the reaction-diffusion model to higher dimensions is straightforward. Assume we have multiple interacting species $u_1, \ldots u_m$, each of them dependent on the same spatial domain $\mathbf{x}$. We can form a vector of densities or concentrations $\vec{u}$,

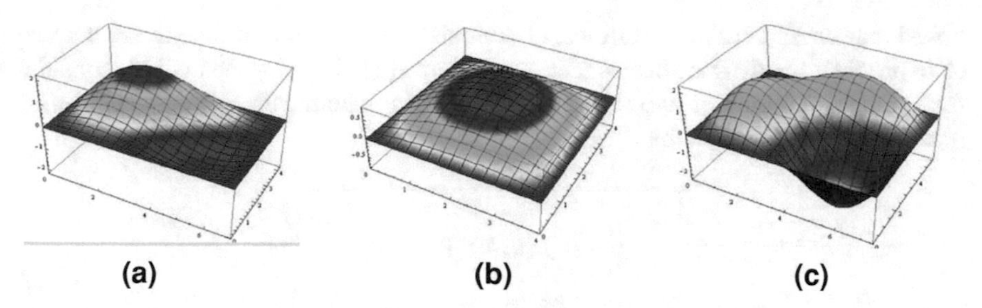

Fig. 8.7 Vibrating membrane on a rectangular domain. *Source* Wolfram alpha

$$\vec{u} = \begin{bmatrix} u_1(\mathbf{x}, t) \\ u_2(\mathbf{x}, t) \\ \vdots \\ u_m(\mathbf{x}, t) \end{bmatrix},$$

where each species diffuses with its own coefficient D_{ii} and diffusion across species i and j is controlled by the coefficient D_{ij}. Then the diffusion matrix D takes the form

$$D = \begin{bmatrix} D_{11} & D_{12} & \dots D_{1m} \\ D_{21} & D_{22} & \dots D_{2m} \\ \vdots & & \vdots \\ D_{m1} & D_{m2} & \dots D_{mm} \end{bmatrix}.$$

Finally, we can write the reaction-diffusion model in higher dimensions as follows

$$\boxed{\frac{\partial \vec{u}}{\partial t} = \vec{f} + \nabla \cdot (D \nabla \vec{u}).} \tag{8.22}$$

We can now consider a two-species reaction diffusion model which can lead, under certain conditions, to the formation of very interesting patterns.

8.2 Turing Patterns

Alan Turing, the British mathematician who became famous for braking German cipher messages encoded by the Enigma machine, was also a philosopher and a well-respected mathematician with strong contributions to the field of pattern forming systems. In 1952, he wrote a seminal manuscript "The Chemical Basis of Morphogenesis" [3], in which he suggested a rather novel idea for its time. Turing proposed that under certain conditions, chemicals can react and diffuse in such a way as to produce steady-state heterogeneous patterns [3]. To get some insight, consider, for instance, a model of two reacting and diffusing chemicals, with concentrations $u(\mathbf{x}, t)$ and $v(\mathbf{x}, t)$, given by

$$\begin{aligned} \frac{\partial u}{\partial t} &= f(u, v) + D_u \nabla^2 u \\ \frac{\partial v}{\partial t} &= g(u, v) + D_v \nabla^2 v, \end{aligned} \tag{8.23}$$

where $\mathbf{x} \in \mathbb{R}^2$ or $\mathbf{x} \in \mathbb{R}^3$. If the diffusion coefficients are identical (including zero), the concentrations u and v are expected to tend to a linearly stable uniform steady-state. Turing argued that, under certain conditions, the chemicals, which Turing called *morphogens*, can, however, form spatially inhomogeneous patterns if they evolve or diffuse at different rates, i.e., when $D_u \neq D_v$. If one of the concentrations acts as an *activator* while the other acts as *inhibitor* then the mutual interplay between the two can lead to nonuniform patterns, specially if they do not evolve or change at the same pace.

Experimental validation of Turing patterns remained elusive for almost sixty years. In 2008, two biologists at Nagoya University, Akiko Nakamsu and Shigeru Kondo, reported the first experimental evidence of Turing patterns in zebra fish. Since the markings on these type of fish develop from juvenile spots, Nakamsu and Kondo used lasers to scar the spots of juveniles and then watch how they change or evolve over time. Their observations, see Fig. 8.8, matched very well with predictions of computer simulations of related reaction-diffusion models.

At the time, the suggestion of diffusion causing patterns was considered to be a revolutionary and novel proposition because diffusion was considered to be a stabilizing process. Nowadays, this phenomenon is known as **diffusion driven instability** and it's the underlying mechanism of the many patterns that can be observed throughout nature, see Fig. 8.9.

8.2.1 Diffusion-Driven Instability

In this section we study, mathematically, the underlying mechanism that selects one pattern over another in a diffusion-driven stability. We consider the model (8.23), rewritten in vector form:

$$\frac{\partial \mathbf{w}}{\partial t} = F(\mathbf{w}, \mu) + D\nabla^2 \mathbf{w} \tag{8.24}$$

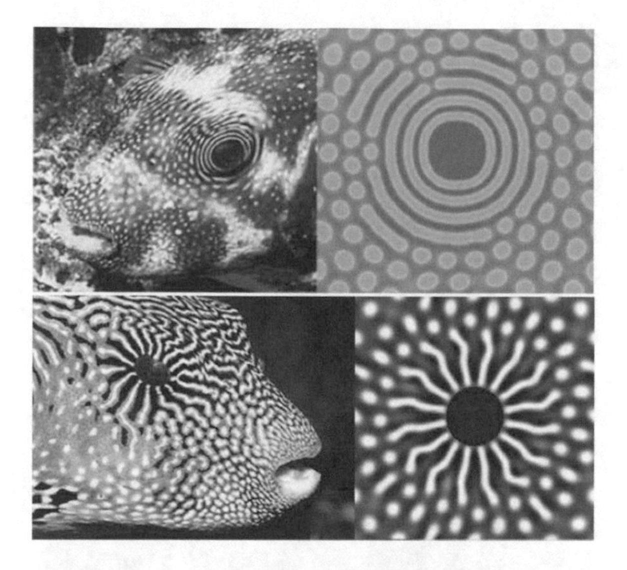

Fig. 8.8 Diffusion drive instability as an underlying mechanism for patterns found in nature. *Source* https://www.wired.com

Fig. 8.9 Turing patterns appear throughout nature. *Source* https://www.wired.com

where $\mathbf{w} = (u, v)$, $F = (f(u, v), g(u, v))$, $D = \text{diag}(D_u, D_v)$, and μ represents a vector of parameters.

Let $\mathbf{w_0} = (u_0, v_0)$ be a homogeneous equilibrium solution, so that

$$\begin{aligned} f(u_0, v_0) + D_u \nabla^2 u_0 &= 0 \\ g(u_0, v_0) + D_u \nabla^2 v_0 &= 0. \end{aligned}$$

Also, let $\mathbf{w} = \mathbf{w_0} + \delta\mathbf{w}$, where $\delta\mathbf{w}$ is a spatio-temporal perturbation given by

$$\delta\mathbf{w} = \sum_j c_j e^{\sigma_j t} e^{ik_j \cdot \mathbf{x}}$$

Substituting into linearized system about $\mathbf{w_0} = (u_0, v_0)$ yields:

$$(J - k_j^2 D - \sigma_j I)\mathbf{w} = 0, \tag{8.25}$$

where $k_j^2 = \vec{k}_j \cdot \vec{k}_j$ and

$$J = \begin{pmatrix} f_u & f_v \\ g_u & g_v \end{pmatrix}_{(u_0,v_0)}.$$

Solving Eq. (8.25) yields:

$$\sigma^2 + ((D_u + D_v)k^2 - f_u - g_v)\sigma + D_u D_v k^4 - (D_v f_u + D_u g_v)k^2 + f_u g_v - f_v g_u = 0. \tag{8.26}$$

8.2.2 Pattern Selection Mechanism

$\sigma(k)$ predicts the growing wave modes: $We^{i\vec{k}\cdot\vec{r}}e^{\sigma(k)t}$. That is, the steady-state (u_0, v_0) is stable if both eigenvalues $\sigma_{1,2}$ of Eq. (8.26) have negative real parts, i.e., if $\Re\{\sigma_{1,2}\} < 0$. But if both eigenvalues satisfy $\Re\{\sigma_{1,2}\} > 0$ then spatial modes with wave numbers k will grow exponentially until the nonlinearities in the reaction kinetics bound this growth.

To obtain the critical wave number at which the instability occurs, we first start with the assumption that in the absence of diffusion the steady-state (u_0, v_0) must be stable. So we must impose that $\Re\{\sigma_{1,2}\} < 0$. Then, if $D_u = D_v = 0$, Eq. (8.26) reduces to

$$\sigma^2 - (f_u + g_v)\sigma + f_u g_v - f_v g_u = 0.$$

From Chap. 4 we know $\Re\{\sigma_{1,2}\} < 0$ if the following two conditions are satisfied

$$\text{trace}(J) = f_u + g_v < 0, \qquad \text{and} \qquad \det(J) = f_u g_v - f_v g_u > 0.$$

Thus these two conditions will guarantee the homogenous steady-state (u_0, v_0) to be linearly stable, and they will restrict the region of parameter space where we look for solutions of Eq. (8.26). Now, in the full model with diffusion, $\Re\{\sigma_{1,2}\} > 0$ in Eq. (8.26) if either

$$(D_u + D_v)k^2 - f_u - g_v < 0, \quad \text{or} \quad D_u D_v k^4 - (D_v f_u + D_u g_v)k^2 + f_u g_v - f_v g_u < 0.$$

The first option cannot happen since $D_u, D_v > 0$, so $(D_u + D_v)k^2 > 0$, and we just imposed $f_u + g_v < 0$. Thus, the only possibility is for second option, which we rewrite as

$$p(s) = D_u D_v s^2 - (D_v f_u + D_u g_v)s + f_u g_v - f_v g_u,$$

where $s = k^2$. Direct computations of $p'(s)$ and solving $p'(s) = 0$ yields

$$k_c^2 = \frac{1}{2}\left(\frac{f_u}{D_u} + \frac{g_v}{D_v}\right).$$

Direct computations show $p''(s) > 0$, so $p(s)$ is concave up. Its minimum is

$$p_{min} = p(k_c^2) = |J| - \frac{1}{4D_u D_v}(g_v D_u + f_u D_v)^2.$$

The instability occurs at the bifurcation point (D_u^c, D_v^c) at which

$$\frac{1}{4D_u^c D_v^c}(g_v D_u^c + f_u D_v^c)^2 = |J|,$$

so that $p_{min} = 0$. We can the re-write the critical wave number as

$$\mathbf{k_c^2} = \frac{D_v^c f_u + D_u^c g_v}{2D_u^c D_v^c} = \sqrt{\frac{|J|}{D_u^c D_v^c}} = \sqrt{\frac{f_u g_v - f_v g_u}{D_u^c D_v^c}}$$

Thus, we have arrived at the following theorem.

Theorem 8.1 (Turing Instability) *Consider a reaction-diffusion model of the form*

$$\frac{\partial u}{\partial t} = f(u, v) + D_u \nabla^2 u$$
$$\frac{\partial v}{\partial t} = g(u, v) + D_v \nabla^2 v.$$

Let (u_0, v_0) be a steady-state, homogeneous, solution. The conditions for the formation of a spatio-temporal pattern via Turing instabilities are:

$$f_u + g_v < 0, \tag{8.27}$$
$$f_u g_v - f_v g_u > 0, \tag{8.28}$$
$$D_v f_u + D_u g_v > 0, \tag{8.29}$$
$$(D_v f_u + D_u g_v)^2 - 4D_u D_v(f_u g_v - f_v g_u) > 0, \tag{8.30}$$

where all partial derivatives are evaluated at the steady-state (u_0, v_0). If these conditions are satisfied then the reaction-diffusion model will undergo a bifurcation at the critical point (D_u^c, D_v^c), which satisfies

$$\frac{1}{4D_u^c D_v^c}(g_v D_u^c + f_u D_v^c)^2 = f_u g_v - f_v g_u,$$

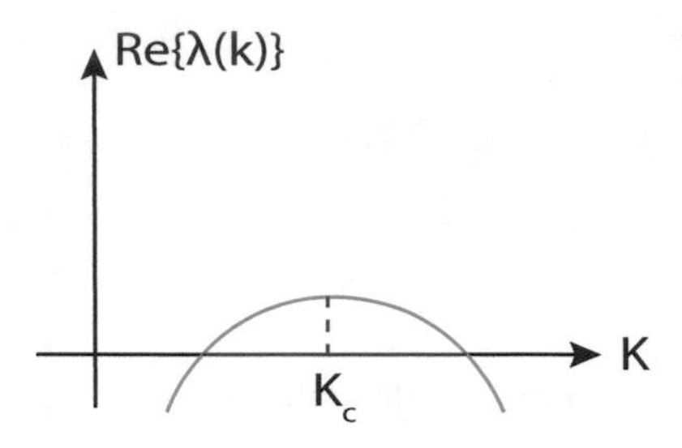

Fig. 8.10 Dispersion relation

to a non-homogenous pattern with wave form given by $\Phi_k = e^{ik_c \cdot \mathbf{x}}$*. The function* Φ_k *are also known as the eigenfunctions or eigenmodes.*

Figure 8.10 illustrates the relation between the sign of the real part of the eigenvalues as a function of the wave vector **k**. This relation is known as the **dispersion relation**. It shows that at the bifurcation point, i.e, when

$$Re\{\sigma(k_c)\} = 0,$$

a spatio-temporal pattern emerges in which the spatial mode that makes up the pattern has a wave vector $\mathbf{k_c}$. In other words, the dispersion relation illustrates the selection mechanism that leads to one particular pattern over many others.

8.3 The Brusselator Model with Diffusion

In Chap. 5 we introduced the Brusselator model for describing an autocatalytic reaction through a system of Ordinary Differential Equations (5.9). If we now consider the effects of the spatial diffusion of the concentrations, the model equations can be cast as a system of Partial Differential Equations (PDEs) of the form:

$$\begin{aligned} \frac{\partial u}{\partial t} &= \kappa_1 \nabla^2 u + (B-1)u + A^2 v - \eta u^3 - \nu_1 (\nabla u)^2 \\ \frac{\partial v}{\partial t} &= \kappa_2 \nabla^2 v - Bu - A^2 v - \eta v^3 - \nu_2 (\nabla v)^2. \end{aligned} \tag{8.31}$$

In this revised form, the model Eq. (8.31) describes the evolution of two coupled, diffusive spatiotemporal fields $u(\mathbf{x}, t)$ and $v(\mathbf{x}, t)$, where κ_1 and κ_2 are the diffusion coefficients of the two linearly coupled fields. The cubic terms control the growth of the linearly unstable modes.

8.3.1 *Linear Stability Analysis*

We now conduct a linear stability analysis of the model (8.31) using the ideas laid out in Sect. 8.2.1. Let $\mathbf{w_0} = (0, 0)$ be a homogeneous sol. and let $\mathbf{w} = \mathbf{w_0} + \delta\mathbf{w}$, where $\delta\mathbf{w}$ is a spatio-temporal perturbation given by

$$\delta\mathbf{w} = \begin{bmatrix} \delta u \\ \delta v \end{bmatrix} \Psi_{nm}.$$

Using the fact that $\nabla^2\Psi_{nm} - (j_{nm}/R)^2\Psi_{nm}$ and substituting into Eq. (8.31) gives

$$\frac{\partial}{\partial t}\begin{bmatrix} \delta u(t) \\ \delta v(t) \end{bmatrix} = \begin{bmatrix} m_{11} \; m_{12} \\ m_{21} \; m_{22} \end{bmatrix}\begin{bmatrix} \delta u(t) \\ \delta v(t) \end{bmatrix},$$

where $m_{11} = B - 1 - \kappa_1(j_{nm}/R)^2$, $m_{12} = A^2$, $m_{21} = -B$, $m_{22} = -A^2 - \kappa_2(j_{nm}/R)^2$. Then, the uniform state destabilizes to $\Psi_{nm}(r, \phi)$ when the real part of the eigenvalues are zero, which yields the following marginal stability curves

$$B_{nm}^M = 1 + \frac{\kappa_1}{\kappa_2}A^2 + \kappa_1\left(\frac{j_{nm}}{R}\right)^2 + \frac{A^2}{\kappa_2}\left(\frac{R}{j_{nm}}\right)^2. \tag{8.32}$$

For a given value of A, the marginal stability curve B_{nm} reaches a minimum of

$$B_0 = 1 + \frac{\kappa_1}{\kappa_2}A^2 + 2A\sqrt{\frac{\kappa_1}{\kappa_2}}.$$

Figure 8.11 illustrates some of the marginal stability curves evaluated at fixed values of $\kappa_1 = -0.2$, $\kappa_2 = 2$, and $A = 5.0$. B and the radius of the domain R are used as control parameters.

8.3.2 *Simulations*

Rotating states of a single cell obtained from the numerical integration of Eq. (8.31) are presented in Fig. 8.12 along with the analogous experimental states. The chiral asymmetry is demonstrated by contrasting Fig. 8.12a with Fig. 8.12b and also Fig. 8.12c with Fig. 8.12d. The computed and experimental cell shapes are similar as seen by comparing Fig. 8.12a, and also Fig. 8.12b and d.

A Fourier–Bessel expansion of the rotating cell confirms that the modes with largest amplitude are Ψ_{01}, Ψ_{11} and Ψ_{21}. The real coefficients z_{0m} are constants of the motion. The rest of the time-dependent coefficients can be computed as follows. We assume the field $u(x, t)$ to be smooth and vanishing on the boundary of a circular domain of radius R. Then u can be expanded in a Fourier–Bessel series as

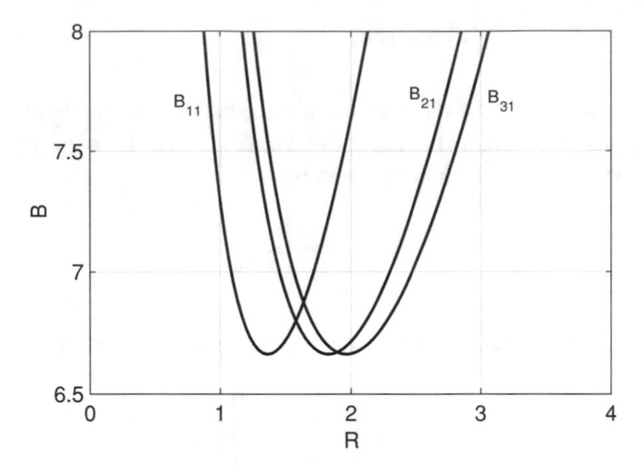

Fig. 8.11 Marginal stability curves along which the uniform state $(u, v) = (0, 0)$ of the Brusselator model (8.31) destabilizes to the Fourier–Bessel modes Ψ_{nm}. The curves are evaluated for fixed values of $\kappa_1 = -0.2$, $\kappa_2 = 2$, and $A = 5.0$. B and the radius of the domain R are used as control parameters

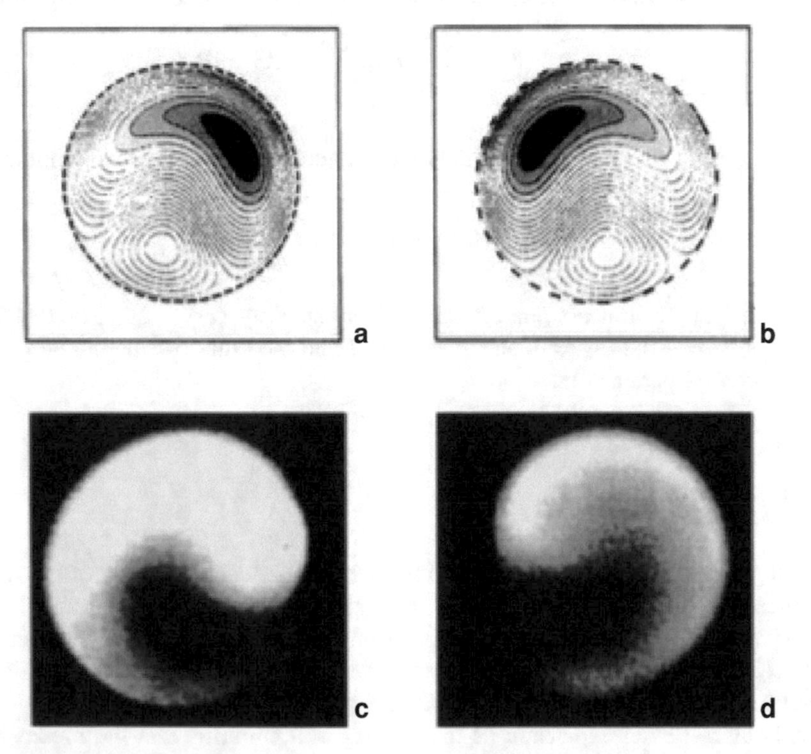

Fig. 8.12 **a** Clockwise and **b** counter-clockwise rotating states of a single cell from the model, and the analogous states (**c**) and (**d**) of the experiment. Observe the qualitative similarity of the cell shape in the two cases. The parameters generating the rotating state are $\eta = 2.0$, $\nu_1 = 0.5$, $\nu_2 = 1.0$, $B = 6.8$ and $R = 1.35$

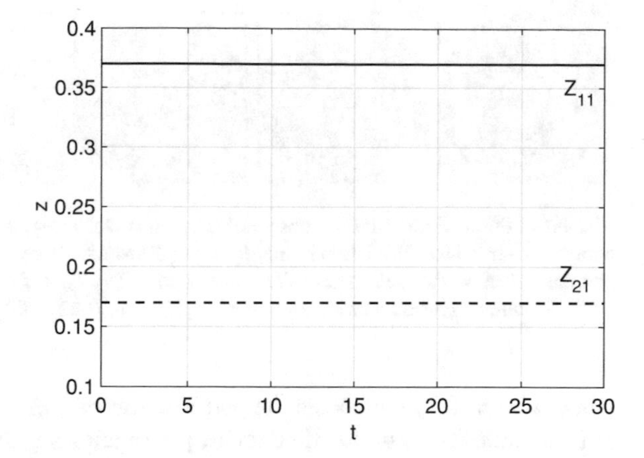

Fig. 8.13 Constant time-dependent coefficients associated with the one-cell dynamic state of Fig. 8.12 confirm the cell is rotating uniformly

$$u(x,t) = \sum_{n=0,m=0}^{\infty,\infty} z_{n,m}(t)\Psi_{nm}(r,\phi) + \text{c.c.},$$

where $\Psi_{nm}(r,\phi) = J_n(j_{nm}r/R)e^{n\phi i}$, $m >, n \geq 0$ and c.c. denotes the complex conjugate. Here $J_n(r)$ is the nth order Bessel function of the first kind and α_{nm} is its mth nontrivial zero. $z_{nm}(t)$ are complex coefficients, except for z_{0m} which are real. A similar expansion can be applied to $v(x,t)$. The orthonormality and completeness of the function $\{\Psi_{nm}\}$ gives

$$z_{nm} = \frac{1}{\pi R^2 J_{n+1}^2(j_{nm})} \int_0^{2\pi} \int_0^R r u(r,\phi)\bar{\Psi}_{nm}(r,\phi) d\phi\, dr,$$

with the proviso that the coefficients are half of the value given when $n = 0$.

For the particular case of the cell pattern of Fig. 8.12, the amplitude of the coefficients z_{11} and z_{21} is constant in time, see Fig. 8.13, which indicates that the state undergoes uniform rotation.

8.4 A Model of Flame Instability

A generic example of a cellular-pattern-forming dynamical system is described by the Kuramoto–Sivashinsky (KS) equation, which can be written in the form

$$\frac{\partial u}{\partial t} = \eta_1 u - (1+\nabla^2)^2 u - \eta_2(\nabla u)^2 - \eta_3 u^3, \tag{8.33}$$

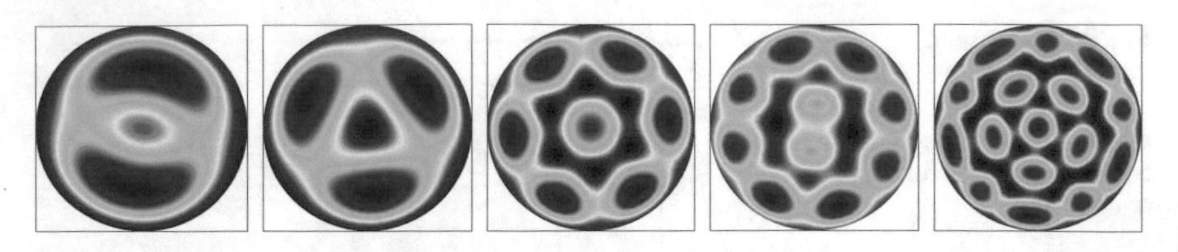

Fig. 8.14 Some static patterns observed using a novel integration scheme. In PANEL#1: 2-cell pattern, observed for simulation radius $R = 5.0$; PANEL#2: 3-cell pattern, $R = 6.0$; PANEL#3: 6/1-cell pattern, $R = 10.0$; PANEL#4: 8/2-cell pattern, $R = 12.0$; PANEL#5: 10/5/1-cell pattern, $R = 14.5$; *Common simulation parameters:* $(\eta_1, \eta_2, \eta_3) = (0.32, 1.00, 0.017)$

where $u = u(\vec{x}, t)$ represents the perturbation of a planar front (normally assumed to be a flame front) in the direction of propagation, η_1 measures the strength of the perturbation force, η_2 is a parameter associated with growth in the direction normal to the domain (burner) of the front, $\eta_3 u^3$ is a term that has been added to help stabilize its numerical integration. $\vec{x} \in \Omega$, where Ω is the domain of integration. Since we are interested in cellular patterns, Ω is assumed to be a polar grid, as it appears in Fig. 8.5. The KS equation describes the perturbations of a uniform wave front by thermo-diffusive instabilities. It has been studied in different contexts, including the existence of heteroclinic connections, by Cross and Hohenberg (1993), Armbruster, Guckenheimer, and Holmes (1988), Holmes, Lumley, and Berkooz (1996), and by Hyman and Nicolaenko (1986). Gassner, Blomgren, and Palacios (2007) have also conducted numerical explorations of the effects of noise on the KS equation in various regions of parameter space.

Figure 8.14 shows five examples of stationary patterns observed through numerical integration of the model (8.33). The computer simulations indicate a greater tendency towards stationary states (as opposed to dynamic states). Stationary states are patterns with petal-like cellular structures and well-defined spatial symmetries. Dynamic states are patterns in which the cells move, either individually or in ring configurations.

As the radius of the circular domain increases, the typical ordered state that appears changes from a single ring of cells to concentric rings of cells. Occasionally, dynamic states are also observed in the transition from one stationary pattern to another.

A linear stability analysis (see exercises) leads to the following marginal stability curves

$$\varepsilon_{nm}(R) = 1 - 2\left(\frac{j_{nm}}{R}\right)^2 + \left(\frac{j_{nm}}{R}\right)^4, \tag{8.34}$$

where j_{nm} are the zeroes of the Fourier–Bessel modes. Figure 8.15 depicts a few of the marginal stability curves.

A critical observation is the fact that beyond the curve, ε_{nm}, on increasing ε, the uniform state $u_0 = 0$ (uniform flame front) destabilizes and a cellular pattern with the shape of the Fourier–Bessel mode, $\Psi_{nm}(r, \theta)$, emerges. More importantly, the marginal stability curves provide a tool to systematically search for the right type

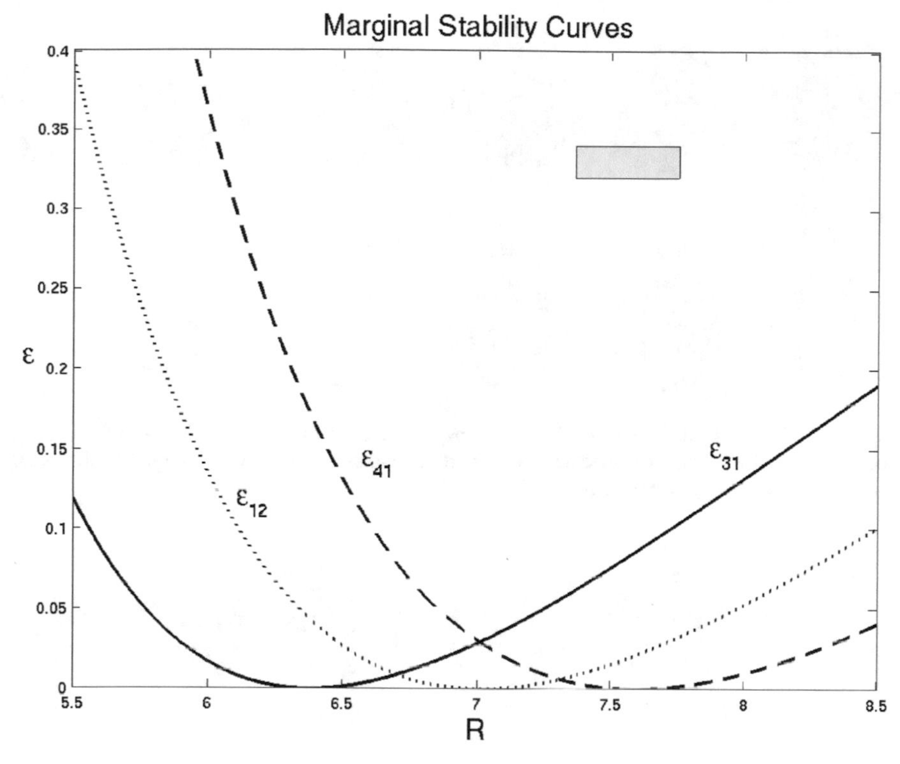

Fig. 8.15 Marginal stability curves associated with the Kuramoto–Sivashinsky model of flame instability. The curves outline the stability domains where the trivial solution $u_0 = 0$ (representing a uniform flame front) bifurcates to Fourier–Bessel modes Ψ_{nm}

of pattern, at least with the right number of cells. For instance, let's assume we are interested in finding a region where the evolution of a single-ring stationary pattern with three cells, as is shown in Fig. 8.14 panel #2,, can be traced. Such region can be found in a neighborhood of the minimum of the marginal stability curve ε_{31} shown in Fig. 8.15. As the curved is crossed, on increasing R, a stationary pattern of three cells with purely spatial D_3-symmetry emerges via a symmetry-breaking bifurcation from the $\mathbf{O(2)}$-invariant trivial solution. Increasing R further, and upon crossing the left edge of the shaded region, the three-cells pattern loses stability, the D_3–symmetry of the ring is broken, and a dynamic pattern of three cells rotating "almost" uniformly and counter-clockwise bifurcates subcritically. Figure 8.16 depicts a sample of snapshots of the space and time evolution of the $u(r, \theta, t)$ field obtained at $R = 7.36$.

Near $R = 7.74$, in particular, the cells repeatedly make abrupt changes in their angular position while they rotate around the ring; in a manner that resembles a *hopping pattern*. Figure 8.17 depicts a few representative snapshots of the spatio-temporal dynamics at $R = 7.7475$. Observe that changes in cell shape are more noticeable. In fact, a hopping cell changes its shape more than the other two and

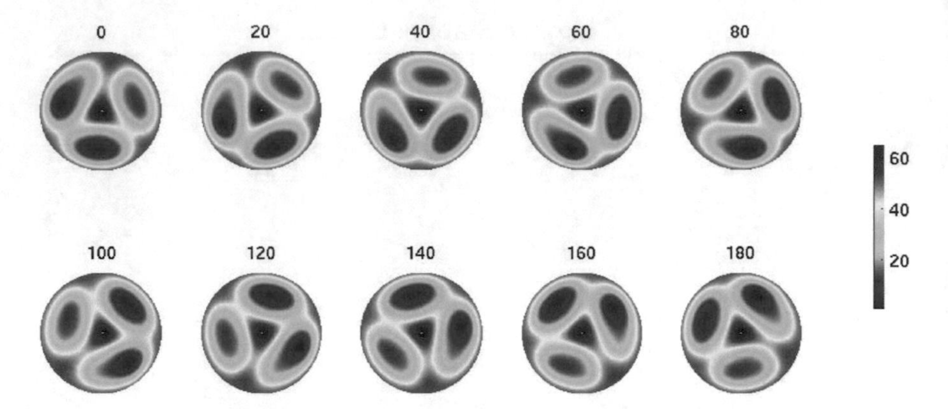

Fig. 8.16 Ten sequential time (top index) snapshots of a dynamic state of three cells rotating counter-clockwise, with small modulations, found in simulations of (8.33). Parameter values are: $\varepsilon = 0.32$, $\eta_1 = 1.0$, $\eta_2 = 0.013$, and $R = 7.36$

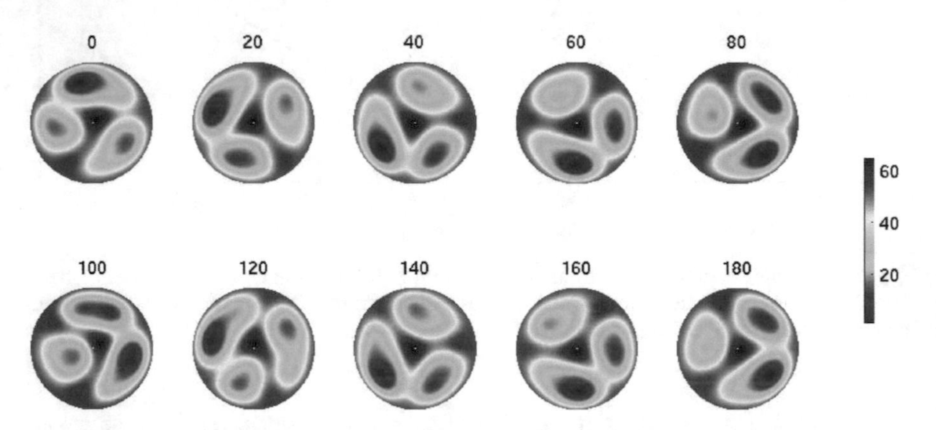

Fig. 8.17 Space and time evolution of a three-cells hopping state found in simulations of (8.33). The cells move nonuniformly and their shapes change periodically. Parameter values are: $\varepsilon = 0.32$, $\eta_1 = 1.0$, $\eta_2 = 0.013$, and $R = 7.7475$

also appears more asymmetric. The hops are small in comparison with experimental states but, up to a time-scale factor, the overall characteristics of the dynamics appear to be in good agreement with experiments.

8.5 Pattern Formation in Butterflies

Gierer and Meinhardt [4] suggested the following set of equations as a model for the activator-inhibitor dynamics that characterizes many reaction-diffusion systems

$$
\begin{aligned}
f(u, v) &= \kappa_1 - \kappa_2 u + \frac{\kappa_3 u^2}{v(\kappa_6 + \kappa_7 u^2)} \\
g(u, v) &= \kappa_4 u^2 - \kappa_5 v.
\end{aligned}
\tag{8.35}
$$

Sekimura et al. [5] combined Gierer-Meinhardt kinetics dynamics with diffusion to arrive at the following model

$$
\boxed{
\begin{aligned}
\frac{\partial u}{\partial t} &= \gamma \left(a - bu + \frac{u^2}{v(1 + \kappa u^2)} \right) + \nabla^2 u \\
\frac{\partial v}{\partial t} &= \gamma (u^2 - v) + d \nabla^2 v,
\end{aligned}
}
\tag{8.36}
$$

after non-dimensionalizing the spatial variables, letting $d = D_v / D_u$ and re-scaling time accordingly.

Explicit analytical solutions of steady-states are too cumbersome to compute. However, they can be found numerically by solving the following system of equations

$$
\begin{aligned}
a - bu + \frac{u^2}{v(1 + \kappa u^2)} &= 0 \\
u^2 - v &= 0.
\end{aligned}
$$

Similarly, a linear stability analysis can be carried out numerically while looking for parameter values that satisfy the following conditions

$$
\begin{aligned}
&f_u + g_v < 0, \quad f_u g_v - f_v g_u > 0, \\
&d f_u + g_v > 0, \ (d f_u + g_v)^2 - 4d(f_u g_v - f_v g_u) > 0.
\end{aligned}
$$

Madzvamuse et al. [6] have conducted a complete numerical exploration of the model behavior and their work has produced the parameter values shown in Fig. 8.18 as guidelines for pattern-forming in a wide variety of species of butterflies.

Figure 8.19 shows results of numerical simulations of the model (8.36) with parameter values drawn from Fig. 8.18. The simulations are then mapped over the wings of the butterflies for comparison purposes.

Thus far we have studied spatio-temporal behavior through PDE models. However, it is also possible for ODEs to exhibit spatial-temporal dynamics even though they are not defined over extended domains. How can this be possible, would be an obvious question to ask. The explanation is, however, straightforward. A system of ODEs can be used to study collective or aggregate behavior. For instance, a single ODE can serve to model the intensity of light emitted by a single firefly in reaction to stimuli. If a large system of ODEs is then used to study the collective behavior of thousands to fireflies then the solutions can form spatial-temporal behavior.

Another example consists of the motion of thousands of bubbles in fluidization processes. Each individual bubble in a fluidization bed can be described by a single

Pattern	Forewing	Hindwing
Niobe	$d = 70.8473$, $\gamma = 619.45$ $\alpha = -0.111$, $\beta = -0.025$, $c_0 = 0.69$	Same as forewing except $\alpha = 0.111$, $\beta = -0.025$ $c_0 = 0.9$
Salaami	Same as for *Niobe* except $c_0 = 0.695$	$c_0 = 0.89$
Trophonius	Same as for *Niobe* except $c_0 = 0.697$	$c_0 = 0.91$
Hippocoonides	Same as for *Niobe* except $c_0 = 0.701$	$c_0 = 0.87$
Planemoides	Same as for *Niobe* with $c_0 = 0.67$,	$c_0 = 0.7$
Natalica	Same as for *Niobe* except $\alpha = -0.0555$, $c_0 = 0.673$	$c_0 = 0.75$
Cenea	Same as *Niobe* except $\alpha = -0.0111$, $\beta = -0.025$, $c_0 = 0.653$	$c_0 = 0.6$
Leighi	Same as *Cenea* except $c_0 = 0.656$	$c_0 = 0.8$
Male–like	Same as *Cenea* except $c_0 = 0.95$	$c_0 = 0.75$

Fig. 8.18 Parameter values for Turing pattern-instabilities associated with a wide range of species of butterflies

ODE while the interaction of thousands of bubbles is governed through a coupled system of ODEs. The entire system evolves according to a predefined set of rules for how the bubbles interact. This is an example of an **agent based model**. These type of models have been gained popularity in recent years, as they can provide useful insight into *self-organization* processes with relatively simple equations. We have chosen the bubble dynamics as an example to introduce agent-based models. Next we provide details.

8.6 Agent-Based Model of Bubbles in Fluidization

Fluidization is a process in which solid particles behave like liquid in a vessel due to some constant flowing medium such as gas or air. Fluidization was introduced in fluid catalytic cracking process to convert heavier petroleum cuts into gasoline in the early 1940s as the first large scale commercial application. Today, fluidization processes have many important industrial applications, especially in chemical fossil and petrochemical industries where good gas-solid mixing is required. Typical industrial applications include coal gasification, solid transportation, polymerization of olefins, heat exchange, polyethylene synthesis, cracking of hydrocarbon, catalytic reaction, water treatment, and nanotubes [7–10]. Figure 8.20 depicts a schematic picture of a large circulating fluidized combustor in Florida, in which upward blowing air lifts solid fuels, providing a turbulent mixing of gas and solids.

The behavior of a fluidization process can have numerous regimes based on size of fluidized bed, flowing medium, flow velocity, physical property of solid particles,

Fig. 8.19 Computer simulations of the reaction-diffusion model (8.36) over a nonuniform grid. Four different species are simulated and rendered on the right wing of each butterfly. For comparison purposes, the left wing shows the original species. Courtesy of Anotida Madzvamuse

and operating conditions [10, 11]. Bubbling is an important phenomenon existing in most fluidization processes in which bubbles are generated continuously, move upward vigorously, coalesce and interact with flowing medium and particles [7]. In the applications in chemical, fossil and petrochemical industries, excellent gas-solid mixing are even achieved through bubbles that are spontaneously formed during the fluidization processes. Therefore, it is necessary to have good understanding of fluidization processes, especially of the bubble dynamics, to provide reliable control mechanisms for the fluidization applications in industries.

The difficulty of modeling three phase gas-solid-bubble dynamics in a fluidized bed lies mainly in modeling bubbles due to their complex dynamical behaviors including coalescence and splitting [12]. Determining the velocity is the essential part in

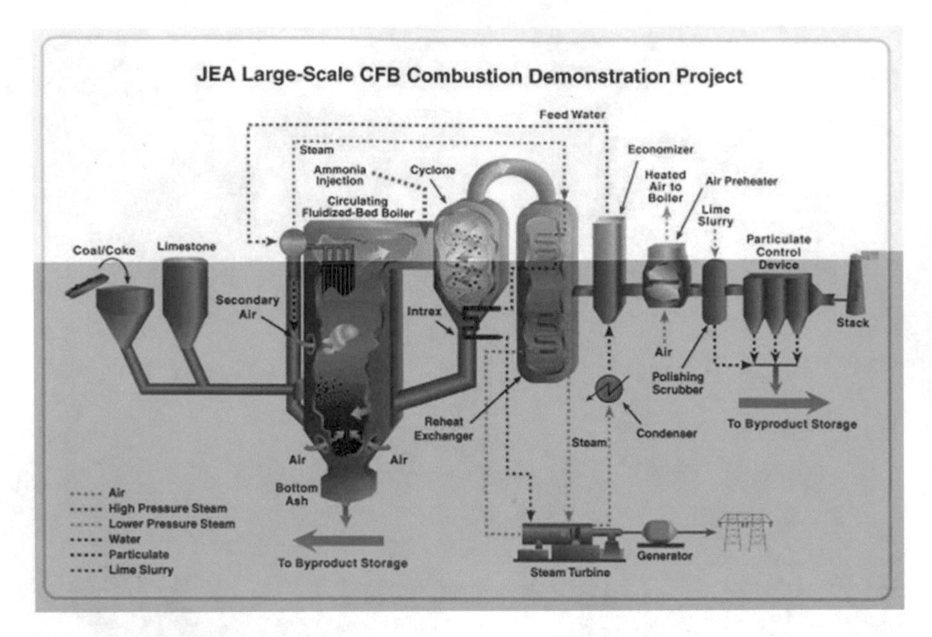

Fig. 8.20 Application of a Circulating Fluidized Bed combustor to generate electric energy (courtesy of DOE/NETL). Upward blowing air lifts solid fuels, providing a turbulent mixing of gas and solids

modeling rising bubbles. In this section, we introduce a low-dimensional, agent-based bubble model, **Dynamic Interacting Bubble Simulation** (DIBS), which focuses on describing main bubble-bubble interactions among rising bubbles based on empirical observations and data fitting technique via imaging apparatus. We also summarize other approaches for modeling bubbles and give critical reviews for these models compared with the DIBS bubble model.

8.6.1 Fluidization Processes

A fluidized bed, regardless of its application, normally consists of a vessel that contains solids and has a porous bottom plate for injecting flowing medium upward. When the flow rate is low, the flowing medium percolates through the gaps among particles. The particles remain packed and are in a steady state as is shown in Fig. 8.21a. As the flowing speed of the medium keeps increasing and reaches a threshold at which the forces from the flow exerted on the particles overcome gravitational forces, particles start to suspend in the flowing medium inside the vessel. Further increasing the flowing speed will cause particles to behave like fluid, a state called fluidization (see Fig. 8.21b). This threshold of flowing velocity for the carrying medium is called the minimum fluidization velocity (U_{mf}). Many efforts have been made to find a

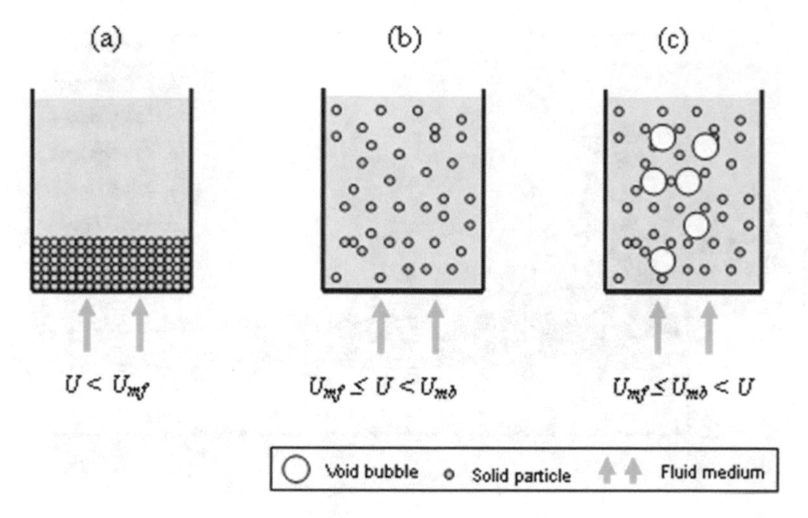

Fig. 8.21 Fluidization Processes. **a** Fluidized bed is in a steady state when the upward flow rate, U, is less than the minimum fluidization velocity (U_{mf}). **b** Fluidization happens when the flow rate crosses U_{mf}. **c** When the flow rate reaches the minimum bubbling velocity (U_{mb}), bubbles are created and the fluidized bed exhibits three phases, bubbles phase, solid phase and gas (or liquid) phase

formula for the correlation between minimum fluidization and physical properties of the flowing medium and particles for the purpose of providing accurate design in building a fluidized bed.

As the flow rate increases further towards a second threshold, called the minimum bubbling velocity (U_{mb}), bubble voids are formed thus creating a dramatical change in the dynamics as bubbles move upwards vigorously and coalesce when bubbles touch. Figure 8.21c illustrates a fluidized bed with medium flow, moving particles, and bubbles. This particular state is the bubbling fluidized regime and is the focus of this section.

8.6.2 *Bubble Dynamics*

Davies and Harrison [13] presented in 1950 the following equation to model the velocity of a free rising bubble in a bubbling fluidized bed

$$V_{b\infty} = (2/3)\sqrt{gR}, \tag{8.37}$$

where g is the gravitational acceleration and R is the bubble diameter. Since then, many attempts have been made for modeling bubble velocities. Earlier work focused mainly on explaining bubble formation and the physical properties such as bubble diameter, size and shape. Harrison and Leung [14], Zenz [15], and Caram and Hsu

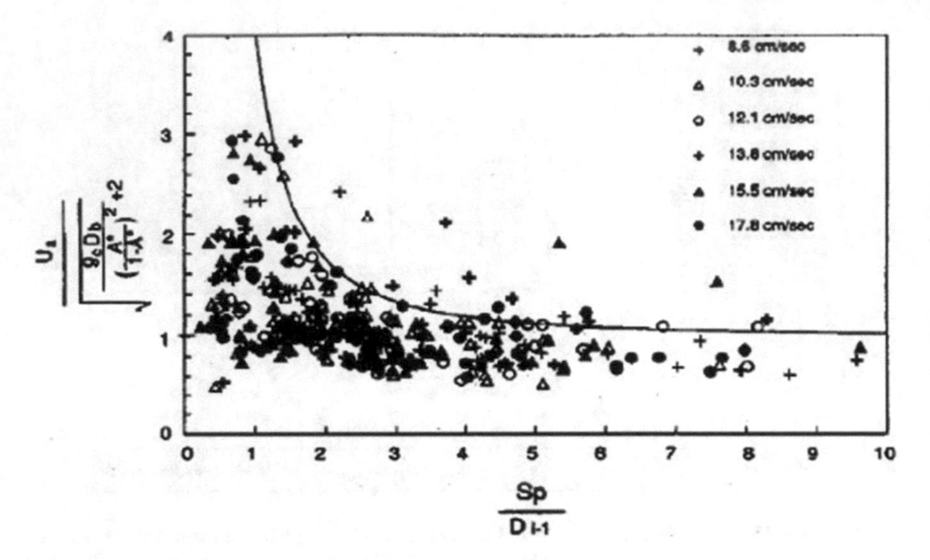

Fig. 8.22 Collected velocity data of a trailing bubble from imaging system and fitting curve from Eq. (8.38) by Halow and Nicoletti. The final Eq. (8.39) used in DIBS model is empirically derived

[16] developed various models that described the growth of bubbles due to gas injection at a single orifice. Nieuwland et al. [17] provided a complete review of existing models in 1996. However, none of these efforts addressed the collective behavior of bubbles nor the complete interactions, in space and time, which were commonly observed in related experiments.

With the arrival of modern computers, scientists and engineers started to develop computational models and numerical simulations of bubble dynamics. The CHEM-FLUB software developed by Systems Science and Software Inc. in 1980 presented one modeling approach to simulate gas and solids flow in fluidized gasifiers [18–21]. In its approach, bubbles are treated as a continuum flow and thus the continuity equation is applied to model the bubble column. Yet the complicated phenomena of bubbles in bubbling fluidized bed, for example, the bubble-bubble interactions including bubble coalescence, are missing in such a computational model.

In recent years, an imaging system was used on a bubbling fluidized bed to capture bubble wake behavior that exerts some pulling force and accelerates the rise of a trailing bubble. This is an additional velocity for a bubble that is trailing another bubble and is related to the diameter of its leading bubble. Figure 8.22 is a plot by Halow and Nicoletti illustrating actual experimental data and fitting curve from the following empirical formula for a trailing bubble:

$$U_b = \sqrt{\frac{g l_b}{2 + \left(\frac{A^*}{1 - A^*}\right)^2}} \left[1 + 3\left(\frac{D_{i-1}}{S_p}\right)^2\right], \tag{8.38}$$

where D_{i-1} is the diameter of the leading bubble and S_p is the distance between a trailing bubble and its leading bubble.

In 1993, Halow and Fasching [22] further examined their fitting model. By comparing Eq. (8.38) with those by Farrokhalaee [23] and Lord [24], they suggested that the square term be replaced by a cubic form. This established the bubble velocity formula of Eq. (8.39) adopted by S. Pannala, C. S. Daw and J. S. Halow who later developed the computational DIBS model [25] for bubble dynamics in bubbling fluidized bed.

8.6.3 Computational DIBS Model

The DIBS model was introduced in 2004 by Pannala et al. [26]. In the DIBS model, each bubble is treated as a single agent and is described by a low-order ordinary differential equation (ODE):

$$\left\|\frac{dX_i}{dt}\right\| = \|V_i\| = \sqrt{\frac{gl_i}{2+\left(\frac{A^*}{1-A_i^*}\right)^2}}\left[1+3\left(\frac{D_{Lj}}{X_{i-j}}\right)^3\right], \tag{8.39}$$

referring Fig. 8.23, X_i is the position, V_i is the velocity, l_i is the length of ith bubble, A_i^* is the fraction of cross section area of ith bubble divided by testbed area, D_{Lj} is the diameter of the leading bubble, and X_{i-j} is the distance between ith bubble and its leading bubble, jth bubble.

In order to simulate the agent-based DIBS model, some important assumptions need to be made. We list them all.

(i) If a bubble does not have a leading bubble, its equation will be the one having no j related term, i.e. the cubic term, in Eq. (8.39).
(ii) Each bubble is spherically shaped if the diameter of the bubble is less than 85% of the bed diameter. If the diameter of a bubble is larger or equal to 85% of the bed diameter, the bubble will be cylindrically shaped with a hemispherical end cap.
(iii) At any moment, the movement of a bubble is affected by its leading bubble through a pulling force. A bubble j is called a leading bubble for a bubble i if j has vertical position above i and has the shortest distance with the bubble i (See Fig. 8.23).
(iv) Two bubbles coalescence when they touch in 3-dimensional space.
(v) When reaching testbed surface, a bubble disappears.
(vi) The bubble rise velocity given by Eq. (8.39) is relative to the solids flow in fluidized bed.

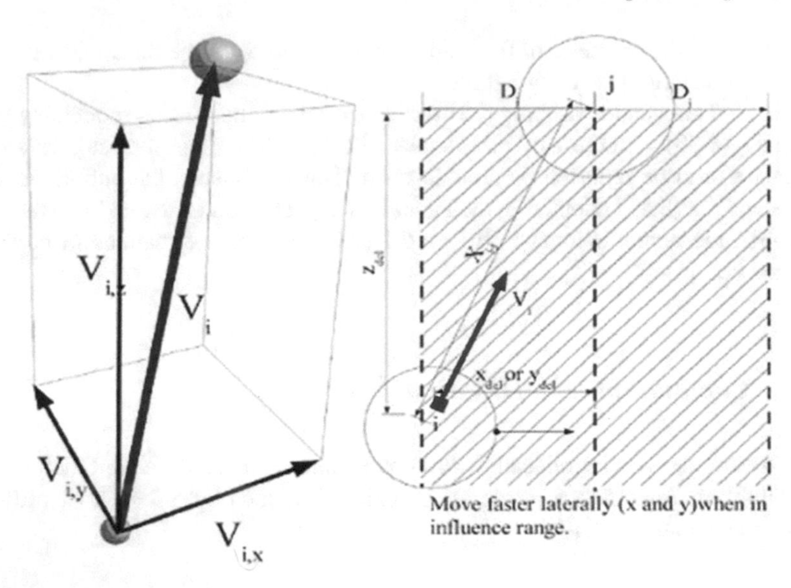

Fig. 8.23 The leading bubble j exerts a pulling force on a trailing bubble i. Bubble j is the leading bubble for bubble i if bubble j is above bubble i and has the shortest distance between two bubble, X_{i-j}

Notice that by assumption (iv) and (v), the number of ODEs varies as the number of bubbles changes with the fluidized bed. Thus it is very difficult to carry analytic study of this model purely by looking at a cluster of coupled bubbles modeled by Eq. (8.39). On the other hand, the use of powerful computers allows such computational model to be implemented numerically and simulated graphically. Figure 8.24 is a modified version of flowchart for the implementation of DIBS model originally designed and implemented by Pannala et al. at Oak Ridge National Laboratory [25].

In assumption (i), the DIBS model stresses the influence of an immediate leading bubble for bubble velocity. It is also worth mentioning that such approach were sought in early 1970s by Orcutt and Carpenter [27] in 1971 and Allahwla [28] in 1975.

In the actual computational simulation, bubbles are generated periodically from a fixed number of bubble injectors through a porous medium located at the bottom of a fluidized bed. The bubble injection frequency will play an important role in our bifurcation analysis. The initial critical values for the DIBS simulation are the bed shape (either cylinder shape or a rectangular test bed), bed size, the number and positions of bubble injectors, bubble injection frequency, minimum fluidization velocity, and superficial velocity of the gas flow.

The operation to update locations of bubbles is to solve numerically Eq. (8.39) for each bubble. The first order Euler method can be used with small value of time step size in integral implementation to acquire better accuracy in finding numerical solutions. After the positions of bubbles are updated, the program checks the bound-

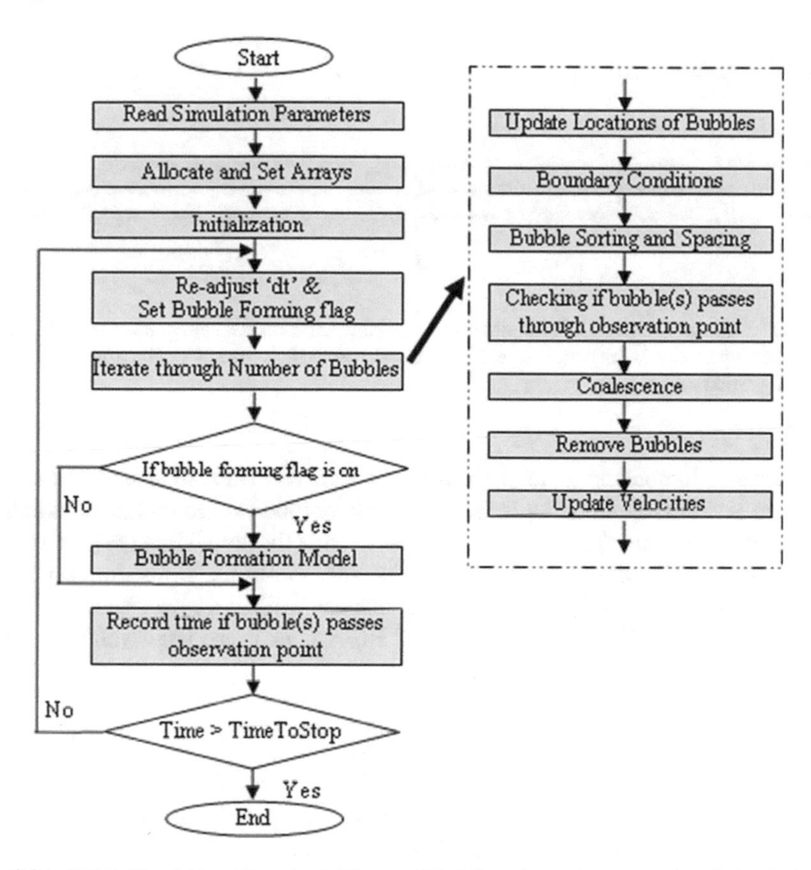

Fig. 8.24 DIBS Simulation Flowchart. The modified flowchart of computational simulation of DIBS model with adaptive integration step size. The operation to update the locations of bubbles is actually to numerically solve the ODE of Eq. (8.39)

ary condition in the sense that a bubble is bounced fully when it meets the side wall of the vessel. The program then checks if there are coalescing bubbles by checking if there are two bubbles that are physically in contact with one another. Bubbles that touch in 3-dimensional space are merged into one bubble with volume equal to all merged ones. Bubbles that surpass the top of the fluidized bed are removed from the bubble computational array.

Bubbles are generated periodically in a real-time simulation in which the software program checks at each loop if it is time to generate new bubbles. New bubbles are formed with a fixed bubble diameter, or with an initial size related to gas velocity, from the bottom of the fluidized bed and are then added into the bubble array. Each bubble has an initial color that is changed once it passes a fixed observation point. If there is a new bubble or bubbles passing the observation point, the passage time is recorded for a bifurcation analysis.

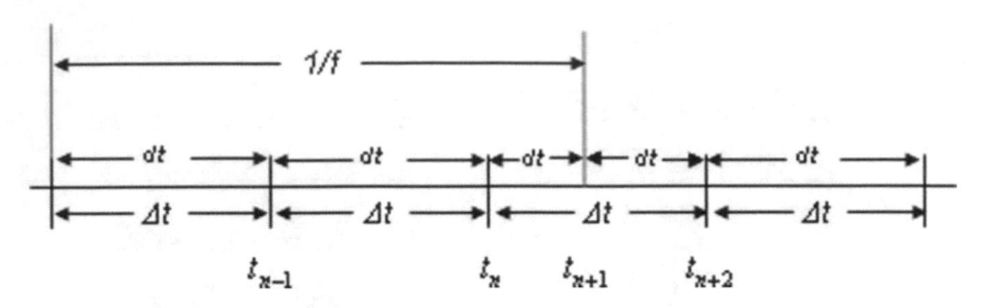

Fig. 8.25 Adjustable dt in DIBS Simulation. Adjustable 'dt' for ODE integration. Δt is the initial fixed time step size. f is the bubble injection frequency. And dt is the adjustable time step size. An extra integration step is taken at t_{n+1}

After each computational loop, the time is advanced to next time step. Usually the time step size is fixed in using finite difference method to find numerical solutions of differential equation(s). However, we realized that the actual integration time step size, dt, has to be adjusted in real-time to be able to run the simulation for evenly distributed values of bubble injection frequencies (BIF) in a given interval. This is because there would be limited choices of BIF values if the fixed value of 'dt' is used due to the following relationship.

$$f = \frac{1}{N \cdot dt} \tag{8.40}$$

where N is the number of time steps between two bubbles to be generated and f is the bubble injection frequency (BIF). If a fixed Δt is used, the values of f from Eq. (8.40) would not be evenly distributed on any given interval by varying N as positive integers. To overcome this problem, a variable integrating time step, dt, is used in actual implementation. The value of 'dt' is checked and adjusted as needed in each loop as illustrated in Fig. 8.25. With this approach, values of BIFs for the simulations can be evenly distributed in any given interval, thus allowing us to carry bifurcation analysis for bubble dynamics in response to changes in injection frequencies.

8.6.4 *Bifurcation Analysis of Single-Bubble Injector*

The goal of the bifurcation analysis is to seek a guide for engineering fluidized beds through numerical simulation and data analysis on some identified key element(s). To achieve this goal, some data analysis methods are needed to understand well the expected nature of the dynamical information such as the generation and transmission of information and self-organization into spatial-temporal systems.

To get insight into the bubble dynamics, we employ a simulated experimental measurement. It consists of a hypothetical laser device that detects passing bubbles

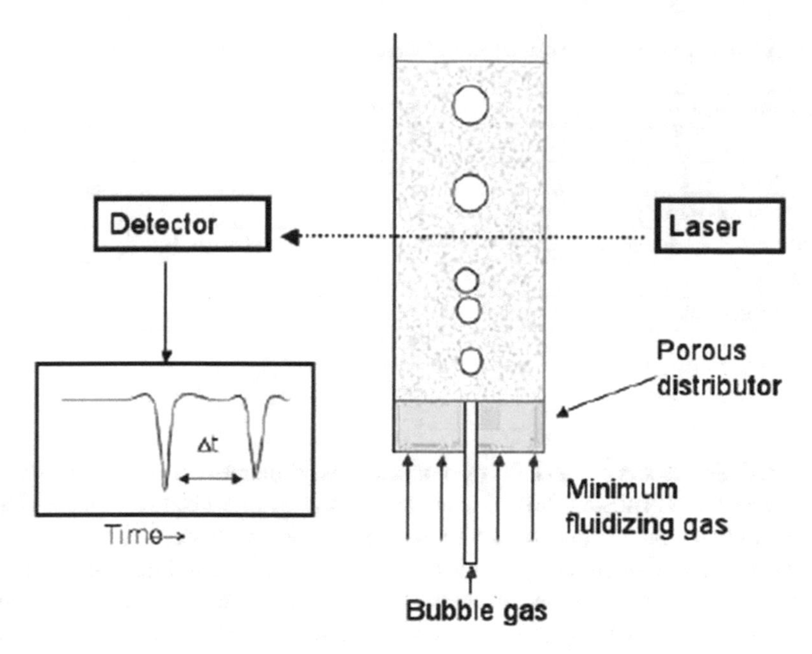

Fig. 8.26 Experiment with Single Bubble Injector. A schematic diagram depicts the devices for detecting and recording the time of bubbles passing the observation point. The time interval between two passing bubbles, Δt, is calculated offline to form a time series to which the bifurcation analysis is carried for a single bubble injector case

and records time series for rising bubbles that pass through a fixed observation point for a single injector case, as is depicted in Fig. 8.26.

Nguyen et al. [29] and Tufaile and Sartorelli [30, 31] successfully used this type of measurement to study bubble-train dynamics. Notice that the actual formation of bubbles is related to the physical properties of flow medium and minimum fluidization velocity (U_{mf}). Therefore, reasonable BIF values for numerical simulations are initially chosen from near 0 to 10 Hz. Here n Hz is defined to be n bubbles to be generated per second by the simulated bubble injector.

The time interval Δt_i between bubbles passing through the laser detector is collected and incorporated into a time series of crossing times. By eliminating the transients, a bifurcation diagram is constructed and shows complex bubble dynamical behavior with fixed points, chaotic attractors, periodic solutions and intermittency behaviors and these will be analyzed in detail later.

To capture the long term behavior, the program runs for a minimum 700 s and in some cases it runs up to 2400 s. The computationally intensive jobs demand use of high performance computing. Table 8.1 shows the main parameters used during the simulations.

An initial real-time visualization was developed using native graphic functions. This first version of bubble visualization software which displays rich visual information showing bubble movement (rising upward rapidly) and bubble coalescence.

Table 8.1 Main parameters for computational simulation of bubbling fluidized bed using DIBS model

Initial bed height	40.0 cm
Bed diameter	22.9 cm
Bed shape	Circular
Superficial gas velocity	1.7127 cm/s
Bubble injection frequencies	a 2000-point grid on (0, 10]
Initial bubble size	1.0 cm
Observation point	20.0 cm
Δt	0.001 s

Figure 8.27 shows some snapshots of numerical experiments with different BIF values. A more sophisticated three-dimensional visualization display was developed later to study the bubble dynamics with multiple bubble injectors.

A bifurcation diagram is constructed by varying the bubble injection frequency (f). The computationally based bifurcation analysis shows that the bubble dynamics transits among different regimes such as fixed point, chaotic attractors and intermittent behavior. For each BIF value, the DIBS simulation runs for 700 s to record the time series, $\{t_i\}$, bubble passage time through a laser detector. The actual signals, $\{\Delta t_i\}$, the time interval between two consecutive data points, are calculated offline. The first 200 points of computed $\{\Delta t_i\}$ are treated as transients and thus are discarded. A bifurcation diagram is constructed by plotting the resulting time series against the BIF values. This computational method has been widely used for constructing bifurcation diagram. One typical example is the bifurcation diagram for logistic equation by Alligood et al. [32]. Figure 8.28 shows the bifurcation diagrams from the DIBS simulations. The diagram reveaks the following observations.

At low injection frequencies, bubbles in the fluidized beds are significantly separate and the bubble-bubble interactions have very little impact on rising bubbles. Thus bubbles rise as a stream with almost fixed gap between two consecutive bubbles. This results in a fixed value in $\{\Delta t_i\}$. This demonstrates that the global dynamics is attracted to a fixed point with low BIF values. As BIF value increases beyond $f = 4$ Hz, bubbles start to coalesce and the global dynamics rapidly changes from a fixed point to a region of quasi-periodic behavior that eventually becomes chaotic. The global dynamics enters a more organized region of period-4 oscillation for $4.3\,\text{Hz} < f < 4.7\,\text{Hz}$, roughly. After this region, the period-4 region bifurcates into a chaotic region with four disjoint attractors. Near $f = 5$ Hz, the dynamics again enters a region with period-3 oscillation that changes into a period-2 orbit. The period-doubling bifurcation then leads to a period-4 orbit.

Beyond $f = 5.5$ Hz, the system displays intermittent behavior in which the underlying dynamics randomly changes between a high period orbit, period-6 and a chaotic attractor.

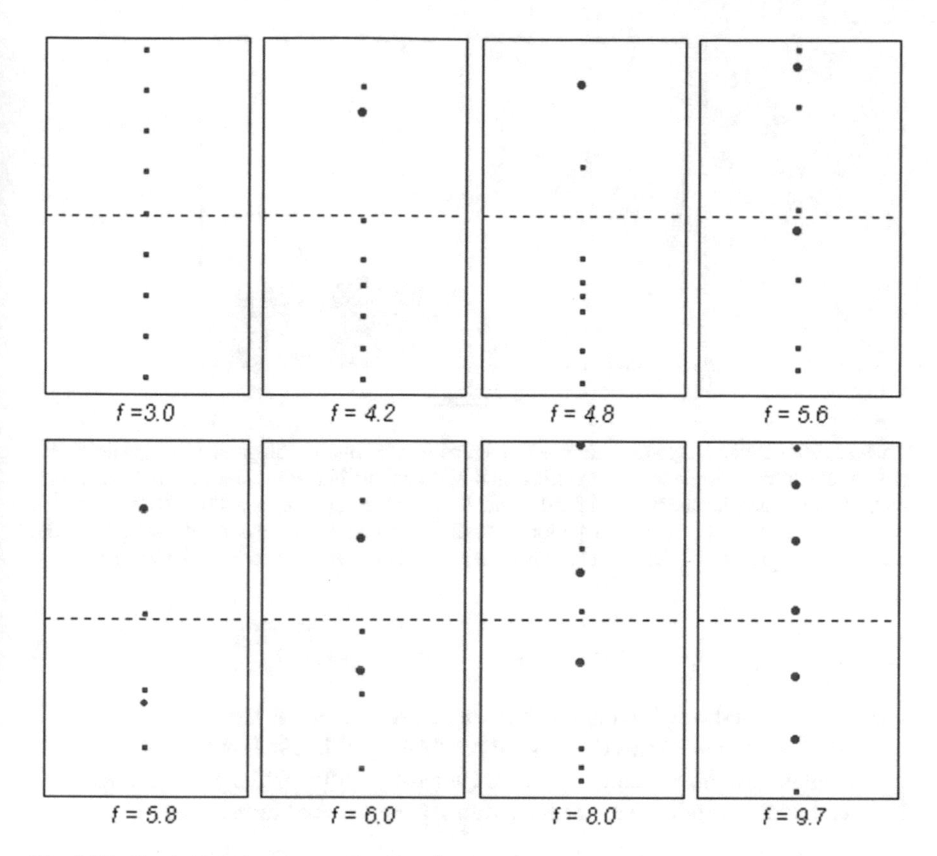

Fig. 8.27 Single Bubble Injector Simulations. Bubble simulations with various bubble injection frequencies. The bubbles below the observation line (laser detector) are colored in blue and bubbles passing the observation line are painted in red. The bed size is 20 cm × 40 cm. The initial diameter of a bubble being injected from nozzle is 1 cm. The observation point is right in the middle height of the bed

8.6.5 Phase Space Embeddings

Phase space embedding is an important method in studying nonlinear time series. In a deterministic system, the states of all future times are determined once an initial state is given. Thus the idea of phase space reconstruction is to use a one-to-one and continuous function, more precisely a topological mapping, to embed the one dimensional time series from a deterministic dynamical process into a multi-dimensional space to disclose the real manifold in which the dynamics takes place and to allow prediction for future evolution of the dynamical system. A typical technical solution is the method of delays. Assume $\{s_n\}$ is a given time series. New vectors are formed by defining the delay coordinates as:

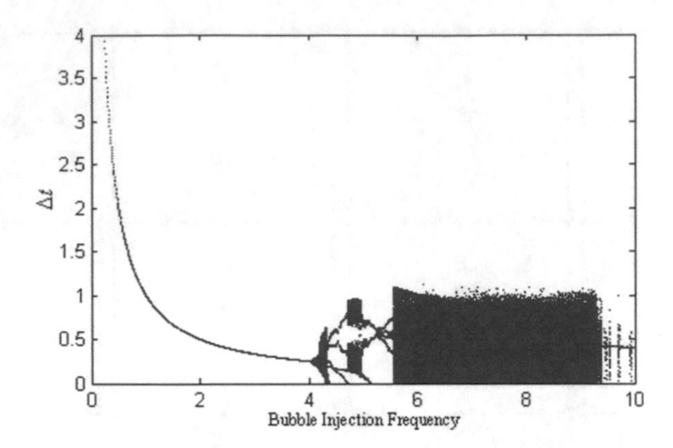

Fig. 8.28 Bifurcation Diagram of Bubble Dynamics. The bifurcation diagram of bubble dynamics with the experiment configuration described in 4.3. As bubble injection frequency (BIF) is small, the global dynamics is attracted to a fixed point. As BIF value goes beyond 4.0 to 10.0, the global dynamics changes to chaotic region, period-4 oscillation, four separate chaotic attractors, periodic oscillation, a region with intermittent behavior, and then into a region with nearly a fixed point

$$\vec{s} = \left(s_{n-(d-1)\tau}, s_{n-(d-2)\tau}, \ldots, s_n\right), \tag{8.41}$$

where d is called the embedding dimension and τ is delay or lag.

The embedding theorems established by Taken [33] in 1981 and Sauer et al. [34] in 1991 guarantee the existence of such d (when d is sufficient large) so that the time delay embedding produces a true embedding from original time series to the space R^d.

The embedding dimension d reveals there are d independent measurements within a given time series. It is then naturally to ask how to find the minimum embedding dimension as the existence of such numbers is guaranteed. The computational method adopted by this study to find the minimum embedding dimension is the false nearest neighbor (FNN) method originally proposed by Kennel et al. [35] and was implemented by Hegger et al. [36, 37] in their TISEAN software package. The FNN method is based on the idea that neighboring points of a given point are also mapped to neighbors in delay space by a true embedding. For a delay map with embedding dimension $d < d_{min}$, such topological properties would no longer be preserved and would produce false neighbors after mapping into to delay space.

In the DIBS model, the local dynamics of a single bubble injector will be dominated by pairwise interactions between leading and trailing bubbles. The bubble stream also tends to collapse toward the bed center as the bubbles rise. This should tend to reduce the effective local dynamical dimension considerably. Based on these, it is reasonable to expect the crossing dynamics to be described by a map of the form $\Delta t_{n+1} = G(\Delta t_n, \Delta t_{n-1})$ for the time series from DIBM simulation. Namely, the anticipated embedding dimension required to resolve the local dynamics to be close to a value of $d = 3$. Figure 8.29 shows the resulting plots after applying FNN algorithm from TISEAN package for the time series from DIBS simulation with BIF

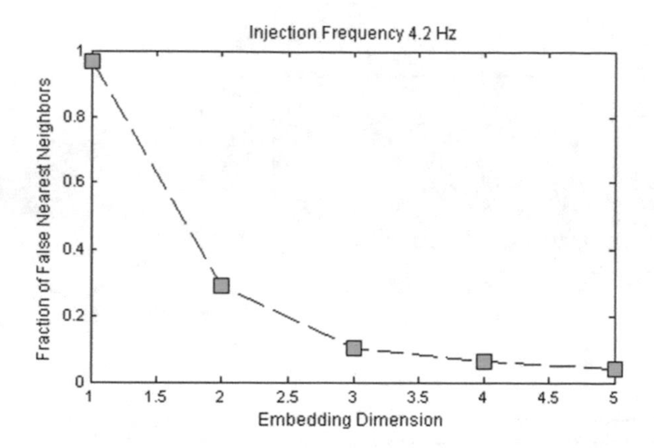

Fig. 8.29 Estimation of embedding dimension through the false nearest neighbor (FNN) method, which was implemented in the TISEAN software for the time series at $f = 4.2$. The plot shows that a good estimate for the embedding dimension is $d = 3$. This is because there is a significant drop in the percentage of false nearest neighbors when d changes from 2 to 3. The embedding dimension $d = 3$ is empirically determined from this graph

value at $f = 4.2$ Hz. Notice there is less than 0.1% change of the false nearest neighbors when d changes from 3 to 4. It is then reasonable to assume that the required embedding dimension to be $d = 3$. This result matches the expected embedding dimension.

Proceeding with an assumed dimension of $d = 3$, the time delay map for the DIBS model can be written as: $\Delta T_n = (\Delta t_n, \Delta t_{n-1}, \Delta t_{n-2})$. Figure 8.30 shows two time series plots and their phase space embedding portraits for two BIF values, $f = 4.2$ Hz and $f = 4.6$ Hz. The phase portrait for $f = 4.2$ shows a successful embedding of an attractor.

8.6.6 *Model Fitting*

With the embedding dimension to be $d = 3$, the embedding function from time series to phase space is $\Delta T_n = (\Delta t_n, \Delta t_{n-1}, \Delta t_{n-2})$. With the new embedded points on the trajectory in phase space, a typical question would be to find an map, at least computationally, to model the deterministic evolution of the new time series $\{\Delta T_n\}$, namely, to seek a nonlinear map $\Delta T_{n+1} = F(\Delta T_n)$ for the purpose of forecasting future trajectory in phase space, eventually for original time series, as is depicted in Fig. 8.31. Many nonlinear prediction models for time series have been studied since early 1990s. Well-studied models are global polynomial model fitting and local linear model fitting. The resulting fitting map then can be used computationally to forecast the future trajectory and to evaluate Lyapunov Exponent spectrum.

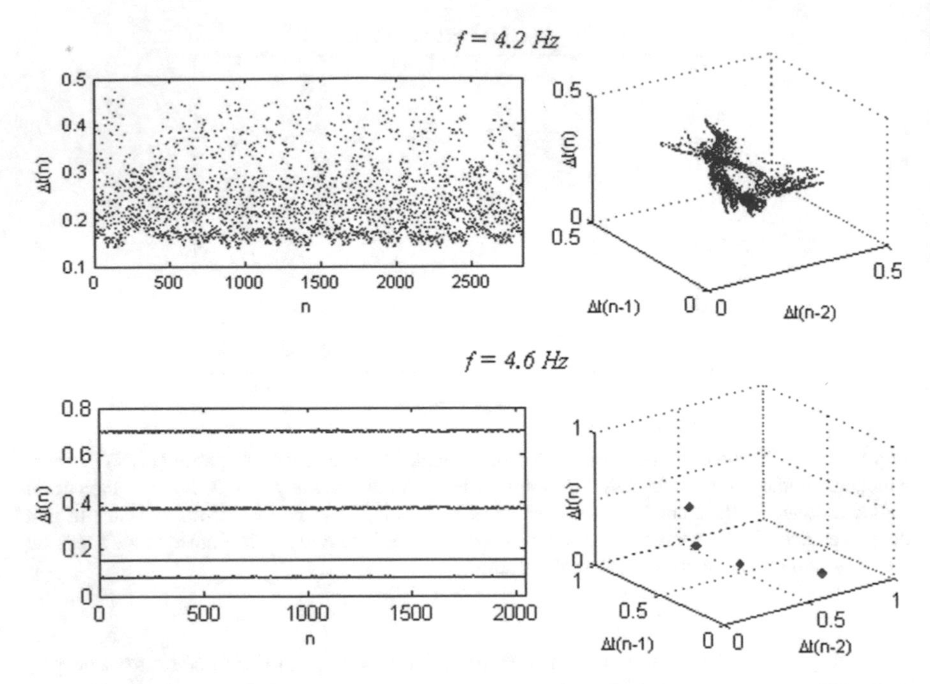

Fig. 8.30 Time series plots and their corresponding phase portraits for the data from DIBS simulations. Two pictures on left side are the time series plots for $f = 4.2$ and $f = 4.6$. The pictures on the right side are the phase portraits using embedding dimension $d = 3$

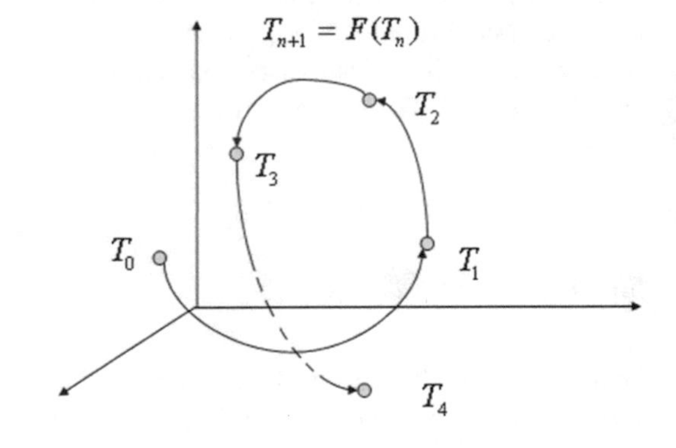

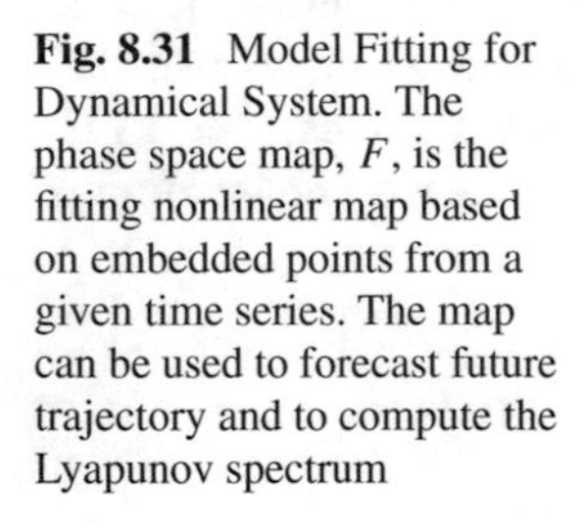
Fig. 8.31 Model Fitting for Dynamical System. The phase space map, F, is the fitting nonlinear map based on embedded points from a given time series. The map can be used to forecast future trajectory and to compute the Lyapunov spectrum

The local linear model was introduced by Eckmann et al. [38] in 1986 and Farmer et al. [39] in 1987. This model is first chosen to do model fitting for the time series of the bubble dynamics. With the embedding vector as $\Delta T_n = (\Delta t_n, \Delta t_{n-1}, \Delta t_{n-2})$, both T_n and T_{n+1} have same Δt_n and Δt_{n-1}. Thus it suffices to find a map f for the last component such that $\Delta t_{n+1} = f(\Delta T_n)$. The simplified case leads to approximate f by a linear function of the form $f(\Delta T_n) = \vec{a_n} \cdot \Delta T_n + b_n$. The vector $\vec{a_n}$ and the scalar b_n are then found by minimizing:

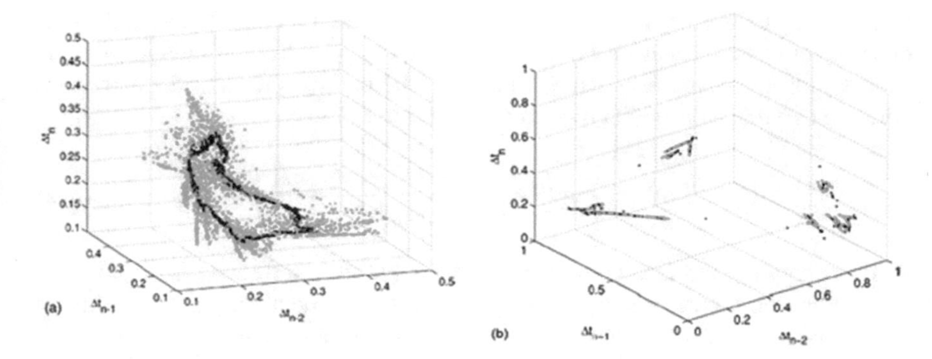

Fig. 8.32 Local Linear Fitting for Bubble Dynamics. The phase space map, F, is the fitting nonlinear map based on embedded points from a given time series. The map can be used to forecast future trajectory and to compute the Lyapunov spectrum

$$\sum_{i=1}^{M} \| \Delta t_{n+1} - \vec{a}_n \cdot \Delta T_i - b_i \| \tag{8.42}$$

where M is the total number of time delay vectors.

The TISEAN software provides two routines, *onestep* and *nstep*, that implement local linear fitting algorithm for a given time series. The *onestep* takes input of time series data and outputs the forecasted error with the local linear fitting model. The important arguments in using the routine are the embedding dimension, delay and neighborhood size. The routine, *nstep*, using the same inputs as those for *onestep*, produces predicted trajectory that is computed by applying the fitting algorithm with the given time series data. Detailed information to use TISEAN software can be found in its online manual by Hegger et al. [40].

For the time series of bubble dynamics from DIBS simulation, the local linear fitting works very well for a wide range of BIF values up to $f = 5.55$ Hz. Figure 8.32 shows the predicted trajectories (in black) of 2000 iterations and the input time series from DIBS simulation (in green) at two representative values of BIF values at 4.2 and 4.8 Hz.

Beyond $f = 5.55$ Hz, where the system exhibits intermittency between periodic orbits and chaotic attractors, both approximation methods, local linear fitting and global nonlinear multivariate polynomial fitting, failed to produce adequate long-term prediction.

8.6.7 Lyapunov Exponent

In Chap. 3 we introduced Lyapunov Exponents as a measure of the future growth and decay rate for a small initial perturbation. Recall the exponents to be a quantitative measure for the sensitivity of a dynamical system on initial conditions, and for the presence of chaotic behavior. For a discrete map, $x_{n+1} = F(x_n)$, and a given initial state x_0, the Lyapunov exponent is defined to be the log value for the divergence between two trajectories with a small perturbation ϵ_0 :

$$\lambda(n) = \frac{1}{n} \ln \left(\frac{\| f^n(x_0 + \epsilon_0) - f^n(x_0) \|}{\| \epsilon_0 \|} \right) \tag{8.43}$$

By computing the Lyapunov exponents for the time series, $\{\Delta T_n\}$ from DIBS simulation, the divergence (or convergence) rates at which neighboring orbits on each individual attractor, for each bubble injection frequency, can be quantified as the time-passage dynamics evolves in time. Many computational algorithms have been proposed for computing an estimate of Lyapunov exponent. The TISEAN software package implements two main algorithms for computing Lyapunov exponent. One is the algorithm to compute a Lyapunov exponent by Rosensetein et al. [41] and by Kantz [42], independently, which test directly the exponential divergence of nearby trajectories. Another algorithm is to compute the Lyapunov spectrum by estimating the local *Jacobians* from a fitting model in embedding space. Unfortunately none of these works well with the time series from the DIBS simulation due to the presence of a fixed point when the bubble injection frequency is very small. To circumvent these problems, we chose a simpler approach based on small perturbations. A small perturbation can be achieved by delaying the bubble injection with a small time δt at some time t_1 in the DIBS simulation. An estimate of largest Lyapunov exponent is computed by averaging the λ_n in Eq. (8.43), for the purpose of obtaining a more statistically meaningful measure of Lyapunov exponent, with the formula:

$$\lambda = \frac{1}{N} \sum_{i=1}^{N} \lambda(i) \tag{8.44}$$

In practice, a small perturbation of $\delta t = 0.05$ generates very good results of Lyapunov exponents for a wide range of injection frequencies, except for a small area between 4.5 and 5.6 Hz, as is shown in Fig. 8.33. Observe that the sign of the exponent agrees with the attractor depicted by the bifurcation diagram. That is, for low frequencies, the largest exponent is negative, indicating convergence towards the fixed point, as is normally observed in laboratory experiments as well as in the DIBS simulations. A positive Lyapunov exponent is indicative of deterministic chaotic behavior in the passage-time dynamics.

For frequencies larger than $f = 5.55$ Hz, the sign of the largest exponent is mainly positive. In this region, however, the time-passing dynamics is not only chaotic but rather intermittent, randomly switching between chaotic attractors and high-period orbits. Observe also that the largest Lyapunov exponent is zero at the points of bifurcation where the system dynamics changes behavior. The presence of the intermittency in the bifurcation sequence is likely to be difficult to ever be observed experimen-

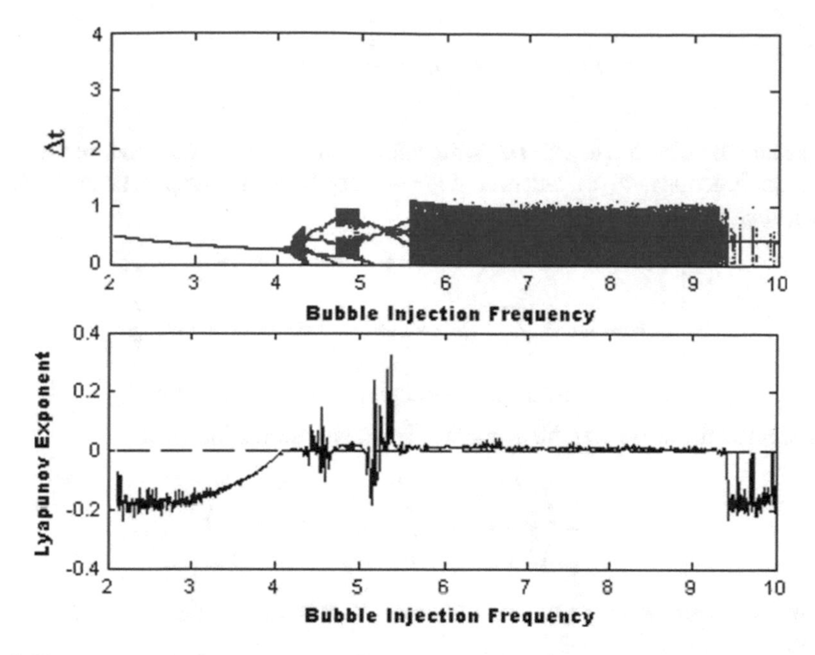

Fig. 8.33 Lyapunov Exponents of Bubble Dynamics. Largest Lyapunov exponent of time-series bubble dynamics for various values of injection frequencies. A positive exponent is indicative of chaotic behavior in the system's dynamics. Fixed points in time series match well with negative Lyapunov exponents

tally because of the presence of parametric noise, e.g. from gas-flow turbulence or granular particle flow.

8.7 Proper Orthogonal Decomposition

Proper Orthogonal Decomposition (POD), also known as Karhunen–Loève Decomposition (KLD), is a well-known technique for determining an optimal basis for a data set [43–48]. This section reviews the definitions and properties of POD decomposition relevant to modeling spatio-temporal systems and discusses how the method can be applied to image data in order to separate spatial and temporal behavior. Additional properties and technical details of the POD decomposition can be found in Appendix C.

Consider a sequence of observations represented by the scalar functions $u(\mathbf{x}, t_i)$, $i = 1 \ldots M$. The functions u are assumed to be L^2 on a domain D which is a bounded subset of R^n. The functions are parametrized by t_i, which represents time in this application. The (time) average of the sequence, defined as

$$\bar{u}(\mathbf{x}) = \langle u(\mathbf{x}, t_i) \rangle = \frac{1}{M} \sum_{i=1}^{M} u(\mathbf{x}, t_i),$$

is assumed to be zero. The POD decomposition extracts time-independent orthonormal basis functions, $\Phi_k(\mathbf{x})$, and time-dependent orthonormal amplitude coefficients, $a_k(t_i)$, such that the reconstruction

$$\boxed{u(\mathbf{x}, t_i) = \sum_{k=1}^{\infty} a_k(t_i)\, \Phi_k(\mathbf{x})\,, \quad i = 1, \ldots, M} \tag{8.45}$$

is optimal in the sense that the average least square truncation error

$$\varepsilon_N = \left\langle \left\| u(\mathbf{x}, t_i) - \sum_{k=1}^{N} a_k(t_i)\, \Phi_k(\mathbf{x}) \right\|^2 \right\rangle \tag{8.46}$$

is always a minimum for any given number N of basis functions over all possible sets of orthogonal functions.

The functions $\Phi_k(\mathbf{x})$, called *empirical eigenfunctions*, *coherent structures*, or *POD modes*, are the eigenvectors of the *two-point spatial correlation* function

$$r(\mathbf{x}, \mathbf{y}) = \frac{1}{M} \sum_{i=1}^{M} u(\mathbf{x},\ t_i)\, u^T(\mathbf{y}, t_i) \tag{8.47}$$

8.7.1 Computational Implementation

POD decomposition can be generally applied to find the optimal basis of a data set. To separate spatial and time behavior for a physical system, each point in the data set should represent an observation of the spatial state of the system at a particular time. POD decomposition is applied to the observations to find an optimal basis for the spatial observations. The data set is projected on the resulting POD basis functions to obtain the time behavior in much the same way as normal mode expansions are used for partial differential equations. The POD technique is based purely on the observations and thus has the advantage of not requiring knowledge of an underlying model equation or normal modes.

In practice the state of a numerical model is only available at discrete spatial grid points, and so the observations that form the data set are vectors rather than continuous functions. In other words, $D = (x_1, x_2, \ldots, x_N)$, where x_j is the jth grid point and $u(\mathbf{x}, t_i)$ is the vector

$$\mathbf{u}_i = [u(x_1, t_i), u(x_2, t_i), \ldots, u(x_N, t_i)]^T.$$

Experimental data also undergoes a discretization process when it is acquired for processing. In the case of the combustion experiment, images of the flame front were digitized to obtain the observations at different times. Each image is a $w \times h = N$ array of pixels. A pixel is a scalar value in the range $[0, 255]$. An image can be converted to a vector by ordering the pixel values in row major form (e.g. pixel (j, k) in the image is stored in position $n = j \times w + k$ in the vector).

8.7.2 The Method of Snapshots

A popular computational technique for finding the eigenvectors of (8.47) is the *method of snapshots* developed by Sirovich [47]. It was introduced as an efficient method when the resolution of the spatial domain (N) is higher than the number of observations (M). The method of snapshots is based on the fact that the data vectors, $\mathbf{u}_i$, and the eigenvectors Φ_k, span the same linear space (see [43, 47] for details). This implies that the eigenvectors can be written as a linear combination of the data vectors

$$\Phi_k = \sum_{i=1}^{M} v_i^k \mathbf{u}_i \tag{8.48}$$

After substitution in the eigenvalue problem, $r(\mathbf{x}, \mathbf{y})\Phi(\mathbf{y}) = \lambda\Phi(\mathbf{x})$, the coefficients v_i^k are obtained from the solution of

$$C\,\mathbf{v} = \lambda\,\mathbf{v} \tag{8.49}$$

where $\mathbf{v^k} = (v_1^k, \ldots, v_N^k)$ is the kth eigenvector of (8.49), and C is a symmetric $M \times M$ matrix defined by

$$\left[c_{ij}\right] = \frac{1}{M}(\mathbf{u}_i, \mathbf{u}_j),$$

where $(\cdot, \cdot)$ denotes the standard vector inner product,

$$(\mathbf{u}_i, \mathbf{u}_j) = u(x_1, t_i)u(x_1, t_j) + \cdots + u(x_N, t_i)u(x_N, t_j).$$

In this way an $N \times N$ eigenvalue problem (the eigenvectors of (8.47)) is reduced to computing the eigenvectors of an $M \times M$ matrix, a preferable task if $N \gg M$. Throughout the remaining of this section, M will denote the number of measurements of a laboratory or numerical experiment, and N will represent the maximum number of POD eigenfunctions employed in a particular reconstruction of an experiment.

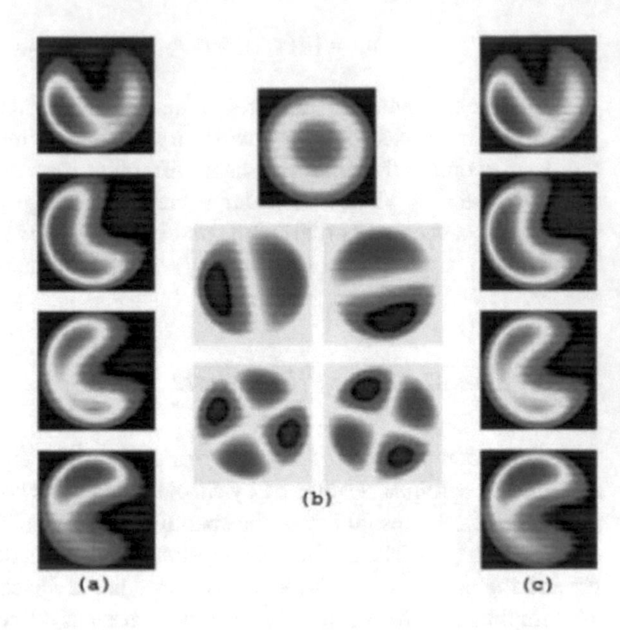

Fig. 8.34 **a** Four snapshots of a uniformly rotating one-cell state produced from simulations of Eq. (8.51); **b** the time-average and (from left-to-right and top-to-bottom) the four most energetic POD modes; **c** reconstruction of the dynamics using the four most energetic POD modes

After computing the POD basis functions, the temporal coefficients $a_k(t)$, are calculated by projecting the data set onto each of the eigenfunctions. This operation is carried out using the inner product defined above, and it leads to

$$a_k(t) = \frac{(u(\mathbf{x}, t), \Phi_k(\mathbf{x}))}{(\Phi_k(\mathbf{x}), \Phi_k(\mathbf{x}))}. \tag{8.50}$$

Example 8.4 (*Flame Dynamics*) We consider in this example the Brusselator model of two chemical reactions, see Eq. (8.31). For completeness purposes, we re-write the equations

$$\boxed{\begin{aligned} u_t &= \kappa_1 \nabla^2 u + (B-1)u + A^2 v - \eta u^3 - \nu_1 (\nabla u)^2 \\ v_t &= \kappa_2 \nabla^2 v - Bu - A^2 v - \eta v^3 - \nu_2 (\nabla v)^2. \end{aligned}} \tag{8.51}$$

Numerical simulations of the Brusselator model have demonstrated the formation of both stationary and nonstationary states. These states emerge as a result of symmetry-breaking bifurcations in which several spatial modes couple and compete for existence. Recall that these equations describe the evolution of two coupled, diffusive spatiotemporal fields $u(\mathbf{x}, t)$ and $v(\mathbf{x}, t)$, where κ_1 and κ_2 are the diffusion coefficients of the two linearly coupled fields.

In order to simulate the circular geometry of the experimental burner, the integration of Eq. (8.51) is carried out in polar coordinates over a circular grid of radius R. Small changes in the radius, R, can produce qualitatively different flame patterns. This observation leads us to consider the radius as a distinguished bifurcation parameter.

Figure 8.34 shows four snapshots of a single cell state simulated with Eq. (8.51) for parameter values ($\eta = 2.0, \nu_1 = 0.5, \nu_2 = 1.0, \kappa_1 = 0.2, \kappa_2 = 2.0, A = 5.0, B = 6.8, R = 1.35$). It consists of a single cell rotating clockwise that does not change its shape. The POD decomposition of a complete period produces an O(2) invariant time-average pattern, Fig. 8.34b and four POD modes, Φ_1, Φ_2, Φ_3 and Φ_4, (depicted from left-to-right and top-to-bottom, respectively).

A similar POD spectrum indicates that the first two modes, Φ_1 and Φ_2, capture about 94% of the energy. Only two modes are necessary to reconstruct the dynamics, and the remaining modes affect other aspects such as cell shape. Observe that nearly 100% of the energy is captured by the first four modes. The reconstruction with these four POD modes is shown in Fig. 8.7c.

The details of the temporal behavior of the cell are extracted from the POD projections. Figure 8.8a shows phase plots of $a_1(t)$ vs $a_2(t)$, $a_3(t)$ vs $a_4(t)$ and $a_1(t)$ vs $a_3(t)$. The first pair indicates the uniform rotation of the cell. The second pair indicates a periodic oscillation at twice the frequency of the dominant pair. The plots of the relative phases for each pair of POD modes shown in Fig. 8.8b confirm this relationship.

Example 8.5 (*Bubble Dynamics*) With multiple bubble injectors, the bubble dynamics becomes more complicated. Bubble coalescence does not only happen in vertical direction but occur in all directions. As bubble injection frequency increases, a large number of bubbles are present in fluidized bed, moving upward and coalescing with each other. To capture the bubble dynamics with multiple injectors, bubble dynamical process is digitized into frames to construct a data matrix, see Fig. 8.35. To generate one frame of data, a grid with pre-determined size is set on the projected rectangular area. For each block inside the grid, the area covered by projected bubbles is computed. Then the fraction of covered area versus the area of the block is evaluated and is saved as the value for the block, or one component for the output vector of final multi-dimensional time series.

When a fine grid is set to digitize the dynamical information, each small block can be viewed as a pixel in the whole image. Each block value composes one cell value in the matrix for a digitized video frame. Then the data in matrix is aligned into one column in the final data matrix for POD analysis.

The computing power must be considered in choosing the grid size. Although a finer grid may create more accurate digitized information, the resulting data matrix can easily have the size beyond the computing capability of a computer. In the case of a 22×70 grid, 3502 frames were collected from bubble simulation and the final data matrix has the size of 1540×3520.

Numerical experiments are first conducted with nine (chosen as an example) injectors that are sparsely distributed on the bottom plate with simultaneous bubble injection mode. The motivation is based on intuition that the behavior of the bubble dynamics should have similar patterns as those observed in a single bubble injector case since nine injectors are sparsely distributed. Several data points are picked from

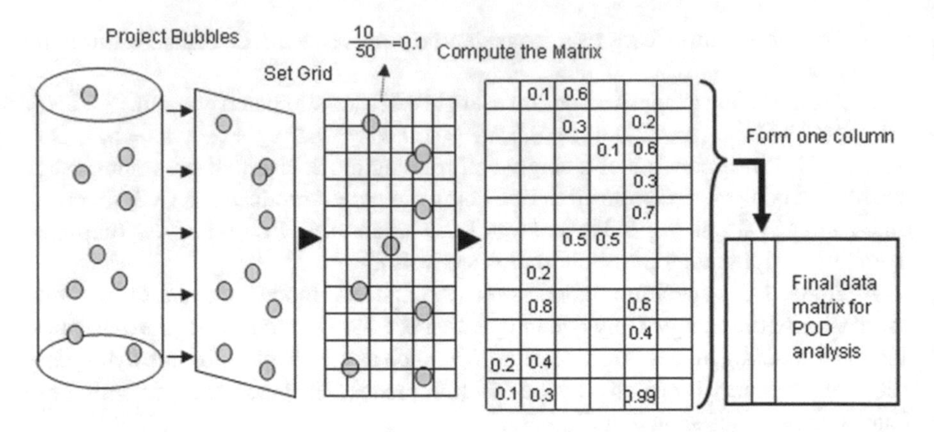

Fig. 8.35 Digitization of Bubble Dynamics. A schematic diagram for constructing one frame of digitized data for bubble dynamics. Rising bubbles are projected onto a plane. A grid is set on the rectangle area and the data in each block of the grid is computed based on the area covered by bubbles. A frame of digitized data is created and is aligned to form one column in the final matrix for POD analysis

(0, 10] for BIF values to conduct our research. A 3D visualization for the DIBS simulation is developed to monitor bubble dynamics in real-time simulation. The top four images in Fig. 8.36 shows a group of snapshots from simulations with the BIF values taken at $f = 0.5$, $f = 3.5$, $f = 6.5$ and $f = 9.5$ Hz. Four pictures in the bottom of Fig. 8.36 are the snapshots for the digitized frames with the chosen BIF values.

Nine bubbles are generated simultaneously and move upward in parallel and form a layer of bubbles in the testbed. With small BIF value, there is a large distance between two layers of bubbles and thus there is no interaction or bubble coalescence among bubbles. As BIF value gets to larger over a threshold around 1.0 Hz, bubbles start to interact with each other and coalesce. One common phenomenon with large BIF value is the channeling in which bubbles tends to collapse into middle of the testbed and thus form a channel. The pictures of the 3D visual display in the top of Fig. 8.36 show such channeling behavior for BIF values at 3.5, 6.5 and 9.5 Hz.

To carry POD analysis, a grid is used to digitize each frame in running the DIBS simulation with multiple bubble injectors. A total of 3502 frames are collected for each BIF value. This generates a data matrix of size 1540×3502 for each BIF value. Our POD analysis program is encoded in Matlab and uses a built-in routine for SVD decomposition. Figure 8.37 shows the first eight modes obtained directly from the simulations.

The energy distribution plots (not shown for brevity) indicate that it requires at least 200 modes for $f = 0.5$ and at least 400 modes for $f = 3.5$, $f = 6.5$ and $f = 9.5$ to capture 80% of the total energy.

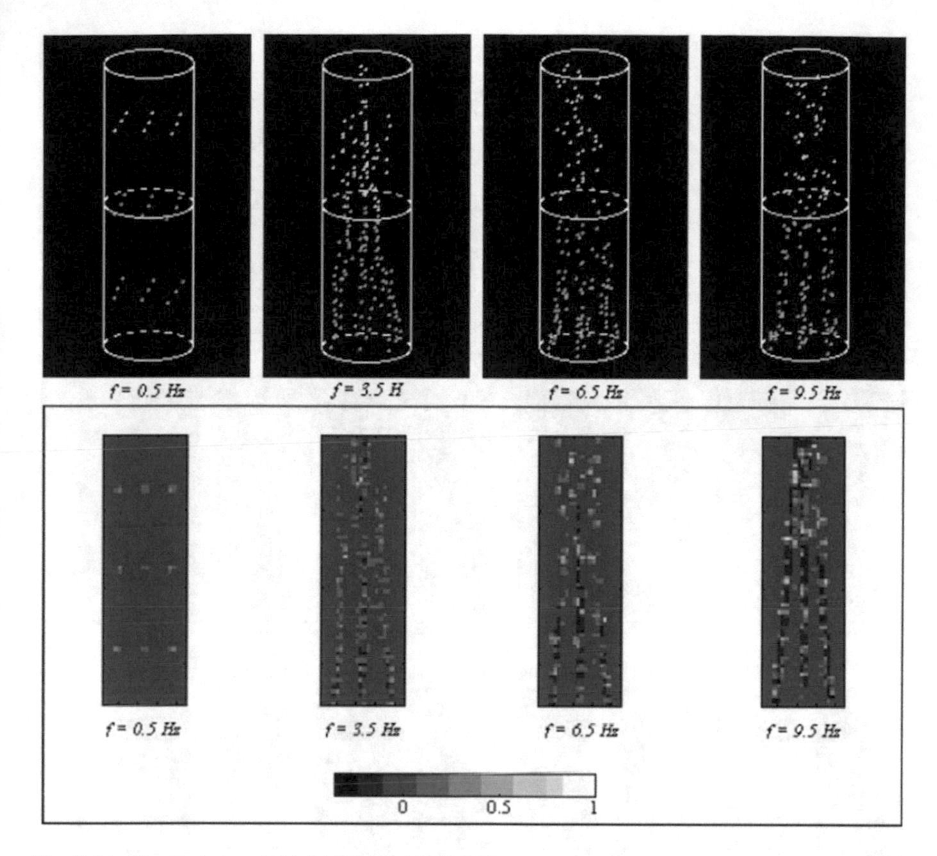

Fig. 8.36 Simulation and Digitization of Bubble Dynamics. DIBS simulations with nine bubble injectors and their digitized frames. The images on the top are snapshots taken from the 3D real-time visualization for bubble simulation with multiple injectors. The images in the bottom are their corresponding images of digitized frames. The values in the matrix of a digitized frame are normalized

8.8 The Symmetry Perspective

In Chap. 5, Sect. 5.7, we introduced the idea of using the underlying group of symmetries of a model to directly predict its behavior, i.e., without using numerical simulations or experiments. Later on, in Sect. 5.8 the concept of *symmetry-breaking bifurcations*, and the relevant methods, were formalized. In particular, two types of symmetry-breaking bifurcations were discussed, steady-state and Hopf bifurcations. It turns out that the same theory, and same methods, can be used to study and predict the emergence of patterns in spatial-temporal systems. The only difference is that this time steady-state bifurcations will be associated with steady-state patterns and Hopf bifurcations with patterns that change periodically (or quasi-periodically) in

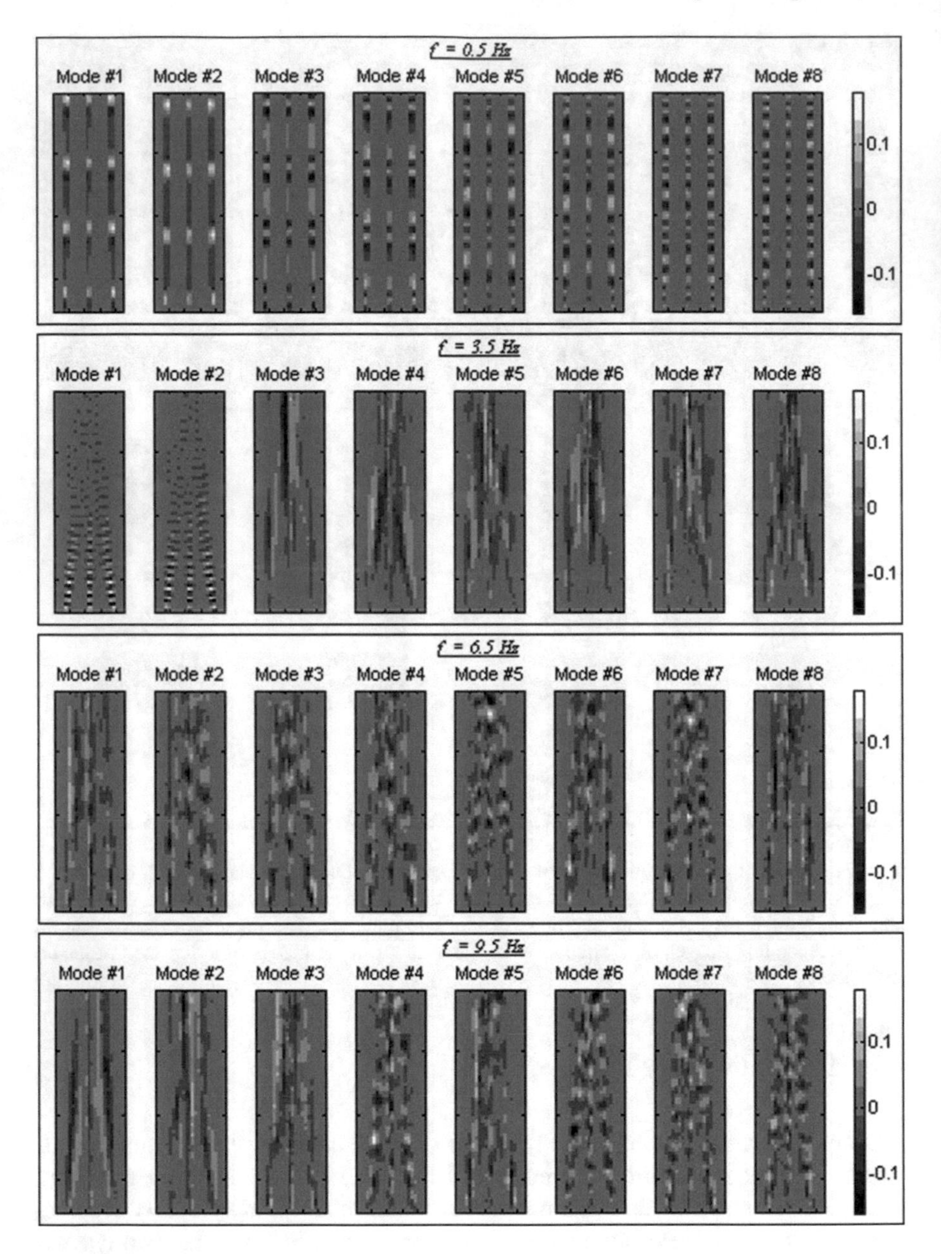

Fig. 8.37 POD modes of bubble dynamics. Plots of the first eight modes for the DIBS simulations with 4 bubble injection frequencies. All modes form an orthogonal basis

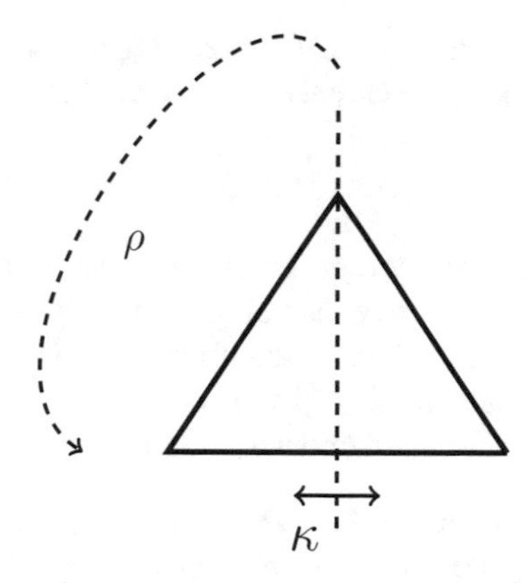

Fig. 8.38 Symmetries of a triangle are described by the dihedral $\mathbf{D}_3$ group

space and time. As an example, we consider next the case of a pattern forming system over a triangular domain.

8.8.1 *Steady-State Bifurcation in a Triangular Domain*

Consider, for instance, a pattern-forming system in a triangular domain. The triangular domain has symmetry group $\mathbf{D}_N$–the dihedral group of symmetries of an N-gon, $N = 3$ in this case. The symmetry group $\mathbf{D}_3$ is generated by a rotation ρ by $2\pi/3$, and a reflection κ across the line that connects a vertex to the middle point of the opposite side, just as is shown in Fig. 8.38.

In Chap. 5, we discussed the concept of the representation of a group. Recall that group elements themselves do not transform an object. Instead, they only describe the abstract geometry of an object. The actual transformation of an object is done through the matrices that are associated with group elements, i.e., the representation of the group. A group will typically have many representations. Each representation will lead to a new bifurcation problem. And when the theory is applied to pattern-forming systems, each representation will lead to the emergence of a new pattern with less symmetry than that of the original domain, i.e., symmetry-breaking bifurcations of patterns. We continue next with our example of a pattern-forming over a triangular domain.

8.8.2 *Irreducible Representations*

The elements of the dihedral group $\mathbf{D}_3$ are: $\{e, \rho, \rho^2, \kappa, \kappa\rho, \kappa\rho^2\}$. Observe that each element can be obtained through the combination of one or two other elements, ρ and κ. For this reason, we can say that $\mathbf{D}_3$ is generated by $\{\rho, \kappa\}$. Two group elements,

γ_1 and γ_2, in Γ are *conjugate*, or in the same *class* or *conjugacy class* if there is a group element $\sigma \in \Gamma$, such that

$$\gamma_1 = \sigma\, \gamma_2\, \sigma^{-1}. \tag{8.52}$$

For the group $\mathbf{D}_3$, one can verify that $\rho^2 = \kappa\rho\kappa^{-1}$, so that ρ and ρ^2 are in the same conjugate class. Similarly, κ, $\kappa\rho$ and $\kappa\rho^2$ are all three in their own conjugate class. Finally, the single identity element e is in its own conjugacy class since it satisfies Eq. (8.52) for any $\sigma \in \Gamma$. In conclusion, $\mathbf{D}_3$ has three conjugacy classes. This number determines the number of irreducible representations, see next theorem.

Theorem 8.2 *The number of inequivalent irreducible representations of a finite group Γ is equal to the number of conjugacy classes of Γ.*

This last theorem means that $\mathbf{D}_3$ has three irreducible representations. To find them, we now employ the following theorem.

Theorem 8.3 *The sum of the squares of the dimensions d_i of the n inequivalent irreducible representations of a finite group Γ, is equal to the order, $|\Gamma|$, of Γ. That is*

$$\sum_{i=1}^{n} d_i^2 = |\Gamma|.$$

According to Theorem 8.3, the only possibility for the group $\mathbf{D}_3$ is for it to have two one-dimensional irreducible representations and one two-dimensional irreducible representation. Since the generators, ρ and κ are of order 3 and 2, respectively, this means that the associated (1×1) matrices must satisfy

$$A_\rho^3 = 1, \quad A_\kappa^2 = 1,$$

Thus, $A_\rho = \{1, -\frac{1}{2} \pm \frac{\sqrt{3}}{2}i\}$ and $A_\kappa = \pm 1$. The complex roots have to be ruled out because $(\kappa\rho)^2 = e$ implies $A_{\kappa\rho}^2 = 1$, which cannot be satisfied. The only root we can use is $A_\rho = 1$, which combined with $A_\kappa = \pm 1$ yields two representations: $A_\rho = 1$ and $A_\kappa = 1$ leads to the trivial representation; $A_\rho = 1$ and $A_\kappa = -1$ leads to the alternating representation. The three representations of the group $\mathbf{D}_3$ are now listed in Table 8.2.

Table 8.2 Character table for the symmetry group $\mathbf{D}_3$

Rep. \ D_3	e	ρ	ρ^2	κ	$\kappa\rho$	$\kappa\rho^2$
Trivial	1	1	1	1	1	1
Non-trivial	1	1	1	−1	−1	−1
Natural	2	−1	−1	0	0	0

As for the *natural representation*, recall from Sect. 5.8 that this is a two-dimensional representation given by six matrices $\{A_e, A_\rho, A_{\rho^2}, A_\kappa, A_{\kappa\rho}, A_{\kappa\rho^2}\}$, where

$$A_e = \begin{bmatrix} 1 & 0 \\ 0 & 1 \end{bmatrix}, \quad A_\rho = \begin{bmatrix} -\frac{1}{2} & -\frac{\sqrt{3}}{2} \\ \frac{\sqrt{3}}{2} & -\frac{1}{2} \end{bmatrix}, \quad A_{\rho^2} = \begin{bmatrix} -\frac{1}{2} & \frac{\sqrt{3}}{2} \\ -\frac{\sqrt{3}}{2} & -\frac{1}{2} \end{bmatrix},$$

$$A_\kappa = \begin{bmatrix} -1 & 0 \\ 0 & 1 \end{bmatrix}, \quad A_{\kappa\rho} = \begin{bmatrix} \frac{1}{2} & \frac{\sqrt{3}}{2} \\ \frac{\sqrt{3}}{2} & -\frac{1}{2} \end{bmatrix}, \quad A_{\kappa\rho^2} = \begin{bmatrix} \frac{1}{2} & -\frac{\sqrt{3}}{2} \\ -\frac{\sqrt{3}}{2} & -\frac{1}{2} \end{bmatrix}.$$

8.8.3 Eigenmodes

We have emphasized throughout the text that each irreducible representation of the group of symmetries Γ leads to a bifurcation problem of the form

$$\frac{dx}{dt} = f(x, \lambda), \tag{8.53}$$

where f is Γ-equivariant. That is,

$$A_\gamma f(x, \lambda) = f(A_\gamma x, \lambda), \quad \text{for all} \quad \gamma \in \Gamma.$$

Assume (0, 0) to be an equilibrium of Eq. (8.53), which undergoes a steady-state bifurcation. From Chap. 5, Eq. (5.11), we know the bifurcation condition to be

$$\begin{aligned} f(x, \lambda) &= 0 \\ f_x(x, \lambda) &= 0. \end{aligned}$$

Next, we address each representation independently and attempt to derive the mathematical model Eq. (8.53) associated with each case.

Trivial Representation

In this case, all matrices are $+1$, so that Γ-equivariance of f implies

$$1 \cdot f(x, \lambda) = f(1 \cdot x, \lambda).$$

In words, the trivial representation does not impose any restrictions on the form of the vector field $f(x, \lambda)$. Hence, a Taylor expansion around the zero equilibrium $(x, \lambda) = (0, 0)$ yields:

$$f(x, \lambda) = f(0, 0) + f_x(0, 0) + f_\lambda(0, 0)\lambda + \frac{1}{2} f_{xx}(0, 0)x^2 + f_{x\lambda} x\lambda + \frac{1}{2} f_{\lambda\lambda}\lambda^2 + \ldots$$

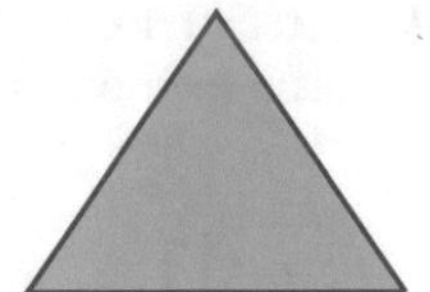
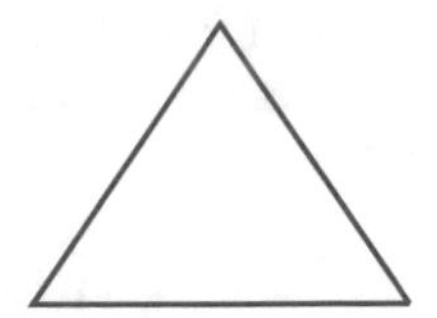

Fig. 8.39 Solutions predicted by trivial irreducible representation of the group $\mathbf{D}_3$

The first two terms vanish due to the bifurcation condition. Rescaling time by $\tau = f_\lambda(0, 0)$, and relabeling the time variable τ, we can rewrite the model equations as

$$\frac{dx}{dt} = \lambda + a_{x^2}(0, 0)x^2 + a_{x\lambda}(0, 0)x\lambda + a_{\lambda^2}(0, 0)\lambda^2 + \dots$$

Now, if we consider the neighborhood of the equilibrium point $(x, \lambda) = (0, 0)$ to be $|x| < \varepsilon$ and $|\lambda| < \varepsilon^2$, where $0 < \varepsilon \ll 1$, then the last equation can be written (after a suitable re-scaling) as

$$\frac{dx}{dt} = \lambda \pm x^2 + O(x\lambda, \lambda^2, x^3). \tag{8.54}$$

We recognize Eq. (8.54) as the normal form for the saddle-node bifurcation of Sect. 5.5, see Eq. (5.27). From the standpoint of a pattern-forming system, and under the trivial representation, every point x is fixed by the group of symmetries $\mathbf{D}_3$. It follows that the bifurcation branch must also share the $\mathbf{D}_3$-symmetry. This means that the emerging pattern does not change, it is still the homogeneous pattern that appears right before the bifurcation, as is shown in Fig. 8.39.

Alternating Representation

In the non-trivial representation, also known as the *alternating representation*, matrices $\{A_e, A_\rho, A_{\rho^2}\}$ are $+1$, while matrices $\{A_\kappa, A_{\kappa\rho}, A_{\kappa\rho^2}\}$ are -1. The restriction imposed on f by these three last matrices implies

$$-f(x, \lambda) = f(-x, \lambda).$$

It follows that f must be an odd function in x. A Taylor expansion of f, similar to the previous case, yields the normal form associated with this bifurcation

$$\frac{dx}{dt} = \lambda x \pm x^3 + O(x\lambda, \lambda^2, x^3). \tag{8.55}$$

Again, we recognize Eq. (8.55) as the normal form for the pitchfork bifurcation, see Eq. (5.29). The "+" case corresponds to a subcritical pitchfork bifurcation and the "−" one to a supercritical pitchfork bifurcation. The pattern solution that emerges from this bifurcation must remain unchanged under rotations by matrices $\{A_e, A_\rho, A_{\rho^2}\}$ because these matrices are all $+1$ under the alternating representation.

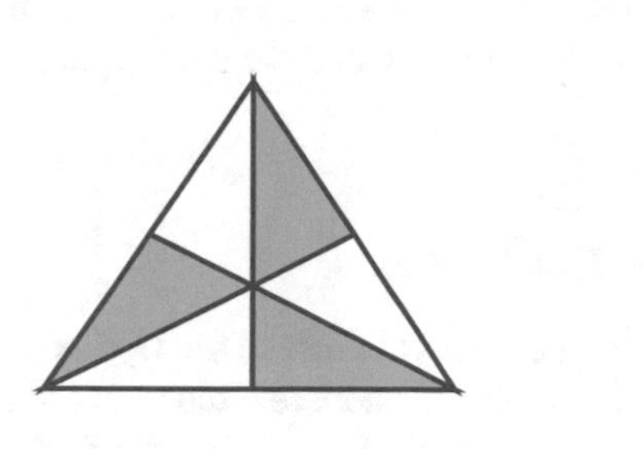

Fig. 8.40 (Left) Pattern solution predicted by the alternating irreducible representation of the group $\mathbf{D}_3$. (Right) Lattice of isotropy subgroup confirms the emerging pattern has $\mathbf{Z}_3$-symmetry, i.e., cyclic rotations by multiples of $\rho = 2\pi/3$

This means that the isotropy subgroup, Σ_x, of the emerging pattern is equivalent to the cyclic group of three elements, that is $\Sigma_x = \mathbf{Z}_3$. Similarly, the matrices associated with reflections, i.e., $\{A_\kappa, A_{\kappa\rho}, A_{\kappa\rho^2}\}$, act as -1, thus sending the pattern to its mirror image. If we represent $+1$ with a light gray color and -1 as a white color then the bifurcation pattern would have the form shown in Fig. 8.40(left). On the right, is the lattice of isotropy subgroup, which describes, schematically, the underlying symmetry-breaking bifurcations that lead to the new pattern solution.

Natural Representation

To derive the mathematical model Eq. (8.53), under the natural representation of $\mathbf{D}_3$, we will employ a little bit of *invariant theory*.

Definition 8.1 Let Γ be a compact Lie group acting on a vector space V. A real-valued function $f : V \to \mathbb{R}$ is Γ-invariant if

$$f(\gamma x) = f(x),$$

for all $\gamma \in \Gamma$ and $x \in V$.

Under the natural representation, the action of the elements γ of the group $\Gamma = \mathbf{D}_3$ on $\mathbf{C}$, are given by

$$\begin{aligned} \rho \cdot z &= e^{2\pi i/3} z \\ \kappa \cdot z &= \bar{z}. \end{aligned} \tag{8.56}$$

Thus, we need to find a function f that will remain $\mathbf{D}_3$-invariant under the action above. It is easy to check that $\{z\bar{z}, z^3 + \bar{z}^3\}$ is a basis for all $\mathbf{D}_3$-invariant functions under the natural representation. Details of the derivation will be left as an exercise.

Next, we need to consider mapping that commute with the action of the group of symmetries.

Definition 8.2 Let Γ be a compact Lie group acting on a vector space V. A mapping $f : V \to V$ *commutes* with Γ or is Γ-equinvariant if

$$f(\gamma x) = \gamma f(x),$$

for all $\gamma \in \Gamma$ and $x \in V$.

Similar calculations to those of the $\mathbf{D}_3$-invariants, yields a basis $\{z, \bar{z}^2\}$, for the $\mathbf{D}_3$-equivariant mappings. And a few more calculations show that every $\mathbf{D}_3$-equivariant polynomial mapping $f : \mathbf{C} \to \mathbf{C}$ has the form

$$f(z, \lambda) = p(z\bar{z}, z^3 + \bar{z}^3, \lambda)z + q(z\bar{z}, z^3 + \bar{z}^3, \lambda)\bar{z}^2, \tag{8.57}$$

where p and q are $\mathbf{D}_3$-invariant real-valued polynomials. Hence, the model equations (written in complex coordinates) for the bifurcation problem associated with the natural representation can be expressed as

$$\frac{dz}{dt} = p(z\bar{z}, z^3 + \bar{z}^3, \lambda)z + q(z\bar{z}, z^3 + \bar{z}^3, \lambda)\bar{z}^2. \tag{8.58}$$

Of course, one can also re-write the model Eq. (8.58) by substituting $z = x + yi$ and solving for dx/dt and dy/dt. We leave this task as an exercise.

Now, observe that vertices of the triangle shown in Fig. 8.38 are mapped into each other by $\Gamma = \mathbf{D}_3$. More importantly, the isotropy subgroup of a vertex on the real axis is the subgroup $\mathbf{Z}_2$ generated by κ. Formally, $\Sigma_x = \mathbf{Z}_2(\kappa)$. Since $\text{fix}(\mathbf{Z}_2) = \mathbb{R}$ then we conclude that the bifurcation problem occurs along the line $y = 0$. The actual pattern solution that emerges at the bifurcation is illustrated in Fig. 8.41.

Since $\text{fix}(\mathbf{Z}_2) = \mathbb{R}$, we can set $z = x$, so Eq. (8.58) becomes

$$\frac{dx}{dt} = p(x^2, 2x^3, \lambda)x + q(x^2, 2x^3, \lambda)x^2.$$

Equilibrium points are found by solving

$$(p + qx)x = 0.$$

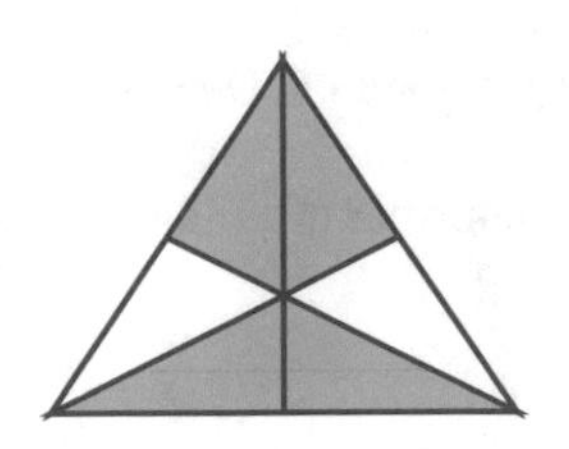

Fig. 8.41 (Left) Pattern solution predicted by the natural irreducible representation of the group $\mathbf{D}_3$. (Right) Lattice of isotropy subgroup confirms the emerging pattern has $\mathbf{Z}_2$-symmetry, i.e., reflections across the real axis

Fig. 8.42 One-dimensional lattice

We recognize this last equation as the normal form for a transcritical bifurcation, see Eq. (5.28). The nontrivial solution corresponds to $p + qx = 0$. In addition, $p(0, 0, 0) = 0$ since there is a bifurcation of $x = 0$ at $\lambda = 0$. A Taylor series expansion yields an expression for the bifurcating branch

$$\lambda = -\frac{q(0)}{p_\lambda(0)}x + \ldots$$

Finally, this last expression shows that if $q(0)/p_\lambda(0) < 0$ then the bifurcation pattern with isotropy subgroup $\mathbf{Z}_2(\kappa)$ is stable for $\lambda > 0$. Otherwise, if $q(0)/p_\lambda(0) > 0$ then it is stable for $\lambda < 0$.

8.8.4 Traveling Wave and Standing Wave Patterns

It is also possible to use the theory of Hopf bifurcation with symmetry to predict the emergence of periodic patterns that oscillate in space and time. In this section we consider those type of patterns. The spatial domain will consist of a one-dimensional lattice of the form as is shown in Fig. 8.42.

Mathematically speaking, the lattice can be defined as

$$\mathcal{L} = \{n \cdot \mathbf{k_1} : n \in \mathcal{Z}\},$$

where $\mathbf{k_1} = (1, 0)$. We are particularly interested in finding traveling waves, which (to lower-order) can be written as

$$u(x, t) = z_1(t)e^{(x-t)i} + z_1(t)e^{-(x+t)i} + c.c. \tag{8.59}$$

The term "*c.c.*" denotes complex conjugate and it is include to guarantee that the pattern $u(x, t)$ is real-valued. Both terms, $e^{(x-t)i}$ and $e^{-(x+t)i}$ are included in order to capture all possible types of spatio-temporal patterns. The subspace $(z_1, z_2) = (z_1, 0)$, in particular, accounts for a wave traveling to the right, in the direction of positive values of x. Similarly, the subspace $(z_1, z_2) = (0, z_2)$ accounts for a wave traveling in the opposite direction, i.e., to the left in the direction of negative values of x. The case where $z_1 = \pm z_2$ corresponds to a standing wave.

In all three cases, the waves should remain the same whether we look at the positive x-axis or the negative one. That is, the lattice has reflectional symmetry

$$\kappa \cdot x = -x.$$

Hoyle [49] shows that this reflection symmetry implies

$$(z_1, z_2) \rightarrow (z_2, z_1) \tag{8.60}$$

Similarly, the lattice can be translated to the right or to the left by any amount p. That is

$$p \cdot x = x + p.$$

Again, Hoyle [49] shows that translation symmetry leads to the following action

$$(z_1, z_2) \rightarrow (e^{-pi} z_1, e^{pi} z_2). \tag{8.61}$$

Observe that the actions described by Eqs. (8.60) and (8.61) are the same as those of a Hopf bifurcation in a system with $\mathbf{O(2)}$-symmetry. This type of symmetry-breaking bifurcation was studied in great detail in Chap. 5, through the example of flame dynamics on a circular domain, see Eq. (5.58). When the system is written in Birkhoff normal form [50, 51], an additional symmetry under phase shift is introduced. Together, the action of the $\mathbf{O(2)}$ group is given by

$$\begin{aligned}
\rho \cdot (z_1, z_2) &= (e^{-\rho i} z_1, e^{\rho i} z_2) \\
\kappa \cdot (z_1, z_2) &= (z_2, z_1) \\
\theta \cdot (z_1, z_2) &= (e^{\theta i} z_1, e^{\theta i} z_2).
\end{aligned}$$

It can be shown that, to cubic order, the mathematical model that governs the amplitudes of the oscillating waves is given by

$$\boxed{\begin{aligned}
\frac{dz_1}{dt} &= \mu z_1 - (\alpha + \beta i)|z_1|^2 z_1 - (\gamma + \delta i)|z_2|^2 z_1 \\
\frac{dz_2}{dt} &= \mu z_2 - (\alpha + \beta i)|z_2|^2 z_2 - (\gamma + \delta i)|z_1|^2 z_2.
\end{aligned}} \tag{8.62}$$

where μ, α, β, γ and δ are all real-valued parameters.

Figure 8.43 illustrates a left-traveling wave and a standing wave found in computer simulations of the amplitude equations (8.62). The waves are reconstructed using Eq. (8.59).

8.9 Exercises

Exercise 8.1 Solve the Sturm–Liouville problem

$$\frac{d^2\Psi}{d\theta^2} + \mu\Psi(\theta) = 0, \qquad \Psi(-\pi) = \Psi(\pi), \quad \Psi_\theta(-\pi) = \Psi_\theta(\pi)$$

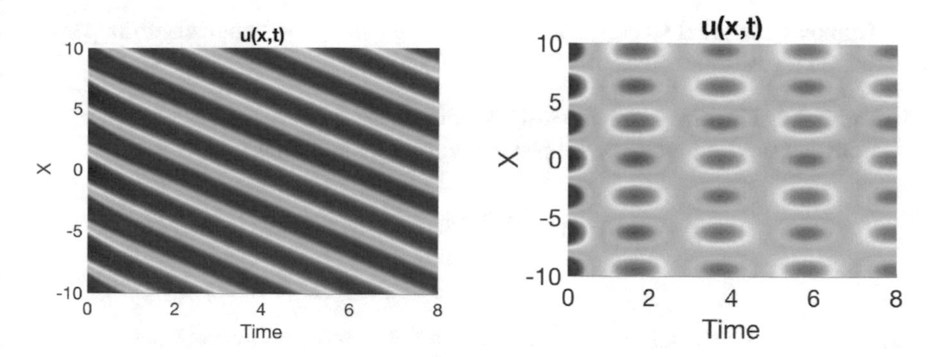

Fig. 8.43 Computer simulations of amplitude equations (8.62) over a one-dimensional lattice show a (Left) Left-traveling wave and (Right) Standing wave

when $\mu > 0$.

Exercise 8.2 Consider the following version of the Brusselator model

$$\frac{\partial X}{\partial t} = X^2Y - (1+\beta)X + \alpha + D_X\nabla^2 X$$
$$\frac{\partial Y}{\partial t} = -X^2Y + \beta X + D_Y\nabla^2 Y.$$

Find a nontrivial equilibrium solution and perform a linear stability analysis around that equilibrium point.

Exercise 8.3 The Swift–Hohenberg Model on a Finite Line

The Swift–Hohenberg equation (1977) was originally proposed as a simplified model of convective instability in a one-dimensional system. It takes the form

$$\frac{\partial u}{\partial t} = \left[\mu - (\nabla^2 + k_c^2)^2\right] u - u^3. \tag{8.63}$$

Assume $k_c^2 = 1$ and the following boundary conditions

$$u = \frac{\partial^2 u}{\partial x^2}, \qquad x = 0,\ L.$$

Study the stability properties of the homogeneous trivial solution $u_0 = 0$ by performing the following tasks:

1. Linearize (8.63) about $u = u_0$: Set $u = u_0 + \tilde{u}$ and write the linearized equation for $\tilde{u}$.
2. Solve linearized equation for $\tilde{u}$ using separation of variables: Set $\tilde{u}(x,t) = G(t)\phi(x)$, substitute and divide across by $G(t)\phi(x)$. Set separated variables to a common constant σ and then solve the resulting eigenvalue problems for $G(t)$ and $\phi(x)$.

3. Compute marginal stability curves $\sigma_n = 0$ and study the behavior of the perturbation for various values of n in the (μ, L) plane.

Exercise 8.4 Kuramoto–Sivashinsly Model.
Consider the Kuramoto–Sivashinsky equation without noise:

$$\frac{\partial u}{\partial t} = \eta_1 u - (1 + \nabla^2)^2 u - \eta_2 (\nabla u)^2 - \eta_3 u^3.$$

(a) Perform a linear stability analysis around the $u = 0$ equilibrium state.
(b) Compute and validate the marginal stability curves with those from Eq. (8.34) in Sect. 8.4.
(c) Plot a few of the marginal stability curves as functions of the radius R.

Exercise 8.5 A rod of length 10 cm is kept at a temperature of 0 °C at one end and at 100 °C at the other end. Assuming both ends are insulated, the model that governs the temperature along the rod can be written as

$$\begin{aligned} \frac{\partial u}{\partial t} &= c^2 \frac{\partial^2 u}{\partial x^2}, \qquad 0 \le x \le 10, \\ u(0, t) &= 0, \quad u(10, t) = 100. \end{aligned}$$

(a) Find a steady-state temperature, $u_s(x)$, by solving $\frac{\partial^2 u}{\partial x^2}$ with the initial conditions.
(b) Use the method of separation of variables to find a solution for the temperature at time t, i.e., solve for $u(x, t)$. Write a general solution as $u_g(x, t) = u_s(x) + u)x, t)$.
(c) Assume that at some time $t > 0$, the temperature at $x = 0$ is raised to 20 °C, while the temperature at $x = 10$ is decreased to 60 °C. Find a solution for $u_g(x, t)$.

Exercise 8.6 Schnakenberg (1979) Model.
Schnakenberg proposed the following kinematic-reaction model while searching for a chemically sensible model that can capture, with a minimum number of terms and equations, periodic behavior.

$$\begin{aligned} \frac{\partial X}{\partial t} &= \kappa_4 \alpha - \kappa_1 X + \kappa_3 X^2 Y + D_X \nabla X^2 \\ \frac{\partial Y}{\partial t} &= \kappa_2 \beta - \kappa_3 X^2 Y + D_Y \nabla Y^2. \end{aligned}$$

(a) Rewrite the model equation in dimensionless form by finding a suitable change of coordinates for X and Y as well a re-scaling of the time variable t.
(b) Calculate the unique nonzero or nontrivial equilibrium.
(c) Perform a linear stability analysis around the nontrivial equilibrium state.

Exercise 8.7 Gray–Scott Model.

The Gray–Scott model describes two irreversible chemical reactions through the following set of equations

$$\frac{\partial U}{\partial t} = -UV^2 + \kappa_1(1 - U) + D_U \nabla^2 U$$
$$\frac{\partial V}{\partial t} = UV^2 - (\kappa_1 + \kappa_2)V + D_V \nabla^2 V,$$

where κ_1 and κ_2 are reaction parameters and D_U and D_V are diffusion coefficients.

(a) Calculate the unique nonzero or nontrivial equilibrium.
(b) Perform a linear stability analysis around the nontrivial equilibrium state.

Exercise 8.8 Predator-Prey Model.

A spatially distributed predator-prey model introduced by Segal and Jackson [52, 53] has the form

$$\frac{\partial X}{\partial t} = (k_0 + k_1 X) - aXY + \mu \nabla^2 X$$
$$\frac{\partial Y}{\partial t} = bXY - mY - cY^2 + \mu \nabla^2 Y,$$

where k_0, k_1, a, b, c, m, and μ are parameters.

(a) Consider the special case $m = 0$. Find an appropriate scaling of the state variables X and Y to derive a dimensionless form

$$\frac{\partial x}{\partial t} = (1 + kx)x - axy + \delta^2 \nabla^2 x$$
$$\frac{\partial y}{\partial t} = xy - y^2 + \nabla^2 y,$$

where k, a, and δ^2 are to be determined constants.
(b) Calculate the nontrivial homogeneous steady-states.
(c) Show that the condition for diffusive instability is $k - \delta^2 > 2\sqrt{a - k}$.

Exercise 8.9 Consider the following model for a reaction-diffusion process

$$\frac{\partial u}{\partial t} = r(a - u + u^2 v) + \frac{\partial^2 u}{\partial x^2}$$
$$\frac{\partial v}{\partial t} = r(b - u^2 v) + D\frac{\partial^2 u}{\partial x^2},$$

where a, b and r are parameters.

(a) Calculate the nontrivial homogeneous steady state.
(b) Assume there is no diffusion. Then determine the stability condition of the homogeneous steady-state.
(c) Repeat part (b) but with diffusion.

Exercise 8.10 Traffic Flow Model.

A mathematical model for traffic flow has the form

$$\frac{\partial \rho}{\partial t} + \frac{\partial(\rho u)}{\partial x} = 0, \tag{8.64}$$

where $\rho = \rho(x, t)$ represents the density of cars, and $u = u(x, t)$ describes the velocity of cars.

Assuming the velocity of cars is a constant, Eq. (8.64) becomes

$$\frac{\partial \rho}{\partial t} + u_0 \frac{\partial \rho}{\partial x} = 0.$$

Apply the method of characteristics to solve the traffic flow model with constant velocity. Hint: the characteristics for this case are solutions of the ode

$$\frac{dx}{dt} = u_0.$$

Exercise 8.11 In a more realistic situation, one can assume that the velocity of cars is a function of traffic density. Consider, for instance the case where the traffic flow model becomes

$$\frac{\partial \rho}{\partial t} + u(\rho) \frac{\partial \rho}{\partial x} = 0.$$

Let $u(\rho) = \sqrt{\rho}$ and consider the initial condition: $\rho(x, 0) = x$, $x > 0$. Use the method of characteristics to find a solution for $\rho(x, t)$.

Exercise 8.12 Consider the following model of traffic flow

$$\frac{\partial \rho}{\partial t} + (x \sin t) \frac{\partial \rho}{\partial x} = 0.$$

Assume initial conditions to be:

$$\rho(x, 0) = 1 + \frac{1}{1 + x^2}.$$

Use the method of characteristics to solve for $\rho(x, t)$.

Exercise 8.13 A homogeneous flexible string, of length L, in a guitar is stretched between two end points, (0, 0) and L, 0). Assume the string to be initially pulled from rest from a position $\mu x(L - x)$. The mathematical model that governs the evolution of the string is the wave equation:

$$\frac{\partial^2 u}{\partial t^2} = c^2 \frac{\partial^2 u}{\partial x^2}, \qquad 0 \le x \le L, \quad t > 0.$$

Find the displacement $u(x, t)$ of the string of the guitar at time t.

Exercise 8.14 Find all absolutely irreducible real representations of $\mathbf{D}_4$, the symmetry group of a square. Then, work out all the possible solutions that are guaranteed at a steady-state bifurcation with $\mathbf{D}_4$ symmetry under one or other of these representations. Draw examples of the eigenmodes in an appropriate square box, and work out the relevant normal form equations.

Exercise 8.15 Consider a coupled cell system with three identical cells coupled unidirectionally, so that $\mathbf{Z}_3$–cyclic group of permutations of three objects–is the underlying group of symmetries. Analyze the Hopf bifurcation in the three-cell coupled system, where the action of the group $\mathbf{Z}_3 \times \mathbf{S}^1$ is given by

$$(e, \theta) \cdot z = e^{i\theta} z,$$
$$(\rho, 0) \cdot z = -z,$$

where $z \in \mathbf{C}$, (e, θ) is the phase shift and $(\rho, 0)$ is the rotation of the cells by $2\pi/3$. Describe the spatio-temporal symmetry of emergent collective patterns of solutions.

Exercise 8.16 Repeat the previous exercise but now with a four-cell system. This time, the action of the group $\mathbf{Z}_4 \times \mathbf{S}^1$ is given by

$$(e, \theta) \cdot z = e^{i\theta} z,$$
$$(\rho, 0) \cdot z = -z,$$

where $z \in \mathbf{C}$, (e, θ) is the phase shift and $(\rho, 0)$ is the rotation of the cells by $\pi/2$.

Exercise 8.17 Let $p : \mathbf{C} \to \mathbb{R}$ be a real-valued function expressed as

$$p(z) = \sum_{\alpha,\beta} a_{\alpha\beta} z^{\alpha} \bar{z}^{\beta}.$$

Apply the invariance conditions

$$\begin{aligned} p(z) &= p(\bar{z}) \\ p(e^{2\pi i/3} z) &= p(z) \\ p(z) &= \overline{p(z)}, \quad \text{since } p \text{ must be real} \end{aligned}$$

imposed by the action of the $\mathbf{D}_3$ group given by Eq. (8.56) to show that $a_{\alpha\beta} \in \mathbb{R}$, $a_{\alpha\beta} = a_{\beta\alpha}$, and $a_{\alpha\beta} = 0$ unless $\alpha = \beta \bmod 3$.

Exercise 8.18 Let $f : \mathbf{C} \to \mathbf{C}$ be a complex-valued mapping expressed as

$$f(z) = \sum_{\alpha,\beta} b_{\alpha\beta} z^{\alpha} \bar{z}^{\beta}.$$

Apply the equivariance conditions

$$\begin{aligned} f(e^{2\pi i/3}z) &= e^{2\pi i/3} f(z) \\ f(z) &= \overline{f(\bar{z})}, \end{aligned}$$

imposed by the action of the $\mathbf{D}_3$ group given by Eq. (8.56) to show that $b_{\alpha\beta} \in \mathbb{R}$, and $b_{\alpha\beta} = 0$ unless $\alpha = \beta + 1 \bmod 3$.

Exercise 8.19 Write a MATLAB (or equivalent software) code to integrate the amplitude equations (8.62) over a 1D grid of the form $x \in [a, b]$. For instance, $a = -50, b = 50$. Turn in a copy of the code. Note: It's easier and more convenient to write the code using complex-valued variables, i.e., $(z_1,\ z_2) \in \mathbf{C}^2$.

Exercise 8.20 Illustrate the stability properties of traveling wave solutions (both left-traveling and right traveling waves) and of standing waves obtained through Eq. (8.62). To do this, assign appropriate values to parameters: $(\mu, \alpha, \beta, \gamma, \delta)$ and initial conditions. For each case, plot the emerging pattern $u(x, t)$, where:

$$u(x, t) = z_1(t)e^{(x-t)i} + z_2(t)e^{-(x+t)i} + c.c.$$

References

1. A. Kandler, R. Unger, Population dispersal via diffusion-reaction equations. Technical Report, University College London and Technical University Chemnitz (2010)
2. M. Abramowitz, I.A. Stegun, *Handbook of Mathematical Functions* (Dover Publications, New York, 2012)
3. A. Turing, The chemical basis of morphogenesis. Philos. Trans. R. Soc. Lond. B **237**(641), 37–72 (1952)
4. A. Geirer, H. Meinhardt, A theory of biological pattern formation. Kybernetik **12**, 30–39 (1972)
5. T. Sekimura, A. Madzvamuse, A.J. Wathen, P.K. Maini, A model for colour pattern formation in butterfly wing of papilio dardanus. Proc. R. Soc. Lond. B **267**, 851–859 (2000)
6. A. Madzvamuse, T. Sekimura, A.J. Wathen, P.K. Maini, A predictive model for color pattern formation in butterfly wing of papilio dardanus. Hiroshima Math. J. **32**, 325–336 (2002)
7. P. Blomgren, A. Palacios, B. Zhu, S. Daw, C. Finney, J. Halow, S. Pannala, Bifurcation analysis of bubble dynamics in fluidized beds. Chaos **17**(2), 509–520 (2007)
8. N. Cheremisinoff, P. Cheremisinoff, *Hydrodynamics of gas-solids fluidization* (Golf Publishing company, Houston, 1984), pp. 194–195
9. J.F. Davidson, R. Clift, D. Harrison, *Fluidization*, 2nd edn. (Academic, London, 1985)
10. V.V. Ranade, *Computational Flow Modeling for Chemical Reactor Engineering* (Academic, San Diego, 2004)
11. L.S. Fan, C. Zhu, *Principles of Gas-Sold Flows* (Cambridge University Press, New York, 1998)
12. O. Borchers, G. Eigenberger, Detailed experimental studies on gas-solid bubble flow in bubble columns with and without recycle, in *Bubbly Flows: Analysis, Modeling and Calculation* (Springer, Berlin, 2004), pp. 1–10
13. R.M. Davies, G.I. Taylor, The mechanics of large bubbles rising through extended liquids and through liquids in tubes. Proc. R. Soc. Lond. A **200**, 375–390 (1950)

14. D. Harrison, L. Leung, Bubble formation at an orifice in a fluidized bed. Trans. Inst. Chem. Eng. **39**, 409–414 (1961)
15. F. Zenz, Bubble formation and grid design. Inst. Chem. Eng. Symp. **30**, 136–139 (1968)
16. H. Caram, K. Hsu, Bubble formation and gas leakage in fluidized beds. Chem. Eng. Sci. **41**(6), 1445–1453 (1986)
17. J. Nieuwland, M. Veenendaal, J. Kuipers, W. Vanswaaij, Bubble formation at a single orifice in gas-fluidized beds. Chem. Eng. Sci. **51**(17), 4087–4102 (1996)
18. T. Blake, P. Chen, Computer modeling of fluidized bed coal gasification reactors. Amer. Chem. Soc. Symp. Ser. **168**, 157–183 (1981)
19. S. Garg, J. Prichett, Dynamics of gas-fluidized beds. J. Appl. Phys. **46**, 4493–4500 (1975)
20. D. Richner, T. Minoura, J. Prichett, T. Blake, A numerical model of gas fluidized beds. AIChE J. **36**, 361–369 (1990)
21. G. Schneyer, E. Peterson, P. Chen, D. Brownell, T. Blake, Computer modeling of coal gasification reactors. Technical Report, Department of Energy ET/10247 (1981)
22. J.S. Halow, G. Fasching, P. Nicolletti, J. Spenik, Observations of a fluidized bed using capacitance imaging. Chem. Eng. Sci. **48**(4), 643–659 (1993)
23. T. Farrokhalaee, R. Clift, Mechanistic prediction of bubble properties in freely-bubbling fluidized beds, in *Proceddings of the 1980 International Fluidization Conference* (New Hampshire, 1980), pp. 135–142
24. W.K. Lord, Bubbly flow in fluidized beds of large particle. Ph.D. Thesis, MIT (1983)
25. S. Pannala, C.S. Daw, J. Halow, Simulations of reacting fluidized beds using agent-based bubble model. Int. J. Chem. React. Eng. **1**, A20 (2003)
26. S. Pannala, C.S. Daw, J. Halow, Dynamic interacting bubble simulation (dibs): an agent-based bubble model for reacting fluidized beds. Chaos **14**(2), 487–498 (2004)
27. J.C. Orcutt, B.H. Carpenter, Bubble coalescence and the simulation of mass transport and chemical reaction in gas fluidized beds. Chem. Eng. Sci. **26**, 1049–1064 (1971)
28. S.A. Allahwla, Bubble frequency and distribution in fluidized beds. Ph.D. Thesis, Monash University (1981)
29. P. Chakka, K. Nguyen, M. Cheng, C.E.A. Finney, D.D. Bruns, C.S. Daw, M.B. Kennel, Spatio-temporal dynamics in a train of rising bubbles. Chem. Eng. J. **64**, 191–197 (1996)
30. A. Tufaile, J.C. Sartorelli, Chaotic behavior in bubble formation dynamics. Physica A **275**, 336–346 (2000)
31. A. Tufaile, J.C. Sartorelli, Hénon-like attractor in air bubble formation. Phys. Lett. A **275**, 211–217 (2000)
32. K.T. Alligood, T.D. Sauer, J.A. Yorke, *Chaos: An Introduction to Dynamical Systems* (Springer, New York, 1996)
33. F. Taken, Detecting strange attractors in turbulence (1981)
34. T. Sauer, J. Yorke, M. Casdagli, Embedelogy. J. Stat. Phys. **65**, 579 (1991)
35. M.B. Kennel, R. Brown, H.D.I. Abarbanel, Determining embedding dimension for phase-space reconstruction using a geometrical construction. Phys. Rev. A **45**, 3403 (1992)
36. H. Kantz, T. Schreiber, *Nonlinear Time Series Analysis* (Cambridge University Press, Cambridge, 1997)
37. R. Hegger, H. Kantz, T. Schreiber, Practical implementation of nonlinear time series methods: the tisean package. Chaos **9**, 413–435 (1999)
38. J.-P. Eckman, S.O. Kamphorst, D. Ruelle, S. Ciliberto, Lyapunov exponents from a time series. Phys. Rev. A **34**, 4971 (1986)
39. J.D. Farmer, J. Sidorowich, Predicting chaotic time series. Phys. Rev. Lett. **59**, 845 (1987)
40. R. Hegger, H. Kantz, T. Schreiber, Nonlinear time series analysis
41. J.J. Collins, M.T. Rosenstein, C.J. De Luca, A practical method for calculating largest lyapunov exponents from small data sets. Physica D **65**, 117–134 (1993)
42. H. Kantz, A robust method to estimate the maximal lyapunov exponent of a time series. Phys. Lett. A **185**(1), 77–87 (1994)
43. P. Holmes, J. Lumley, G. Berkooz, *Turbulence, Coherenet Structures, Dynamical Systems and Symmetry* (Cambridege University Press, Cambridege, 1996)

44. K. Karhunen, Uber lineare methoden in der wahrscheinlichkeitsrechnung. Ann. Acad. Sci. Fennicae Ser. A Math-Phys **37**, 1–79 (1947)
45. M. Loève, *Probability Theory* (Van Nostrand, New York, 1955)
46. J.L. Lumley, The structure of inhomogeneous turbulence, in *Atmospheric Turbulence and Radio Wave Propagation*. ed. by A.M. Yaglom, V.I. Tatarsk (Nauka, Moscow, 1967), pp. 166–178
47. L. Sirovich, Turbulence and the dynamics of coherent structures. Part i: Coherent structures. Q. Appl. Maths, **5**, 561 (1987)
48. M. Dellnitz, M. Golubitsky, M. Nicol, Symmetry of attractors and the karhunen-loeve decomposition, in *Trends and Perspectives in Applied Mathematics*, ed. by L. Sirovich, vol. 100 (Springer, Berlin, 1990), p. 73
49. R. Hoyle, *Pattern Formation. An Introduction to Methods* (Cambridge University Press, Cambridge, 2006)
50. J. Guckenheimer, P.J. Holmes, *Nonlinear Oscillations, Dynamical Systems and Bifurcations of Vector Fields* (Springer, New York, 1993)
51. S. Wiggins, *Introduction to Applied Nonlinear Dynamical Systems* (Springer, New York, 1990)
52. L.A. Segel, J.L. Jackson, Dissipative structure: an explanation and an ecological example. J. Theor. Biol. **37**, 545–559 (1972)
53. L. Edelstein-Keshet, *Mathematical Models in Biology* (Society for Industrial and Applied Mathematics (SIAM), 1988)

Chapter 9
Stochastic Models

Many physical or biological phenomena often have stochastic or random effects, such as Brownian motion or a birth event. These phenomena rarely are completely deterministic, so cannot be accurately modeled with difference or differential equations. One method of introducing these variations into models is by adding some type of noise, such as white and colored noise. This chapter examines stochastic modeling methods, including methods from *stochastic differential equations*, and the *Fokker-Planck equation* for describing the probability that a system is in some state at a given time.

9.1 Definitions

In this section we introduce some basic definitions from probability theory and stochastic processes relevant to modeling. We start off with the concept of a **random variable**.

Definition 9.1 (*Random Variable*) A *random variable*, usually written X, is a variable whose possible values are numerical outcomes of a random phenomenon.

The random variable is either *discrete* or *continuous*, which are stated in the following definitions.

Definition 9.2 (Discrete Random Variable) A *discrete random variable* X has a countable number of possible values.

Definition 9.3 (Continuous Random Variable) A *continuous random variable* X takes all values in a given interval of numbers.

A. Palacios, *Mathematical Modeling*, Mathematical Engineering,
https://doi.org/10.1007/978-3-031-04729-9_9

For instance, rolling out a fair die can lead to six equally possible outcomes of getting an integer number, i.e., a *discrete random variable*, which can take values from one through six. An example of a *continuous random variable* would be the life expectancy of a population of individuals from a particular country, in which individuals are known to (typically) die between age X_1 and age X_2.

For probabilistic modeling one needs knowledge of how the random variables are distributed. For the example of the fair die, we assumed that each integer between one and six has the same chance of appearing as the outcome of rolling out a die. Thus, each value from one to six has exactly the sixth chance of occurring in a given roll. This information is encapsulated in the form of **probability distributions**.

Definition 9.4 (Probability Distribution) A *probability distribution* of a random variable X indicates the possible values of X and how the probabilities are assigned to those values.

For continuous random variables one needs the concept of a **probability density function**.

Definition 9.5 (Probability Density Function) A *probability density function*, $f(x)$, of a continuous random variable is a function whose integral across an interval gives the probability that the value of the variable lies within the same interval:

$$Pr\{x_1 \leq X \leq x_2\} = \int_{x_1}^{x_2} f(x)dx.$$

Note that for $f(x)$ to be a probability density function, it must satisfy $f(x) \geq 0$ for all x and

$$\int_{-\infty}^{\infty} f(x)\, dx = 1.$$

Now that we know what a probability density function is, we can define the concept of a **probability space**.

Definition 9.6 (Probability Space) A three-tuple (S, F, P), where S is the sample space, F is the event space, and P is the probability function, is known as the *probability space*.

In our running example of rolling a die, the sample space is $S = \{1, 2, 3, 4, 5, 6\}$. The event space is the collection of subsets of S, which can be any singleton set, the empty set, and the sample space S itself.

Phenomena that involve the operation of chance are also known as **stochastic processes**. More formally, we have the following definition.

Definition 9.7 (Stochastic Process) A stochastic process is a collection of random variables $\{X_t, t \in T\}$, defined on some probability space (S, F, P).

9.2 Stochastic Differential Equations

Random fluctuations are inherent to many natural and artificial systems and they have led to mathematical models in the form of *Stochastic Differential Equations* (SDE). These type of models incorporate fluctuations as *noise*, and they appear in many applications that include: chemical physics, laser systems, electronic oscillators, combustion, mathematical biology, antennas and radars, gyroscope systems, precision timing devices, and sensors.

From a dynamical systems standpoint, a physical or biological system subject to random noise can be modeled through the following stochastic differential equation

$$\boxed{\frac{dx(t)}{dt} = f(x(t)) + g(x(t))\,\xi(t),} \tag{9.1}$$

where $x(t) \in \mathbb{R}^n$ represents the state of the system at time $t \geq 0$, $f(x) \in \mathbb{R}^n$ describes the deterministic dynamics, $g(x) \in \mathbb{R}^n$ is a smooth vector-valued function, and $\xi(t)$ is a scalar function describing random fluctuations–internal and external to the system. When g is constant, the noise is called *additive*, and when it depends explicitly on the state variable, x, is called *multiplicative*.

The simplest approximation of real-life fluctuations, $\xi(t)$, are in the form of Gaussian *white noise*. White noise can be interpreted as a series of independent pulses that act on a system in random directions, so that the average of all perturbations is zero. This leads to a uniform power spectral density over equal intervals of frequencies. Gaussian noise exhibits zero mean, zero correlation time and infinite variance, so that the autocorrelation function follows a Dirac delta function. That is,

$$\begin{aligned} \langle \xi(t) \rangle \quad &= 0, \\ \langle \xi(t)\,\xi(s) \rangle &= 2D\delta(t-s), \end{aligned} \tag{9.2}$$

where $D = \lambda k_B T$ is Einstein diffusion coefficient, which represents the noise intensity, k_B is Boltzmann's constant, and T is absolute temperature, $\xi(t)$ is the derivative of a Wiener (i.e., Brownian motion) process [1, 2]. That is,

$$\xi(t) = \frac{dW}{dt}.$$

However, this derivative does not exist since the Wiener process, $W(t)$, is nowhere differentiable. It is used, nevertheless, as a heuristic representation. With this in mind, we can write the stochastic model as

$$\frac{dx(t)}{dt} = f(x(t)) + g(x(t))\,\frac{dW}{dt}.$$

A common way of writing a stochastic differential equation is to multiply by "dt" the previous expression, which leads to the more commonly used form of a stochastic differential equation:

$$\boxed{\begin{aligned} dx(t) &= f(x(t))\,dt + g(x(t))\,dW \\ x(0) &= x_0. \end{aligned}} \tag{9.3}$$

The solution to the SED Eq. (9.3) is given by

$$x(t) = x_0 + \int_0^t f(x)s))\,ds + \int_0^t g(x(s))\,dW, \qquad \forall t > 0. \tag{9.4}$$

The second term in the right-hand side of Eq. (9.4) is a standard integral over time. The last term is a *stochastic integral*, known as an Itô's integral [2, 3]. This leads to identifying the stochastic process x as an Itô process.

9.2.1 Itô's Formula

Now that we have defined a stochastic differential equation, we can address a more subtle issue of whether the SDE Eq. (9.3) is a faithful model of the physical phenomenon. This issue is subtle because in stochastic calculus the chain rule of standard calculus takes a different form. To explore this issue in more detail, we will investigate which stochastic differential equation is satisfied by a smooth function, $u(x)$, twice differentiable, of an Itô process, $x(t)$. Thus, let

$$Y(t) = u(x(t), t).$$

A Taylor expansion of $u(x(t), t)$ yields

$$dy = \frac{\partial u}{\partial t}\,dt + \frac{\partial u}{\partial x}\,dx + \frac{1}{2}\frac{\partial^2 u}{\partial x^2}(dx)^2 + \cdots .$$

Substituting dx from Eq. (9.3) we get

$$\begin{aligned} dy = {} & \frac{\partial u}{\partial t}\,dt + \frac{\partial u}{\partial x}\big(f(x(t))\,dt + g(x(t))\,dW\big) + \\ & \frac{1}{2}\frac{\partial^2 u}{\partial x^2}\big(f^2(x(t))(dt)^2 + 2f(x(t))g(x(t))dt dW + g^2(x(t))(dW)^2\big) + \cdots . \end{aligned}$$

In stochastic calculus, the following rules apply:

$$(dW)^2 = dt, \qquad dt\,dW = 0, \qquad (dt)^2 = 0.$$

These rules hold because due to the quadratic variance of a Wiener process, W, in the limit $dt \to 0$, the terms $(dt)^2$ and $dt\,dW$ tend to zero faster than $(dW)^2$, which is $\mathcal{O}(dt)$. If we apply these rules to the previous equation, replace $(dW)^2$ with dt, and collect terms in dt and dW, we arrive at Itô's formula:

$$dy(t) = \left(\frac{\partial u}{\partial t} + f(x(t))\frac{\partial u}{\partial x} + \frac{1}{2}g^2(x(t))\frac{\partial^2 u}{\partial x^2}\right)dt + g(x(t))\frac{\partial u}{\partial x}\,dW. \tag{9.5}$$

The main conclusions here are two. First, Itô's calculus rules lead to an additional term in the stochastic version of the chain rule, mainly

$$\frac{1}{2}g^2(x(t))\frac{\partial^2 u}{\partial x^2}.$$

Second, a smooth function, $y(t)$, of an Itô process, x, which itself is a solution of Eq. (9.3), satisfies an associated stochastic differential equation of the form

$$dy = f_y\,dt + g_y\,dW, \tag{9.6}$$

where

$$f_y = \frac{\partial u}{\partial t} + f(x(t))\frac{\partial u}{\partial x} + \frac{1}{2}g^2(x(t))\frac{\partial^2 u}{\partial x^2}, \qquad g_y = g(x(t))\frac{\partial u}{\partial x}\,dW.$$

9.2.2 Examples

Example 9.1 (*Stochastic Model for Stock Prices*). Let $P(t)$ represent the price of a stock at time t. The evolution of this stock in time can be studied by considering a SDE model for the relative change of price, $\frac{dP}{P}$, of the form

$$\frac{dP}{P} = \mu dt + \sigma dW, \tag{9.7}$$

where the parameter $\mu > 0$ represents the drift of the stock, and σ can be interpreted as the volatility of the market, and W is a Wiener process. We can rewrite Eq. (9.7) in the familiar form

$$\begin{aligned} dP &= \mu P dt + \sigma P dW \\ P(0) &= P_0. \end{aligned} \tag{9.8}$$

The deterministic version of Eq. (9.8), i.e., with zero volatility, $\sigma = 0$, corresponds to Malthusian growth, which we know from Chap. 4 to have as a solution

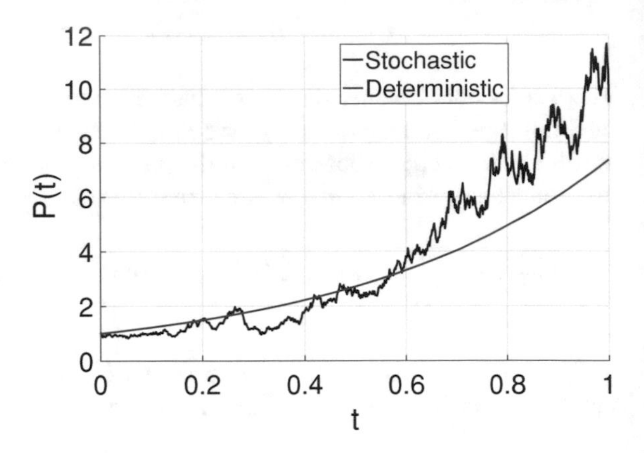

Fig. 9.1 Deterministic and stochastic solutions of a stock price model obtained through the Euler-Maruyama method. Parameters are: $\mu = 2.0$ and $\sigma = 1.0$. (MATLAB code in Appendix)

$$P(t) = P_0 e^{\mu t}.$$

To solve the stochastic version, we let

$$y = u(P) = \ln P.$$

Applying Itô's formula, see Eq. (9.5), we get

$$dy = (\mu - \frac{\sigma^2}{2})dt + \sigma dW.$$

Since $y = \ln P$, it follows that a solution of the original stock price model Eq. (9.7) is

$$P(t) = P_0\, e^{\sigma W(t) + (\mu - \frac{\sigma^2}{2})t}.$$

Assuming that the initial price P_0 is always positive, then the stock price will always be positive. Figure 9.1 illustrates the numerical solutions obtained from the original SDE model of a stock price evolution. For comparison purposes, both solutions, deterministic and stochastic are shown.

Example 9.2 Suppose we want to solve the following SDE:

$$\begin{aligned} dy &= \lambda y\, dW \\ y(0) &= 1. \end{aligned} \tag{9.9}$$

The obvious guess for a solution is $y(t) = e^{\lambda W(t)}$.

But let us find out what this guess is missing. Consider the following SDE:

$$dx = dW, \tag{9.10}$$

which is in the form of Eq. (9.3) with $f = 0$ and $g = 1$.

Let

$$y = u(x,t) = e^{\lambda x - \frac{\lambda^2 t}{2}}.$$

Applying Itô's formula we get

$$dy = \left(-\frac{\lambda^2}{2}e^{\lambda x - \frac{\lambda^2 t}{2}} + \frac{\lambda^2}{2}e^{\lambda x - \frac{\lambda^2 t}{2}}\right) dt + \lambda e^{\lambda x - \frac{\lambda^2 t}{2}}\, dW,$$

which reduces to our original Eq. (9.9). Since $x(t) = W(t)$, the solution to Eq. (9.9) is

$$y = e^{\lambda W(t) - \frac{\lambda^2 t}{2}}.$$

Heuristically, in Itô's calculus the solution $e^{\lambda W(t) - \frac{\lambda^2 t}{2}}$ plays the role of $e^{\lambda t}$ in ordinary calculus.

Example 9.3 Consider the following SDE:

$$\begin{aligned} dx &= h(x)x\, dW, \\ x(0) &= 1, \end{aligned} \tag{9.11}$$

where $h(x)$ is a smooth function of x.

This equation is also in the form of Eq. (9.3) with $f = 0$ and $g = h(x)x$. Let

$$y = u(x,t) = \ln x.$$

Applying Itô's formula we get

$$\begin{aligned} dy &= -\frac{1}{2x^2}h^2x^2\, dt + \frac{1}{x}hx\, dW \\ &= -\frac{1}{2}h^2\, dt + h\, dW. \end{aligned}$$

Solving for $y(t)$ we get

$$y(t) = -\frac{1}{2}\int_0^t h^2\, ds + \int_0^t h\, dW.$$

Since $y = \ln x$, it follows that a solution to the original SDE Eq. (9.11) is

$$x(t) = e^{-\frac{1}{2}\int_0^t h^2\, ds + \int_0^t h\, dW}.$$

9.3 Colored Noise

So far we have considered white noise in our calculations of solutions of models described by stochastic differential equations. Recall that white noise is characterized by a uniform frequency spectral. On the contrary, *colored noise*, such as *pink*, *red*, or *blue*, have their own spectral profile, which usually (but not always) is related to the color of light with similar spectral. Thus, in general, the *color of noise* is a property that is directly related to the power spectrum of the signal or the fluctuations.

In the next sections we will discuss in more detail the formulation of the stochastic differential Eq. (9.1), the advantages of using colored noise in $\xi(t)$, and how to numerically solve related models.

9.3.1 Langevin Equation

The Langevin [4] equation

$$\boxed{m\frac{dv}{dt} = -\lambda v + \xi(t),} \tag{9.12}$$

is a stochastic differential equation that was introduced [4] to describe Brownian motion. That is, the motion of small particles suspended in a fluid and moving under the influence of random forces that result from collisions with molecules of the fluid induced by thermal fluctuations. m is the mass of the particles and λ describes mobility or the ratio of the particle's terminal drift velocity to the collision forces. Langevin introduced this formulation to describe the motion into two parts, a slow varying deterministic part for the velocity, $v(t)$, of the particles, and a rapidly varying random part, $\xi(t)$, which represents Gaussian white noise fluctuations.

The Langevin Eq. (9.12) can be formally solved as if it were a deterministic ordinary differential equation, leading to

$$v(t) = v_0 e^{-t/\tau_B} + \frac{1}{m}\int_0^t e^{-(t-s)/\tau_B}\xi(s)\,ds, \tag{9.13}$$

where $\tau_B = m/\lambda$ is the characteristic time scale for the loss of speed gained in a single collision by the particles. Langevin went on to show that the position, $x(t)$, of the particles can be described by

$$x(t) = x_0 + \int_0^t v(t)\,dt, \tag{9.14}$$

which is differentiable on a short time scale, and, it behaves diffusively on a long time scale with Gaussian probability distribution

$$p(x,t) \approx \frac{1}{(4\pi Dt)^{1/2}} e^{-x^2/(4Dt)}.$$

Example 9.4 Consider the following mathematical model of an electric circuit consisting of a resistor and a capacitor

$$\frac{dV}{dt} = -\frac{1}{RC}V + \eta(t), \tag{9.15}$$

where $V(t)$ represents the voltage across the resistor and $\eta(t)$ represents white noise fluctuations due to electronics.

We can rewrite this equation in the following form

$$C\frac{dV}{dt} = -\frac{1}{R}V + \tilde{\eta}(t),$$

where $\tilde{\eta}(t) = C\eta(t)$. This last equation is in Langevin form (9.12), with $m = C$ and $\lambda = 1/R$. It follows that the correlation of the white noise fluctuations, $\eta(t)$, is

$$\langle \eta(t)\, \eta(s) \rangle = \left\langle \frac{1}{C}\tilde{\eta}(t)\, \frac{1}{C}\tilde{\eta}(s) \right\rangle = \frac{1}{C^2} 2D\delta(t-s) = \frac{1}{C^2} 2\lambda k_B T\, \delta(t-s).$$

Since $\lambda = 1/R$ we get

$$\langle \eta(t),\, \eta(s) \rangle = \frac{2k_B T}{RC^2}\, \delta(t-s).$$

9.3.2 *Ornstein-Uhlenbeck Process*

A difficulty of modeling natural and artificial systems with Gaussian white noise is that this type of noise is a mathematical idealization of fluctuations that do not occur in real-life situations. They do not occur because white noise has no time scale, while the fluctuations that affect physical, biological, and even electronic systems, have an inherent nonzero correlation time. For instance, in the collision of water molecules with Brownian particles there is a time scale beyond which the fluctuations are no longer uncorrelated. But since the time scale of the fluctuations in this problem is much smaller than that of the particles, then the white noise approximation can be used, and, accordingly, the corresponding Langevin equation can be formally solved.

Many electronic systems exhibit fluctuations where the differences in time scales is not as large to justify the use of white noise. For instance, the fluctuations in the fluxgate magnetometer that was described in Chap. 4, can be attributed to signal contamination and noise from the electronics in the coupling circuitry. These fluctuations manifest themselves as fluctuations about the steady states of the cores with a finite time scale and nonzero correlation time.

To overcome this problem, the time scale of the fluctuations and of the physical system must be taken into account. This is done by using colored noise, $\eta(t)$, which has a finite frequency-bandwidth. The simplest example of time-correlated noise is the Ornstein-Uhlenbeck process [5], which exhibits a zero mean and an exponential correlation function. That is,

$$\begin{aligned} \langle \eta(t) \rangle &= 0, \\ \langle \eta(t)\, \eta(s) \rangle &= \frac{D}{\tau_c} e^{(-|t-s|/\tau_c)}, \end{aligned} \tag{9.16}$$

where τ_c is the noise correlation time. Observe that in the limit $\tau_c \to 0$ the exponential in Eq. (9.16) becomes a Dirac delta function, and then the Ornstein-Uhlenbeck process becomes a white noise process. In fact, it can be shown that the Ornstein-Uhlenbeck process is a solution of the Langevin Eq. (9.12)

$$\frac{d\eta(t)}{dt} = -\frac{1}{\tau_c}\eta(t) + \frac{\sqrt{2D}}{\tau_c}\xi(t), \tag{9.17}$$

with initial conditions $\eta(0)$ being a Gaussian random variable of zero mean and variance $\sqrt{2D}/\tau_c$. If we combine the general form of the stochastic model, Eq. (9.1), with the Langevin Eq. (9.17), we arrive to a $(2N)$D formulation of an Ornstein-Uhlenbeck process

$$\boxed{\begin{aligned} \dot{x}(t) &= f(x, \mu) + g(x(t))\, \eta(t), \\ \dot{\eta}(t) &= -\frac{1}{\tau_c}\eta(t) + \frac{\sqrt{2D}}{\tau_c}\xi(t), \end{aligned}} \tag{9.18}$$

where $\xi(t)$ is a Gaussian white noise function.

A common approach to conduct computer simulations of stochastic models subject to colored noise is to solve, numerically, both systems of Eq. (9.18). An alternative approach is to solve, first, analytically, the second equation in Eq. (9.18) for $\eta(t)$, and, then substitute the solution in the numerical integration of the first equation. Next we show details of the derivation of an analytical solution for $\eta(t)$.

First we rewrite Eq. (9.17) as

$$d\eta(t) = -\frac{1}{\tau_c}\eta(t)\, dt + \frac{\sqrt{2D}}{\tau_c}\, dW.$$

Next we apply a suitable change of coordinates by letting $y = e^{-t/\tau_c}\eta(t)$. This change of variables leads to

$$dy = \frac{\sqrt{2D}}{\tau_c} e^{-t/\tau_c} dW.$$

The solution of this equation is

$$e^{-t/\tau_c}\eta(t) = \eta_0 + \int_0^t \frac{\sqrt{2D}}{\tau_c} e^{-s/\tau_c}\, dW(s),$$

which leads to the desired solution for $\eta(t)$ given by:

$$\eta(t) = \eta_0 e^{-t/\tau_c} + \sigma \int_0^t e^{-(s-t)/\tau_c}\, dW(s). \tag{9.19}$$

Example 9.5 Let us consider again the electric circuit modeled by Eq. (9.15). In our previous example, we found the correlation function:

$$\langle \eta(t)\, \eta(s) \rangle = \frac{2k_B T}{RC^2}\, \delta(t-s),$$

which shows that $D = k_B T/(RC^2)$. Thus, if we rewrite Eq. (9.15) as

$$\frac{dV}{dt} = -\frac{1}{RC} V + \frac{\sqrt{2R^2C^2D}}{RC}\tilde{\eta}(t).$$

then we can use D to obtain the correlation function

$$\langle V(t)\ V(s) \rangle = \frac{R^2C^2 \frac{k_B T}{RC^2}}{RC} e^{(-|t-s|/RC)} = \frac{k_B T}{C} e^{(-|t-s|/RC)}.$$

Observe that this correlation function leads to that of a white noise (Johnson noise) function when the capacitance, C, becomes negligibly small.

Next we introduce a popular numerical technique, known as the Euler-Maruyama method, to numerically solve stochastic differential equations.

9.3.3 *Euler-Maruyama Numerical Algorithm*

Consider a stochastic differential equation of the form

$$\dot{x}(t) = a(x(t), t)dt + b(x(t), t)\dot{W}(t), \tag{9.20}$$

with initial condition $x(0) = x_0$ and $W(t)$ being a Wiener process. Let us assume that we are interested in seeking a numerical solution to Eq. (9.20) on the interval $[0, T]$. The Euler-Murayam algorithm follows the following steps.

Step 1: Partition the interval $[0, T]$ into N equal subintervals of width $\Delta t > 0$. That is

$$0 = t_0 < t_1 < \cdots < t_N = T, \quad \text{and,} \quad \Delta t = \frac{T}{N}.$$

Step 2: $y_0 = x_0$.

Step 3: Apply the recursive relation

$$y_{n+1} = y_n + a(y_n, t_n)\Delta t + b(y_n, t_n)\Delta W_n,$$

where $\Delta W_n = W_{t_{n+1}} - W_{t_n} = \sqrt{\Delta t}\, Z_n$, with $Z_n \sim N(0, 1)$, for all n.

The random variables ΔW_n are independent and identically distributed normal random variables with zero mean and variance Δt.

Example 9.6 Consider the Langevin model

$$\dot{\eta}(t) = -\frac{1}{\tau_c}\eta(t) + \frac{\sqrt{2D}}{\tau_c}\xi(t), \tag{9.21}$$

with $D = 0.5$. Below is a MATLAB code to solve this model using the Euler-Maruyama algorithm.

```
1  function ou
2
3  close all;
4  clear all;
5  clc
6
7  randn('state',100)
8  tau  = 1;
9  xi0  = 1;
10 dt   = 0.01;
11 D    = 0.5;
12 N    = 1000;
13 T    = N*dt;
14
15 pd  = makedist('Normal',0,sqrt(dt));
16 dW  = random(pd);
17
18 xi       = zeros(1,N);      % preallocate for ...
      efficiency
19 xi(1)  = xi0 - dt*xi0/tau + sqrt(2*D)*dW/tau;
20
```

```
21  for j=2:N
22      dW     = random(pd);
23      xi(j)  = xi(j-1)  - dt*xi(j-1)/tau + ...
            sqrt(2*D)*dW/tau;
24  end
25
26  plot([0:dt:T],[xi0,xi],'k-', 'LineWidth',3);
27  xlabel('t','FontSize',12)
28  ylabel('\eta(t)','FontSize',16,'Rotation',0,...
29  'HorizontalAlignment','right');
30  set(gca,'FontSize',40);
31  grid on;
```

Figure 9.2 illustrates the result of numerically solving Eq. (9.21) for $\eta(t)$ through the Euler-Mayurama algorithm.

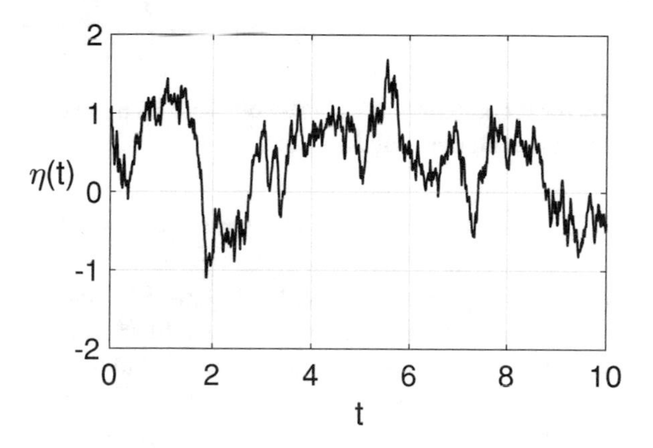

Fig. 9.2 Numerical solution, $\eta(t)$, of the Langevin model Eq. (9.21) obtained through the Euler-Mayurama algorithm

9.4 Colored Noise in Bistable Systems

While in most practical applications, e.g., electronic systems, noise can degrade performance and, thus, it must be avoided, under certain circumstances, however, small fluctuations can help improve performance [6, 7]. This phenomenon, in which noise can have positive effects, is known as *stochastic resonance* [8–13]. In dynamic sensors, for instance, weak signals that need to be detected can be strengthened or amplified by adding a little bit of white noise. This paradigm works as follows. Since white noise contains a wide spectrum of frequencies, those that are associated with the natural frequencies of the weak signal will resonate with one another. Then the resonant effect will lead to the original signal being amplified while the remaining white noise frequencies remain at low-power levels. In this way, the signal-to-noise ratio is increased, thus making the original signal stronger and, consequently, more detectable.

To get more insight, recall our introduction to dynamic sensors from Chap. 4. In that chapter we discussed the fact that dynamic sensors are typically governed by overdamped bistable dynamics of the generic form

$$\frac{dx}{dt} = -\nabla U(x) + \eta(t), \tag{9.22}$$

where $x(t)$ is the state variable of the device, i.e., magnetization state or electric field, and $U(t)$ is the bistable potential function. $\eta(t)$ is a zero mean and an exponential correlation function. That is,

$$\begin{aligned} \langle \eta(t) \rangle &= 0, \\ \langle \eta(t)\ \eta(s) \rangle &= \frac{D}{\tau_c} e^{(-|t-s|/\tau_c)}, \end{aligned}$$

The most common model is actually that of an overdamped Duffing oscillator with a double-well potential function

$$U(x) = -\frac{1}{2}ax^2 + \frac{1}{4}bx^4.$$

with positive parameters a and b. Figure 9.3(left) illustrates the potential function, whose minima are located at $\pm x_m = (a/b)^{1/2}$ and the height of the potential barrier between the two minima is $U_0 = a^2/(4b)$ and it is located at $x = 0$.

Introducing the scale variables $\tilde{x} = x\sqrt{b/a}$, $\tilde{t} = at$, $\tilde{\eta}(t) = \eta\sqrt{b/a^3}$, $\tilde{\tau} = a\tau$, $\tilde{D} = (b/a^2)D$, we arrive at a normalized Langevin version

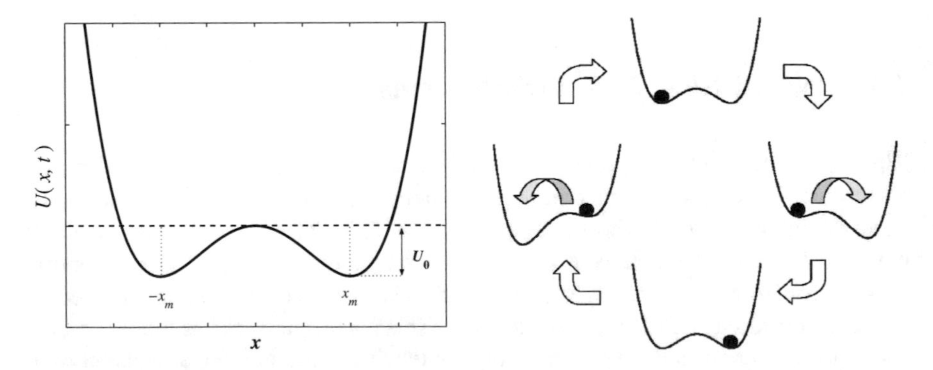

Fig. 9.3 (Top) Bistable Potential $U(x) = (1/4)bx^4 - (1/2)ax^2$. (Bottom) Noise-induced switching between wells of a potential function occurs when the period of a weak driving force approximately equals twice the noise-induced inter-well transition time

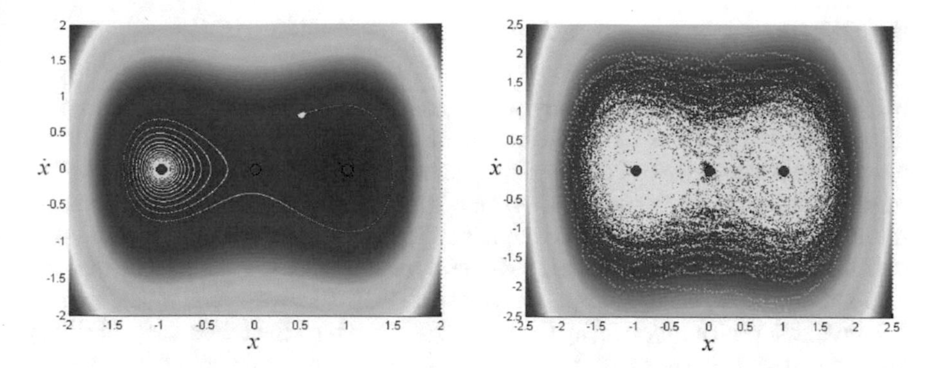

Fig. 9.4 (Left) Motion of a particle on a double-well potential function. In the absence of noise, the system dynamics quickly settles into an equilibrium point. Which equilibrium point is selected depends on the initial conditions. (Right) In the presence of noise, the system dynamics now lingers intermittently between the two equilibrium states of the deterministic system, independently of initial conditions

$$\frac{d\tilde{x}}{dt} = \tilde{x} - \tilde{x}^3 + \tilde{\eta}(t),$$

which satisfies the autocorrelation function

$$\langle \tilde{\eta}(t)\ \tilde{\eta}(s) \rangle = \frac{\tilde{D}}{\tilde{\tau}_c} e^{(-|t-s|/\tilde{\tau}_c)}.$$

In the absence of noise, and in the presence of an external weak periodic force $f(t)$, the state variable in $U(x + f(t))$ can be induced to rock back and forth as the double-well potential is tilted asymmetrically up and down, periodically raising and lowering the potential barrier, as is shown in Fig. 9.3(right).

The forcing term $f(t)$ was assumed to be too weak to induce the state variable to oscillate freely between the two minima. Consequently, the state variable will relax to one of the two equilibrium points or minima of the double-well potential function, as is seen in Fig. 9.4(left). Which equilibrium point is observed depends on initial conditions. A small amount of noise can, however, allow the system to escape the *basin of attraction* and hope back and forth between the two equilibrium points, as is now shown in Fig. 9.4(right). In this way, noise-induced oscillations can be achieved due to the presence of *stochastic resonance*.

That is, stochastic resonance refers to noise-induced transitions which, due to the nonlinear nature of a system dynamics, become synchronized with the period of an external force. In fact, the transitions occur when the driving force approximately equals twice the noise-induced inter-well transition time [6, 7]. A distinctive feature of stochastic resonance is a rapid increase in a system's output Signal-to-Noise Ratio (SNR) under weak coupling followed by a slower decrease in SNR for stronger coupling. At intermediate noise intensities, the system exhibits maximum SNR. This feature is inherently due to the nonlinear nature of a system's dynamics and cannot be reproduced by linear systems.

9.5 Fokker-Planck Equation

In Chap. 4 we studied, among many other features, the local behavior of deterministic continuous dynamical models based on the analysis of the existence and stability of equilibrium points. Such equilibrium points were found by setting the derivative of the right-hand side of the model to zero and then solving for the state variable. In the case of stochastic dynamic models, this is no longer possible because random fluctuations render the model equations nonautonomous. Thus, instead of using equilibrium points, the long-term behavior of a stochastic dynamical model is studied through a probability density function, $p(x, t)$, which measures the probability that the system is in state x at time t. The time evolution of this probability density function is governed by the Fokker-Planck equation. There are two versions of the Fokker-Planck equation. One due to Stratonovich and one due to Itô. We discuss both next.

Stratonovich Version

For a one-dimensional stochastic model described by Eq. (9.1), i.e., $x \in \mathbb{R}$, with Gaussian white noise, $\xi(t)$, with correlation given by Eq. (9.2), the Fokker-Planck equation is

$$\frac{\partial}{\partial t} p(x,t) = -\frac{\partial}{\partial x}\left[f(x,t)p(x,t)\right] + D\frac{\partial}{\partial x}\left[g(x)\frac{\partial}{\partial x}\left(g(x)p(x,t)\right)\right]. \quad (9.23)$$

In higher dimensions, where the state variable is $\mathbf{x} = [x_1, \ldots, x_N]^T \in \mathbb{R}^N$, and the internal dynamics is $f = [f_1(\mathbf{x}), \ldots, f_N(\mathbf{x})]^T \in R^N$, the Fokker-Planck equation becomes

$$\begin{aligned} \frac{\partial}{\partial t} p(\mathbf{x},t) = &-\sum_{i=1}^{N} \frac{\partial}{\partial x_i}\left[f_i(\mathbf{x},t)p(\mathbf{x},t)\right] + \\ &D\sum_{i,j,k=1}^{N} \frac{\partial}{\partial x_i}\left[g_{i,k}(\mathbf{x})\frac{\partial}{\partial x_j}\left(g_{jk}(x)p(\mathbf{x},t)\right)\right]. \end{aligned} \quad (9.24)$$

Itô's Version

An alternative interpretation due to Itô yields the following 1D version of the Fokker-Planck equation

$$\frac{\partial}{\partial t} p(x,t) = -\frac{\partial}{\partial x}\left[f(x,t)p(x,t)\right] + D\frac{\partial^2}{\partial x^2}\left[g^2(x)p(x,t)\right]. \quad (9.25)$$

And in higher dimensions, Itô's version of the Fokker-Planck equation becomes

$$\frac{\partial}{\partial t} p(\mathbf{x},t) = -\sum_{i=1}^{N} \frac{\partial}{\partial x_i}\left[f_i(\mathbf{x},t)p(\mathbf{x},t)\right] + D\sum_{i,j=1}^{N} \frac{\partial^2}{\partial x_i \partial x_j}\left[g_{ij}^2(x)p(\mathbf{x},t)\right]. \quad (9.26)$$

Notice that both interpretations, Stratonovich and Itô, coincide when $g(x)$ is a constant, which corresponds to the case of additive noise. We also note that it is common practice to use $\langle \xi(t)\, \xi(s) \rangle = \delta(t-s)$, in which case $D = 1/2$ must be substituted in the corresponding Fokker-Planck equations.

Next, we discuss a few examples, starting with the case where g is a constant, so that both approaches, Stratonovich and Itô, lead to the same Fokker-Planck equation.

Example 9.7 Consider an overdamped Brownian motion modeled by a 1D stochastic equation driven purely by a Wiener process, i.e., $f = 0$ and $g = 1$. Assume the diffusion coefficient to be D, so that Eq. (9.27) becomes

$$\frac{dx}{dt} = \frac{dW(t)}{dt}.$$

In this case, the Fokker-Planck equation is

$$\frac{\partial p(x,t)}{\partial t} = D \frac{\partial^2 p(x,t)}{\partial x^2}.$$

This last equation is simply the heat equation, which describes the diffusion of a "temperature" field, $p(x,t)$. Assuming the initial condition $p(x,0) = \delta(x)$, and solving for $p(x,t)$ yields

$$p(x,t) = \frac{1}{\sqrt{4\pi D t}} e^{-x^2/(4Dt)}.$$

Example 9.8 For the case of a one-dimensional stochastic model driven by a Wiener process, $W(t)$, in which $\dot{W}(t)$ is a Gaussian white noise function (zero mean and correlation as discussed earlier), with diffusion coefficient $\sigma^2(x,t)/2$, Eq. (9.1) can be written as

$$dx = f(x,t)\, dt + \sigma(x,t)\, dW(t), \tag{9.27}$$

the Fokker-Planck Eq. (9.67) becomes

$$\frac{\partial}{\partial t} p(x,t) = -\frac{\partial}{\partial x}\left[f(x,t) p(x,t)\right] + D \frac{\partial^2}{\partial x^2}\left[\sigma^2(x,t) p(x,t)\right]. \tag{9.28}$$

Example 9.9 Let us consider an Ornstein-Uhlenbeck process described by Eq. (9.18). For completeness purposes, we rewrite the model in Langevin form

$$\dot{x}(t) = -\frac{1}{\tau_c} x(t) + \frac{\sigma}{\tau_c} \xi(t).$$

Using Eq. (9.28), we find the associated Fokker-Planck equation to be

$$\frac{\partial p(x,t)}{\partial t} = \frac{1}{\tau_c}\frac{\partial}{\partial x}(xp(x,t)) + \frac{D\sigma^2}{\tau_c^2}\frac{\partial^2 p(x,t)}{\partial x^2}.$$

The steady-state solution, $p_s(x)$, of the Fokker-Planck equation can be found by setting $\partial_t p = 0$, and then solving for $p(x,t)$, which yields

$$p_s(x) = \frac{1}{\sqrt{\frac{2\pi D}{\tau_c}\sigma}} e^{-x^2/(2D\sigma^2/\tau_c)}.$$

Example 9.10 Let us consider again the bistable system modeled by Eq. (9.22),

$$\frac{dx}{dt} = -\nabla U(x) + \eta(t)$$

with a double well potential potential function

$$U(x) = U(x) = -\frac{1}{2}x^2 + \frac{1}{4}x^4.$$

and $\eta(t)$ assumed to be a Gauss white noise function with correlation

$$\langle \tilde{\eta}(t)\, \tilde{\eta}(s)\rangle = 2D\delta(t-s).$$

The Langevin equation becomes

$$\frac{dx}{dt} = x - x^3 + \eta(t). \tag{9.29}$$

The corresponding Fokker-Planck equation is

$$\frac{\partial p(x,t)}{\partial t} = \frac{\partial}{\partial x}\big(U'(x)p(x,t)\big) + D\frac{\partial^2 p(x,t)}{\partial x^2}.$$

Once again, we seek a steady-state solution, $p_s(x)$, by setting $\partial_t p = 0$:

$$D\frac{d^2 p_s}{dx^2} + \frac{d}{dx}\big(U'(x)p_s\big) = 0.$$

Integrating once, we get

$$\frac{dp_s}{dx} = -\frac{U'}{D}p_s + c_1,$$

where c_1 is an arbitrary constant. Assuming the boundary condition $p_s(\infty) = 0$, leads to $c_1 = 0$. Integrating a second time we find the desired steady-state solution:

$$p_s(x) = c_2 e^{-\frac{1}{D}U(x)}.$$

Substituting the potential function $U(x)$ defined above, we get

$$p_s(x) = c_2 \exp\left[-\frac{1}{D}\left(\frac{x^4}{4} - \frac{x^2}{2}\right)\right],$$

where c_2 is also an arbitrary constant.

For small values of D, this steady-state solution, $p_s(x)$, is a bimodal distribution. In fact, the time-evolution of the bistable system is spent, mostly, near the two equilibrium points, $x = -1$ and $x = 1$, of the deterministic system, see Fig. 9.4.

Example 9.11 Let us consider now a 2D stochastic system of the form

$$\begin{aligned}\frac{dx_1}{dt} &= \mu(1 - x_1^2 - x_2^2)x_1 - \omega x_2 + \eta_1(t)\\ \frac{dx_2}{dt} &= \mu(1 - x_1^2 - x_2^2)x_2 + \omega x_1 + \eta_2(t),\end{aligned} \tag{9.30}$$

where μ and ω are parameters, and

$$\langle \tilde{\eta}_i(t)\, \tilde{\eta}_j(s)\rangle = 2D\delta_{ij}\delta(t - s).$$

Let us examine first the deterministic system. Using polar coordinates, with $x_1 = r\cos\theta$ and $x_2 = r\sin\theta$, the deterministic system leads to the following amplitude-phase dynamics

$$\begin{aligned}\frac{dr}{dt} &= \mu(1 - r^2)r\\ \frac{d\theta}{dt} &= \omega.\end{aligned}$$

Observe that in polar coordinates the amplitude and phase equations decouple from one another. Furthermore, the equilibrium $r = 1$ corresponds to a limit cycle. A linear stability analysis shows that when $\mu > 0$, this limit cycle is locally asymptotically stable. Next, we investigate what happens to this limit cycle under the presence of noise.

The Fokker-Planck equation for the joint probability distribution function, $p(x_1, x_2, t)$ is

$$\frac{\partial p(x,t)}{\partial t} = -\frac{\partial}{\partial x_1}(f_1 p) - \frac{\partial}{\partial x_1}(f_2 p) + D\left(\frac{\partial^2 p(x,t)}{\partial x_1^2} + \frac{\partial^2 p(x,t)}{\partial x_2^2}\right). \tag{9.31}$$

To find a steady-state solution, it is more convenient to rewrite Eq. (9.31) in polar coordinates

$$\frac{\partial p(x,t)}{\partial t} = -\frac{1}{r}\frac{\partial}{\partial r}\left[\mu(1-r^2)r^2 p\right] - \frac{\partial}{\partial \theta}(\omega p) + D\left(\frac{1}{r}\frac{\partial}{\partial r}\left(r\frac{\partial p}{\partial r}\right) + \frac{1}{r^2}\frac{\partial^2 p}{\partial \theta^2}\right).$$

It is convenient to assume $p_s(x)$ to be independent of θ. That is,

$$\frac{\partial p_s(x)}{\partial \theta} = 0.$$

Then, setting $\partial_t p = 0$ we get

$$-\frac{1}{r}\frac{d}{dr}\left[\mu(1-r^2)r^2 p\right] - D\frac{1}{r}\frac{d}{dr}\left(r\frac{dp_s}{dr}\right) = 0.$$

Solving this ordinary differential equation produces the desired steady-state solution:

$$p_s(x) = c\exp\left[\frac{\mu}{D}\left(\frac{r^2}{2} - \frac{r^4}{4}\right)\right],$$

where c is a constant.

Figure 9.5 illustrates the computer simulation of the model Eq. (9.30) obtained through the Euler-Maruyama method. For comparison purposes, both, the deterministic and stochastic solutions are shown.

Example 9.12 We now consider a case where the function $g(x)$ is no longer constant, and discuss the differences between the Stratonovich's and the Itô's version of the Fokker-Planck equation. Consider the Langevin SDE

$$\begin{aligned} \frac{dx}{dt} &= x\,\eta(t) \\ x(0) &= x_0, \end{aligned} \tag{9.32}$$

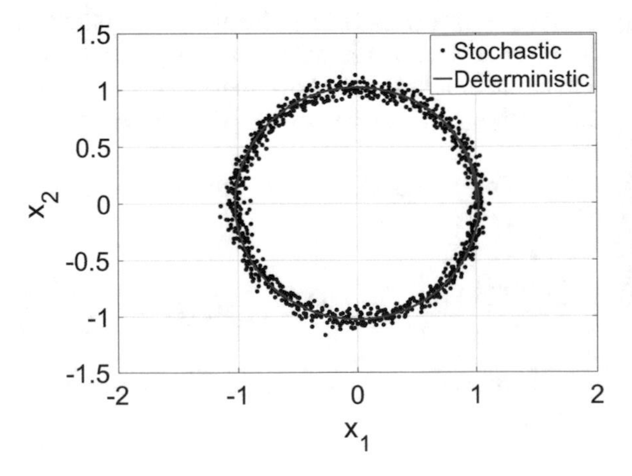

Fig. 9.5 Numerical solution of the 2D stochastic model Eq. (9.30) obtained through the Euler-Mayurama algorithm

where $\eta(t)$ is a Gaussian white noise function with zero mean and correlation $\langle \eta(t)\, \eta(s) \rangle = 2D\delta(t-s)$. The Fokker-Planck equations produced by these two approaches are

$$\begin{aligned} \frac{\partial p(x,t)}{\partial t} &= D\frac{\partial}{\partial x}\left(x\frac{\partial}{\partial x}(xp)\right) \qquad \text{(Stratonovich)}, \\ \frac{\partial p(x,t)}{\partial t} &= D\frac{\partial^2}{\partial x^2}\left(x^2 p\right) \qquad \text{(Itô)}. \end{aligned} \tag{9.33}$$

We will explore the differences in these two approaches through the mean of the state variable $x(t)$, which is defined as

$$\langle x(t) \rangle = \int_{-\infty}^{\infty} x\, p(x,t)\, dx.$$

To find an expression for $\langle x(t) \rangle$ we can multiply the Fokker-Planck equations by x and then integrate. Let's do this first with the Stratonovich's version:

$$\frac{\partial}{\partial t}\int_{-\infty}^{\infty} x\, p(x,t)\, dx = D\int_{-\infty}^{\infty} x\frac{\partial}{\partial x}\left(x\frac{\partial}{\partial x}(xp)\right) dx.$$

Integrating by parts once we get

$$\frac{d}{dt}\langle x(t) \rangle = -D\int_{-\infty}^{\infty} x\frac{\partial}{\partial x}(xp)dx.$$

Integrating by parts, again, we arrive at

$$\frac{d}{dt}\langle x(t) \rangle = D\int_{-\infty}^{\infty} x\, p\, dx = D\langle x(t) \rangle.$$

Solving this ODE we obtain the desired mean value

$$\langle x(t) \rangle = x_0\, e^{Dt}. \tag{9.34}$$

We now apply a similar process to Itô's version. Start by multiplying by x and then integrate:

$$\frac{\partial}{\partial t}\int_{-\infty}^{\infty} x\, p(x,t)\, dx = D\int_{-\infty}^{\infty} x\frac{\partial^2}{\partial x^2}\left(x^2 p\right) dx.$$

Integrating by parts we arrive at

$$\frac{d}{dt}\langle x(t) \rangle = 0,$$

which leads to the solution

$$\langle x(t) \rangle = x_0. \tag{9.35}$$

Now, if we average the original SDE model Eq. (9.32) we get

$$\frac{d}{dt}\langle x(t) \rangle = \langle x(t)\, \eta(t) \rangle.$$

Thus, comparing Eqs. (9.34) and (9.35), we see that Itô's interpretation of the Fokker-Planck equation leads to $\langle x(t)\, \eta(t) \rangle = 0$, while Stratonovich's approach yields $\langle x(t)\, \eta(t) \rangle \neq 0$. This happens because in Itô's approach, $x(t)$ depends on $\eta(s)$ only for $s < t$. Beyond $s > t$, $x(t)$ is independent, so $x(t)$ and $\eta(t)$ are uncorrelated. In Stratonovich's approach, $x(t)$ and $\eta(t)$ are, however, correlated, so the average is nonzero.

Some authors point out that Stratonovich's approach is more suitable for modeling stochastic process from physics and engineering, while Itô's interpretation is more suitable for mathematical and financial models.

9.6 Phase Drift in a Network of Gyroscopes

The French mathematician, mechanical engineer, and scientist, Gaspard-Gustave de Coriolis (1792–1843) is best known for the discovery of the "Coriolis" effect: "an apparent deflection and acceleration of moving objects from a straight path when viewed from a rotating frame of reference" [14]. The observed inertial acceleration of the object, also known as Coriolis acceleration, serves nowadays as the basic principle of operation of many inertial navigation systems, including gyroscopes. Vibratory gyroscopes, in particular, are sensor devices that can measure absolute angles of rotation (type I gyroscope) or rates of angular rotation (type II). All vibratory gyroscopes operate on the basis of energy transferred between two vibration modes, a driving mode and a sensing mode, by Coriolis force [15–17]. The conventional model of a vibratory gyroscope consists of a mass-spring system, see Fig. 9.6. A change in the acceleration around the driving axis caused by the presence of Coriolis force induces a vibration in the sensing axis which can be converted to measure angular rate output or absolute angles of rotation.

The accuracy of most gyroscope systems depend on three parameters: quality factor, phase drift, and robustness. The quality factor is the linear deviation of the measured rate from the true rate (normally given as a percentage of full scale). It characterizes the capability of a gyroscope to accurately sense angular velocity at different angular rates, including the sensitivity of the angular rate sensor and its ability to convert voltage output into angular rate, so its units are in $(deg/s)/V$. The phase drift is the offset error output that appears as an additive term on the gyroscope output due, mainly, to temperature fluctuations. It characterizes the ability of a gyroscope to reference all rate measurements to the nominal zero rate output, so its units are in deg/sec or deg/h. Robustness is the deviation of the measured rate due to noise influence or parameter variations and it is very important because signal processing of the gyroscope output can introduce noise. The units of measurement for the effect of noise are generally $deg/\sqrt{D}$, where D is the intensity of noise.

9.6.1 Equations of Motion

The configuration of the vibratory gyroscope of Fig. 9.6 contains two vibration modes: the *primary* mode (x-direction) and the *secondary* mode (y-direction). Both modes are coupled to one another by Coriolis force through the term $F_{cx} = |2m\vec{\Omega}_z \times \dot{y}| = 2m\Omega_z\dot{y}$ and $F_{cy} = |2m\vec{\Omega}_z \times \dot{x}| = -2m\Omega_z\dot{x}$, respectively, where m is mass and Ω_z is the angular rate of rotation along a perpendicular direction (z-axis). The governing equations for the entire spring-mass system can then be written in the following form

$$\begin{aligned} m\ddot{x} + c_x\dot{x} + F_r(x) &= F_e(t) + 2m\Omega_z\dot{y} \quad \text{(drive)} \\ m\ddot{y} + c_y\dot{y} + F_r(y) &= \qquad\quad - 2m\Omega_z\dot{x}, \quad \text{(sense)} \end{aligned} \tag{9.36}$$

where c_x (c_y) is the damping coefficient along the x-direction (y-direction), $F_r(\cdot)$ is the elastic restoring force of the springs. A typical model for the restoring force along the x-direction, for instance, has the form: $F_r(x) = \kappa_x x + \mu_x x^3$, where κ_x and μ_x

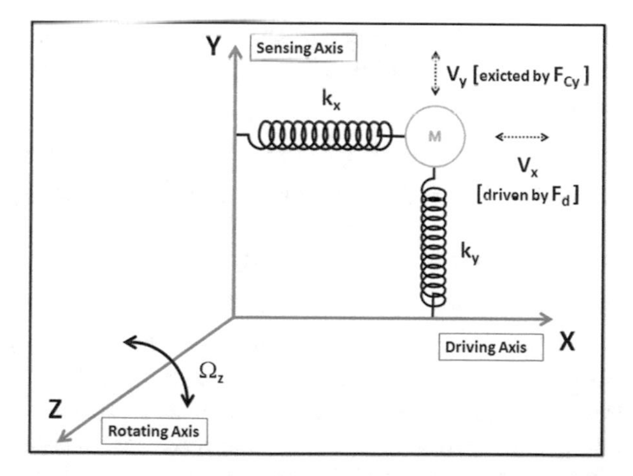

Fig. 9.6 Schematic diagram of a model for a vibratory gyroscope system. An internal driving force induces the spring-mass system to vibrate in one direction, the x-axis in this case. An external rotating force, perpendicular to the plane of the spring-mass system, induces, on the other hand, the spring-mass system to oscillate in the y-direction by transferring energy through Coriolis force. The oscillations along the y-axis can be used to detect and quantify the rate of rotation

Table 9.1 System parameters for a vibratory gyroscope

Parameter	Value	Unit
m	1.0E-09	Kg
c_x, c_y	5.1472E-07	N s/meter
κ_x, κ_y	2.6494	N/meter
μ_x, μ_y	2.933	N/meter^3
A_d	1.0E-03	N
w_d	5.165E+04	rad/sec

are constant parameters. The same model applies to the y-axis, just replace x by y. The x-axis mode, which is also known as the drive axis, is also excited by a reference driving force, typically a periodic signal of the form $F_e = A_d \cos w_d t$, where A_d is the amplitude and w_d is the frequency of the excitation. Typical parameter values, which we will consider in this work, are shown in Table 9.1.

Under these conditions, the gyroscope of Fig. 9.6 can detect an applied angular rate Ω_z by measuring the displacements along the y-axis (also known as sensing axis) caused by the transfer of energy by Coriolis force. If there is no external rotation, i.e., $\Omega_z = 0$, the motion Eqs. (9.36) along the two axes become uncoupled from one another.

9.6.2 Bi-Directionally Coupled Ring

We now consider an array of N vibratory gyroscopes arranged in a ring configuration, coupled bidirectionally along the drive axis, so that the equations of motion can be written in the general form

$$
\begin{aligned}
m_j\ddot{x}_j + c_{xj}\dot{x}_j + F_r(x_j) &= F_{ej}(t) + 2m_j\Omega_z\dot{y}_j + \sum_{k\to j} c_{jk}h(x_j, x_k) \\
m_j\ddot{y}_j + c_{yj}\dot{y}_j + F_r(y_j) &= \qquad - 2m_j\Omega_z\dot{x}_j,
\end{aligned}
\tag{9.37}
$$

where h is the coupling function between gyroscopes j and k, the summation is taken over those gyroscopes k that are coupled to gyroscopes j and c_{jk} is a matrix of coupling strengths. Parameter values are the same as those shown in Table 9.1. We choose to couple the gyroscopes through the drive axis because this type of coupling is the most natural way to add signals on top of the already existing external drive signal. One may also choose to couple through the sense axis but that may involve more design changes and added circuitry to accommodate the input signal. In this section we will consider, in particular, a diffusive coupling function of the form $h(x_j, x_k) = x_k - x_j$. Here we consider the response of the coupled gyroscope system to a weak periodic force, so we apply the transformation $A_d \to \varepsilon$. In an attempt to understand the collective behavior of the network, we make the simplifying assumption of the mass-spring-dampers to be identical and set all coefficients equal to the mean value for a typical ensemble of gyroscopes. In addition, we assume each gyroscope to be excited by the same external harmonic sine-wave signal with one driving frequency in the drive coordinate axis, i.e., $F_{ei} = F_d \sin w_d t$. Further assuming the coupling strength to be identical, i.e., $c_{jk} = \lambda$, the equations of motion take the form

$$\begin{aligned} m\ddot{x}_j + c\dot{x}_j + \kappa x_j + \mu x_j^3 &= \varepsilon \sin w_d t + 2m\Omega_z \dot{y}_j + \lambda(x_{j+1} - 2x_j + x_{j-1}) \\ m\ddot{y}_j + c\dot{y}_j + \kappa y_j + \mu y_j^3 &= \qquad\qquad - 2m\Omega_z \dot{x}_j. \end{aligned} \tag{9.38}$$

9.6.3 Computational Bifurcation Analysis

Computer simulations and the continuation software package AUTO [18] confirm the existence of all three solution classes predicted by the lattice of isotropy subgroups for the special case $n = 3$, see Fig. 9.7, including the IP transition $d_1 \to d_n$.

The onset of oscillations in the model Eqs. (9.38) occurs when the coupling strength exceeds a critical value, which we denote by λ_c. When $\lambda < \lambda_c$, there are two stable periodic solutions and one unstable periodic solution. The stable solutions correspond to (a, a, b)—two patterns of oscillation in which two of the driving modes oscillate in synchrony but with non-zero mean (one positive and one negative) while the third mode oscillates with a different non-zero mean. The unstable solution represents the complete-synchronization state (a, a, a). As λ increases towards

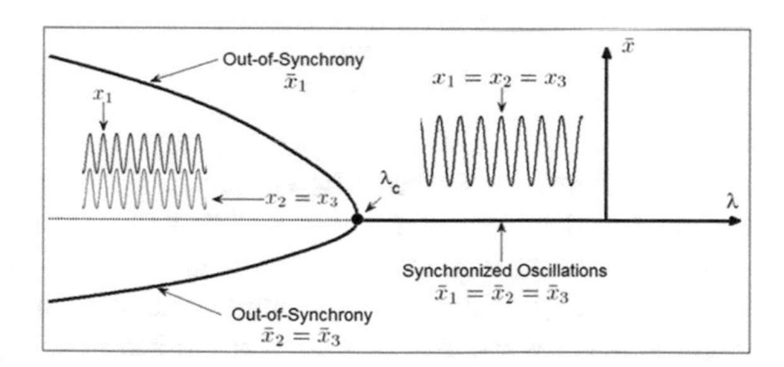

Fig. 9.7 One-parameter bifurcation diagram illustrating the existence and stability properties of synchronized periodic oscillations in a ring of three vibratory gyroscopes bi-directionally coupled. As λ approaches (from the left) a critical coupling strength, λ_c, three periodic solutions merge in a supercritical pitchfork bifurcation. The stable solutions are periodic solutions with non-zero mean while the unstable solution is the synchronized state in which the driving modes oscillate with the same amplitude and the same phase. Past λ_c the synchronized state becomes globally asymptotically stable, as is supported by numerical calculation of eigenvalues of the linearized vector field

λ_c, the two non-zero mean periodic solution and the zero-mean periodic solution merge in a supercritical pitchfork bifurcation. Past λ_c, only the zero-mean periodic solution exists and becomes globally asymptotically stable (as is determined from the eigenvalues obtained numerically with the aid of AUTO). The oscillations along the sensing axis are, however, unaffected by the change in coupling. They are always stable and completely synchronized with one another though they are out-of-phase by π with those of the driving axis due to the sign difference in the Coriolis force terms.

9.6.4 Robustness

We expect noise in a coupled gyroscope system to arise from two main sources: fluctuations in the mass of each individual gyroscope and contamination of a target signal. In the former case, we need to replace m in the motion equations by m_i. Experimental data suggest that the range $m_i = 1.0E - 09 \pm 10\%$ is actually reasonable. In the latter case, we consider a target signal contaminated by noise, assumed to be Gaussian band-limited noise having zero mean, correlation time τ_c (usually $\tau_F << \tau_c$, where τ_F is the time constant of each individual gyroscope, so that noise does not drive its response), and variance σ^2. This type of noise is a good approximation (except for a small $1/f$ component at very low frequencies) to what is actually expected in an experimental setup. From a modeling point of view, colored noise $\eta(t)$ that contaminates the signal should appear as an additive term in the sensing axis, leading to a stochastic (Langevin) version of the model equations, which for the ring configuration with bidirectional coupling we get

$$\begin{aligned} m_j\ddot{x}_j + c\dot{x}_j + \kappa x_j + \mu x_j^3 &= \varepsilon \sin w_d t + 2m_j\Omega_z\dot{y}_j + \lambda(x_{j+1} - 2x_j + x_{j-1}) \\ m_j\ddot{y}_j + c\dot{y}_j + \kappa y_j + \mu y_j^3 &= -2m_j\Omega_z\dot{x}_j + \eta_j(t), \\ \frac{d\eta_j}{dt} &= -\frac{\eta_j}{\tau_c} + \frac{\sqrt{2D}}{\tau_c}\xi(t). \end{aligned} \tag{9.39}$$

In general, we would expect somewhat different noise in each equation, since, realistically, the reading of the external signal is slightly different in each sensing axis. This is due to non-identical circuit elements, mainly. In this work we will consider, therefore, the situation wherein the different noise terms $\eta_i(t)$ are uncorrelated; however, for simplicity, we will assume them to have the same intensity D. Each (colored) noise $\eta_i(t)$ is characterized by $\langle\eta_i(t)\rangle = 0$ and $\langle\eta_i(t)\eta_i(s)\rangle = (D/\tau_c) \times \exp[-|t-s|/\tau_c]$, where $D = \sigma^2\tau_c^2/2$ is the noise intensity, $\xi(t)$ is a gaussian white noise function of zero mean, and the "white" limit is obtained for vanishing τ_c.

Computer simulations of ensembles of various network sizes N of uncoupled and coupled gyroscopes were conducted for comparison purposes of phase drifts. Each ensemble consisted of $M = 100$ simulation samples with random fluctuations in

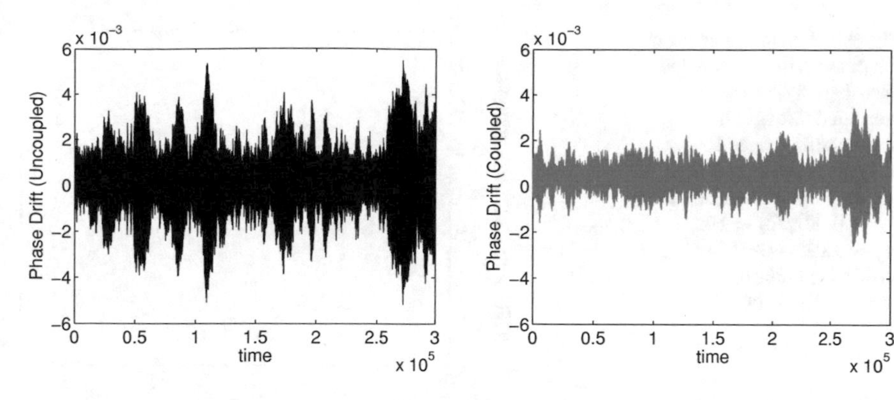

Fig. 9.8 Comparison of phase drift between (left) an ensemble of three uncoupled gyroscopes and (right) a three-gyroscope coupled system. Parameters are: $A_d = 0.001$, $\Omega_z = 100$, and $\lambda = -0.65$, $m_j = 1.0E - 09 \pm 10\%$, and noise intensities $D = \pm 1.0E - 09$

mass and noise intensities. The phase of each individual j gyroscope was calculated through $\alpha_j = \arctan\left(-\dot{y}_j / w_d y_j\right)$. Then the phase drift on that individual gyroscope was obtained as the difference between its phase with noise and its phase without noise, i.e., $\theta_j = \alpha_j^{\text{noise}} - \alpha_j^{\text{no noise}}$. Finally, the average phase drift

$$\theta(t) = \frac{1}{MN} \sum_{j=1}^{MN} \theta_j,$$

of the entire ensemble was calculated for both cases, uncoupled and coupled ensembles. Figure 9.8 shows, in particular, the phase drift of an ensemble of three individual gyroscopes and the phase drift of a similar ensemble but with coupling. The reduction in the phase drift of the sensing axis of the coupled system is, approximately, by a factor of 1.7 times that of the uncoupled system.

To calculate the actual reduction factor we first compute the interquartile range (IQR) of both uncoupled and coupled ensembles. The IQR measures the phase drift variation from the 25% percentile to the 75% percentile. The reduction factor is then the ratio IQR(θ^c) / IQR(θ^u), where the superscript indicates whether the gyroscopes are coupled or uncoupled, respectively. Figure 9.9 shows the resulting reduction factors for various network sizes.

For small N the reduction factor of a coupled vs. uncoupled ensemble appears to decrease steadily as N increases but it then increases for networks larger than $N = 8$ gyroscopes, approximately. Careful examination of the average amplitude response of an ensemble of coupled gyroscopes reveals that the amplitude of the sensing axis is dynamically dependent on the number N of gyroscopes and the coupling strength λ, see Fig. 9.10. In fact, the largest amplitudes are achieved in the vicinity of $N = 8$. Larger amplitudes, in turn, can better attenuate the effects of noise and

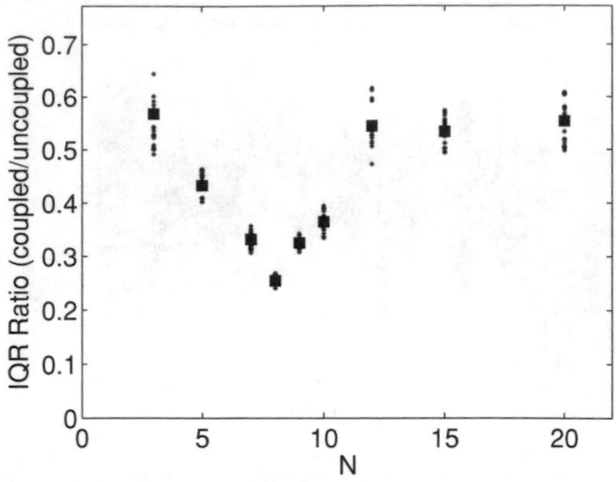

Fig. 9.9 Reduction factor in the phase drift of a coupled gyroscope system as measured through the interquartile range of ensembles between 80 and 100 samples. Parameters are: $A_d = 0.001$, $\Omega_z = 100$, $m_j = 1.0E - 09 \pm 10\%$ with noise intensities $D = \pm 1.0E - 09$

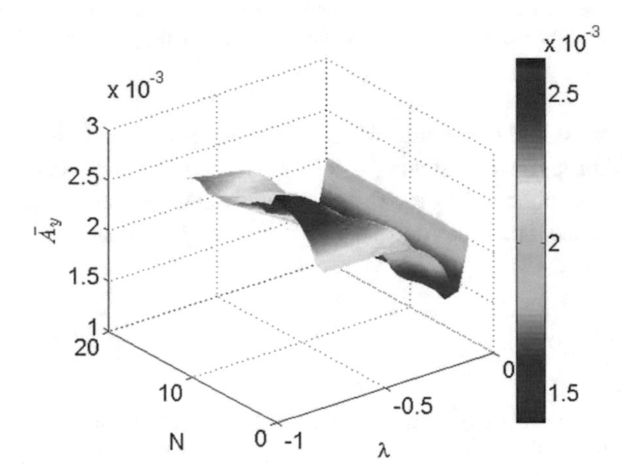

Fig. 9.10 Average amplitude response of the sensing axis of ensembles of coupled gyroscopes with various network sizes and coupling strengths. Parameters are: $A_d = 0.001$, $\Omega_z = 100$, $m_j = 1.0E - 09 \pm 10\%$ without noise

mass fluctuations, and thus, this explains why the bidirectionally coupled gyroscope system yields an optimal phase drift around $N = 8$.

9.7 Phase Drift in a Model for Precision Timing

Precise time is crucial to a variety of economic activities around the world. Communication systems, electrical power grids, and financial networks all rely on precision timing for synchronization and operational efficiency. The free availability of GPS [19] time has enabled cost savings for industrial and scientific developments that depend on precise time and has led to significant advances in capability. For example, wireless telephone and data networks use GPS time to keep all of their base stations in synchronization.

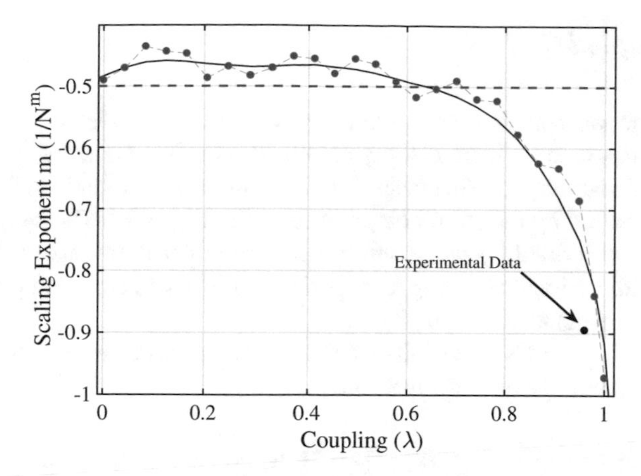

Fig. 9.11 Phase error scaling in an ensemble of N crystal oscillators. When the crystal oscillators are uncoupled, the scaling is about $1/\sqrt{N}$, see red line. When the oscillators are coupled, the scaling can be (for certain values of the coupling strength, λ, and for certain patterns such as rotating wave solutions) about $1/N$. Blue markers represent experimental measurements. Solid black curve shows predictions from a relevant mathematical model

The *standard* practice in precision timing is to average out the timing of multiple (uncoupled) clocks. For instance, at the United States Naval Observatory time is measured by averaging the time of an ensemble of (uncoupled) atomic clocks, in which phase error scales as $1/\sqrt{N}$, see red line in Fig. 9.11.

In this section, we show proof of concept that an ensemble of coupled crystal oscillators can produce (through certain patterns of oscillations, such as rotating waves with constant phase differences) a phase error that scales, at least, as $1/N$, see experimental data shown as blue markers in Fig. 9.11. Predictions from a mathematical model (see Appendix B) for details of computational work) of a network of coupled crystal oscillators are shown by a solid black curve in Fig. 9.11. Crystal oscillators were chosen because they are readily available, inexpensive and require low power for operation. However, the fundamental idea of performance enhancement via collective behavior is a model-independent feature which should apply to any network of coupled nonlinear oscillators, provided that the collective oscillations are, qualitatively, the same.

The success of the averaging technique in the analysis of a single crystal oscillator model, see Eq. (4.62) in Chap. 4, has lead us to consider a similar approach for the analysis of a network of coupled crystal oscillators. We discuss next the coupled system.

9.7.1 Coupled System

In this section we consider a Coupled Crystal Oscillator System (CCOST) made up of N, assumed to be identical, crystal oscillators. We consider first the case of unidirectional coupling in a ring fashion, as is shown schematically in Fig. 9.12. Each node is represented by the circuit diagram found in Fig. 4.35. The spatial symmetry of the ring is described by the group $\mathbf{Z}_N$ of cyclic permutations of N objects. In the case of bidirectional coupling, the spatial symmetry is captured by the dihedral group $\mathbf{D}_N$ of permutations of an N-gon.

Applying Kirchhoff's law to the CCOST network with unidirectional coupling yields the following governing equations

$$L_{k,j}\frac{d^2 i_{k,j}}{dt^2} + R_{k,j}\frac{di_{k,j}}{dt} + \frac{1}{C_{k,j}} i_{k,j} = \left[a - 3b\big(i_{k,1} + i_{k,2} - \lambda\left[i_{k+1,1} + i_{k+1,2}\right]\big)^2\right] \left[\frac{di_{k,1}}{dt} + \frac{di_{k,2}}{dt} - \lambda\left(\frac{di_{k+1,1}}{dt} + \frac{di_{k+1,2}}{dt}\right)\right], \tag{9.40}$$

where $k = 1, 2, \ldots, N \mod N$, $j = 1, 2$. Since we assume identical components in each crystal oscillator, then the set of parameters reduces to: $L_{k,1} = L_1$, $L_{k,2} = L_2$, $R_{k,1} = R_1$, $R_{k,2} = R_2$, $C_{k,1} = C_1$ and $C_{k,2} = C_2$. Letting $t = \sqrt{L_1 C_1}\tau$, $\Omega_1^2 = 1$, $\Omega_2^2 = \frac{L_1}{L_2}\frac{C_1}{C_2}$, $L_r = \frac{L_1}{L_2}$, $\varepsilon = \sqrt{\frac{C_1}{L_1}}$, and relabeling τ as time t, we write Eq. (9.40) in dimensionless form

$$\begin{aligned}
\frac{d^2 i_{k,1}}{dt^2} + \Omega_1^2 i_{k,1} &= \varepsilon\Bigg\{-R_1\frac{di_{k,1}}{dt} + \left[a - 3b\big(i_{k,1} + i_{k,2} - \lambda\left[i_{k+1,1} + i_{k+1,2}\right]\big)^2\right] \\
&\quad \left[\frac{di_{k,1}}{dt} + \frac{di_{k,2}}{dt} - \lambda\left(\frac{di_{k+1,1}}{dt} + \frac{di_{k+1,2}}{dt}\right)\right]\Bigg\} \\
\frac{d^2 i_{k,2}}{dt^2} + \Omega_2^2 i_{k,2} &= \varepsilon L_r\Bigg\{-R_2\frac{di_{k,2}}{dt} + \left[a - 3b\big(i_{k,1} + i_{k,2} - \lambda\left[i_{k+1,1} + i_{k+1,2}\right]\big)^2\right] \\
&\quad \left[\frac{di_{k,1}}{dt} + \frac{di_{k,2}}{dt} - \lambda\left(\frac{di_{k+1,1}}{dt} + \frac{di_{k+1,2}}{dt}\right)\right]\Bigg\}.
\end{aligned} \tag{9.41}$$

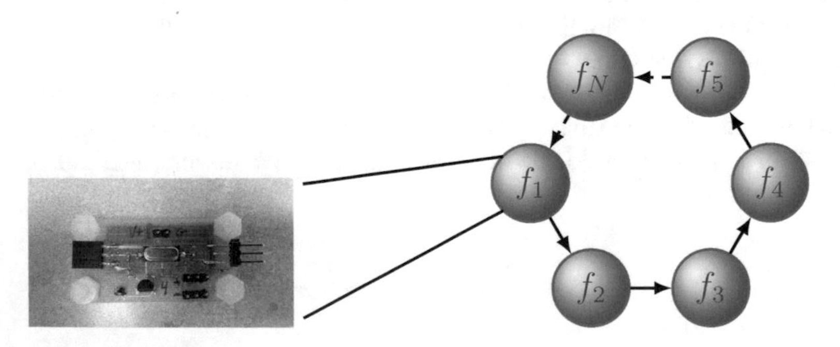

Fig. 9.12 CCOST concept with unidirectionally coupled crystal oscillators

After applying the following set of invertible coordinates transformations

$$\begin{aligned} i_{kj} &= x_{kj}\cos\phi_{kj}; \\ i'_{kj} &= -\Omega_j x_{kj}\sin\phi_{kj}; \\ i''_{kj} &= -\Omega_j x'_{kj}\sin\phi_{kj} - \Omega_j^2 x_{kj}\cos\phi_{kj} - \Omega_j x_{kj}\psi'_{kj}\cos\phi_{kj}; \\ \phi_{kj} &= \Omega_j t + \psi_{kj}; \end{aligned} \tag{9.42}$$

for $j = 1, 2$ we arrive at the following set of equations, written symbolically as:

$$\begin{bmatrix} \mathbf{x}'_k \\ \boldsymbol{\phi}'_k \end{bmatrix} = \begin{bmatrix} 0 \\ \Omega^0 \end{bmatrix} + \varepsilon \begin{bmatrix} \mathbf{X}^{[1]}(\mathbf{x}_k, \boldsymbol{\phi}_k, \boldsymbol{\phi}_{k+1}, \varepsilon) \\ \boldsymbol{\Omega}^{[1]}(\mathbf{x}_k, \boldsymbol{\phi}_k, \boldsymbol{\phi}_{k+1}, \varepsilon) \end{bmatrix}, \tag{9.43}$$

where $\mathbf{x}_k = (x_{k1}, x_{k2})$, $\boldsymbol{\phi}_k = (\phi_{k1}, \phi_{k2})$ and $\boldsymbol{\Omega}^0 = (\Omega_1, \Omega_2)$. The vector $\mathbf{X}^{[1]}$ has polynomial functions containing linear and cubic terms in $x_{k1}, x_{k2}, x_{k+1,1}$ and $x_{k+1,2}$ while $\boldsymbol{\Omega}^{[1]}$ has terms only dependent on $\boldsymbol{\phi}_k$ and at most quadratic terms in $\mathbf{x}_{k+1}$ divided by $\mathbf{x}_k$.

Next we remove the $O(\varepsilon)$ dependence in the equation for ϕ_k by using coordinates $\boldsymbol{\phi}_k \mapsto \boldsymbol{\phi}_k + \boldsymbol{\phi}_s$ and $\boldsymbol{\phi}_{k+1} \mapsto \boldsymbol{\phi}_{k+1} + \boldsymbol{\phi}_s$, where $\boldsymbol{\phi}_s = (\phi_{s1}, \phi_{s2})$. Then Eq. (9.43) becomes

$$\begin{bmatrix} \mathbf{x}'_k \\ \boldsymbol{\phi}'_k \\ \boldsymbol{\phi}'_s \end{bmatrix} = \begin{bmatrix} 0 \\ 0 \\ \boldsymbol{\Omega}^0 \end{bmatrix} + \varepsilon \begin{bmatrix} \mathbf{X}^{[1]}(\mathbf{x}_k, \boldsymbol{\phi}_k + \boldsymbol{\phi}_s, \boldsymbol{\phi}_{k+1} + \boldsymbol{\phi}_s, \varepsilon) \\ \boldsymbol{\Omega}^{[1]}(\mathbf{x}_k, \boldsymbol{\phi}_k + \boldsymbol{\phi}_s, \boldsymbol{\phi}_{k+1} + \boldsymbol{\phi}_s, \varepsilon) \\ 0 \end{bmatrix}. \tag{9.44}$$

The explicit form of these equations is not shown for brevity. In the bidirectional case, the dimensionless equations are

$$\begin{aligned} &\frac{d^2 i_{k,1}}{dt^2} + \Omega_1^2 i_{k,1} = \varepsilon\Big\{-R_1\frac{di_{k,1}}{dt} + \big[a - 3b\big(i_{k,1} + i_{k,2} - \\ &\lambda\big[i_{k+1,1} + i_{k+1,2} + i_{k-1,1} + i_{k-1,2}\big]\big)^2\big] \\ &\left[\frac{di_{k,1}}{dt} + \frac{di_{k,2}}{dt} - \lambda\left(\frac{di_{k+1,1}}{dt} + \frac{di_{k+1,2}}{dt} + \frac{di_{k-1,1}}{dt} + \frac{di_{k-1,2}}{dt}\right)\right]\Big\} \\ &\frac{d^2 i_{k,2}}{dt^2} + \Omega_2^2 i_{k,2} = \varepsilon L_r\Big\{-R_2\frac{di_{k,2}}{dt} + \big[a - 3b\big(i_{k,1} + i_{k,2} - \\ &\lambda\big[i_{k+1,1} + i_{k+1,2} + i_{k-1,1} + i_{k-1,2}\big]\big)^2\big] \\ &\left[\frac{di_{k,1}}{dt} + \frac{di_{k,2}}{dt} - \lambda\left(\frac{di_{k+1,1}}{dt} + \frac{di_{k+1,2}}{dt} + \frac{di_{k-1,1}}{dt} + \frac{di_{k-1,2}}{dt}\right)\right]\Big\}. \end{aligned} \tag{9.45}$$

The transformation (9.42) leads to the following network equations

$$\begin{bmatrix} \mathbf{x}'_k \\ \phi'_k \\ \phi'_s \end{bmatrix} = \begin{bmatrix} 0 \\ 0 \\ \boldsymbol{\Omega}^0 \end{bmatrix} + \varepsilon \begin{bmatrix} \mathbf{X}^{[1]}(\mathbf{x}_k, \phi_k + \phi_s, \phi_{k+1} + \phi_s, \phi_{k-1} + \phi_s \varepsilon) \\ \boldsymbol{\Omega}^{[1]}(\mathbf{x}_k, \phi_k + \phi_s, \phi_{k+1} + \phi_s, \phi_{k-1} + \phi_s \varepsilon) \\ 0 \end{bmatrix} \tag{9.46}$$

with $\mathbf{X}^{[1]}$ is a vector of polynomial functions containing linear and cubic terms in $x_{k1}, x_{k2}, x_{k+1,1}, x_{k+1,2}, x_{k-1,1}$ and $x_{k-1,2}$ and $\boldsymbol{\Omega}^{[1]}$ has a similar structure as described above. The complete set of equations is not shown for brevity.

A similar set of equations are obtained for the bidirectional case. The complete equations are omitted for brevity as they are very long, but they are found in [20]. The symmetry of these averaged amplitude-phase equations is captured by the groups $\mathbf{Z}_N \times \mathbf{O(2)} \times \mathbf{O(2)}$ and $\mathbf{D}_N \times \mathbf{O(2)} \times \mathbf{O(2)}$ for the unidirectional and bidirectional coupling cases, respectively. A complete analysis of the equations can be found in [20].

9.7.2 Phase Relations

At the beginning of this section we showed proof of concept, via computer simulations, that certain patterns of collective behavior produced by a network of crystal oscillators, mainly standard rotating waves, can lead to phase drift reduction that follows a $1/N$ scaling law, which is better than the $1/\sqrt{N}$ shown by an uncoupled ensemble [21]. We now provide a mathematical explanation of why the standard wave patterns are the desirable patterns that can lead to the best phase drift reduction. More importantly, we also show, analytically, that $1/N$ is the fundamental limit of phase drift scaling that can be achieved by a network configuration.

But, first, we need to comment on some technical issues that justify the use of the averaged Eqs. (9.46). These averaged amplitude-phase system can be written in complex coordinates, (z_1, z_2), as follows:

$$\begin{aligned}
\dot{z}_{k1} &= (a - R_1)z_{k1} - \frac{3b}{4}(|z_{k1}|^2 + |z_{k2}|^2)z_{k1} - a\lambda z_{k+1,1} + \\
&\quad \frac{3}{4}b(\lambda + \lambda^3)(|z_{k+1,1}|^2 + 2|z_{k+1,2}|^2)z_{k+1,1} - \frac{3}{2}b\lambda^2(|z_{k+1,1}|^2 + |z_{k+1,2}|^2)z_{k,1} - \\
&\quad \tfrac{3}{4}b\lambda^2 z_{k+1,1}^2 \bar{z}_{k,1} - \tfrac{3}{2}b\lambda^2 \bar{z}_{k,2} z_{k+1,1} z_{k+1,2}\left(e^{2i\alpha_{k1}} + e^{2i(\alpha_{k1}+\alpha_{k2})}\right) + \\
&\quad 3b z_{k,1}\bar{z}_{k,2} z_{k+1,2} e^{i\alpha_{k2}} \cos\alpha_{k2} \\
\dot{z}_{k2} &= L_r(a - R_2)z_{k2} - \frac{3}{4}L_r b(|z_{k1}|^2 + |z_{k2}|^2)z_{k2} - L_r a\lambda z_{k+1,2} + \\
&\quad \tfrac{3}{4}L_r b(\lambda + \lambda^3)(|z_{k+1,2}|^2 + 2|z_{k+1,1}|^2)z_{k+1,2} - \\
&\quad \tfrac{3}{2}L_r b\lambda^2(|z_{k+1,2}|^2 + |z_{k+1,1}|^2)z_{k,2} - \tfrac{3}{4}L_r b\lambda^2 z_{k+1,2}^2 \bar{z}_{k,2} - \\
&\quad \tfrac{3}{2}L_r b\lambda^2 \bar{z}_{k,1} z_{k+1,1} z_{k+1,2}\left(e^{2i\alpha_{k1}} + e^{2i(\alpha_{k1}+\alpha_{k2})}\right) + \\
&\quad 3L_r b z_{k,2}\bar{z}_{k,1} z_{k+1,1} e^{i\alpha_{k1}} \cos\alpha_{k1},
\end{aligned} \tag{9.47}$$

where $z_{kj} = x_{kj}e^{i\phi_{kj}}$ and $\alpha_{kj} = \phi_{kj} - \phi_{k+1,j}$, $j = 1, 2$. These equations are now symmetric under the group $\mathbf{Z}_N \times \mathbf{O(2)} \times \mathbf{O(2)}$.

A similar set of equations are obtained for bidirectional coupling but they have been omitted for brevity. Their symmetries are described by the group $\mathbf{D}_N \times \mathbf{O}(2) \times \mathbf{O}(2)$. We discussed above the fact that the averaging process rendered the amplitude-phase equations of the precision-timing network with $\Gamma \subset \mathbf{S}_N$ symmetry into a new set of equations with $\Gamma \times \mathbf{O}(2)$ symmetry. In the case of crystal oscillators, there are two modes of oscillations, which adds an additional $\mathbf{O}(2)$ symmetry, i.e., $\Gamma \times \mathbf{O}(2) \times \mathbf{O}(2)$. We considered two types of coupling, unidirectional, in which case $\Gamma = \mathbf{Z}_N$, and bidirectional, where $\Gamma = \mathbf{D}_N$. In both cases, the averaged system decouples along fixed point subspaces of the two $\mathbf{SO}(2) \subset \mathbf{O}(2)$ actions. The decoupling guarantees that we can work with a simpler system corresponding to a unique mode in $\mathrm{Fix}(\mathbf{SO}(2))$.

Another way to observe the decoupling is by setting z_{k2} to zero as an initial condition, and then notice that the right hand side of the second equation in (9.47) vanishes completely. Thus, z_{k2} must be constant, remaining at the initial condition, i.e., zero at all times. It follows that $z_{k2} = 0$ is an invariant subspace of the network dynamics. Similarly, $z_{k1} = 0$ is also an invariant subspace, since the right hand side of the first equation in (9.47) vanishes when z_{k1} is set to zero (as an initial condition). Consequently, Eq. (9.47) decouples over two invariant subspaces, $z_{k2} = 0$ or $z_{k1} = 0$. This means that we can focus the analysis of the network dynamics on one mode of oscillation at a time. This is also true for the case of an array of crystals coupled bidirectionally. In both cases we choose the invariant subspace $z_{k2} = 0$ to study phase drift under the influence of noise because the other invariant subspace, $z_{k1} = 0$, corresponds to the spurious mode of oscillation.

Although one of the principal justifications for using the averaged system is to work with equations that are more manageable, i.e., in terms of modes being decoupled and having computationally manageable eigenvalues, one must be a little careful about relating the phase dynamics of the averaged system to that of the original system. We argue next that the phases of equilibrium points of the averaged system correspond, indeed, to the phases of the periodic solutions of the original system. In fact, we have shown in [22], first, that steady-state branches of the averaged system lead to periodic solutions of the original system. More importantly, we showed, secondly, that the symmetry group $\Sigma \subset \Gamma \times \mathbf{SO}(2)$ of steady-state solutions of the averaged system transfers to spatio-temporal symmetry groups of the corresponding solutions. Furthermore, the projection of the isotropy subgroup Σ into the $\mathbf{SO}(2)$ symmetry becomes the desired temporal shift of the phase dynamics of the original system. This provides the necessary justification for using the phase dynamics of the averaged system in the analysis of phase drift, which explain the computational results shown in Fig. 9.11.

9.7.3 Analysis of Uncoupled Network

In the uncoupled case, $\alpha_{kj} = 0$. Using the explicit expressions for the averaged system (9.47), we obtain the linearization at the trivial equilibrium $z_k = 0$,

$$\dot{z}_k = (a - R_1)z_k + (\eta_k^A(t) + \eta_k^p(t)i)z_k. \tag{9.48}$$

Rewriting Eq. (9.48) in amplitude-phase form we get

$$\begin{aligned}
\dot{x}_k &= (a - R_1)x_k + x_k\eta_k^A(t) \\
\dot{\phi}_k &= \eta_k^p(t) \\
\dot{\eta}_k^A &= -\frac{\eta_k^A}{\tau_c} + \frac{\sqrt{2D}}{\tau_c}\xi_k^A \\
\dot{\eta}_k^p &= -\frac{\eta_k^p}{\tau_c} + \frac{\sqrt{2D}}{\tau_c}\xi_k^p.
\end{aligned} \tag{9.49}$$

Observe that in this special case the phase dynamics decouples from the amplitude dynamics, as it is driven purely by noise. We introduce phase drift $\delta\phi_k$ as a perturbation of the trivial equilibrium $\phi_k^e = 0$, i.e., set $\phi_k = \phi_k^e + \delta\phi_k$, and get

$$\delta\dot{\phi}_k = \eta_k^p(t). \tag{9.50}$$

The Fourier transform of the phase component in Eq. (9.50) yields

$$(iw)\widehat{\delta\phi_k}(w) = \widehat{\eta}_k(w).$$

To compute the power spectral of the oscillators, we define first an inner product operator over the complex vector space $\mathbf{C}^N$ through

$$\langle \vec{v}, \vec{w}^* \rangle = \frac{1}{N}[v_1 w_1^* + \cdots + v_N w_N^*],$$

where * denotes complex conjugation. We assume all noise functions η_k to have an identical power spectral, $|\eta_k|^2 = |\eta|^2$, for $k = 1, \ldots, N$. Then the power spectral per oscillator k is given by

$$\left|\widehat{\delta\phi_k}\right|^2 = \langle \widehat{\delta\phi_k}, \widehat{\delta\phi_k^*} \rangle = \langle \frac{i}{w}\widehat{\eta}_k, -\frac{i}{w}\widehat{\eta}_k^* \rangle = \frac{1}{N}\frac{1}{w^2}|\widehat{\eta}|^2,$$

which leads to

$$\left|\widehat{\delta\phi_k}\right| = \frac{1}{\sqrt{N}}\frac{1}{w}|\widehat{\eta}|. \tag{9.51}$$

Consequently, the phase drift for the uncoupled network scales as $1/\sqrt{N}$.

9.7.4 Analysis of Unidirectional Coupling

Using the explicit expressions for the averaged system (9.47), we obtain the linearization at the trivial equilibrium $z_k = 0$, given by

$$\dot{z}_k = (a - R_1)z_k - a\lambda z_{k+1} + z_k(\eta_k^A(t) + \eta_k^P(t)i). \tag{9.52}$$

Rewriting Eq. (9.52) in amplitude-phase form we get

$$\begin{aligned}
\dot{x}_k &= (a - R_1)x_k - a\lambda x_{k+1}\cos\alpha_k + x_k\eta_k^A(t) \\
\dot{\phi}_k &= a\lambda\frac{x_{k+1}}{x_k}\sin\alpha_k + \eta_k^P(t) \\
\dot{\eta}_k^A &= -\frac{\eta_k^A}{\tau_c} + \frac{\sqrt{2D}}{\tau_c}\xi_k^A \\
\dot{\eta}_k^P &= -\frac{\eta_k^P}{\tau_c} + \frac{\sqrt{2D}}{\tau_c}\xi_k^P,
\end{aligned} \tag{9.53}$$

where $\alpha_k = \phi_k - \phi_{k+1}$. The analysis in [22] shows that all symmetry-breaking patterns of collective behavior that are generated by unidirectional coupling, i.e., with $\mathbf{Z}_N \times \mathbf{O(2)}$ symmetry (there is only one $\mathbf{O(2)}$ symmetry since the two modes decouple) are discrete rotating wave periodic solutions that have the same amplitude, i.e., $x_k = x_{k+1}$. It follows that the phase equation in (9.53) decouples from the amplitude one. The symmetry-preserving case, which leads to complete synchronization, will be discussed later on in Sect. 9.7.5. We introduce again the phase drift $\delta\phi_k$ as a perturbation of the equilibrium ϕ_k^e by substituting $\phi_k = \phi_k^e + \delta\phi_k$ into Eq. (9.53) and writing the right hand side of the phase equation as a Taylor series expansion in ϕ_k and ϕ_{k+1}, we get

$$\delta\dot{\phi}_k = a\lambda(\delta\phi_k - \delta\phi_{k+1})\cos\alpha_k + \eta_k^P(t). \tag{9.54}$$

Once again, the classification of the patterns of oscillations, conducted in [22], shows that the phase difference, $\alpha_k = \phi_k - \phi_{k+1}$, of the discrete rotating waves is constant among consecutive oscillators. Consequently, $\cos\alpha_k$ can be replaced by a constant as well. Thus, let $\mu = \cos\alpha_k$. In general, this constant will be nonzero. However, there are cases where it can actually be zero. For instance, consider a standard traveling wave, in which $\alpha_k = T/N$, where T represents the period of the wave. If $T = 2\pi$ and for the special case $N = 4$ oscillators then $\mu = 0$.

We proceed to solve Eq. (9.54) by applying the Fourier Transform $\mathcal{F}$ to get

$$(iw)\widehat{\delta\phi}_k(w) = a\lambda\mu(\widehat{\delta\phi}_k(w) - \widehat{\delta\phi}_{k+1}(w)) + \widehat{\eta}_k(w),$$

where $\widehat{\delta\phi}_k(w) = \mathcal{F}(\delta\phi_k(t))$ and $\widehat{\eta}_k(w) = \mathcal{F}(\eta_k(t))$. This last equation can be rewritten in matrix form

$$A_{\text{unidir}}\,\vec{\widehat{\delta\phi}} = \vec{\widehat{\eta}}, \tag{9.55}$$

where $\widehat{\vec{\delta\phi}} = [\widehat{\delta\phi}_1, \ldots, \widehat{\delta\phi}_N]^T$, $\widehat{\vec{\eta}} = [\widehat{\eta}_1, \ldots, \widehat{\eta}_N]^T$, and the matrix A_{unidir} is the circulant matrix

$$A_{\text{unidir}} = \begin{bmatrix} iw - a\lambda\mu & a\lambda\mu & 0 & 0 & \ldots & 0 \\ 0 & iw - a\lambda\mu & a\lambda\mu & 0 & \ldots & 0 \\ \vdots & \vdots & \ddots & \ddots & \ldots & \vdots \\ \vdots & \ldots & & \ddots & \ddots & 0 \\ 0 & \ldots & \ldots & & iw - a\lambda\mu & a\lambda\mu \\ a\lambda\mu & 0 & \ldots & & \ldots & iw - a\lambda\mu \end{bmatrix}.$$

Matrix A_{unidir} is invertible [23], so we can solve for the phase drift (in the frequency domain) to get

$$\widehat{\vec{\delta\phi}} = A_{\text{unidir}}^{-1} \widehat{\vec{\eta}}. \tag{9.56}$$

To compute the power spectral of phase drift of each individual oscillator k, we write

$$\widehat{\delta\phi}_k = \sum_{j=1}^{N} a_{kj}^{-1} \widehat{\eta}_j,$$

where a_{kj}^{-1} are the entries of the inverse matrix A_{unidir}^{-1}, which depend on the parameters a, λ, μ and on the frequency variable w.

The power spectral per oscillator k is then given by

$$\left|\widehat{\delta\phi}_k\right|^2 = \langle \widehat{\delta\phi}_k, \widehat{\delta\phi}_k^* \rangle = \frac{1}{N}[a_{k1}^{-1}\widehat{\eta}_1(a_{k1}^{-1}\widehat{\eta}_1)^* + \cdots + a_{kN}^{-1}\widehat{\eta}_N(a_{kN}^{-1}\widehat{\eta}_N)^*]. \tag{9.57}$$

Assuming again all noise functions, η_k, to have an identical power spectral, Eq. (9.57) reduces to

$$\left|\widehat{\delta\phi}_k\right|^2 = \frac{1}{N}|\widehat{\eta}|^2 \sum_{j=1}^{N} |a_{kj}^{-1}|^2. \tag{9.58}$$

The power spectral of phase drift in the network is computed by averaging the power spectral of phase drift of all individual oscillators. That is,

$$\left|\widehat{\delta\phi}_{\text{network}}\right|^2 = \frac{1}{N} \sum_{k=1}^{N} \left|\widehat{\delta\phi}_k\right|^2. \tag{9.59}$$

Substituting Eq. (9.58) into Eq. (9.59) yields

$$|\widehat{\delta\phi}_{\text{network}}| = \frac{1}{N}\sqrt{\sum_{k=1} |\widehat{\eta}|^2 \left(\sum_{j=1}^{N} |a_{kj}^{-1}|^2\right)}. \tag{9.60}$$

The square root term in (9.60) is a scalar-valued term. Consequently, phase drift in the network scales down as $1/N$.

Observe that if the coupling strength is set to $\lambda = 0$ then the matrix A_{unidir} becomes purely diagonal, i.e., $b_{kj} = 0$ if $k \neq j$, and $b_{kk} = iw$. Substitution into Eq. (9.58) leads to the $1/\sqrt{N}$ scaling law that was previously obtained for the uncoupled network. Also, the case where $\mu = 0$ will reduce the matrix A_{unidir} to that of the uncoupled network and will yield, again, the $1/\sqrt{N}$ scaling power. These cases are, however, the exception to the rule, so the $1/N$ scale holds, in general, with unidirectional coupling.

In the bidirectional coupling case we get a similar version of Eq. (9.55) for phase drift:

$$A_{\text{bidir}}\,\widehat{\vec{\delta\phi}} = \widehat{\vec{\eta}}, \tag{9.61}$$

where the matrix A_{bidir} is the circulant matrix

$$A_{\text{bidir}} = \begin{bmatrix} iw - a\lambda\mu & a\lambda\mu & 0 & 0 & \dots & 0 \\ a\lambda\mu & iw - a\lambda\mu & a\lambda\mu & 0 & \dots & 0 \\ \vdots & \vdots & \ddots & \ddots & \dots & \vdots \\ \vdots & \dots & & \ddots & \ddots & 0 \\ 0 & \dots & a\lambda\mu & iw - a\lambda\mu & & a\lambda\mu \\ a\lambda\mu & 0 & \dots & a\lambda\mu & & iw - a\lambda\mu \end{bmatrix}.$$

This matrix is also invertible [23] and the computation (not shown for brevity) of the power spectral of phase drift per oscillator takes exactly the same form as in Eq. (9.58), except that the terms a_{kj}^{-1} correspond to the entries of the new inverse matrix A_{bidir}^{-1}. Thus, the phase drift of the network is also given by Eq. (9.60). It follows that phase drift in the bidirectional case also scales as $1/N$.

9.7.5 Fundamental Limit

In this section we address the more transcendental question of why the standard traveling wave pattern, in which adjacent cells oscillate out-of-phase by a constant amount, mainly T/N, yield the best scaling in terms of phase drift. We also address the issue of whether $1/N$ is the fundamental limit of phase drift reduction that can be achieved by a network configuration.

To get insight into the answers to these questions, we first explore other coupling configurations. For instant, consider all-to-all coupling, which leads to a network with $\mathbf{S}_N$ symmetry, where $\mathbf{S}_N$ is the group of permutations of N objects. In this case, there are also standing waves and discrete rotating waves that emerge via bifurcations that break the $\mathbf{S}_N$ symmetry. As before, in the case of discrete rotating waves (same wave forms and constant phase differences) one can show that the matrix associated with the solution of the phase drift takes the form

$$A_{\text{all-to-all}} = \begin{bmatrix} iw - a\lambda\mu & a\lambda\mu & a\lambda\mu \; a\lambda\mu & \dots & a\lambda\mu \\ a\lambda\mu & iw - a\lambda\mu & a\lambda\mu \; a\lambda\mu & \dots & a\lambda\mu \\ \vdots & \vdots & \ddots \;\; \ddots & \dots & \vdots \\ \vdots & \dots & \ddots & \ddots & a\lambda\mu \\ a\lambda\mu & \dots & a\lambda\mu & iw - a\lambda\mu & a\lambda\mu \\ a\lambda\mu & a\lambda\mu & \dots & a\lambda\mu & iw - a\lambda\mu \end{bmatrix}.$$

This matrix is invertible [23] and the computation of the power spectral of phase drift per oscillator leads, once again, to a solution that takes the same form as in Eq. (9.58), except that the terms a_{kj}^{-1} correspond to the entries of the new inverse matrix $A_{\text{all-to-all}}^{-1}$. Consequently, the phase drift of the network follows a $1/N$ scaling given by Eq. (9.60).

Other linear coupling configurations will produce variations of the A matrix. Those variations will appear, however, only in the scalar terms a_{kj}^{-1} of the power spectral of phase drift per oscillator given by Eq. (9.58). For instance, a coupling configuration with nearest neighbors and second-nearest neighbor coupling will lead to a banded matrix that will only change a few of the scalar values in Eq. (9.58). For these reasons, the phase drift of the network will continue to follow a $1/N$ scaling law.

Synchronization State. Two of the most common mechanisms that can lead to oscillatory behavior in a network system include either symmetry-breaking Hopf bifurcations, which can be associated with collective patterns of oscillations, or symmetry-preserving Hopf bifurcations, which lead to complete synchronization states. So far we have considered the former case in all calculations of phase drifts. We now consider the latter case. Our aim is to explain why the synchronization state of an ensemble of N oscillators does not perform any better, in terms of phase drift reduction, than a single one. At first glance one may think the contrary. That is, that the synchronization state could produce a better phase drift reduction because the coherence of the collective pattern could, in principle, attenuate more the negative effects of noise. However, the simulations in [21] show us otherwise. If we examine the phase Eq. (9.53) for unidirectional coupling (or for bidirectional coupling), we can see that when the oscillators are synchronized, i.e., $\alpha_k = 0$ and $\beta_k = 0$, then the corresponding phase equations reduce to that of a single uncoupled oscillator Eq. (9.49), which we already showed earlier to yield a $1/\sqrt{N}$ scaling in phase drift reduction. For this reason alone the synchronization state cannot improve performance.

Finally, we have now arrived at the most critical question of whether $1/N$ is the fundamental limit of phase-drift reduction or whether there is a network configuration that can produce better results. To address this question, we observe first that the previous computations of phase drift show the parameter μ entering as a multiplicative factor of the coupling strength λ in the related matrices A. This happens for all different linear-coupling topologies. Furthermore, as the size N of the network tends to infinity, μ tends to unity. It follows that the coupling strength reaches a maximum as N approaches infinity. The actual value of the term $a\lambda\mu$ in the matrix A only affects the scalar component of the power spectral per oscillator, as can be seen in Eq. (9.58). In other words, the structure of the matrix A may change for different linear-coupling topologies but the changes do not affect the scaling law. Hence, $1/N$ is indeed the fundamental limit that can be achieved by a network approach with linear coupling.

9.8 Stochastic Model of Flame Instability

In this section we investigate the effects of noise on a model of cellular pattern formation. As an example, we consider a stochastic (Langevin) version of a generic example of a cellular-pattern forming dynamical system, known as the Kuramoto-Sivashinsky (KS) equation

$$\frac{\partial u}{\partial t} = \eta_1 u - (1 + \nabla^2)^2 u - \eta_2 (\nabla u)^2 - \eta_3 u^3 + \xi(\vec{x}, t), \tag{9.62}$$

which is the same as Eq. (8.33), except for the stochastic term $\xi(\vec{x}, t)$ that is now included. Recall from Chap. 8 that $u = u(\vec{x}, t)$ represents the perturbation of a planar front (which is normally assumed to be a flame front) in the direction of propagation, η_1 measures the strength of the perturbation force, η_2 is a parameter associated with growth in the direction normal to the domain (burner) of the front, $\eta_3 u^3$ is a term that is added to help stabilize the numerical integration. This time, the term $\xi(\vec{x}, t)$ represents Gaussian white noise, which models thermal fluctuations, dimensionless in space and time. We assume $\xi(\vec{x}, t)$ to be distributed with zero mean $\langle \xi(\vec{x}, t) \rangle = 0$, and to be uncorrelated over space and time, i.e., $\langle \xi(\vec{x}, t) \xi(\vec{x}', t') \rangle = 2D\delta(\vec{x} - \vec{x}')\delta(t - t')$, where D is a measure of the intensity of the noise, $\langle \cdot \rangle$ represents the time-average over a range of observations.

9.8.1 *Computer Simulations*

Numerical integration of Eq. (9.62) shows that the typical ordered state that appears changes (as the radius of the circular domain increases) from a single ring of cells to concentric rings of cells, see Fig. 9.13.

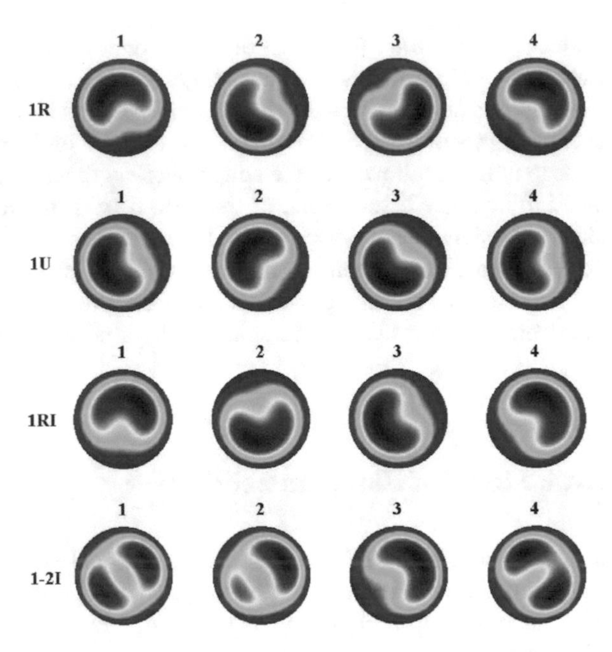

Fig. 9.13 Generic behavior of the KS model for various parameter values of radius and noise Intensity. Notation: S = Stationary, U = Unsteady, I = Intermittent State, R = Rotation. Noise intensity, $D = \sigma^2/2$, is in the range [0.00, 2.5E-04]. This range represents low noise levels, relative to the dynamic range of u, which in the Kuramoto-Sivashinsky model is of order 10. As the noise intensity increases, the radius-parameter range of complex dynamic patterns is extended; when the intensity reaches $D = 1.25 \times 10^{-3}$ no static patterns are observed. For each of these simulations $\eta_1 = 0.32$, $\eta_2 = 1.0$, $\eta_2 = 0.017$, $4.1 \leq radius \leq 4.35$, and $D \leq 0.5$

Occasionally, dynamic states are also observed in the transition from from one stationary pattern to another. Related studies [24] have shown that uniformly rotating and modulated rotating single-ring states with k cells are typically generated by the interaction of two steady-state modes with Fourier wave numbers in a $k : 2k$ ratio. For simplicity, let's focus our attention around a $1 : 2$ mode interaction, though the analysis we are about to conduct still captures many essential features of the effects of noise on larger patterns. Figure 9.13 illustrates the patterns, and transitions between them, which occur near a radius $R = 4.35$.

Without noise, i.e., noise amplitude $D = 0$, a one-cell rotating state (1R) appears in the transition from a one-cell stationary state to a two-cell stationary state, just as predicted by the corresponding $1 : 2$ mode interaction. As the noise intensity increases, the domain of existence of the 1R-state increases and additional patterns emerge. For very weak noise, an unsteady dynamic pattern (1U) appears between the 1S and 1R states. The 1U pattern does not sustain rotations; instead, the pattern rocks back and forth. With increased noise intensity, a one-cell rotating pattern (1RI), which intermittently changes its direction of rotation, is observed between the 1U and 1R patterns. Near the bifurcation point, where the 1RI state forms, there are two bistable branches of rotating states created by symmetry, one branch for each direction of rotation. Which branch is observed depends mainly on initial conditions. Noise appears to act as a switch, inducing recurrent transitions between these two branches. Between the 1R and 2S (or, for higher noise intensities, 2U) patterns, an intermittent 1–2 cell pattern forms. This dynamic pattern is very peculiar: one of the two cells in the 2S state is extinguished; the remaining one-cell state is short-lived, the pattern immediately splits into a new 2S-state, the orientation of which is roughly a quarter-rotation of the previous 2S-state. Each appearance of the 2S state lasts an irregular amount of time, ranging from a few to several hundreds of frames. This is qualitative evidence of a heteroclinic connection where the stable (unstable) manifold of a two-cell equilibrium is also the unstable (stable) manifold of another two-cell equilibrium.

9.8.2 *Mode Decomposition*

In order to explain, quantitatively, the origin and formation mechanisms of the noise-induced intermittent pattern shown in Fig. 9.13, a Proper Orthogonal Decomposition analysis of an ensemble, made up of about 4000 computer-simulated spatio-temporal data points (frames), is performed for each individual case. The method of snapshots (see Chap. 8) is employed to extract time-independent orthonormal basis functions, $\Phi_k(\mathbf{x})$, and time-dependent orthonormal amplitude coefficients, $a_k(t_i)$, such that the reconstruction

$$u(\mathbf{x}, t_i) = \sum_{k=1}^{\infty} a_k(t_i)\, \Phi_k(\mathbf{x})\,, \quad i = 1, \ldots, M$$

is optimal in the sense that the average least square truncation error. To ensure that the POD steady-state modes contain the correct symmetry properties, we have taken special care of including the average over the symmetry group, $\mathbf{O(2)}$, of the numerical simulations, in the ensemble average. In all four cases, shown in Fig. 9.13, the POD analysis reveals that two pairs of modes with wave numbers in a 1:2 ratio capture most of the dynamics, see Fig. 9.14.

The time-average (considered mode Φ_0) is shown first followed by four POD modes, Φ_1–Φ_4, with the highest percentage of energy (see Appendix for an exact definition). The actual amount of energy in each mode is indicated below each mode. Each mode shows some amount of symmetry. The symmetry of the time-average, in particular, reflects the $\mathbf{O(2)}$-symmetry of the burner, even though none of the instantaneous snapshots has this symmetry. This feature is studied in more detail in. Φ_1 and Φ_2 show D_1-symmetry, meaning that one complete revolution leaves them unchanged, while Φ_3 and Φ_4 show D_2-symmetry, i.e., the patterns are restored after half a revolution. Observe also that the energy is equally distributed among these two pairs of modes, which together capture almost 90% of the original behavior. It follows that intermittent behavior in all three cases is created from the mutual interaction of two invariant eigenspaces,

$$V_1 = \text{span}\{\Phi_1, \Phi_2\}, \qquad V_2 = \text{span}\{\Phi_3, \Phi_4\},$$

whose dihedral symmetries are in a $1 : 2$ ratio, just as expected from direct inspection of the transition diagram of Fig. 9.13.

Next we examine results of the POD decomposition of the 1-2I intermittent state. Figure 9.15 shows the time-dependent coefficients associated with each individual POD mode. To help visualize the actual transitions, we have added two markers, a green circle and a red circle. The time between the green (red) circle and the red (green) define the beginning and end of a 2-cell (1-cell) pattern, respectively. Observe that when the oscillations in $a_1(t)$ and $a_2(t)$ have large amplitudes relative to those of a_3 and a_4, the 1-cell pattern shows up. The opposite relation, small amplitude in a_1, a_2 and large amplitude in a_3, a_4, leads to the appearance of the 2-cell pattern.

The heteroclinic saddle-node connections that underlie the transitions between the 1-cell pattern and the 2-cell state, can be observed better in the phase-space portrait of Fig. 9.16. Black arrows indicate the approximate direction of the flow around the two saddle-nodes that are associated with a 2-cell state, while there are four saddle-nodes that correspond to the 1-cell state. This difference deserves an explanation. Once a 2-cell state appears in the simulations, there is only one distinct orthogonal position in which the same pattern can reappear. On the other hand, a 1-cell state has four orthogonally distinct positions where it can reappear. These geometric facts determine the structure of the phase portrait of Fig. 9.16.

9.8.3 Amplitude Equations

As it was mentioned before, all three intermittent patterns, 1RI, 1U, and 1-2I, emerge from the mutual interaction of two pairs of spatial modes, $\{\Phi_1, \Phi_2\} - \{\Phi_3, \Phi_4\}$, with wave numbers in a 1:2 ratio, while the time evolution of each individual pattern is determined by the amplitude coefficients $a_1(t)$-$a_4(t)$ that are associated with the

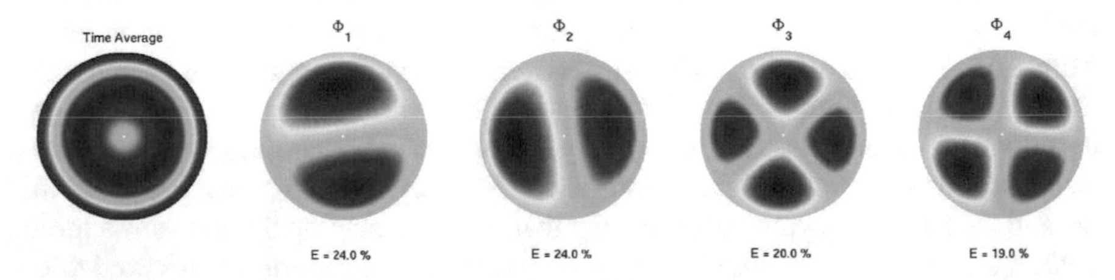

Fig. 9.14 A proper orthogonal decomposition analysis reveals that all four patterns of Fig. 9.13 are created from the mutual interaction of two pairs of spatial modes whose wave numbers are in a 1:2 ratio. These modes were obtained using computer-simulated ensembles of 4000 data set points of each individual pattern

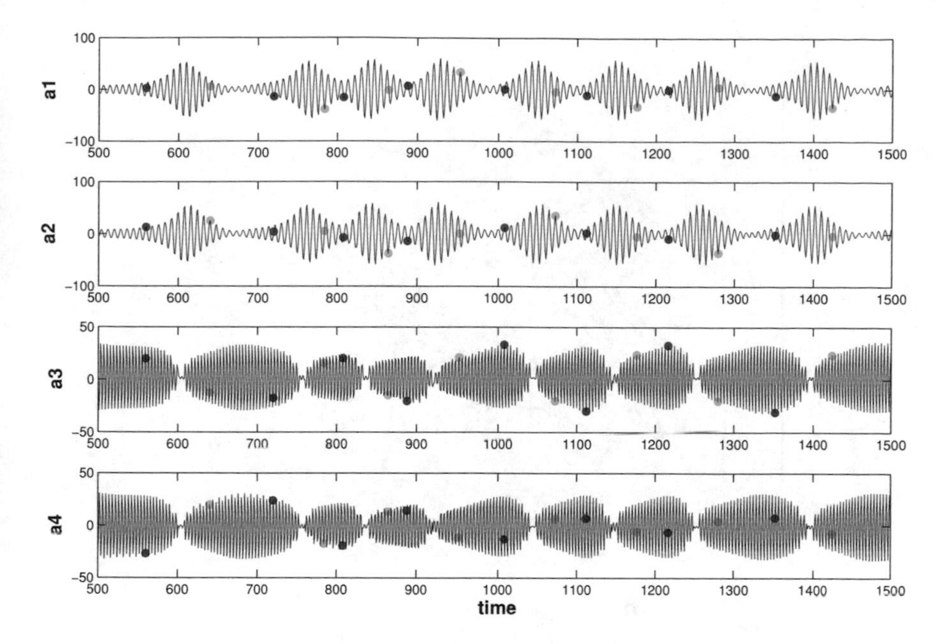

Fig. 9.15 Amplitude coefficients associated with the POD modes of the intermittent state 1-2I of Fig. 9.13. Horizontal axis denotes time. Markers indicate beginning (green) and end (red) of a 2-cell pattern

spatial modes $\Phi_1 - \Phi_4$, respectively. The amplitude equations that govern the evolution of the time-dependent coefficients are derived from the 1-to-2 Fourier-mode interaction in a system with **O(2)**–symmetry, i.e., the symmetry group of rotations and reflections of the circular domain. The deterministic version of these amplitude equations in Birkhoff Normal Form has been thoroughly studied by Armbruster et al. [25–27]. The Langevin version below

$$\begin{aligned} \dot{z}_1 &= \bar{z}_1 z_2 + z_1(\mu_1 + e_{11}|z_1|^2 + e_{12}|z_2|^2) + \varepsilon\eta_1(t) \\ \dot{z}_2 &= \pm z_1^2 + z_2(\mu_2 + e_{21}|z_1|^2 + e_{22}|z_2|^2) + \varepsilon\eta_2(t), \end{aligned} \tag{9.63}$$

where $\eta_1(t)$ and $\eta_2(t)$ are Gaussian white noise functions, uncorrelated with zero mean and with amplitude ε. This model has also been considered by Stone and Holmes [28] in a study of the effects of noise on heteroclinic cycles. Let us start with the 1R pattern. According to the transitions seen in Fig. 9.13, it is reasonable to associate the temporal evolution of the 1R pattern with that of a traveling wave of the deterministic normal forms, i.e., $\eta_1 = 0$ and $\eta_2 = 0$ in (9.63). For convenience, we let $z_j = re^{\theta_j i}$ and $\phi = 2\theta_1 - \theta_2$, so that we can rewrite (9.63) in polar coordinates

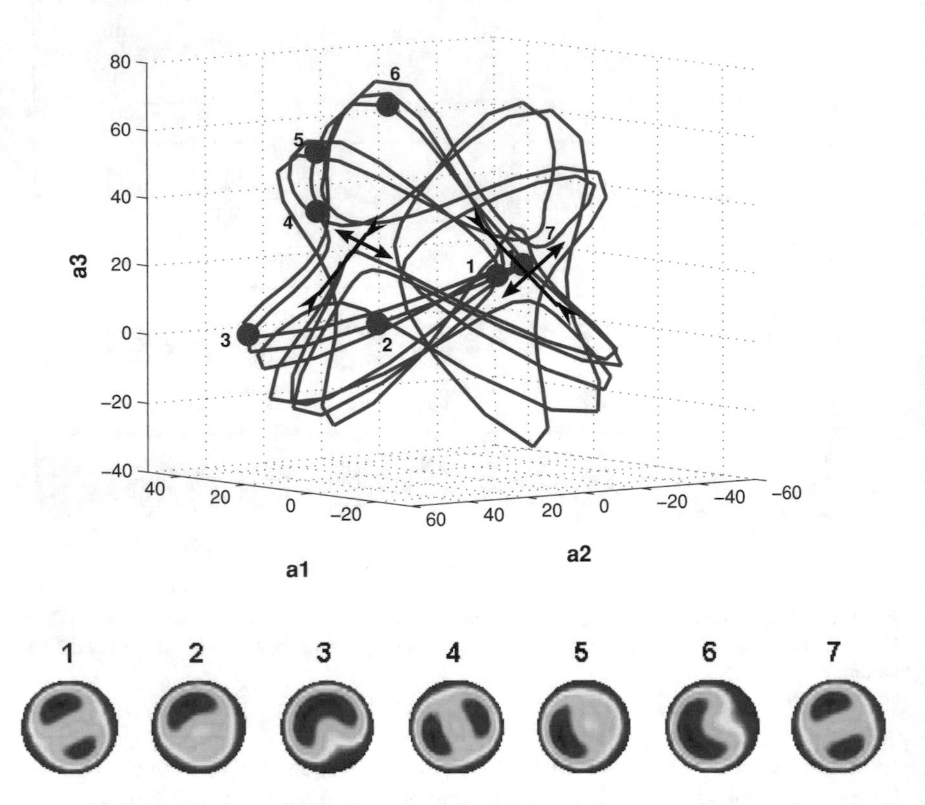

Fig. 9.16 Phase-space portrait from time-dependent POD coefficients for an intermittent state 1-2I capture saddle-node connections between the stable and unstable manifolds associated with each individual ordered pattern, one with one cell and one with two cells

$$\begin{aligned} \dot{r}_1 &= r_1 r_2 \cos\phi + r_1(\mu_2 + e_{11} r_1^2 + e_{12} r_2^2) + \varepsilon\eta_1(t) \\ \dot{r}_2 &= \pm r_1^2 \cos\phi + r_2(\mu_4 + e_{21} r_1^2 + e_{22} r_2^2) + \varepsilon\eta_2(t) \\ \dot{\phi} &= -\left(2r_2 \pm \frac{r_1^2}{r_2}\right)\sin\phi. \end{aligned} \tag{9.64}$$

Observe that the noise functions η_1 and η_2 do not appear, explicitly, in the last equation in (9.64), which governs the evolution of the phase-difference variable. We will show that noise can, however, change the evolution of the phase difference through the radial components r_1 and r_2. Consider the noise-free system: $\eta_1 = 0$ and $\eta_2 = 0$. Traveling Waves (TW) are equilibria of (9.63) in which the phase difference remains constant, though $\phi_2 \neq 0, \pi$. In physical space, TWs correspond to uniformly rotating patterns produced by evolution equations; e.g., the 1R pattern that appears in simulations of the KS model (9.62). Following Armbruster et al. [25], traveling waves (of the deterministic system) are created via a pitchfork bifurcation from the

π-mixed mode solution ($r_1 \neq 0, r_2 \neq 0, \phi = \pi$) when $2r_4 \pm r_2^2/r_4 = 0$, and $\phi_2 = \pi$, so that they only exist in the "$-$" case or when $r_1^2 = 2r_2^2$. Letting $e = 4e_{11} + 2e_{12} + 2e_{21} + e_{22}$, it can be shown that TW solutions of (9.63), without noise, exist and are stable for

$$-2\mu_1 - e\mu_1^2 + O(\mu_1^3) < \mu_2 < \mu_1 \left(1 + \frac{9(e_{22} - e_{12})}{e - 3(e_{22} - e_{12})}\right) + O(\mu_1^2). \quad (9.65)$$

Consider now the noisy system. Direct calculations of the equilibria of (9.64) lead to

$$\lambda r_2 + e\, r_2^3 + \sigma\eta_3(t) = 0, \quad (9.66)$$

where $\lambda = 2\mu_1 + \mu_2$ and η_3 is also a Gaussian white noise function, uncorrelated with zero mean, but with amplitude $\sigma = \sqrt{3}\varepsilon$. When $\sigma = 0$, Eq. (9.66) reduces to the normal form equation for the pitchfork bifurcation that underlie the birth of the TWs of the deterministic system. A more critical observation is the fact that *additive white noise does not modify qualitatively the solution set of a codimension-one, perfect, pitchfork bifurcation* [29]. It follows that TW solutions, and their stability properties, of the noisy system (9.63) necessarily coincide with those of the deterministic, $\sigma = 0$, system; and Eq. (9.65) is still valid for the noisy system. But if the 1RI pattern is indeed a noise-perturbed TW, then we seem to have an apparent contradiction: how can noise change the direction of rotation of the 1RI state if noise cannot modify the qualitative properties of the pitchfork bifurcations that lead to traveling waves? To clarify this subtle issue, we need to take into account that equilibria of (9.64) are now described by a probability density function $p(r_i, t)$. In particular, $p(r_2, t)$ is governed by the following Fokker-Planck equation

$$\frac{\partial}{\partial t} p(r_2, t) = -\frac{\partial}{\partial r}\left[(\lambda r_2 + e r_2^3) p(r_2, t)\right] + \frac{\sigma^2}{2}\frac{\partial^2}{\partial r^2} p(r_2, t). \quad (9.67)$$

Traveling waves solutions are described by stationary solutions of (9.67), i.e., solutions of $\partial_t p(r_2, t) = 0$, which, in turn, yields the stationary probability density function

$$p_s(r_2) = N \exp\left[\left(\frac{2}{\sigma^2}\right)\left(\lambda\frac{r_2^2}{2} + e\frac{r_2^4}{4}\right)\right]. \quad (9.68)$$

Computer simulations, see Fig. 9.17, show that this function changes from single to double peaked as λ increases across zero.

In both cases, $\lambda < 0$ and $\lambda > 0$, the location of the peaks always coincide with the steady states of the deterministic system. As predicted by theory, noise does not modify the qualitative characteristics of the underlying pitchfork bifurcation. However, noise can change the probability distribution around the steady-state $r_2 = 0$. Assuming $\lambda > 0$, we notice that as noise intensity increases from zero, the proportion of time spent by a typical solution of (9.64) around $r_2 = 0$ increases continuously until it reaches a maximum, at which time the phase-difference angle is no longer

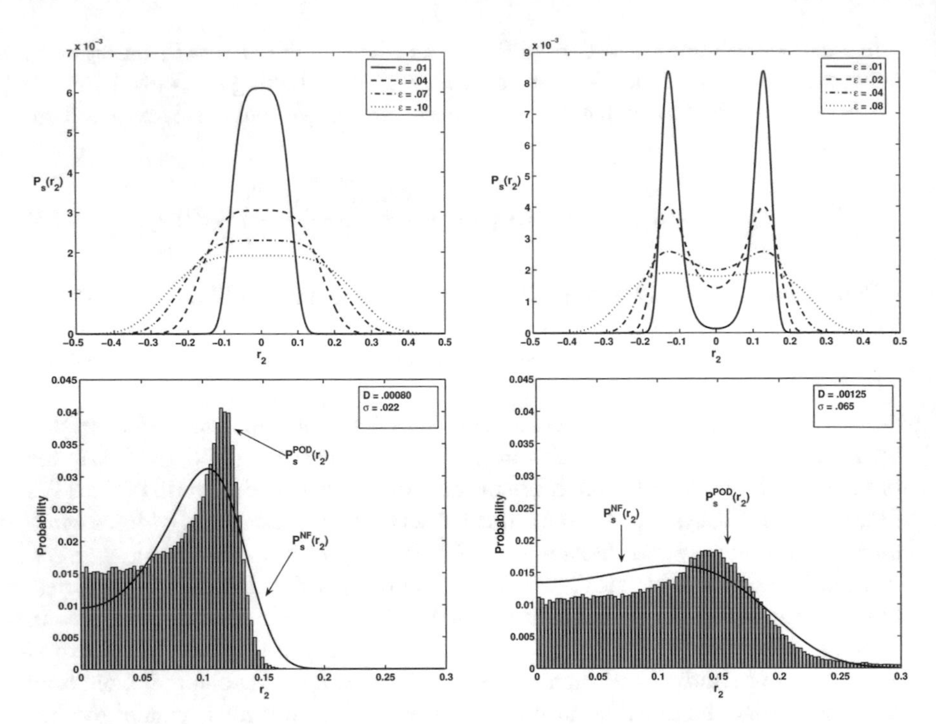

Fig. 9.17 Stationary probability density of the radial component of the pitchfork bifurcation that leads to TW solutions

at an equilibrium, thus triggering a transition that changes the sign of the phase-difference angle, and ultimately, the direction of rotation of the wave. This cycle of events repeats itself at random time-intervals as the system dynamics in r_2 change back and forth between zero and the values of the deterministic system. As for the 1U pattern, since standing waves lie on the invariant subspace $\phi = 0$, or $\phi = \pi$, noise perturbations of the radial components r_1 and r_2 cannot destroy the invariance of the subspaces because they do not enter, explicitly, into the dynamics of the phase angle. Consequently, the only possible effect of noise variations in r_1 and r_2 is to create small oscillations in the phase-angle variable ϕ, thus rocking the wave back and forth.

We now turn our attention to the 1-2I pattern, studied by Stone and Holmes [28]. Among their findings, most relevant to the analysis of the 1RI pattern, is the realization that certain intermittent states can be described as noise-induced "stochastic limit cycles" that are created from the perturbation of heteroclinic orbits connecting saddle-node equilibria of the deterministic ($\varepsilon = 0$) normal forms.

Figure 9.18 depicts the phase-space projection of a typical trajectory of (9.63) onto the first two components of z_1 and the x-component of z_2, which are the analogous of the POD amplitude coefficients a_1, a_2 and a_3, respectively. The reconstructed pattern, calculated through the following equation,

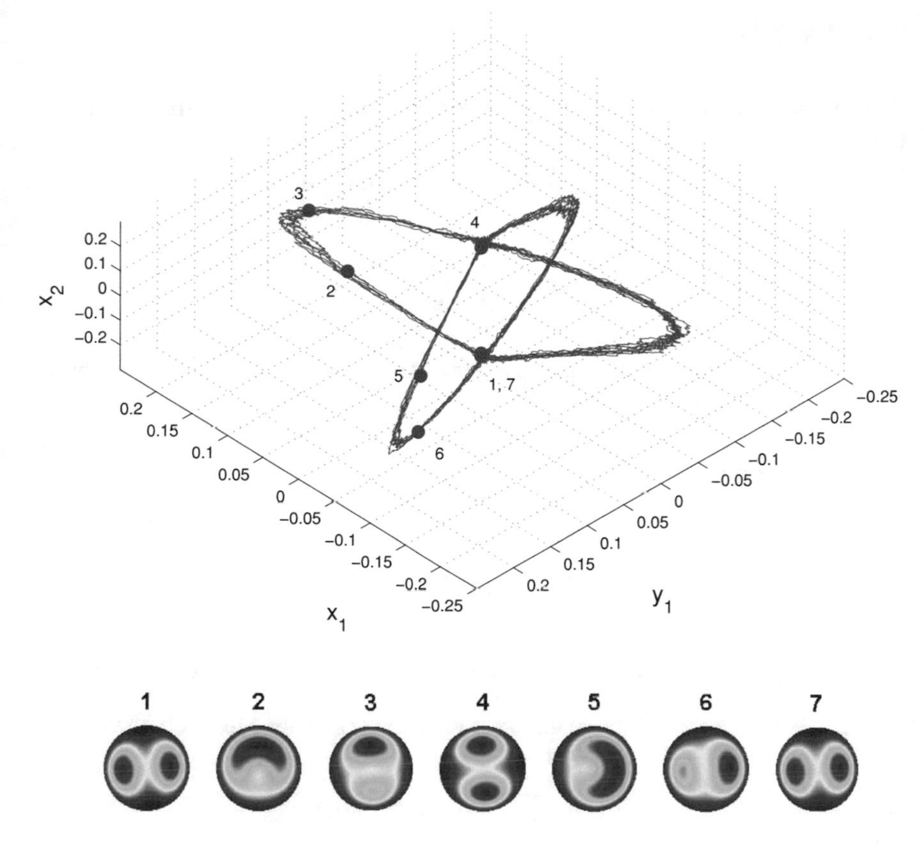

Fig. 9.18 (Top) Phase-space depiction of a heteroclinic cycle found in the Kuramoto-Sivashinsky. (Bottom) Phase-space reconstruction from normal form equations. Parameters are: $R = 4.285$ (radius of domain of integration), $D = 0.0008$ (noise intensity), and $(\eta_1, \eta_2, \eta_3) = (0.32, 1.0, 0.17)$

$$U_{rec}(\mathbf{x}, t_i) = \sum_{k=1}^{4} z_k(t_i)\, \Phi_k(\mathbf{x}), \quad i = 1, \ldots, M,$$

where M is the size of the ensemble, 4000 frames in this case, is also shown immediately below the phase-space projection. The resemblance of the phase space with the POD phase-space projection of Fig. 9.16 is clear. More importantly, the reconstructed intermittent state is qualitatively similar, up to a rotation, to the PDE simulations. The cell rotates uniformly and, intermittently changes direction of rotation. In summary, numerical calculations and the phase-space reconstruction of Fig. 9.18, are strong evidence that the 1-2I intermittent state is indeed a stochastic limit cycle created from the perturbation of a heteroclinic connection. Such connections would be unobservable under noise-free conditions.

9.9 Exercises

Exercise 9.1 A stochastic birth only process could be given by the differential equation

$$\frac{dP_n}{dt} + \lambda N P_n = \lambda(N-1)P_{N-1},$$

with

$$P_N(0) = \begin{cases} 0 & N \neq N_0 \\ 1 & N = N_0 \end{cases},$$

where $P_N(t)$ is the probability that there are exactly N individuals in a population. The solution was given by

$$\begin{aligned} P_{N_0}(t) &= e^{-\lambda N_0 t} \\ P_{N_0+1}(t) &= N_0 e^{-\lambda N_0 t}(1 - e^{-\lambda t}) \\ &\cdots \\ P_{N_0+j}(t) &= \frac{N_0(N_0+1)\cdots(N_0+j-1)}{j!} e^{-\lambda N_0 t}(1 - e^{-\lambda t})^j. \end{aligned}$$

Use mathematical induction to prove that this last formula holds for all j.

Exercise 9.2 The expected population at time t, $E(t)$, is obtained from the formula

$$E(t) = \sum_{j=0}^{\infty} (N_0 + j) P_{N_0+j}(t).$$

a. Can you explain why this is valid using probabilistic ideas?

b. Show that $E(t) = N_0 e^{\lambda t}$.

c. Explain the significance of Part b.

Exercise 9.3 Consider the equation above for $P_{N_0+j}(t)$. Assume that $N_0 = 10{,}000$. Estimate λ, if it is observed that a total of 4500 births occur in 20 days. [Hint: See the previous exercise.]

Exercise 9.4 Since 1973, the British Forestry Commission has surveyed for the presence of the American gray squirrel (*Sciurus carolinensis Gmelin*) and the native red squirrel (*Sciurus vulgaris L.*). From two consecutive years of data for 10 km square regions across Great Britain, data were collected on movement of the two types of squirrels. The transition matrix for red squirrels, gray squirrels, both, or neither in that order was given by

$$T = \begin{pmatrix} 0.8797 & 0.0382 & 0.0527 & 0.0008 \\ 0.0212 & 0.8002 & 0.0041 & 0.0143 \\ 0.0981 & 0.0273 & 0.8802 & 0.0527 \\ 0.0010 & 0.1343 & 0.0630 & 0.9322 \end{pmatrix}.$$

Find the equilibrium distribution of squirrels based on this transition matrix. Does this model suggest that the invasive gray species will significantly displace the native red squirrel over long periods of time?

Exercise 9.5 An enclosed area is divided into four regions with varying habitats. One hundred tagged frogs are released into the first region. Earlier experiments found that on average the movement of frogs each day about the four regions satisfied the transition model given by

$$\begin{pmatrix} f_1(n+1) \\ f_2(n+1) \\ f_3(n+1) \\ f_4(n+1) \end{pmatrix} = \begin{pmatrix} 0.42 & 0.16 & 0.19 & 0.16 \\ 0.07 & 0.38 & 0.24 & 0.13 \\ 0.34 & 0.19 & 0.51 & 0.27 \\ 0.17 & 0.27 & 0.06 & 0.44 \end{pmatrix} \begin{pmatrix} f_1(n) \\ f_2(n) \\ f_3(n) \\ f_4(n) \end{pmatrix}$$

a. Give the expected distribution of the tagged frogs after 1, 2, 5, and 10 days.

b. What is the expected distribution of the frogs after a long period of time? Which of the four regions is the most suitable habitat and which is the least suitable for these frogs?

Exercise 9.6 The following equation serves as a model for the concentration of cGMP, which is a common regulator of ion channel conductance, glycogenolysis, and cellular apoptosis. It serves to relax smooth muscle tissues.

$$\frac{d[\text{cGMP}]}{dt} = -k_{\text{PDE}}[\text{cGMP}] + \gamma, \tag{9.69}$$

where k_{PDE} represents the maximum cGMP hydrolysis rate, in which PDE stands for "phosphodiesterases", which hydrilyze the cyclic nucleotide into GMP, and γ is a constant.

(a) Let

$$k_{\text{PDE}} = \langle k_{\text{PDE}} \rangle + \xi k_{\text{PDE}}(t),$$

where $\xi k_{\text{PDE}}(t)$ represents a small colored noise perturbation off of the mean value $\langle k_{\text{PDE}} \rangle$, with autocorrelation function given by

$$\langle \xi k_{\text{PDE}}(t)\ \xi k_{\text{PDE}}(s) \rangle = A e^{(-k|t-s|)}.$$

Show that the model Eq. (9.69) can be rewritten in Langevin form

$$\frac{dx}{dt} = \alpha x + \gamma + \mu\, x\, \eta(t), \tag{9.70}$$

where α and μ are constant parameters, and

$$\langle \eta(t)\, \eta(s) \rangle = A e^{(-|t-s|/\tau_c)}.$$

(b) Rewrite Eq. (9.70) as

$$\frac{dx}{dt} = \alpha x + \gamma + \mu\, x\, \eta(t),$$
$$\frac{d\eta}{dt} = -\frac{\eta(t)}{\tau_c} + \frac{\sqrt{2D}}{\tau_c}\xi(t),$$

where $\xi(t)$ is Gaussian white noise function of zero mean. Then apply the Euler-Maruyama algorithm to solve Eq. (9.70). Plot the results.

Exercise 9.7 Let $C(t)$ denote the concentration of a drug present at time t in a pharmokinetic, one-compartment, system. A stochastic first-order linear equation describing the loss of the substance from the system is

$$\frac{dC}{dt} = -\rho(t)\, C(t),$$

where $\eta(t)$ is the transfer rate subject to stochastic perturbations.

(a) Split the transfer rate, $\rho(t)$ into deterministic, k, and stochastic components, $\eta(t)$, by letting $\rho(t) = k + \gamma\eta(t)$, where $\eta(t)$ is assumed to be a Gaussian white noise function scaled by the constant parameter γ. Write the resulting equation in Langevin form

$$\frac{dC(t)}{dt} = -kC(t) + \gamma C(t)\, \eta(t).$$

Let $k = 4$ and $\gamma = 0$. Assume the initial condition $C(0) = 1$ and find an analytical solution. Hint: Apply Itô's formula.

(b) Let $\gamma = 2$ and solve this time the stochastic differential equation. Plot and compare the results of part (a) and (b).

Exercise 9.8 A more realistic version of the pharmokinetic, one-compartment, model introduced above, is when the deterministic component, k is randomly perturbed by a Gaussian white noise function $\xi(t)$. This situation leads to a model of the form

$$\frac{dC}{dt} = -k\, C(t),$$
$$\frac{dk}{dt} = k_e - k(t) + \gamma\sqrt{k(t)}\, \xi(t),$$

where k_e is a mean value for the constant k. Let $k_e = 4$ and $\gamma = 2$. Assume initial conditions $C(0) = 1$ and $k(0) = 3.5$. Find a numerical solution and compare the results with those of the previous model.

Exercise 9.9 The Vasicek interest rate model [30, 31] can be expressed in Langevin form

$$\frac{dr(t)}{dt} = a(b - r(t)) + \sigma\,\xi(t)$$

where a represents speed of the reversion towards a mean, b is the long-term level of the mean, and σ describes volatility. Assume $\xi(t)$ is Gaussian white noise of zero mean.

(a) Find an analytical solution as follows.
Step 1: Apply the change of variables, $x(t) = r(t) - b$, noting that $dx = dr$, and rewrite a new stochastic equation in x.
Step 2: Let $y = e^{at}x$ and rewrite the SDE in x in terms of y.
Step 3: Solve the resulting equation for y, then solve for x by reversing the change of variables, i.e., let $r(t) = x(t) + b$.
(b) Solve numerically for $r(t)$.
(c) Compare analytical and numerical solutions.

Exercise 9.10 The Cox Ingersoll-Ross [30, 31] interest rate model can also be expressed in Langevin form

$$\frac{dr(t)}{dt} = a(b - r(t)) + \sigma\sqrt{r(t)}\,\xi(t)$$

where a represents speed of adjustment towards a mean, b is the long-term level of the mean, and σ describes volatility. Assume $\xi(t)$ is Gaussian white noise of zero mean.

(a) Find an analytical solution. Hint: apply a suitable change of coordinates as in the previous problem.
(b) Solve numerically for $r(t)$.
(c) Compare the results against those of the Vasicek model.

Exercise 9.11 The state equations for a Stuart-Landau oscillator with colored noise are

$$\begin{aligned}\frac{d\phi}{dt} &= \alpha - \beta + \rho^2 + \rho\eta(t)\\ \frac{d\rho}{dt} &= \rho - \rho^3 + \rho^2\eta(t)\\ \frac{d\eta}{dt} &= -\frac{\eta(t)}{\tau_c} + \frac{\sqrt{2D}}{\tau_c}\xi(t),\end{aligned} \tag{9.71}$$

where α and β are parameters that define the oscillator free running frequency. Apply the Euler-Maruyama algorithm to numerically solve for $\phi(t)$ and $\rho(t)$. Plot time series for $\phi(t)$ and $\rho(t)$, as well as phase plots on the plane $(\phi(t), \rho(t))$.

Exercise 9.12 A stochastic model for a van der Pol oscillator subject to environmental and internal noise takes the form

$$\begin{aligned}\frac{dv}{d\tilde{t}} &= u \\ \frac{du}{d\tilde{t}} &= -\frac{1}{LC}v - \frac{1}{C}g'(v)u - \frac{1}{C}\nu(\tilde{t}),\end{aligned} \tag{9.72}$$

where L and C are, respectively, the inductance and capacitors in the circuit, $g(v) = -v + v^3/3$ is the characteristic of a nonlinear resistor, and $\nu(t)$ is a source of colored noise, which is modulated by the current through the capacitor. That is

$$\nu(t) = \sqrt{\frac{C}{L}}\, u\, \eta(t)$$

(a) Apply the following change of coordinates

$$t = \frac{1}{\sqrt{LC}}\tilde{t}, \quad x_1 = v, \quad x_2 = \sqrt{LC}\, u,$$

to show that the original model Eq. (9.72) can be transformed to Langevin form

$$\begin{aligned}\frac{dx_1}{dt} &= x_2 \\ \frac{dx_2}{dt} &= -x_1 + \alpha(1 - x_1^2)x_2 + x_2\eta(t), \\ \frac{d\eta}{dt} &= -\frac{\eta(t)}{\tau_c} + \frac{\sqrt{2D}}{\tau_c}\xi(t),\end{aligned} \tag{9.73}$$

where $\alpha = \sqrt{L/C}$.

(b) Solve Eq. (9.73) numerically by using the Euler-Maruyama method and plot the phase-space dynamics on the plane (x_1, x_2).

Exercise 9.13 Transform the stochastic model for the van der Pol oscillator, Eq. (9.73), to an equivalent system with white noise

$$\begin{aligned}\frac{dx_1}{dt} &= x_2 \\ \frac{dx_2}{dt} &= -x_1 + \alpha(1 - x_1^2)x_2 + \frac{D^2}{2}x_2 + Dx_2\xi(t).\end{aligned}$$

Then compute the Fokker-Plank equation, $p(x_1, x_2, t)$, and solve it numerically.

Exercise 9.14 The stochastic version of a single spring-mass system, of mass m and spring constant k, is given by

$$\frac{dx}{dt} = v(t)$$
$$m\frac{dv}{dt} = -kx(t) - b(v(t) + \sqrt{2\gamma^2\lambda}\,\eta(t)$$
$$\frac{d\eta}{dt} = -\frac{\eta(t)}{\tau_c} + \frac{\sqrt{2D}}{\tau_c}\xi(t),$$

where $x(t)$ measures the displacement of the mass from equilibrium, $v(t)$ is the velocity. Assume $k = 1, b = 0.5, \gamma^2 = 0.25, \lambda = 0.4, m = 20, x(0) = 6,$ and $v(0) = 1$. Find a numerical solution and plot the results.

Exercise 9.15 A stochastic model for the phase-amplitude equations of a nonlinear oscillator is

$$\frac{d\phi}{dt} = \alpha + \left(\frac{D^2}{2} - \beta\right)\rho^2 + D\eta(t)$$
$$\frac{d\rho}{dt} = \rho + (D-1)\rho^3 + D\rho^2\eta(t),$$

where ϕ is phase and ρ is amplitude, α and β are parameters, D is the diffusion coefficient (i.e., noise intensity), and η is Gaussian white noise.

Observe that the amplitude equation decouples from the phase equations. Thus, compute the 1D Fokker-Planck equation, $p(\rho, t)$, and calculate its stationary solution.

Exercise 9.16 The Langevin equation for the freely diffusing ions in a microelectrode recessed into a surface is

$$\frac{dx}{dt} = \eta(t),$$

where $\eta(t)$ is a Gaussian white noise function with zero mean and correlation

$$\langle \eta(t)\,\eta(s)\rangle = 2D\delta(t-s).$$

(a) Write the Fokker-Planck equation for the for the probability distribution of the concentration of ions. That is, assume $c(x, t) = p(x, t)$, where $x(t)$ represents the vertical position of the electrode.
(b) Assume the concentration of ions in bulk to be c_b. Consider the electrode to be at $x = 0$, with the flat surface at $x = L$, so that the boundary conditions for the concentration are

$$c(0, t) = 0, \qquad c(L, t) = c_b.$$

Find the steady-state solution that satisfies these boundary conditions.

Exercise 9.17 Consider the Langevin equation

$$\frac{dx(t)}{dt} = f(x(t)) + g(x(t))\,\xi(t),$$

with $f(x) = ax - bx^3$ and $g(x) = Qx$, where $a, b, Q > 0$.

(a) Calculate Itô's version of the Fokker-Planck equation.
(b) Find an analytical expression for the steady-state solution, $p_s(x)$. Hint: set $y(x) = g^2(x)p_s(x)$ and solve for $y(x)$.

Exercise 9.18 Consider the two spring-mass system studied in Chap. 6.

(a) Write a stochastic version of the Model Eq. (6.4) as a first-order system of equations on the state variables $[x_1, x_2, x_3, x_4]^T$.
(b) Assume $m = 1, c = 0.1, k = 1$. Numerically solve the resulting stochastic model and plot the results on various projections of the phase space, i.e., (x_1, x_2), (x_2, x_4), and (x_1, x_3).
(c) Change the damping coefficient to $c = 0$ and repeat part (b).

References

1. D. Cox, H. Miller, *Stochastic Processes* (Chapman and Hall, London, 1996)
2. G.F. Lawler, *Introduction to Stochastic Processes*, 2nd edn. (Chapman & Hall, 2006)
3. L. Evans, *An Introduction to Stochastic Differential Equations* (American Mathematical Society, 2013)
4. P. Langevin, On the theory of brownian motion. C.R. Acad. Sci. (Paris) **146**, 530–533 (1908)
5. G.E. Uhlenbeck, L.S. Ornstein, On the theory of brownian motion. Phys. Rev. **36**, 823–841 (1930)
6. L. Gammaitoni, P. Hanggi, P. Jung, F. Marchesoni, Stochastic resonance. Rev. Mod. Phys. **70**(1), 223–287 (1998)
7. R. Bartussek, P. Hanggi, P. Jung, Stochastic resonance in optical bistable systems. Phys. Rev. E **49**, 3930–3939 (1994)
8. M. Inchiosa, A. Bulsara, Noise-mediated cooperative behavior and signal detection in DC squids, in *Stochastic and Chaotic Dynamics in the Lakes*. ed. by D. Broomhead, E. Luchinskaya, P. McClintock, T. Mullin (Melville, New York, 2000), pp. 583–595
9. M. Inchiosa, V. In, A. Bulsara, K. Wiesenfeld, T. Heath, M. Choi, Stochastic dynamics in a two-dimensional oscillator near a saddle-node bifurcation. Phys. Rev. E **63**, 066114–1–066114–10 (2001)
10. L. Gammaitoni, A.R. Bulsara, Nonlinear sensors activated by noise. Phys. A **325**, 8–14 (2003)
11. L. Gammaitoni, A.R. Bulsara, Noise activated nonlinear dynamic sensors. Phys. Rev. Lett. **88**, 230601 (2002)
12. A. Hibbs, B. Whitecotton, New regime in the stochastic resonance dynamics of squids in *Applied Nonlinear Dynamics and Stochastic Systems Near the Millenium*, ed. by J. Kadtke, A. Bulsara (AIP, New York, 1997)
13. B. McNamara, K. Wiesenfeld, Theory of stochastic resonance. Phys. Rev. A **39**, 4854–4869 (1989)
14. Online Source, Coriolis effect. http://en.wikipedia.org/wiki/Coriolis_effect
15. V. Apostolyuk, *MEMS/NEMS Handbook*, vol. 1 (Springer, New York, 2006)
16. V. Apostolyuk, F. Tay, Dynamics of micromechanical coriolis vibratory gyroscopes. Sensor Lett. **2**, 252–259 (2004)
17. A. Shkel, Type i and type ii micromachined vibratory gyroscopes, in *Proceedings of IEEE/ION PLANS* (San Diego, CA, 2006), pp. 586–593
18. E. Doedel, X. Wang, *Auto94: Software for Continuation and Bifurcation Problems in Ordinary Differential Equations. Applied Mathematics Report, California Institute of Technology* (1994)
19. D.W. Allan, The science of timekeeping. Technical Report 1289 (Hewlett Packard, 1997)

20. P.-L. Buono, B. Chan, J. Ferreira, A. Palacios, S. Reeves, V. In, P. Longhini, Collective patterns of oscillations in networks of crystals for precision timing. SIAM J. Appl. Dyn. Syst. In Print (2018)
21. P.-L. Buono, V. In, P. Longhini, L. Olender, A. Palacios, S. Reeves, Phase drift on networks of coupled of crystal oscillators for precision timing. Phys. Rev. E **98**, 012203 (2018)
22. P.-L. Buono, B. Chan, J. Ferreira, A. Palacios, S. Reeves, P. Longhini, V. In, Symmetry-breaking bifurcations and patterns of oscillations in rings of crystal oscillators. SIAM J. Appl. Dyn. Syst. **17**(2), 1310–1352 (2018)
23. P.J. Davis, *Circulant Matrices* (Wiley, New York, 1970)
24. A. Palacios, G. Gunaratne, M. Gorman, K. Robbins, Cellular pattern formation in circular domains. Chaos **7**(3), 463–475 (1997)
25. D. Armbruster, J. Guckenheimer, P. Holmes, Heteroclinic cycles and modulated traveling waves in systems with o(2) symmetry. Phys. D **29**, 257–282 (1988)
26. Emily Stone, Dieter Armbruster, Noise and o(1) amplitude effects on heteroclinic cycles. Chaos **9**(2), 499–506 (1999)
27. Dieter Armbruster, Emily Stone, Vivien Kirk, Noisy heteroclinic networks. Chaos **13**(1), 71–79 (2003)
28. Emily Stone, Philip Holmes, Random perturbations of heteroclinic attractors. J. SIAM Appl. Math. **50**(3), 726–743 (1990)
29. A. Juel, A.G. Darbyshire, T. Mullin, The effect of noise on pitchfork and hopf bifurcations. Proc. R. Soc. A: Math., Phys. Eng. Sci. **453**, 2627–2647 (1997)
30. E. Allen, *Modeling with Itò Stochastic Differential Equations* (Springer, Netherlands, 2007)
31. J.H. Barret, J.S. Bradley, *Ordinary Differential Equations* (International Text Book Company, 1972)

Chapter 10
Model Reduction and Simplification

The process of deriving a model, either through first principles or through data fitting, or through any other valid technique, typically leads to many more equations, variables or parameters, that may be needed to describe the behavior of a given phenomenon. Thus, it is not surprising that one of the most commonly asked question in mathematical modeling is

> How can we reduce a mathematical model to a simpler form that can be more amenable to analysis?

The words **reduction** and **simplification** may have similar connotations but, from a mathematical modeling standpoint, there is a clear distinction among them, which we should discuss in more detail. In any generic version of a model, either in the form of ODEs or PDEs, certain variables may approach, asymptotically, zero or a constant value or they may vary slowly compared to other variables. In these cases, it might be possible to capture the spatio-temporal evolution of the phenomenon with fewer equations. Consider, for example, the following model in the form of a system of ODEs

$$\begin{aligned} \dot{x} &= \mu - x^2 \\ \dot{y} &= -y. \end{aligned} \tag{10.1}$$

Without doing any complicated calculations, we can see that the two equations decoupled. That is, the values of x do not affect those of y and vice versa. In this sense, we can think of the 2D model (10.1) as being, effectively, a 1D system with two equations that can be treated separately. Furthermore, the solution for the second equation is $y(t) = y_0 e^{-t}$. Thus, $y(t) \to 0$ as $t \to \infty$, while $x(t)$ goes through a saddle-node bifurcation at $\mu = 0$. When $\mu < 0$ there are no solutions but for $\mu \geq 0$ there are two equilibrium points at $(x = \pm\sqrt{\mu}, y = 0)$. The one at $x = -\sqrt{\mu}$ is unstable while the other equilibrium at $x = \sqrt{\mu}$ is stable. Figure 10.1 illustrates the various phase portraits as μ changes values.

A. Palacios, *Mathematical Modeling*, Mathematical Engineering,
https://doi.org/10.1007/978-3-031-04729-9_10

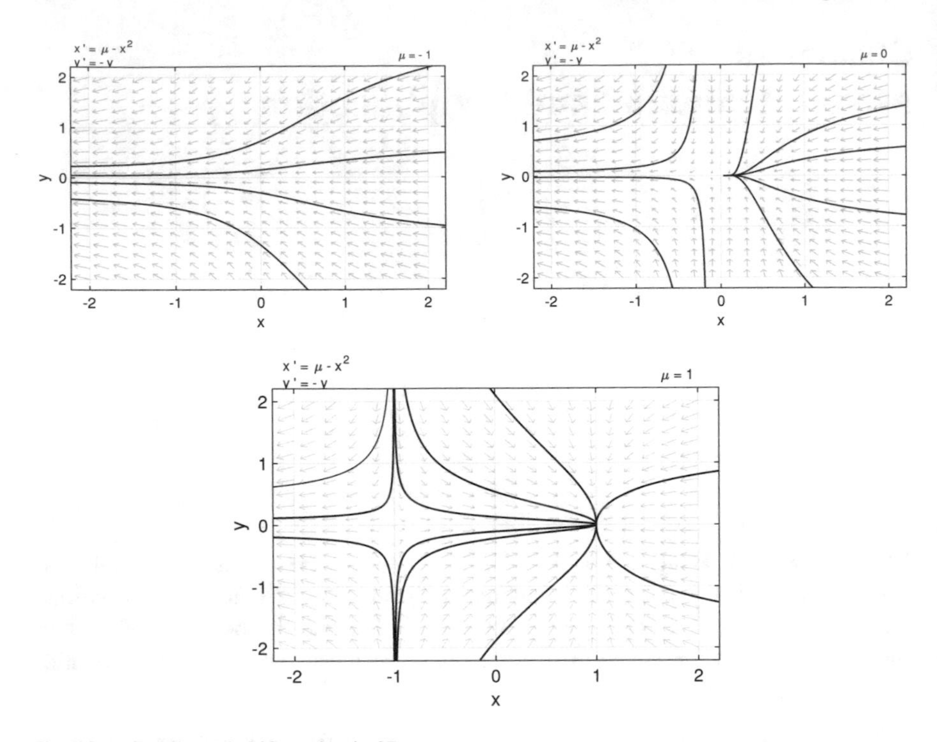

Fig. 10.1 Saddle node bifurcation in 2D

It is then straightforward to see that the original model (10.1) can be reduced to a single equation, $\dot{x} = \mu - x^2$. However obvious this might be, in general, it can be a daunting task to identify which equations can be eliminated from the original model. And often times the reduction of a model may not even involve eliminating equations. But, rather, reduction to *invariant subspaces* of lower dimension than the original phase space. On the other hand, simplification of a model may only involve reducing the number of parameters by rewriting the model equations in dimensionless form or eliminating certain terms, e.g., higher order terms, that may be negligible near the origin or near an equilibrium point.

A popular technique for deriving a reduced order model is the *Center Manifold*. This technique is broader since it can be applied to infinite-dimensional (evolution) models or to a vector field. In this book we focus on the latter case and consider a vector field of the form

$$\frac{dX}{dt} = F(X), \tag{10.2}$$

where $X \in \mathbb{R}^n$. Next, we describe the technical details of the reduction.

10.1 Center Manifold Reduction

In a nutshell, this technique is based on the existence of an *invariant manifold*, e.g., the *Center Manifold*, identified as W^c, and assumed to be located near a fixed point of (10.2). The manifold is invariant in the sense that all solution trajectories with initial conditions on it will remain on the manifold for all times. But if we assume all solution trajectories with initial conditions transverse to the manifold to be attracted towards (or repelled from) the manifold then we could reduce the dynamics of the original model to that of a lower-dimensional model by restricting the vector field (10.2) to the state variables that span the manifold. Figure 10.2 illustrates this scenario. On the manifold itself, there is a limit cycle solution. The reduced order model should capture, in principle, only the dynamics of the limit cycle solution.

The opposite scenario, in which the eigenvalues transverse to the manifold have positive real part is also possible. In that case, the manifold would be unstable so nearby solutions will move away from it. In both cases, however, the invariant properties of the manifold guarantee that solution trajectories with initial conditions on the manifold will remain on it at all times. Nevertheless, we wish to emphasize that this is a local procedure because it works only in the neighborhood of an equilibrium point. Next we describe the procedure to derive the reduced order model.

We start by assuming the n-dimensional phase-space $X \in \mathbb{R}^n$ to be decomposed as $n = n_c + n_s + n_u$, where n_c, n_s and n_u denote, respectively, the dimensions of the subspaces in which the real-part of the eigenvalues are zero, negative and positive. We consider the case $n_s > 0$ and $n_u = 0$, so that the vector field can be split in the form

$$\begin{aligned} \dot{x}_1 &= Ax_1 + f(x_1, x_2) \\ \dot{x}_2 &= Bx_2 + g(x_1, x_2), \end{aligned} \tag{10.3}$$

where $x_1 \in \mathbb{R}^{n_c}$ and $x_2 \in \mathbb{R}^{n_s}$.

We also assume Eq. (10.3) to have a *nonhyperbolic* fixed point, i.e., a fixed point at which, at least, one eigenvalue has zero real part with a corresponding eigenvector

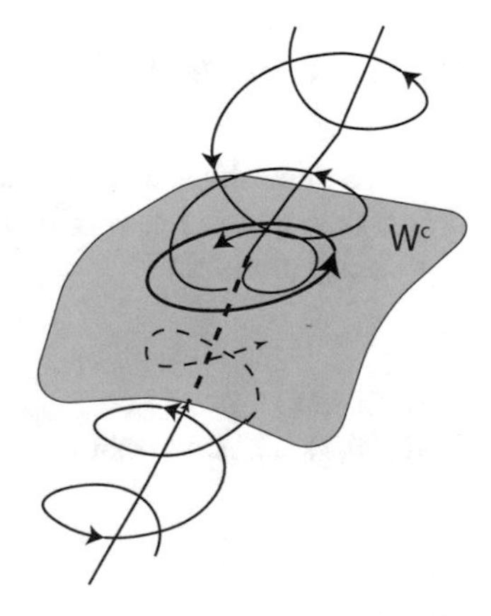

Fig. 10.2 Schematic representation of a Center Manifold containing a limit cycle. In this scenario, all solution trajectories with initial conditions outside the manifold collapse onto the manifold and onto the limit cycle solution

or center eigenspace E^c. Without loss of generality, assume the fixed point to be at $(x_1, x_2) = (0, 0)$, so that

$$\begin{aligned} f(0,0) &= 0, \quad Df(0,0) = 0 \\ g(0,0) &= 0, \quad Dg(0,0) = 0, \end{aligned}$$

where Df and Dg are the Jacobian matrix of $\mathbf{C}^r$ $(r \geq 2)$ continuous functions, f and g. In the above formulation, $A \in \mathbb{R}^{n_c \times n_c}$ is the matrix that contains the eigenvalues with zero real parts, in fact there are n_c of such eigenvalues. Since $n_u = 0$ there are no eigenvalues with positive real part, only eigenvalues with negative real part, which are those of $B \in \mathbb{R}^{n_s \times n_s}$.

Under the above conditions, the *Center Manifold Theorem* [1, 2] predicts the existence of an invariant center manifold $W^c(0)$, which is tangent to E^c. The manifold is formally defined as follows.

Definition 10.1 A center manifold, $W^c(0)$, of the vector field (10.3) is an invariant manifold, which can be locally described as follows

$$W^c(0) = \{(x_1, x_2) \in (\mathbb{R}^{n_c} \times \mathbb{R}^{n_s} | x_2 = h(x_1), |x_1| < \delta, h(0) = 0, Dh(0) = 0\},$$

where δ is small.

The requirements $h(0) = 0$ and $Dh(0) = 0$ guarantee that $W^c(0)$ is tangent to the center eigenspace E^c at the fixed point $(0, 0)$. Transverse to $W^c(0)$, the flow will move towards it since the eigenvalues have negative real part. The case $n_u > 0$ and $n_s = 0$ is similar, the main difference being the center manifold to be unstable, so that solution trajectories with initial conditions near the manifold will asymptotically diverge away.

The emphasis on this book is to describe, mainly, the procedure to derive the reduced order model. Readers interested in the mathematical theory of the existence of the center manifold $W^c(0)$ are referred to [3].

10.1.1 Computing the Center Manifold

Next we show how to compute the center manifold $x_2 = h(x_1)$. Since $(x_1, x_2) \in W^c(0)$ then $(\dot{x}_1, \dot{x}_2) \in W^c(0)$. This last point is obtained by differentiating directly $x_2 = h(x_1)$ to obtain

$$\dot{x}_2 = Dh(x_1)\dot{x}_1. \tag{10.4}$$

Both sets of points (x_1, x_2) and $(\dot{x}_1, \dot{x}_2)$ are still governed by the original vector field (10.3). Thus, we can write

$$\begin{aligned} \dot{x}_1 &= Ax_1 + f(x_1, h(x_1)) \\ \dot{x}_2 &= Bh(x_1) + g(x_1, h(x_1)). \end{aligned}$$

Substituting this last set of equations into Eq. (10.4) we get

$$Dh(x_1)\,[Ax_1 + f(x_1, h(x_1))] - Bh(x_1) - g(x_1, h(x_1)) = 0. \tag{10.5}$$

Solving Eq. (10.5) for $h(x)$ leads directly to the graph of the Center Manifold. And the actual form of the reduced order model is given by the existence theorem [2] of the Center Manifold. Here we quote the theorem.

Theorem 10.1 (Existence) *Under the conditions described above, there exists a* $\mathbf{C}^r$ *center manifold for the vector field (10.3). Furthermore, the dynamics restricted to this manifold leads to the reduced-order,* n_c*-dimensional, model of the form*

$$\boxed{\dot{u} = Au + f(u, h(u)), \quad u \in \mathbb{R}^{n_c}.} \tag{10.6}$$

Proof See Carr [3].

Another theorem by Carr also shows that as time evolves the solution $(x_1(t), x_2(t))$ of the original model converges to $(u(t), h(u(t)))$. More importantly, the stability properties of the solution of the original model are preserved by those of the reduced, low-dimensional, model. These conclusions can stated formally through the following theorem. Again, details can be found in [3].

Theorem 10.2 (Stability)

(i) Suppose the trivial solution $u = 0$ *of (10.6) is asymptotically stable (unstable); then the trivial solution* $(x_1, x_2) = (0, 0)$ *of (10.3) is alo asymptotically stable (unstable).*

(ii) Suppose the trivial solution $u = 0$ *of (10.6) is stable. Then if* $(x_1(t), x_2(t))$ *is a solution of (10.3) with* $(x_1(0), x_2(0))$ *sufficiently small, there is a solution of* $u(t)$ *of (10.6) such that as* $t \to \infty$

$$\begin{aligned} x_1(t) &= u(t) + O(e^{-\gamma t}) \\ x_2(t) &= h(u(t)) + O(e^{-\gamma t}), \end{aligned}$$

where $\gamma > 0$ *is a constant.*

10.1.2 Examples

Let us start with a simple two-dimensional problem.

Example 10.1

$$\begin{aligned} \dot{x}_1 &= x_1^2 - x_1^5 \\ \dot{x}_2 &= -x_2 + x_1^2. \end{aligned} \tag{10.7}$$

If we rewrite this system as

$$\begin{bmatrix} \dot{x}_1 \\ \dot{x}_2 \end{bmatrix} = \begin{bmatrix} 0 & 0 \\ 0 & -1 \end{bmatrix} \begin{bmatrix} x_1 \\ x_2 \end{bmatrix} + \begin{bmatrix} x_1^2 x_2 - x_1^5 \\ x_1^2 \end{bmatrix}.$$

We can then see that

$$A = 0, \quad B = -1, \quad f = x_1^2 x_2 - x^5, \quad g = x_1^2.$$

Thus, we can assume the center manifold to be of the form

$$x_2 = h(x_1) = ax_1^2 + bx_1^3 + \mathcal{O}(x_1^4).$$

We can then write

$$\begin{aligned} h(x_1) &= ax_1^2 + bx_1^3 + \cdots \\ Dh(x) &= 2ax_1 + 3bx_1^2 + \cdots \\ Dh(x)\,[Ax_1 + f(x_1, h(x_1))] &= (2ax_1 + 3bx_1^2 + \cdots)(ax_1^4 + bx_1^5 - x_1^5 + \cdots) \\ Bh(x_1) + g(x_1, h(x_1)) &= -(ax_1^2 + bx_1^3 + \cdots) + x_1^2. \end{aligned}$$

Substituting into Eq. (10.5) and then collecting like powers of terms up to order three, we get

$$\begin{aligned} \mathcal{O}(x_1^2): &\quad a - 1 = 0 \\ \mathcal{O}(x_1^3): &\quad b \quad\;\; = 0. \end{aligned}$$

It follows that the center manifold is

$$x_2 = h(x_1) = x_1^2 + \mathcal{O}(x_1^4).$$

Consequently, the flow on the center manifold is given by

$$\dot{x}_1 = x_1^4 + \mathcal{O}(x_1^5). \tag{10.8}$$

Equation (10.8) shows that for small values, $x_1 \approx 0$, the zero equilibrium $x_1 = 0$ is unstable, which implies that the zero equilibrium, $(0, 0)$, of the original system, is also unstable. Notice, however, that if we had considered that x_2 appears to decay exponentially fast, we could have assumed $x_2 \approx 0$, and obtained the approximation

$$\dot{x}_1 = -x^5,$$

which it would have, erroneously, indicated the opposite stability result of $(0, 0)$ being stable. This example highlights the importance of studying the dynamics and stability properties via the center manifold reduction.

We now illustrate the process of computing a center manifold through an example in which a limit cycle solution lies on the invariant manifold while the solution trajectories, transverse to the manifold, decay towards it. This is a similar situation the illustration shown in Fig. 10.2.

Example 10.2 Consider the following nonlinear system of equations

$$\begin{aligned}\dot{x}_1 &= -x_2 + x_1(1-x_3)x_3 \\ \dot{x}_2 &= x_1 + x_2(1-x_3)x_3 \\ \dot{x}_3 &= -x_3 + x_1^2 + x_2^2 + 2(1-x_3)x_3^2.\end{aligned} \tag{10.9}$$

Figure 10.3 illustrates the long-term behavior of the model Eqs. (10.9). On the left, the time-series solutions indicate that x_3 remains constant at $x_3 = 1$ while x_1 and x_2 oscillate periodically in a limit cycle trajectory. By inspection, we can see that $(0, 0, 0)$ is an equilibrium solution of (10.9). Then the plots suggest that this equilibrium is unstable and the long-term dynamics appears to be captured by a center manifold in which x_3 remains constant while x_1 and x_2 oscillate in a limit cycle solution.

To compute the center manifold, the right-hand side of Eq. (10.9) can be split first between linear and nonlinear terms as follows

$$\begin{bmatrix}\dot{x}_1\\ \dot{x}_2\\ \dot{x}_3\end{bmatrix} = \begin{bmatrix}0 & -1 & 0\\ 1 & 0 & 0\\ 0 & 0 & -1\end{bmatrix}\begin{bmatrix}x_1\\ x_2\\ x_3\end{bmatrix} + \begin{bmatrix}x_1(1-x_3)x_3\\ x_2(1-x_3)x_3\\ x_1^2 + x_2^2 + 2(1-x_3)x_3^2\end{bmatrix}.$$

We can then see that

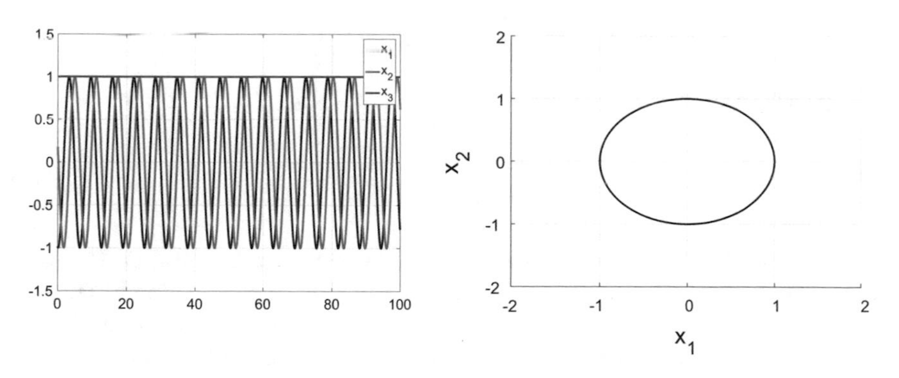

Fig. 10.3 Computer simulations of the dynamics of Eq. (10.7). (Left) Time Series solutions show x_3 remains constant while x_1 and x_2 oscillate periodically forming a limit cycle solution. (Right) Phase space shows the long-term behavior of the dynamics in the form of a limit cycle

$$A = \begin{bmatrix} 0 & -1 \\ 1 & 0 \end{bmatrix}, \quad B = -1, \quad f = \begin{bmatrix} x_1(1-x_3)x_3 \\ x_2(1-x_3)x_3 \end{bmatrix}, \text{ and } g = x_1^2 + x_2^2 + 2(1-x_3)x_3^2.$$

Thus, we can assume the center manifold to be of the form

$$x_3 = h(x_1, x_2) = ax_1^2 + bx_2^2.$$

Substituting into Eq. (10.5) we get

$$[2ax_1 \mid 2bx_2] \left\{ \begin{bmatrix} 0 & -1 \\ 1 & 0 \end{bmatrix} \begin{bmatrix} x_1 \\ x_2 \end{bmatrix} + \begin{bmatrix} x_1(1-h(x_1,x_2))h(x_1,x_2) \\ x_2(1-h(x_1,x_2))h(x_1,x_2) \end{bmatrix} \right\} + h(x_1,x_2) - x_1^2 - x_2^2 - 2(1-h(x_1,x_2))h^2(x_1,x_2) = 0.$$

Expanding this last equation and collecting like powers of terms up to order two, i.e., x_1^2, x_2^2 and x_1x_2, we find:

$$\begin{aligned} \mathcal{O}(x_1^2): &\quad a - 1 = 0 \\ \mathcal{O}(x_1x_2): &\quad b - a = 0 \\ \mathcal{O}(x_2^2): &\quad b - 1 = 0. \end{aligned}$$

It follows that $a = b = 1$, and the center manifold is

$$x_3 = h(x_1, x_2) = x_1^2 + x_2^2,$$

which leads to the flow on the Center Manifold:

$$\begin{bmatrix} \dot{x}_1 \\ \dot{x}_2 \end{bmatrix} = \begin{bmatrix} 0 & -1 \\ 1 & 0 \end{bmatrix} \begin{bmatrix} x_1 \\ x_2 \end{bmatrix} + \begin{bmatrix} x_1(1-x_1^2-x_2^2)(x_1^2+x_2^2) \\ x_2(1-x_1^2-x_2^2)(x_1^2+x_2^2) \end{bmatrix}. \tag{10.10}$$

In polar coordinates, $x_1 = r\cos\theta$ and $x_2 = r\sin\theta$, Eq. (10.10) becomes

$$\begin{aligned} \dot{r} &= r^3(1-r^2) \\ \dot{\theta} &= 1. \end{aligned} \tag{10.11}$$

We can see that $r = 0$ and $r = 1$ are equilibrium points of Eq. (10.11). The latter represents the limit cycle solution observed in Fig. 10.3(right). A little calculation (left as an exercise) can show that $r = 0$ is actually unstable, while the limit cycle at $r = 1$, is stable. It follows from Theorem 10.1 that a limit cycle solution exists in the original Eqs. (10.9) while Theorem 10.2 implies that the limit cycle in (10.9) is also stable.

Example 10.3 (*Parameter Dependency: Fluxgate Magnetometer.*) We now consider the model of a N-dimensional (N odd for negative feddback) unidirectionally coupled ring of overdamped bistable systems, which serves as a model for a fluxgate magnetometer with N fluxgates:

$$\dot{x}_i(t) = -x_i(t) + \tanh\left(c(x_i(t) + \lambda x_{i+1}(t) + \varepsilon)\right), \tag{10.12}$$

where $i = 1, \ldots, N \mod N$. As a representative example, we consider the case $N = 3$. Recall from Fig. 6.24 that a Hopf bifurcation occurs around $\lambda = 4/3$ from the zero equilibrium $(0, 0, 0)$, which leads to unstable periodic oscillations. Indeed, around the trivial equilibrium, the linearized vector field yields the Jacobian matrix

$$J = \begin{bmatrix} c-1 & c\lambda & 0 \\ 0 & c-1 & c\lambda \\ c\lambda & 0 & c-1 \end{bmatrix},$$

whose eigenvalues are

$$\sigma_1 = c - 1 + c\lambda, \quad \sigma_{2,3} = c - 1 - \frac{1}{2}c\lambda \pm \frac{1}{2}c\lambda\sqrt{3}i.$$

At the critical value of coupling strength $\lambda_c = 2(c-1)/c$, the eigenvalues are:

$$\sigma_1 = 3(c-1), \quad \sigma_{2,3} = 0 \pm (c-1)\sqrt{3}i.$$

It follows that a Hopf bifurcation occurs at λ_c. The bifurcation diagram of Fig. 6.24 was generated with $c = 3$, which corresponds to the critical value $\lambda_c = 4/3$ observed in the figure. In fact, if we assume $c > 1$, so that the cores operate in the ferromagnetic regime, we can see that the system (10.12) always has one negative eigenvalue, $\sigma_1 < 0$. The dynamics is then expected to converge to a two-dimensional manifold. To compute the reduced dynamics, we perform first a Taylor series expansion of Eq. (10.12) around the trivial equilibrium $(x_1, x_2, x_3) = (0, 0, 0)$, yielding

$$\begin{bmatrix} \dot{x}_1 \\ \dot{x}_2 \\ \dot{x}_3 \end{bmatrix} = \begin{bmatrix} c-1 & c\lambda & 0 \\ 0 & c-1 & c\lambda \\ c\lambda & 0 & c-1 \end{bmatrix} \begin{bmatrix} x_1 \\ x_2 \\ x_3 \end{bmatrix} + \begin{bmatrix} -\frac{1}{3}x_1^3 - \lambda x_1^2 x_2 - \lambda^2 x_1 x_2^2 - \frac{1}{3}\lambda^3 x_2^3 \\ -\frac{1}{3}x_2^3 - \lambda x_2^2 x_3 - \lambda^2 x_2 x_3^2 - \frac{1}{3}\lambda^3 x_3^3 \\ -\frac{1}{3}x_3^3 - \lambda x_3^2 x_1 - \lambda^2 x_3 x_1^2 - \frac{1}{3}\lambda^3 x_1^3 \end{bmatrix}.$$

We now rewrite this last system of equations in the form given by Eq. (10.3). We do this by exploiting the cyclic nature of the Jacobian matrix J, which leads to the eigenvectors

$$V_j = \left\{\left[v, \zeta^j v, \zeta^{2j} v\right] : v \in R\right\},$$

where $j = 0, 1, 2$ and $\zeta = e^{2\pi i/3}$. Let

$$P = [\mathrm{Re}(V_1) \mid \mathrm{Im}(V_2) \mid V_0]$$

be the transformation for the change of variables $X = PY$, where $X = (x_1, x_2, x_3)$ and $Y = (y_1, y_2, y_3)$. Substituting into the ODE system in X we arrive at

$$\begin{bmatrix} \dot{y}_1 \\ \dot{y}_2 \\ \dot{y}_3 \end{bmatrix} = \begin{bmatrix} \mu & (c-1)\sqrt{3} & 0 \\ -(c-1)\sqrt{3} & \mu & 0 \\ 0 & 0 & 3(c-1) \end{bmatrix} \begin{bmatrix} y_1 \\ y_2 \\ y_3 \end{bmatrix} + \begin{bmatrix} f_1(y_1, y_2) \\ f_2(y_2, y_3) \\ f_3(y_3, y_1) \end{bmatrix},$$

where $\mu = c - 1 - c\lambda$ is treated as a bifurcation parameter, which at λ_c becomes zero, i.e., $\mu(\lambda_c) = 0$. The vector function $[f_1, f_2, f_3]$ contains cubic terms in y_i and y_j, which can be written as: $f_i(y_i, y_j) = a^i_{iii} y_i^3 + a^i_{iij} y_i^2 y_j + a^i_{ijj} y_i y_j^2 + a^i_{jjj} y_j^3$.

To compute the center manifold with the dependency on the bifurcation parameter μ, the terms μy_1 and μy_2 are treated as nonlinear and the system of ODEs is extended to

$$\dot{\mu} = 0.$$

Then we set

$$A = \begin{bmatrix} 0 & (c-1)\sqrt{3} \\ -(c-1)\sqrt{3} & 0 \end{bmatrix}, \quad B = 3(c-1), \quad f = \begin{bmatrix} f_1(y_1, y_2) \\ f_2(y_2, y_3) \end{bmatrix}, \quad g = f_3(y_3, y_1).$$

We now seek a Center Manifold of the form:

$$y_3 = h(y_1, y_2, \mu) = c_1 y_1^2 + c_2 \mu y_1 + c_3 \mu y_2 + c_4 y_1 y_2 + c_5 y_2^2 + c_6 \mu^2,$$

where $c_{i's}$ are unknown coefficients. Direct substitution into Eq. (10.5) and collecting like powers of terms up to order two yields: $c_2 = c_3 = c_6 = 0$, $c_1 = 3(c_1)$, $c_4 = 4(c-1)\sqrt{3}$, and $c_5 = -3(c_1)$, which yields

$$y_3 = 3(c-1)y_1^2 + 4(c-1)\sqrt{3} y_1 y_2 - 3(c-1) y_2^2.$$

The flow on the Center Manifold can be expressed as

$$\begin{bmatrix} \dot{y}_1 \\ \dot{y}_2 \end{bmatrix} = \begin{bmatrix} \mu & (c-1)\sqrt{3} \\ -(c-1)\sqrt{3} & \mu \end{bmatrix} \begin{bmatrix} y_1 \\ y_2 \end{bmatrix} + \begin{bmatrix} f_1(y_1, y_2) \\ f_2(y_2, h(y_1, y_2)) \end{bmatrix}. \tag{10.13}$$

10.2 Lyapunov-Schmidt Reduction

Consider an n-dimensional mathematical model of the form

$$\frac{dX}{dt} = \Phi(X, \alpha), \tag{10.14}$$

where $X = (X_1, \ldots, X_n)$, $\Phi = (\Phi_1, \ldots, \Phi_n)$, and $\alpha = (\alpha_1, \ldots \alpha_s)$ is a vector of parameters. Assume $\Phi : \mathbb{R}^n \times \mathbb{R}^s \to \mathbb{R}^n$ to be a smooth function satisfying $\Phi(0, 0) = 0$. Let

$$L = (d_X \Phi)_{(0,0)} = \left(\frac{\partial \Phi_i}{\partial X_i}(0, 0) \right)$$

be the associated Jacobian matrix. Observe that $L : \mathbb{R}^n \to \mathbb{R}^n$. Now, if we are interested in computing the equilibrium points of Eq. (10.14), we must solve the algebraic equation

$$\Phi(X, \alpha) = 0, \tag{10.15}$$

for the vector of unknowns $X = (X_1, \ldots, X_n)$ as a function of the parameters, typically just the first parameter, i.e., $\alpha_0 = \lambda$, which is commonly treated as the distinguished bifurcation parameter. According to the Implicit Function Theorem (IFT), see Sect. 5.2, if L is nonsingular, then there exists a unique solution $X(\lambda)$ such that $\Phi(X(\lambda), \lambda) = 0$, $X(0) = 0$ and $\Phi(X) = LX +$ higher order terms.

If L is singular, however, the IFT theorem cannot be applied but the Lyapunov-Schmidt (LS) Reduction shows that solving the high-dimensional Eq. (10.15) can be reduced to that of solving a single equation. That is

$$\boxed{\begin{array}{l} \Phi(X, \alpha) = 0, \\ X \in \mathbb{R}^n, \ \alpha \in \mathbb{R}^s. \end{array}} \quad \Longleftrightarrow \quad \boxed{\begin{array}{l} g(x, \lambda) = 0, \\ x \in \mathbb{R}, \ \lambda \in \mathbb{R}. \end{array}}$$

Next we show how the reduction is performed. We will follow the exposition in [4], which is based on Euclidean spaces. But it should be emphasized that the same procedure applies over Banach spaces.

Assume $\ker L \neq \{0\}$, i.e., $\operatorname{rank} L < n$. This assumption is made because if $\operatorname{rank} L = n$ then the Implicit Function theorem is applicable, there is no degeneracy or bifurcation, so we can solve explicitly for a unique solution $X(\alpha)$. The minimum degenerate case occurs when

$$\operatorname{rank} L = n - 1.$$

This is the case to be considered in the remaining of the derivation. Then we seek to find vector spaces M and N such that

$$\begin{aligned} \mathbb{R}^n &= \ker L \oplus M \\ \mathbb{R}^n &= \operatorname{range} L \oplus N. \end{aligned} \tag{10.16}$$

Since $\operatorname{rank} L = n - 1$, it follows that $\dim \operatorname{range} L = n - 1$ and $\dim \ker L = 1$. This means that $\dim M = n - 1$ and $\dim N = 1$. Next, we seek to solve $\Phi(X, \alpha) = 0$ by splitting X into $\ker L$ and M. That is, let $X = v + w$, and solve

$$\Phi(v + w, \alpha) = 0, \quad v \in \ker L, \; w \in M.$$

Let $E : \mathbb{R}^n \to \text{range}\, L$. That is, E is the projection of $\mathbb{R}^n$ onto range L. Notice that ker $E = N$, which means that $Eu = 0$ if and only if $u = 0$, for all $u \in \mathbb{R}^n$. This also means that the complimentary projection $(I - E) : \mathbb{R}^n \to N$ has range equal to N. Consequently, $\Phi = 0$ if and only if

$$E\,\Phi(X, \alpha) = 0 \tag{10.17a}$$

$$(I - E)\,\Phi(X, \alpha) = 0. \tag{10.17b}$$

The fundamental idea behind the LS reduction is that Eq. (10.17a) can be solved explicitly, for $n - 1$ variables, using the IFT Theorem. Then, those $n - 1$ variables can be substituted into Eq. (10.17b) and solve for the remaining variable.

Claim $E\,\Phi(v + w, \alpha) = 0$ can be solved by the IFT theorem for $w = W(v, \alpha)$.

Proof Since E is linear, we can write

$$d_W(E\,\Phi)_{(0,0,0)} = E(d_W\,\Phi)_{(0,0,0)} = E(L|M) = (L|M) \simeq \text{onto range}\, L,$$

where $L|M : M \to \text{range}\, L$ is the restriction of L onto M. Thus, we can write

$$E\,\Phi(v + W(v, \alpha), \alpha) = 0, \quad W(0, 0) = 0.$$

In fact, all solutions to the equation $E\,\Phi(X, \alpha) = 0$ near the origin, $X = 0$, are parametrized this way. □

Hence, all solutions to the original algebraic equation $\Phi(X, \alpha) = 0$, near $X = 0$ are parametrized by solutions to the equation

$$g(v, \alpha) = (I - E)\,\Phi(v + W(v, \alpha), \alpha) = 0,$$

where $g : \ker L \times \mathbb{R}^s \to N$. Recall that by assumption, $\text{rank}\, L = n - 1$ and $\text{corank}\, L = 1$. Thus, we have reduced the original algebraic problem from $\mathbb{R}^n$ to $\mathbb{R}^1$. That is, $g : \mathbb{R} \times \mathbb{R}^s \to \mathbb{R}$.

10.2.1 Computational Aspects

Let $\ker L = \mathbb{R}\{v\}$. Thus, we can view xv as coordinates on $\ker L$. Likewise, let $N = \mathbb{R}\{v^*\}$. Choose $v^* \perp \text{range}\, L$, where $N = (\text{range}\, L)^{\perp}$. Assume also $s = 1$ and $\alpha = \lambda$. Let $< \cdot, \cdot >$ denote an inner product operation. Then the reduced model $g : \ker L \times \mathbb{R}^s \to N$ is given by

$$g(x, \lambda) = < v^*, \Phi(xv + W(x, \lambda), \lambda) > . \tag{10.18}$$

The following calculations confirm that $g_x(0, 0) = 0$, as is expected from the bifurcation condition.

$$g_x = \langle v^*, \frac{d\Phi}{dx}(xv + W(x, \lambda), \lambda)\rangle = \langle v^*, \Phi_v v + (d_W \Phi) W_x\rangle = \langle v^*, Lv + LW_x\rangle = 0.$$

The last equality holds since $v^* \in (\text{range}\, L)^\perp$ and $LW_x \in \text{range}\, L$. Additional derivative terms are computed as follows:

$$\begin{aligned}
g_\lambda(0, 0) &= \langle v^*, \Phi_\lambda(0, 0)\rangle \\
g_{xx}(0, 0) &= \langle v^*, (d^2\Phi)_0(v, v)\rangle \\
g_{x\lambda} &= \langle v^*, d\,\Phi_\lambda(v) - (d^2\Phi)_0(v, z)\rangle \\
g_{xxx} &= \langle v^*, (d^3\Phi)_0(v, v, v) - 3(d^2\Phi)_0(v, z)\rangle,
\end{aligned}$$

where $z = L^{-1}E\,(d^2\Phi)_0(v, v)$. The first derivative of Φ can be computed as a "directional derivative", i.e.,

$$(d\,\Phi)_{(X,\alpha)} Y = \frac{d}{dt}\Phi(X + tY, \alpha)|_{t=0}.$$

Second-order derivatives can be computed as follows

$$\begin{aligned}
(d^2\,\Phi)(Y_1, Y_2) &= \frac{d}{dt_2}\frac{d}{dt_1}\Phi(X + tY_1 + t_2 Y_2, \alpha)|_{t_1=0,t_2=0} \\
&= \sum_{i,j=1}^{n} \frac{\partial^2\Phi}{\partial X_i \partial X_j}(X, \alpha) Y_1^i Y_2^j \in \mathbb{R}^n.
\end{aligned}$$

We can then re-write the derivatives of g as follows

$$\begin{aligned}
g_x(x, \lambda) &= \langle v^*, (d\,\Phi)_{(xv, W(v,\lambda),\lambda)}(v + W_x)\rangle \\
g_{xx}(x, \lambda) &= \langle v^*, (d^2\,\Phi)(v + W_x, v + W_x) + (d\,\Phi)_{(xv,W,\lambda)}(W_{xx})\rangle \\
g_{xx}(0, 0) &= \langle v^*, (d^2\,\Phi)_0(v + W_x(0), v + W_x(0)) + LW_{xx}(0)\rangle \\
&= \langle v^*, (d^2\,\Phi)_0(v + W_x(0), v + W_x(0))\rangle,
\end{aligned}$$

where the last equality follows from the fact that $LW_{xx}(0) \in \text{range}\, L$, so $LW_{xx}(0) = 0$.

Claim $W_x(0) = 0$.

Proof Remember that W is defined implicitly by $E\,\Phi(xv + W(x, \lambda), \lambda) = 0$. Differentiating with respect to x we get

$$E\,L\,W(v + W_x) = L\,W_x,$$

because $v \in \ker L$ implies $Lv = 0$. Then $LW_x(0) = 0$ and since $W(x, \lambda) \in M$ then $W_x \in M$. It follows that since $L|M : M \to \text{range}\, L$ then $W_x(0) = 0$. $\square$.

Similarly, since $LW_{xxx}(0) \in \text{range}\, L$, so that $LW_{xxx}(0) = 0$, we get

$$\begin{aligned} g_{xxx}(0,0) &= \langle v^*, (d^3\Phi)_0(v,v,v) + 3(d^2\Phi)_0(W_{xx}(0), v) + LW_{xxx}(0)\rangle \\ &= \langle v^*, (d^3\Phi)_0(v,v,v) + 3(d^2\Phi)_0(W_{xx}(0), v)\rangle. \end{aligned}$$

Next, we show how to find $W_{xx}(0)$. We start once again with

$$E\,\Phi(xv + W(x,\lambda), \lambda) = 0, \quad \text{and,} \quad E((d\,\Phi)(v + W_x)) = 0.$$

Differentiating with respect to x we get

$$E[(d^2\Phi)(v + W_x, v + W_x) + (d\,\Phi)W_{xx}] = E[(d^2\Phi)(v,v) + LW_{xx}(0)] = 0.$$

Then, $LW_{xx}(0) = -E(d^2\Phi)(v,v)$, which yields

$$W_{xx}(0) = -L^{-1}E(d^2\Phi)(0,0).$$

An important feature of the LS reduction is that the stability properties of the original model equations are preserved by the reduced model.

Theorem 10.3 (Exchange of Stability) *If $g(x,\alpha) = 0$, then the equilibrium $(xv_0, W(x,\alpha), \alpha)$ to Eq. (10.14) is linearly stable if and only if $g_x(x,\alpha) < 0$.*

Proof See Golubitsky [4].

We would like to emphasize that if $\mathcal{X}$ and $\mathcal{Y}$ are Banach spaces with a bounded Fredholm operator $L : \mathcal{X} \to \mathcal{Y}$, such that $\dim(\ker L) < \infty$ and range$L$ is closed and of finite codimension, the it is still possible to apply the Lyapunov-Schmidt reduction by considering

$$\begin{aligned} \mathcal{X} &= \ker L \oplus M \\ \mathcal{Y} &= \text{range}\, L \oplus N. \end{aligned}$$

10.2.2 Symmetries

In this section we show that if the original model is symmetric under some transformation γ, then the reduced Lyapunov-Schmidt model will retain the same symmetries.

Suppose that $\gamma : \mathcal{X} \to \mathcal{X}$ and $\gamma : \mathcal{Y} \to \mathcal{Y}$ are linear (symmetry) mappings, such that γ acts on the domain of the dirrential operator. That is, γ acts on Ω, and, consequently, it acts on the space of functions. Thus,

$$\Phi(\gamma u, \alpha) = \gamma\Phi(u,\alpha), \quad \forall(u,\alpha).$$

This means that Φ is γ-equivariant, i.e., γ is a symmetry of the original model. Next we show that γ is also a symmetry of the reduced model.

Claim

$$g(\gamma v, \alpha) = \gamma g(v, \alpha).$$

Proof Let $\hat{w}(v, \alpha) = \gamma^{-1} w(\gamma v, \alpha)$. We show that $\hat{w}$ is an implicit function.

$$\begin{aligned} E\Phi(v + \hat{w}(v, \alpha), \alpha) &= E\Phi(v + \gamma^{-1} w(\gamma v, \alpha), \alpha) \\ &= E\Phi(\gamma^{-1}(\gamma v + w(\gamma v, \alpha), \alpha)) \\ &= E(\gamma^{-1}\Phi(\gamma v + w(\gamma v, \alpha), \alpha)) \\ &= \gamma^{-1} E\Phi(\gamma v + w(\gamma v, \alpha), \alpha) \\ &= \gamma^{-1} 0 = 0. \end{aligned}$$

By uniqueness of the Implicit Function Theorem, we conclude $\hat{w} = w$. Thus, $\gamma w(v, \alpha) = w(\gamma v, \alpha)$. In addition, if $\Phi(\gamma v, \alpha) = \gamma\Phi(v, \alpha)$ then

$$\gamma\Phi(\gamma^{-1} u, \alpha) = \Phi(v, \alpha).$$

This shows that $\Phi(\gamma^{-1} v, \alpha) = \gamma^{-1}\Phi(v, \alpha)$. This result implies that if γ is a symmetry of the model then so is its inverse. In coordinates this means that

$$\gamma(n, r) = (\gamma n, \gamma r),$$

where $n \in N$ and $r \in$ range L. In other words

$$E\gamma = \gamma E.$$

Finally, we can write

$$\begin{aligned} g(\gamma v, \alpha) &= (I - E)\Phi(\gamma v + w(\gamma v, \alpha), \alpha) \\ &= (I - E)\Phi(\gamma v + \gamma w(v, \alpha), \alpha) \\ &= \gamma g(v, \alpha). \end{aligned}$$

□

10.2.3 Examples

Example 10.4 (*Euler's Beam Model*) We now consider Euler's Beam experiment whose original model appeared in Eq. (5.7). For completeness, we re-write the model here

$$E_m I\theta''(x) + \lambda \sin\theta(x) = 0, \qquad 0 < x < L_b, \tag{10.19}$$

where x is the material coordinate, E_m is the elastic modulus, I is moment of inertia, λ is the compressive force and L_b is the length of the beam. Notice that some of the labels for parameters have been labeled in order to avoid confusion with the notation used in the derivation of the LS reduction.

Assume Newmann (clamped) boundary conditions

$$\theta'(0) = \theta'(L_b) = 0. \tag{10.20}$$

We start by decomposing the ambient space according to Eq. (10.16). That is, we perform the following steps.

Step 1. Decompose Ambient Space

First, we define the space of two-times, real-valued, continuous and differentiable functions from the interval $[0, L_b]$ as the space for the solution $\theta(x)$. Formally, let

$$X = \left\{\theta \in C^2([0, L_b]; \mathbb{R}) : \theta(0) = \theta(L_b) = 0\right\}.$$

Next, we need to define Φ and compute $L = (d\,\Phi)_{(0,0)}$. Let

$$Y = \left\{\theta \in C^0([0, L_b]; \mathbb{R})\right\}.$$

We can now define $\Phi : X \times \mathbb{R} \to Y$ through

$$\Phi(\theta(x), \lambda) = E_m I \theta''(x) + \lambda \sin \theta(x). \tag{10.21}$$

Then the linearized boundary value problem yields

$$L\,\theta(x) = \Phi_\theta(0, \lambda_0)\theta(x) = E_m I \theta''(x) + \lambda_0 \theta(x).$$

Solving $\Phi_\theta(0, \lambda_0)\theta(x) = 0$, for θ, we get a nontrivial solution

$$\theta(x) = A_n \cos\left(\frac{n\pi x}{L_b}\right),$$

if and only if

$$\lambda_0 = \frac{n^2\pi^2 E_m I}{L_b},$$

where n is an integer. This result implies that in a neighborhood of $(\theta(x), \lambda) = (0, n^2\pi^2 E I_m/L_b^2)$, the boundary value problem Eqs. (10.19) and (10.20), has only the trivial solution when $\lambda < n^2\pi^2 E I_m/L_b^2$, and three solutions when $\lambda > n^2\pi^2 E I_m/L_b^2$. In both cases, the trivial solution corresponds to the unbuckled beam, while in the latter case, the two nontrivial solutions correspond to the buckled states. This observation suggest that the beam problem undergoes a pitchfork bifurcation. We proceed to verify this expectation as follows. Observe that

$$\dim \ker L = \begin{cases} 1 & \text{if } \lambda_0 = n^2\pi^2 E_m I/L_b^2 \\ 0 & \text{otherwise.} \end{cases}$$

Since we are interested in a nonzero kernel, we consider $\lambda_0 = n^2\pi^2 E_m I/L_b^2$, so that

$$\ker L = \left\{\theta \in X : \theta(x) = A_n \cos\left\{\frac{n\pi x}{L_b}\right\},\ A_n \in \mathbb{R}\right\}.$$

We can now write

$$\begin{aligned} X &= \ker L \oplus M \\ Y &= \text{range}\, L \oplus N, \end{aligned}$$

where the subspaces M and N can be computed using the standard inner product, $\langle\cdot,\cdot\rangle$, which is defined by

$$\langle u, v\rangle = \int_0^{L_b} u(x)\, v(x) dx.$$

We can then use this definition of inner product to find the orthogonal complement subspace M of $\ker L$, and N of range L. Let $\xi = \frac{n\pi x}{L_b}$, $w \in M$ and $\phi \in N$

$$\begin{aligned} M &= \left\{\theta \in X : \textstyle\int_0^{L_b} \theta(\xi) \cos\xi d\xi = 0\right\}, \\ N &= \{\phi \in Y : \phi(\xi) = B_n \cos\xi,\ B_n \in \mathbb{R}\}. \end{aligned}$$

In fact, integrating by parts, we can verify that

$$\begin{aligned} \langle \cos\xi, L\theta(\xi)\rangle &= \textstyle\int_0^{L_b} \cos(\xi) L\theta(\xi) d\xi = \int_0^{L_b} \cos(\xi)[E_m I\theta''(\xi) + \lambda_0\theta(\xi)] d\xi \\ &= E_m I \textstyle\int_0^{L_b} \cos(\xi)\theta''(\xi) d\xi + \lambda_0 \int_0^{L_b} \theta(\xi)\cos(\xi) d\xi \\ &= E_m I\ \cos(\xi)\theta'(x)\big|_0^{L_b} - E_m I \textstyle\int_0^{L_b} \sin(\xi)\theta'(\xi) d\xi \\ &= E_m I\ \sin(\xi)\theta(x)|_0^{L_b} - E_m I \textstyle\int_0^{L_b} \cos(\xi)\theta(\xi) d\xi \\ &= 0, \quad \forall\theta \in X. \end{aligned}$$

This means that ϕ satisfies the linearized boundary value problem

$$E_m I\phi''(\xi) + \lambda_0\phi(\xi) = 0, \qquad \phi'(0) = \phi'(L_b) = 0.$$

Thus, we can see that in this case $N = \ker L$ and $M = \text{range}\, L$. Observe that these results can also be derived from the fact that L is a self-adjoint linear operator.

Step 2. Split Algebraic Equations

We now proceed to transfer the decomposition of the ambient space X and Y into the solution of two algebraic equations, as is shown in Eq. (10.22). We start with Eq. (10.17a) as follows. Write $\phi(x) \in Y$ as

$$\phi(x) = (I - E)\,\phi(x) + E\,\phi(x), \tag{10.22}$$

where $(I - E)\,\phi(x) \in \ker L$ and $E\,\phi(x) \in \text{range } L$. We know what $\ker L$ is, so we can write

$$(I - E)\,\phi(x) = A_n \cos\left\{\frac{n\pi x}{L_b}\right\}.$$

Using the orthogonality properties of the sine function, we find

$$A_n = \frac{2}{L_b}\int_0^{L_b} \phi(s)\cos\left\{\frac{n\pi s}{L_b}\right\} ds.$$

Substituting into Eq. (10.22) and solving for $E\,\phi(x)$, we find the projection of Y onto range L to be

$$E\,\phi(x) = \phi(x) - \left[\frac{2}{L_b}\int_0^{L_b} \phi(s)\cos\left\{\frac{n\pi s}{L_b}\right\} ds\right]\cos\left\{\frac{n\pi x}{L_b}\right\}.$$

Solving $E\,\phi(x) = 0$ yields $\phi(x) = \cos\left\{\frac{n\pi s}{L_b}\right\}$.

Step 3. Write Reduced Model

We must now choose coordinates on $\ker L$ and $N = (\text{range } L)^{\perp}$ to obtain the reduced ordered model Eq. (10.18). Observe that the functions that make up both subspaces, $\ker L$ and N, satisfy the same linearized boundary value problem. Thus, we can choose the following coordinates:

$$v = v^* = \cos\left\{\frac{n\pi x}{L_b}\right\}.$$

Using Eq. (10.21), we compute the following Frétchet derivatives of Φ:

$$\Phi_\lambda(\theta, \lambda) = \sin\theta, \quad (d\Phi_\lambda)(v) = \frac{d}{dt}\sin(tv)\bigg|_{t=0} = \cos(tv)v|_{t=0} = v,$$

which leads to $\Phi_{\theta\lambda}(\theta, \lambda) = \cos\theta$. In addition,

$$\begin{aligned} d^2\Phi(-v_1, -v_2) &= \frac{d}{dt_1}\frac{d}{dt_2}\Phi(-t_1 v_1 - t_2 v_2, \lambda)\bigg|_{t=0} \\ &= -\frac{d}{dt_1}\frac{d}{dt_2}\Phi(t_1 v_1 + t_2 v_2, \lambda)\bigg|_{t=0} \\ &= -d^2\Phi(v_1, v_2). \end{aligned}$$

By direct differentiation, we get

$$\left.\frac{d}{dt_1}\frac{d}{dt_2}\Phi(-t_1v_1 - t_2v_2, \lambda)\right|_{t=0} = (-1)^2 d^2\Phi(v_1, v_2) = d^2\Phi(v_1, v_2).$$

Thus, $(d^2\Phi)(v_1, v_2) = 0$, which implies that $\Phi_{\theta\theta} = 0$. Also

$$\begin{aligned} d^3\Phi(v_1, v_2, v_3)\big|_{\theta=0} &= \left.\frac{d}{dt_1}\frac{d}{dt_2}\frac{d}{dt_3}\Phi(t_1v_1 + t_2v_2 + t_3v_3, \lambda_0)\right|_{t=0} \\ &= \frac{d}{dt_1}\frac{d}{dt_2}\frac{d}{dt_3}(t_1v_1'' + t_2v_2'' + t_3v_3'' + \lambda_0 \sin(t_1v_1 + t_2v_2 + t_3v_3)) \\ &= \lambda_0\frac{d}{dt_1}\frac{d}{dt_2}\cos(t_1v_1 + t_2v_2)v_3 = -\lambda_0 v_1v_2v_3, \end{aligned}$$

which implies that $\Phi_{\theta\theta\theta} = -\lambda_0$. Evaluating all derivatives at $(0, \lambda_0)$ we get

$$\Phi_\lambda = 0,\ \Phi_{\theta\theta} = 0,\ \Phi_{\theta\lambda} = 1,\ \Phi_{\theta\theta\theta} = -\lambda_0.$$

Then, direct computations yield

$$\begin{aligned} g_\lambda &= \langle v*, \Phi_\lambda(0, 0)\rangle = 0 \\ g_{xx} &= \langle v*, (d^2\Phi)_0(v, v)\rangle = 0 \\ g_{x\lambda} &= \langle v*, (d\Phi_\lambda)v - d^2\Phi(v, L^{-1}E\Phi\lambda)\rangle = \\ &= \langle v*, (d\Phi_\lambda)v)\rangle = \langle\cos\left\{\tfrac{n\pi x}{L_b}\right\}, \cos\left\{\tfrac{n\pi x}{L_b}\right\}\rangle = \frac{L_b}{2}, \\ g_{xxx} &= \langle v*, (d^3\Phi)(v, v, v) - 3d^2\Phi(v, L^{-1}Ed^2\Phi(v, v))\rangle = \\ &= \langle v*, (d^3\Phi)(v, v, v)\rangle = -\langle\cos\left\{\tfrac{n\pi x}{L_b}\right\}, \cos^3\left\{\tfrac{n\pi x}{L_b}\right\}\rangle = -\frac{3}{8}L_b\lambda_0. \end{aligned}$$

In summary, including the bifurcation problem, $g = g_x = 0$, we have found:

$$g = g_x = g_\lambda = g_{xx} = 0, \qquad g_{x\lambda} = \frac{1}{2}L_b,\quad g_{xxx} = -\frac{3}{8}L_b\lambda_0.$$

The first set of four zero-equalities are known as the *defining conditions* for the bifurcation problem. The last two nonzero-equalities are known as the *degenerate conditions* [4]. The normal form consistent with these conditions is that of a pitchfork bifurcation:

$$\dot{x} = g(x, \lambda) = \lambda x - x^3.$$

Example 10.5 (*Traveling Wave Solutions of PDE Model*) In this example, we study the bifurcations of traveling wave solutions in a PDE model through the Lyapunov-Schmidt reduction. The PDE model is reduced to an algebraic system of equations whose solutions are in one-to-one correspondence with the traveling-wave solutions. The PDE model is

$$\frac{\partial^2 u}{\partial t^2} + \frac{\partial^4 u}{\partial x^4} + \lambda\frac{\partial^2 u}{\partial x^2} + u + u^2\frac{\partial^2 u}{\partial x^2} + u\left(\frac{\partial u}{\partial x}\right)^2 = 0, \qquad (10.23)$$

where $u = u(x, t)$. Traveling wave solutions can be studied by employing a co-rotating frame of reference. That is, let

$$\tilde{u}(y) = u(x, t), \qquad y = x - ct,$$

Substituting into Eq. (10.23) (and re-labeling $\tilde{u}$ as u) we get

$$\frac{\partial^4 u}{\partial y^4} + \lambda \frac{\partial^2 u}{\partial y^2} + u + u^2 \frac{\partial^2 u}{\partial y^2} + u \left(\frac{\partial u}{\partial y} \right)^2 = 0, \tag{10.24}$$

with $\lambda = 1 + c^2$. In this new frame of reference, $u(y)$, is a periodic function of the form $u(y) = u(y + 2\pi)$.

Now, we define the space of four-times, real-valued, continuous and differentiable functions from the interval $[0, 2\pi]$ as the space for the solution $u(y)$. Formally, let

$$X = \left\{ u \in C^4([0, 2\pi]; \mathbb{R}) : u(0) = u(2\pi) = 0 \right\}.$$

Next, we need to define Φ and compute $L = (d\,\Phi)_{(0,0)}$. Let

$$Y = \left\{ u \in C^0([0, 2\pi]; \mathbb{R}) \right\}.$$

We can now define $\Phi : X \times \mathbb{R} \to Y$ through

$$\Phi(u(y), \lambda) = \frac{\partial^4 u}{\partial y^4} + \lambda \frac{\partial^2 u}{\partial y^2} + u + u^2 \frac{\partial^2 u}{\partial y^2} + u \left(\frac{\partial u}{\partial y} \right)^2. \tag{10.25}$$

Then the linearized boundary value problem yields

$$L\,u(y) = \Phi_u(0, \lambda_0) u(y) = \frac{\partial^4 u}{\partial y^4} + \lambda \frac{\partial^2 u}{\partial y^2} + u.$$

Solving $\Phi_u(0, \lambda_0) u(y) = 0$, for u, we get

$$u_n(y) = A_n \sin\left(ny\right) + B_n \cos\left(ny\right), \quad n = 1, 2, \ldots$$

The characteristic polynomial associated with this solution is

$$n^4 - \lambda n^2 + 1 = 0,$$

which indicates that a bifurcation occurs at the value $\lambda = 2$. This, in turn, yields $n = 1$. Thus, if we choose the modes

$$e_1 = A \sin y, \qquad e_2 = B \cos y,$$

where $A = B = \sqrt{2}$, then we can define

$$\ker L = \text{span}\{u_1, u_2\}.$$

Observe that in this case dim ker$L = 2$. The Lyapunov-Schmidt reduction is still applicable. We can now write

$$\begin{aligned} X &= \ker L \oplus M \\ Y &= \text{range}\, L \oplus N, \end{aligned}$$

where the subspaces M and N can be computed using the standard inner product, $\langle \cdot, \cdot \rangle$, which is defined by

$$\langle u, v \rangle = \frac{1}{2\pi} \int_0^{2\pi} u(x)\, v(x) dx.$$

We can then use this definition of inner product to find the orthogonal complement subspace M of ker L, and N of range L. Let $\xi = \frac{n\pi x}{L_b}$, $w \in M$ and $\phi \in N$

$$\begin{aligned} M &= \left\{ u \in X : \tfrac{1}{2\pi} \int_0^{2\pi} u(y)\big(\sin y + \cos y\big) = 0 \right\}, \\ N &= \{ u \in Y : u = A \sin y + B \cos y,\ \ A, B \in \mathbb{R} \}. \end{aligned}$$

In fact, integrating by parts, we can verify that $\langle A \sin y + B \cos y, L\, u \rangle = 0$. We leave this task as an exercise. The reduced equation can now be written as

$$g(x_1, x_2, \lambda) = (I - E)\Phi(v + w, \lambda), \tag{10.26}$$

where $v \in \ker L$ and $w \in M$. Letting $u = v + w$, we can rewrite Eq. (10.25) as

$$\begin{aligned} \Phi(v + w, \lambda) &= L(v + w) + (v + w)^2 (v + w)'' + (v + w)((v + w)')^2 \\ &= Lv + v^2 v'' + v(v')^2 + N(w), \end{aligned}$$

where $N(w)$ represents all the terms that contain w. The projection onto ker L yields

$$g(x_1, x_2, \lambda) = (I - E)\Phi(v + w, \lambda) = \sum_{i=1}^{2} \langle Lv + v^2 v'' + v(v')^2 + N(w), e_i \rangle e_i. \tag{10.27}$$

Since $v \in \ker L$, then we can write $v = x_1 e_1 + x_2 e_2$, where (x_1, x_2) are coordinates of v in ker L. Substituting into Eq. (10.27), and after tedious computations of inner products, which use the fact that $Le_1 = (1 - k^2)e_1$ and $Le_2 = (1 - k^2)e_2$, we arrive at

$$\begin{bmatrix} \dot{x}_1 \\ \dot{x}_2 \end{bmatrix} = g(x_1, x_2, \lambda) = \begin{bmatrix} (1 - c^2)x_1 - (x_1^2 + x_2^2)x_1 \\ (1 - c^2)x_2 - (x_1^2 + x_2^2)x_2. \end{bmatrix} \tag{10.28}$$

The reduced model Eq. (10.27) has five equilibrium points:

$$X_1 = (0, 0), \quad X_{2,3} = \left(0, \pm\sqrt{1-c^2}\right), \quad X_{4,5} = \left(\pm\sqrt{1-c^2}, 0\right).$$

It can be shown that the nontrivial equilibrium points correspond to the following traveling wave solutions:

$$u_{2,3} = \pm\sqrt{2(1-c^2)}\cos(x - ct), \quad u_{4,5} = \pm\sqrt{2(1-c^2)}\sin(x - ct)$$

10.3 Galerkin Projection

The Galerkin method is a convenient and straight-forward method to reduce a high-dimensional model into a low-dimensional one. The former model is usually in the form of an evolution equation or Partial Differential Equation (PDE), which serves to describe a system or phenomenon that changes in space and time. The latter model is usually in the form of a system of Ordinary Differential Equations (ODE), which describes only the temporal evolution of the space-time dynamics. To start, let's consider first the evolution model

$$\frac{\partial u(\mathbf{x}, t)}{\partial t} = F(u), \tag{10.29}$$

in which F is, in general, a nonlinear function that involves spatial derivatives. But since the method is rather general, F may also include integrals. Now, let's assume that the scalar functions $u(\mathbf{x}, t)$ have already been decomposed via the POD into Eq. (8.45). For completeness, we rewrite the decomposition

$$u(\mathbf{x}, t) = \sum_{k=1}^{\infty} a_k(t)\, \Phi_k(\mathbf{x}), \tag{10.30}$$

Computing the derivative of $u(\mathbf{x}, t)$ with respect to t yields

$$\frac{\partial u(\mathbf{x}, t)}{\partial t} = \sum_{k=1}^{\infty} \frac{da_k(t)}{dt} \Phi_k(\mathbf{x}). \tag{10.31}$$

We rewrite Eq. (10.29) in the following form

$$\frac{\partial u(\mathbf{x}, t)}{\partial t} - F(u) = 0.$$

Then we can project the PDE onto each ith POD mode by computing the inner product $(\cdot, \cdot)$. That is,

$$\left(\frac{\partial u(\mathbf{x},t)}{\partial t} - F(u),\ \Phi_i\right) = 0. \tag{10.32}$$

Substituting Eq. (10.31) into Eq. (10.32) we get

$$\left(\frac{da_k(t)}{dt}\Phi_k(\mathbf{x}) - F(u),\ \Phi_i\right) = \sum_{k=1}^{N}\frac{da_k(t)}{dt}\,(\Phi_k(\mathbf{x}), \Phi_i(\mathbf{x})) - (F(u), \Phi_i(\mathbf{x})) = 0.$$

Since the POD modes $\Phi_{i's}$ are orthogonal (actually we can assume they are orthonormal), the only nonvanishing terms in the summation above are those where $k = i$. Thus, we get

$$\frac{da_i(t)}{dt} = \left(F\left(\sum_{k=1}^{N} a_k(t)\,\Phi_k(\mathbf{x})\right),\ \Phi_i(\mathbf{x})\right), \quad i = 1, \ldots, N. \tag{10.33}$$

Equation (10.33) is the desired reduced-order model of our original evolution equation. It is low-dimensional because it contains only one independent variable, time t. Since the POD modes $\Phi_{j's}$ do not change over time, the reduced-order model can be thought as the time-evolution of the amplitudes $a_k(t)$ in the POD decomposition, Eq. (8.45), of the scalar field $u(\mathbf{x}, t)$.

Example 10.6 (*Heat Transfer Model*) A one-dimensional model for heat transfer along a rod of length $l = 1$ is given by

$$\frac{\partial u}{\partial t} = -\nu\frac{\partial u}{\partial x} + \frac{1}{P_e}\frac{\partial^2 u}{\partial x^2}, \tag{10.34}$$

with initial and boundary conditions

$$u(x, 0) = u_0(x), \quad \frac{\partial u}{\partial x}u(0, t) = \frac{\partial u}{\partial x}u(1, t).$$

In Eq. (10.34), the parameter P_e represents the Peclet number. Let us assume that we have already POD decomposed $u(x, t)$ as is shown in Eq. (10.30). Then direct substitution into Eq. (10.33) leads to

$$
\begin{aligned}
\frac{da_i}{dt} &= \left(-\nu \sum_{k=1}^{n} a_k \frac{\partial \Phi_k}{\partial x} + \frac{1}{P_e} \sum_{k=1}^{n} a_k \frac{\partial^2 \Phi_k}{\partial x^2}, \Phi_i\right) \\
&= \sum_{k=1}^{n} \left[-\nu \left(\frac{\partial \Phi_k}{\partial x}, \Phi_i\right) + \frac{1}{P_e}\left(\frac{\partial^2 \Phi_k}{\partial x^2}, \Phi_i\right)\right] a_k.
\end{aligned}
$$

If we let $X = [a_1, a_2, \ldots, a_n]^T$, then we can write the reduced order model in matrix notation

$$
\frac{dX}{dt} = AX, \qquad X(0) = X_0, \tag{10.35}
$$

where

$$
A_{ij} = -\nu \left(\frac{\partial \Phi_j}{\partial x}, \Phi_i\right) + \frac{1}{P_e}\left(\frac{\partial^2 \Phi_j}{\partial x^2}, \Phi_i\right), \quad X_{0_i} = (u_0, \Phi_i), \; i = 1, \ldots, N.
$$

Example 10.7 (*Flame Dynamics.*) Consider the spatio-temporal model of flame instability described by the Kuramoto-Sivashinsky Eq. (8.33). For completeness purposes, we show the model equation again

$$
\boxed{\frac{\partial u}{\partial t} = \eta_1 u - (1 + \nabla^2)^2 u - \eta_2 (\nabla u)^2 - \eta_3 u^3.} \tag{10.36}
$$

Let's assume there is no noise, i.e., $\xi(\mathbf{x}, t) = 0$. Assume also that the POD decomposition of the scalar field $u(\mathbf{x}, t)$ has been performed, so that the spatial modes Ψ_k in Eq. (10.30) are readily available. Now we seek to find a low-dimensional model for the evolution of the corresponding amplitude coefficients $a_k(t)$. Then we can rewrite the right-hand side of Eq. (10.36) in terms of the POD expansion as follows

$$
\begin{aligned}
F(u) = (\eta_1 - 1) \sum_{k=1} a_k \Phi_k - 2 \sum_{k=1} a_k \nabla^2 \Phi_k - \sum_{k=1} a_k \nabla^4 \Phi_k - \\
\eta_2 \left(\sum_{k=1} a_k \nabla \Phi_k\right)^2 - \eta_3 \left(\sum_{k=1} a_k \Phi_k\right)^3.
\end{aligned}
$$

Computing the inner product of $F(u)$ with each individual kth POD mode we get

$$
\begin{aligned}
(F(u), \Phi_i) = (\eta_1 - 1) \sum_{k=1} a_k (\Phi_k, \Phi_i) - 2 \sum_{k=1} a_k \left(\nabla^2 \Phi_k, \Phi_i\right) - \sum_{k=1} a_k \left(\nabla^4 \Phi_k, \Phi_i\right) - \\
\eta_2 \left(\left(\sum_{k=1} a_k \nabla \Phi_k\right)^2, \Phi_i\right) - \eta_3 \left(\left(\sum_{k=1} a_k \Phi_k\right)^3, \Phi_i\right).
\end{aligned}
$$

Simplifying, we arrive at the reduced order model

$$
\begin{aligned}
\frac{da_i}{dt} &= (\eta_1 - 1)a_i - 2\sum_{k=1} \left(\nabla^2 \Phi_k, \Phi_i\right) a_i - \sum_{k=1} \left(\nabla^4 \Phi_k, \Phi_i\right) a_i - \\
&\eta_2 \left(\left(\sum_{k=1} a_k \nabla \Phi_k \right)^2, \Phi_i \right) - \eta_3 \left(\left(\sum_{k=1} a_k \Phi_k \right)^3, \Phi_i \right).
\end{aligned}
\tag{10.37}
$$

For the special case of a 1D domain, i.e., $0 < x < l$, the following POD modes can be used as a basis

$$
\Phi_{k=1}^N = \left\{ \sqrt{2}\cos(kx),\ \sqrt{2}\cos(kx),\ k = 1, \ldots, k_{\max} \right\}
$$

with an inner product

$$
(u(x)\, v(x)) = \frac{1}{2\pi} \int_0^{2\pi} u(x)\, v(x) dx.
$$

Observe that truncating at wave number $k_{\max}$ leads to $N = 2k_{\max}$ POD modes. Using the fact that $\nabla^2 \Phi_k = -k^2 \Phi_k$ and $\nabla^4 \Phi_k = \Phi_k$, direct calculations of inner products, yield the following coefficients for linear terms:

$$
\begin{aligned}
&\left(\nabla^2 \Phi_1, \Phi_1\right) = -k^2,\ \left(\nabla^2 \Phi_2, \Phi_1\right) = 0,\ \left(\nabla^2 \Phi_1, \Phi_2\right) = 0,\ \left(\nabla^2 \Phi_2, \Phi_2\right) = -k^2, \\
&\left(\nabla^4 \Phi_1, \Phi_1\right) = k^4,\ \left(\nabla^4 \Phi_2, \Phi_1\right) = 0,\ \left(\nabla^4 \Phi_2, \Phi_2\right) = k^4,\ \left(\nabla^4 \Phi_1, \Phi_2\right) = 0.
\end{aligned}
$$

For the coefficients of the quadratic terms, we get

$$
\begin{aligned}
&\left((\nabla \Phi_1)^2, \Phi_1\right) = 0,\ (\nabla \Phi_1 \nabla \Phi_2, \Phi_1) = 0,\ \left((\nabla \Phi_2)^2, \Phi_1\right) = 0, \\
&\left((\nabla \Phi_1)^2, \Phi_2\right) = 0,\ (\nabla \Phi_1 \nabla \Phi_2, \Phi_2) = 0,\ \left((\nabla \Phi_2)^2, \Phi_2\right) = 0,
\end{aligned}
$$

Thus, all quadratic terms vanish. For the special case of $k = 1$, the coefficients of the cubic terms lead to

$$
\begin{aligned}
&\left(\Phi_1^3, \Phi_1\right) = \frac{3\pi}{2},\ \left(\Phi_1^2 \Phi_2, \Phi_1\right) = 0,\ \left(\Phi_1 \Phi_2^2, \Phi_1\right) = \frac{\pi}{2},\ \left(\Phi_2^3, \Phi_1\right) = 0, \\
&\left(\Phi_1^3, \Phi_2\right) = 0,\ \left(\Phi_1^2 \Phi_2, \Phi_2\right) = \frac{\pi}{2},\ \left(\Phi_1 \Phi_2^2, \Phi_2\right) = 0,\ \left(\Phi_2^3, \Phi_2\right) = \frac{3\pi}{2}.
\end{aligned}
$$

Substituting into Eq. (10.38) we arrive at the desired reduced order model

$$
\begin{aligned}
\dot{a}_1 &= \eta_1 a_1 - \frac{3\pi}{2} \eta_3 a_1 \left(a_1^2 + a_2^2\right) \\
\dot{a}_2 &= \eta_1 a_2 - \frac{3\pi}{2} \eta_3 a_2 \left(a_1^2 + a_2^2\right).
\end{aligned}
\tag{10.38}
$$

Similar calculations apply for higher-order modes. They are left as an exercise.

10.4 Normal Forms

In this section we focus our attention on the process of simplifying a mathematical model. There are two types of simplifications that can be performed on a given model. One deals with simplifying the number of parameters to a minimum. This type of simplification can be achieved by the process of writing the relevant equations in dimensionless form, and it was already discussed earlier on Chap. 2. The second type of simplification deals with eliminating terms that are negligible or that do not influence the dynamics nor the type of bifurcation that occurs in the model.

10.4.1 Hopf Bifurcation

To illustrate the process, we consider the case of a Hopf bifurcation near a zero equilibrium [5] point $x = 0$, where $x = (x_1, x_2)^T$. Assume the model is written as

$$\frac{dx}{dt} = L(\lambda)x + N(x, \lambda), \tag{10.39}$$

where L is the linear part, i.e., a constant 2×2 matrix that depends on the bifurcation parameter λ, of the form

$$L(\lambda) = \begin{bmatrix} l_{11}(\lambda) & l_{12}(\lambda) \\ l_{21}(\lambda) & l_{22}(\lambda) \end{bmatrix}.$$

$N(x, \lambda)$ is a smooth vector-valued function that contains nonlinear terms. Near $x = 0$ the matrix $L(\lambda)$ represents the Jacobian matrix whose eigenvalues, $\sigma_{1,2}$, are the roots of the characteristic polynomial

$$\sigma^2 - \text{Tr}(L)\sigma + \det(L) = 0,$$

where $\text{Tr}(L)(\lambda) = l_{11}(\lambda) + l_{22}(\lambda)$ and $\det(L)(\lambda) = l_{11}(\lambda)l_{22}(\lambda) - l_{12}(\lambda)l_{21}(\lambda)$.

Direct computation yields

$$\sigma_{1,2} = \frac{1}{2}\text{Tr}(L) \pm \frac{1}{2}\sqrt{\text{Tr}^2(L) - 4\det(L)}.$$

Recall from Chap. 5 that the condition for a Hopf bifurcation requires

$$\text{Tr}(L)(0) = 0, \quad \text{and} \quad \det(L)(0) > 0.$$

Thus, if we set $\det(L)(0) = \omega_0^2 > 0$ and define

$$\tau(\lambda) = \frac{1}{2}\mathrm{Tr}(L)(\lambda), \quad \omega(\lambda) = \frac{1}{2}\sqrt{4\det(L) - \mathrm{Tr}^2(L)},$$

then we can assume the eigenvalues of L to be of the form

$$\sigma(\lambda) = \tau(\lambda) \pm \omega(\lambda)i, \quad \tau(0) = 0, \quad \omega(0) = \omega_0 > 0.$$

Claim The matrix $L(\lambda)$ in the model Eq. (10.39) has the canonical real Jordan form

$$J(\lambda) = \begin{bmatrix} \tau(\lambda) & -\omega(\lambda) \\ \omega(\lambda) & \tau(\lambda) \end{bmatrix}. \tag{10.40}$$

Proof Let $q = v_1 \mp v_2 i$ be the eigenvectors associated with the eigenvalues $\sigma(\lambda)$, and let

$$P = [v_1 | \, v_2]$$

be the matrix whose columns are the real and imaginary parts (vectors) of q. It is known [5] that v_1 and v_2 are linearly independent, then it follows that P is invertible and

$$Pe_1 = v_1, \;\; Pe_2 = v_2, \quad \text{and} \quad P^{-1}v_1 = e_1, \;\; P^{-1}v_2 = e_2.$$

Since q is an eigenvector of the matrix L then

$$L(\lambda)q(\lambda) = \sigma(\lambda)q(\lambda).$$

Consider $\sigma(\lambda) = \tau(\lambda) - \omega(\lambda)i$, for which $q(\lambda) = v_1 + v_2 i$. Then direct computations show

$$L(v_1 + v_2 i) = (\tau - \omega i)(v_1 + v_2 i) = \tau v_1 + \omega v_2 + (\tau v_2 - \omega v_1)i,$$

so that $Lv_1 = \tau v_1 + \omega v_2$ and $Lv_2 = \tau v_2 - \omega v_1$. These results lead to

$$\begin{aligned} P^{-1}LPe_1 &= P^{-1}Lv_1 = \tau P^{-1}v_1 + \omega P^{-1}v_2 = \tau e_1 + \omega e_2, \\ P^{-1}LPe_2 &= P^{-1}Lv_2 = \tau P^{-1}v_2 - \omega P^{-1}v_1 = -\omega e_1 + \tau e_2. \end{aligned}$$

Observe that $P^{-1}LPe_1$ and $P^{-1}LPe_2$ are the first and second column of $P^{-1}LP$, respectively. Consequently, the matrix $P^{-1}LP$ takes the form

$$J(\lambda) = P^{-1}LP = \begin{bmatrix} \tau(\lambda) & -\omega(\lambda) \\ \omega(\lambda) & \tau(\lambda) \end{bmatrix}.$$

□

Let $y = Px$, where $y = (y_1, y_2)$. Substituting into Eq. (10.39) leads to

$$\frac{dy}{dt} = J(\lambda)y + \tilde{N}(y, \lambda), \tag{10.41}$$

where $\tilde{N} = P^{-1}PN(Py, \lambda)$. Next, we complexify by introducing $z = y_1 + y_2 i$, so that Eq. (10.41) can be rewritten as

$$\frac{dz}{dt} = \sigma(\lambda)z + g(z, \bar{z}, \lambda), \tag{10.42}$$

where $g = O(|z|^2)$ is a smooth function of $(z, \bar{z}, \lambda)$.

The process of reducing Eq. (10.42) (or any other model equation for that matter) into normal form consists of eliminating, as much as possible, nonlinear terms, through successive near identity transformations of the form

$$z = w + h(w, \bar{w}), \quad \text{where } h = O(|w|^2). \tag{10.43}$$

Quadratic Terms
We start with quadratic terms, and rewrite Eq. (10.42) as a Taylor series up to order two in z and $\bar{z}$, through

$$\frac{dz}{dt} = \sigma(\lambda)z + \frac{g_{20}}{2}z^2 + g_{11}z\bar{z} + \frac{g_{02}}{2}\bar{z}^2 + O(|z|^3), \tag{10.44}$$

where $\sigma = \sigma(\lambda) = \tau(\lambda) \pm \omega(\lambda)i$, $\tau(0) = 0$, $\omega(0) = \omega_0 > 0$, and $g_{ij} = g_{ij}(\lambda)$. We write the near-identity transformation (10.43) as a Taylor series up to order two:

$$z = w + \frac{h_{20}}{2}w^2 + h_{11}w\bar{w} + \frac{h_{02}}{2}\bar{w}^2 + O(|w|^3). \tag{10.45}$$

The inverse transformation can also be written as a Taylor series of the form

$$w = z - \frac{h_{20}}{2}z^2 - h_{11}z\bar{z} - \frac{h_{02}}{2}\bar{z}^2 + O(|z|^3).$$

Then, direct computations yield

$$\begin{aligned}
\dot{w} &= \dot{z} - h_{20}z\dot{z} - h_{11}(\dot{z}\bar{z} + z\dot{\bar{z}}) - h_{02}\bar{z}\dot{\bar{z}} + \cdots \\
&= \sigma z + \left(\frac{g_{20}}{2} - \sigma h_{20}\right) + (g_{11} - \sigma h_{11} - \bar{\sigma}h_{11}z\bar{z} + \left(\frac{g_{02}}{2} - \bar{\sigma}h_{02}\right) + \cdots \\
&= \sigma w + \frac{1}{2}(g_{20} - \sigma h_{20})w^2 + (g_{11} - \bar{\sigma}h_{11})w\bar{w} + \\
&\quad \tfrac{1}{2}(g_{02} - (2\bar{\sigma} - \sigma)h_{02})\bar{w}^2 + O(|w|^3).
\end{aligned}$$

It follows that all quadratic terms can be eliminated by setting

$$h_{20} = \frac{g_{20}}{\sigma}, \ h_{11} = \frac{g_{11}}{\bar{\sigma}}, \ h_{02} = \frac{g_{02}}{2\bar{\sigma} - \sigma}.$$

Cubic Terms

We can now rewrite Eq. (10.42) as a Taylor series up to order three

$$\frac{dz}{dt} = \sigma(\lambda)z + \frac{g_{30}}{6}z^3 + \frac{g_{21}}{2}z^2\bar{z} + \frac{g_{12}}{2}z\bar{z}^2 + \frac{g_{03}}{6}\bar{z}^3 + O(|z|^4). \tag{10.46}$$

We write the near-identity transformation (10.43) as a Taylor series up to order three:

$$z = w + \frac{h_{30}}{6}w^3 + \frac{h_{21}}{2}w^2\bar{w} + \frac{h_{12}}{2}w\bar{w}^2 + \frac{h_{03}}{6}\bar{w}^3 + O(|w|^4). \tag{10.47}$$

The inverse transformation can also be written as a Taylor series of the form

$$w = z - \frac{h_{30}}{6}z^3 - \frac{h_{21}}{2}z^2\bar{z} - \frac{h_{12}}{2}z\bar{z}^2 - \frac{h_{03}}{6}\bar{z}^3 + O(|z|^4).$$

Then, direct computations yield

$$\begin{aligned}
\dot{w} &= \dot{z} - \frac{h_{30}}{2}z^2\dot{z} - \frac{h_{21}}{2}(2z\bar{z}\dot{z} + z^2\dot{\bar{z}}) - \frac{h_{12}}{2}(\dot{z}\bar{z}^2 + 2z\bar{z}\dot{\bar{z}}) - \frac{h_{03}}{2}\bar{z}^2\dot{\bar{z}} + \cdots \\
&= \sigma z + \left(\frac{g_{30}}{6} - \frac{\sigma h_{30}}{2}\right)z^3 + \left(\frac{g_{21}}{2} - \sigma h_{21} - \frac{\bar{\sigma}h_{21}}{2}\right)z^2\bar{z} + \\
&\quad \left(\frac{g_{12}}{2} - \sigma h_{12} - \frac{\bar{\sigma}h_{12}}{2}\right)z\bar{z}^2 + \left(\frac{g_{03}}{6} - \frac{\sigma h_{03}}{2}\right)\bar{z}^3 + \cdots \\
&= \sigma w + \frac{1}{6}(g_{30} - 2\sigma h_{30})w^3 + \frac{1}{2}(g_{21} - (\sigma + \bar{\sigma})h_{21})w^2\bar{w} + \\
&\quad \frac{1}{2}(g_{12} - 2\bar{\sigma}h_{12})w\bar{w}^2 + \frac{1}{6}(g_{03} + (\sigma - 3\bar{\sigma})\sigma h_{03})\bar{w}^3 + O(|w|^4).
\end{aligned}$$

It follows that all cubic terms, except for the term containing $w^2\bar{w}$, can be eliminated by setting

$$h_{30} = \frac{g_{30}}{2\sigma}, \; h_{12} = \frac{g_{12}}{2\bar{\sigma}}, \; h_{03} = \frac{g_{03}}{3\bar{\sigma} - \sigma}.$$

The reason why the term $w^2\bar{w}$ cannot be eliminated is because it would require setting

$$h_{21} = \frac{g_{21}}{\sigma + \bar{\sigma}},$$

which might be possible for small $\lambda \neq 0$ but not for $\lambda = 0$, at which point the denominator, $\sigma + \bar{\sigma}$, vanishes.

Consequently, the normal form equations for the Hopf bifurcation can be written, up to order three, as

$$\dot{w} = \sigma(\lambda)w + c_1 w^2\bar{w} + O(|w|^4). \tag{10.48}$$

We can rewrite this last equation in the more familiar normal form for a Hopf bifurcation

$$\dot{z} = \left(\sigma(\lambda) + c_1|z|^2\right) z \tag{10.49}$$

Stability

The stability of emerging oscillations is determined by the coefficient c_1, which can be computed as follows. Differentiate the near identity transformation Eq. (10.45) to get

$$\dot{z} = \dot{w} + h_{20} w\dot{w} + h_{11}(w\dot{\bar{w}} + \bar{w}\dot{w}) + h_{02}\bar{w}\dot{\bar{w}}.$$

Substituting Eq. (10.48), and the conjugate version for $\dot{w}$, into this last expression we get (up to order three) the following expression

$$\dot{z} = \sigma w + h_{20}\sigma w^2 + h_{11}(\sigma + \bar{\sigma}) w\bar{w} + h_{02}\bar{\sigma}\bar{w}^2. \tag{10.50}$$

Now, we consider the original equation, which can be written as a Taylor series in z and $\bar{z}$ as

$$\dot{z} = \sigma z + \sum_{2\le j+k\le 3} \frac{1}{j!k!} g_{jk} z^j \bar{z}^k + O(|z|^4). \tag{10.51}$$

Substituting Eq. (10.45) into Eq. (10.51) and expanding up to order three, we get

$$\begin{aligned} \dot{z} &= \sigma w + \tfrac{1}{2}(h_{20}\sigma + g_20)w^2 + \tfrac{1}{2}(h_02\sigma + g_{02})\bar{w}^2 + (h_{11}\sigma + g_{11})w\bar{w} + \\ &\tfrac{1}{6}(3g_{11}\bar{h}_{02} + 3g_{20}h_{20} + g_{30})w^3 + \tfrac{1}{6}(3g_{02}\bar{h}_{20} + 3g_{11}h_{02} + g_{03})\bar{w}^3 + \\ &\tfrac{1}{2}(2g_{02}\bar{h}_{11} + 2g_{11}h_{11} + g_{11}\bar{h}_{20} + g_{20}h_{02} + g_{12})w\bar{w}^2 + \\ &\tfrac{1}{2}(g_{02}\bar{h}_{02} + g_{11}h_{20} + 2g_{11}\bar{h}_{11} + 2g_{20}h_{11} + g_{21})w^2\bar{w}. \end{aligned} \tag{10.52}$$

Comparing the coefficients of the quadratic terms of Eq. (10.50) with those of Eq. (10.52) yields (as expected) the previously found formulas for h_{20}, h_{11}, and h_{02}. Similarly, the coefficient in front of the term $w^2\bar{w}$ corresponds to c_1:

$$c_1 = \frac{1}{2}(g_{02}\bar{h}_{02} + g_{11}h_{20} + 2g_{11}\bar{h}_{11} + 2g_{20}h_{11} + g_{21}).$$

Substituting the formulas for h_{20}, h_{11}, and h_{02} yields the stability condition in terms of the Taylor coefficients g_{ij}, as

$$c_1 = \frac{g_{20}g_{11}(2\sigma + \bar{\sigma})}{2|\sigma|^2} + \frac{|g_{11}|^2}{\bar{\sigma}} + \frac{|g_{02}|^2}{2(2\sigma - \bar{\sigma})} + \frac{1}{2}g_{21}. \tag{10.53}$$

At the bifurcation point $\lambda = 0$, the formula for the coefficient c_1 reduces to

$$c_1(0) = \frac{i}{2w_0^2}\left(g_{20}g_{11} - 2|g_{11}|^2 - \frac{1}{3}|g_{02}|^2\right) + \frac{g_{21}}{2}. \tag{10.54}$$

10.4.2 General Method

Let us now formalize the process of simplifying a model into its normal form. Consider a generic version of a mathematical model written as

$$\dot{x} = f(x), \quad x \in \mathbb{R}^n, \quad f(0) = 0. \tag{10.55}$$

Consider the change of coordinates

$$x = \varphi(y), \quad \varphi(0) = 0,$$

and assume $(d\varphi)_0$ is invertible. By the chain rule $\dot{x} = (d\varphi)_y\dot{y} = f(\varphi(y))$, so that the original model Eq. (10.55) can be rewritten in the new coordinates y as

$$\dot{y} = g(y) = (d\varphi)_y^{-1} f(\varphi(y)). \tag{10.56}$$

The general idea of writing a model in normal form is to make Eq. (10.56) as easy to deal with as possible. To do that, let

$$\dot{x} = Lx + f_2(x) + \cdots + f_{k-1}(x) + h(x), \tag{10.57}$$

where $L(x)$ represents the linear terms, while each function $f_k(x)$ represents terms of order k. To make Eq. (10.56) as easy as possible, we need to find changes of coordinates that will leave the terms through order $k-1$ fixed and simplify terms at order k. This process can be accomplished by using near identity changes of coordinates

$$\varphi(y) = y + \varphi_k(y),$$

where φ_k is a homogeneous polynomial of degree k, and $(d\varphi_k)_0 = I_n$, which implies invertibility:

$$(d\varphi)_y = I + (d\varphi_k)_y \quad \Longleftrightarrow \quad (d\varphi)_y^{-1} = I - (d\varphi_k)_y + \cdots + \mathcal{O}(|y|^k).$$

The invertibility equations can be verified directly by computing:

$$\big(I + (d\varphi_k)_y\big)\big(I - (d\varphi_k)_y\big) = I + (d\varphi_k)_y - (d\varphi_k)_y + (d\varphi_k)_y^2 = \mathcal{O}(|y|^k).$$

Substituting $(d\varphi)_y^{-1} = I - (d\varphi_k)_y$ into Eq. (10.56) we obtain,

$$\dot{y} = \left[I - (d\varphi_k)_y\right]\left[L(y + \varphi_k(y)) + f_2(y + \varphi_k(y)) + \cdots + f_{k-1}(y + \varphi_k(y)) + h(y + \varphi_k(y))\right].$$

Modulo terms of order $k + 1$, represented by M^{k+1}, we get

$$f_i(y + \varphi_k(y)) = f_i(y) + M^{k+1}, \quad i = 1, \ldots, k - 1.$$

Then,

$$\begin{aligned}\dot{y} &= \left[I - (d\varphi_k)_y\right]\left[L\,y + f_2(y) + \cdots + f_{k-1}(y) + h(y) + L\varphi_k(y)\right],\\ \dot{y} &= L\,y + f_2(y) + \cdots + f_{k-1}(y) + h(y) + L\varphi_k(y) - (d\varphi_k)_y L y.\end{aligned}$$

Now, let us introduce the following adjoint map.

Definition 10.2 (*Adjoint Map*) The adjoint map, $\mathrm{ad_L} : p_k \rightarrow p_k$, defined by

$$\mathrm{ad_L}(p(y)) = Lp(y) - (dp)_y L\,y,$$

is linear in p.

We have them arrived at the following theorem.

Theorem 10.4 *Using successive near identity changes of coordinates, through order k, we can assume that a general mathematical model*

$$\dot{x} = L\,x + f_2(x) + \cdots + f_{k-1}(x) + \cdots,$$

can be written in the form

$$\dot{y} = L\,y + g_2(y) + \cdots + g_k(y) + \cdots,$$

where $g_j(y) \in \mathcal{J}_j$ and $p_j = Im\ ad_L \oplus \mathcal{J}_j$.

The main idea is that all terms in Im $\mathrm{ad_L}$ can be discarded, while the terms in $\mathcal{J}_j$ are the ones that make up the normal form. Let us now revisit the case of a Hopf bifurcation.

Example 10.8 (*Hopf Bifurcation Revisited*) Our previous discussion has shown that all that is needed to apply the general methodology of rewriting a mathematical model in normal form is the linear part of the model. Thus, for this example assume

$$L = \begin{bmatrix} 0 & -1 \\ 1 & 0 \end{bmatrix}.$$

Since the eigenvalues are $\pm i$, it is reasonable to expect a Hopf bifurcation. Next, let

$$p(x) = (p_1(x_1, x_2),\ p_2(x_1, x_2)),$$

and proceed to compute

$$\text{ad}_{\text{L}} p = Lp(x) - (dp)_x L x.$$

Direct computations show

$$L p = (-p_2, p_1), \qquad (dp) = \begin{bmatrix} p_{1,x_1} & p_{1,x_2} \\ p_{2,x_1} & p_{2,x_2} \end{bmatrix}.$$

Then

$$\text{ad}_{\text{L}} p = \begin{bmatrix} -p_2 + p_{1,x_1} x_2 - p_{1,x_2} x_1 \\ p_1 + p_{2,x_1} x_2 - p_{2,x_2} x_1 \end{bmatrix}. \tag{10.58}$$

Quadratic Terms

We now seek to find out which quadratic terms can be eliminated. Thus, let

$$\begin{aligned} p_1 &= a_{11} x_1^2 + a_{12} x_1 x_2 + a_{22} x_2^2 \\ p_2 &= b_{11} x_1^2 + b_{12} x_1 x_2 + b_{22} x_2^2. \end{aligned}$$

Substituting into Eq. (10.58) we get

$$\text{ad}_{\text{L}} p = \begin{bmatrix} -(a_{12} + b_{11}) x_1^2 + (2a_{11} - 2a_{22} - b_{12}) x_1 x_2 + (a_{12} - b_{22}) x_2^2 \\ (a_{11} - b_{12}) x_1^2 + (a_{12} + 2b_{11} - 2b_{22}) x_1 x_2 + (a_{22} + b_{12}) x_2^2 \end{bmatrix}.$$

In coordinates, $[a_{11}, a_{12}, a_{22}, b_{11}, b_{12}, b_{22}]$, we can write

$$\text{ad}_{\text{L}} \begin{bmatrix} a_{11} \\ a_{12} \\ a_{22} \\ b_{11} \\ b_{12} \\ b_{22} \end{bmatrix} = T \begin{bmatrix} a_{11} \\ a_{12} \\ a_{22} \\ b_{11} \\ b_{12} \\ b_{22} \end{bmatrix} = \begin{bmatrix} 0 & 2 & 0 & 1 & 0 & 0 \\ -1 & 0 & 1 & 0 & 1 & 0 \\ 0 & -2 & 0 & 0 & 0 & 1 \\ -1 & 0 & 0 & 0 & 2 & 0 \\ 0 & -1 & 0 & -1 & 0 & 1 \\ 0 & 0 & -1 & 0 & -2 & 0 \end{bmatrix} \begin{bmatrix} a_{11} \\ a_{12} \\ a_{22} \\ b_{11} \\ b_{12} \\ b_{22} \end{bmatrix}$$

Direct computations show that rank $T = 6$. This means that T is nonsingular, so T is invertible, so that

$$\text{Im ad}_{\text{L}}|_{\mathcal{P}_2} = \mathcal{P}_2.$$

Consequently, the complement space is $\mathcal{J}_2 = \{0\}$. It follows that all quadratic terms can be eliminated through successive near-identity transformations.

Cubic Terms

We now seek to find out which cubic terms can be eliminated. Thus, let

$$\begin{aligned} p_1 &= a_{111} x_1^3 + a_{112} x_1^2 x_2 + a_{122} x_1 x_2^2 + a_{222} x_2^3 \\ p_2 &= b_{111} x_1^3 + b_{112} x_1^2 x_2 + b_{122} x_1 x_2^2 + b_{222} x_2^3. \end{aligned}$$

Direct computations lead to the matrix representation for ad_L, which we write as follows:

$$\text{ad}_\text{L}\begin{bmatrix} a_{111} \\ a_{112} \\ a_{122} \\ a_{222} \\ b_{111} \\ b_{112} \\ b_{122} \\ b_{222} \end{bmatrix} = T\begin{bmatrix} a_{111} \\ a_{112} \\ a_{122} \\ a_{222} \\ b_{111} \\ b_{112} \\ b_{122} \\ b_{222} \end{bmatrix} = \begin{bmatrix} 0 & 3 & 0 & 0 & 1 & 0 & 0 & 0 \\ -1 & 0 & 2 & 0 & 0 & 1 & 0 & 0 \\ 0 & -2 & 0 & 1 & 0 & 0 & 1 & 0 \\ 0 & 0 & -3 & 0 & 0 & 0 & 0 & 1 \\ -1 & 0 & 0 & 0 & 0 & 3 & 0 & 0 \\ 0 & -1 & 0 & 0 & -1 & 0 & 2 & 0 \\ 0 & 0 & -1 & 0 & 0 & -2 & 0 & 1 \\ 0 & 0 & 0 & -1 & 0 & 0 & -3 & 0 \end{bmatrix}\begin{bmatrix} a_{111} \\ a_{112} \\ a_{122} \\ a_{222} \\ b_{111} \\ b_{112} \\ b_{122} \\ b_{222} \end{bmatrix}$$

This time we find rank $T = 6$. This implies that $\text{ad}_\text{L}(p)$ has a complement space which is two-dimensional. To find this complement, we could try to find two vectors, with eight components each, that are linearly independent and orthogonal to each column of the matrix T. But this is the same as computing two linearly independent left-eigenvectors of the zero eigenvalue. That is, we need to compute v such that $v^T T = 0$. Direct computations yield

$$\begin{aligned} v_1 &= [1\ 0\ 1\ 0\ 0\ 1\ 0\ 1]^T \\ v_2 &= [0\ -1\ 0\ -1\ 1\ 0\ 1\ 0]^T \end{aligned}.$$

Interpreting the components of these eigenvectors as the coordinates of cubic terms, we find the complement space $\mathcal{J}_2$ to be of the form

$$\begin{bmatrix} x_1^3 + x_1 x_2^2 \\ x_1^2 x_2 + x_2^3 \end{bmatrix}, \qquad \begin{bmatrix} -x_1^2 x_2 - x_2^3 \\ x_1^3 + x_1 x_2^2 \end{bmatrix}.$$

We can then write the complement space as

$$\mathcal{J}_2 = \left\{ p(x_1^2 + x_2^2)\begin{bmatrix} x_1 \\ x_2 \end{bmatrix} + q(x_1^2 + x_2^2)\begin{bmatrix} -x_2 \\ x_1 \end{bmatrix} \right\}. \tag{10.59}$$

In fact, the previous result can be generalized to include higher order terms, which lead to all even terms being eliminated from the normal form, while the odd terms can be written as

$$\mathcal{P}_{2k+1} = \text{ad}_\text{L}(\mathcal{P}_{2k+1}) \oplus \left\{ (x_1^2 + x_2^2)^k \begin{bmatrix} x_1 \\ x_2 \end{bmatrix} + (x_1^2 + x_2^2)^k \begin{bmatrix} -x_2 \\ x_1 \end{bmatrix} \right\}.$$

Consequently, by near identity changes of coordinates, we can assume the normal form of a Hopf bifurcation to be

$$\frac{d}{dt}\begin{bmatrix} x_1 \\ x_2 \end{bmatrix} = \begin{bmatrix} 0 & -1 \\ 1 & 0 \end{bmatrix}\begin{bmatrix} x_1 \\ x_2 \end{bmatrix} + \sum_{j=1}^{k} \left\{ a_j (x_1^2 + x_2^2)^j \begin{bmatrix} x_1 \\ x_2 \end{bmatrix} + b_j (x_1^2 + x_2^2)^j \begin{bmatrix} -x_2 \\ x_1 \end{bmatrix} \right\}$$
$$+\mathcal{O}(|x|^{2k+2}).$$

10.5 Exercises

Exercise 10.1 Consider the planar system

$$\begin{aligned}\dot{x} &= xy + x^3\\ \dot{y} &= -y - 2x^2.\end{aligned}$$

Compute the Center Manifold near the (0, 0) equilibrium and determine its stability.

Exercise 10.2 Study the dynamics near the origin via the center manifold for each of the following systems:

(a)

$$\begin{aligned}\dot{x}_1 &= -x_1 + x_2^2\\ \dot{x}_2 &= -\sin x.\end{aligned}$$

(b)

$$\begin{aligned}\dot{x}_1 &= \frac{1}{2}x_1 + x_2 + x_1^2 x_2\\ \dot{x}_2 &= x_1 + 2x_2 + x_2^2.\end{aligned}$$

(c)

$$\begin{aligned}\dot{x}_1 &= -x_1 - x_2 + x_3^2\\ \dot{x}_2 &= 2x_1 + x_2 - x_3^2\\ \dot{x}_3 &= x_1 + 2x_2 - x_3.\end{aligned}$$

Exercise 10.3 Consider the following 3D system

$$\begin{aligned}\dot{x} &= y\\ \dot{y} &= -x - xz\\ \dot{z} &= -z + \alpha x^2.\end{aligned}$$

Compute the Center Manifold near the (0, 0, 0) equilibrium and determine its stability.

Exercise 10.4 Show that $(\cos t, \sin t, 1)$ is a solution of Eq. (10.7).

Exercise 10.5 Show that the zero equilibrium solution of Eq. (10.11) is unstable.

Exercise 10.6 Consider the planar system

$$\begin{aligned}\dot{u} &= v\\ \dot{v} &= \beta u - u^2 - \delta v.\end{aligned}$$

Compute the Center Manifold near the (0, 0) equilibrium and determine its stability.

Exercise 10.7 Consider the following 3D model of a laser system

$$\begin{aligned}\dot{x} &= \frac{g^2}{\gamma+\kappa}\left[\lambda(x+y) - N(x+y)z\right]\\ \dot{y} &= -(\gamma+\kappa)y - \frac{g^2}{\gamma+\kappa}(x+y)(\lambda+z)\\ \dot{z} &= -rz + \frac{2}{N}\left[(\kappa x - \gamma y)(x^* + y*)\right],\end{aligned}$$

where $g, \gamma, \kappa, \lambda$ and r are all parameters. x^* and y^* are mode amplitudes. Compute a center manifold of the form $y, z = f(x^2, \lambda x, \lambda^2)$ near the $(0, 0, 0)$ equilibrium and determine its stability.

Exercise 10.8 The singular perturbation problem

$$\begin{aligned}\dot{y} &= -y + (y+c)z\\ \varepsilon\dot{z} &= y - (y+1)z,\end{aligned} \tag{10.60}$$

where $\varepsilon > 0$ and small, and $0 < c < 1$. For $\varepsilon = 0$ a solution is given by:

$$z = \frac{y}{y+1}.$$

Substitution of this solution into $\dot{y}$ gives

$$\dot{y} = \frac{(c-1)}{(1+y}y. \tag{10.61}$$

Show via center manifold reduction that solutions of (10.60) are close to solutions of (10.61).

Exercise 10.9 Consider the Lorenz equations

$$\begin{aligned}\dot{x} &= \sigma(y-x)\\ \dot{y} &= \tilde{\rho}x + x - y - xz\\ \dot{y} &= -\beta z + xy.\end{aligned} \tag{10.62}$$

Here $\tilde{\rho} = \rho - 1$ is the usual parameter in the Lorenz system. Assume σ and β to be fixed.

(a) Compute the equilibrium points of Eq. (10.62).
(b) Calculate the linearization of Eq. (10.62) near the zero equilibrium $(0, 0, 0)$.
(c) Compute the eigenvalues and eigenvectors of the Jacobian matrix associated with the linearization of Eq. (10.62).
(d) Find a transformation matrix T that allows to rewrite the model equation in the form given by Eq. (10.3).

(e) Consider $\tilde{\rho}$ as the distinguished bifurcation parameter. Study the stability of the zero equilibrium using the center manifold. Describe the nature of the bifurcations near the trivial equilibrium.

Exercise 10.10 Consider the following system of differential equations

$$\begin{aligned} \dot{x} &= \lambda^2 + x - x^2 + y^2 \\ \dot{y} &= \lambda + x^2 - xy. \end{aligned} \tag{10.63}$$

Equilibrium points of the system (10.63) can be though of as being zeros of the function $\Phi : \mathbb{R}^2 \times \mathbb{R} \to \mathbb{R}^2$ given by

$$\Phi(x, y, \lambda) = (\lambda^2 + x - x^2 + y^2, \lambda + x^2 - xy).$$

Apply the Lyapunov-Schmidt procedure to reduce the problem of solving $\Phi(x, y, \lambda) = 0$ into solving an equivalent equation of the form $g(x, \lambda) = 0$.

Exercise 10.11 Repeat the Lyapunov-Schmidt model reduction on the Euler's Beam model Eq. (10.19) but this time with hinged contact boundary conditions:

$$\theta(0) = \theta(L_b) = 0.$$

Hint: this time ker L is spanned by eigenfunctions of the form: $\sin\left\{\frac{n\pi x}{L_b}\right\}$.

Exercise 10.12 Consider the PDE model Eq. (10.23) with traveling wave solutions. Carry out the computations outlined in Eq. (10.28) to show that the PDE model reduces to the algebraic system of Eq. (10.28).

Exercise 10.13 A model for a dispersive long wave has the form

$$\boxed{\begin{aligned} u_{ty} + v_{xx} + \frac{1}{2}(u^2)_{xy} &= 0 \\ v_t + (uv + u + u_{xy})_x &= 0, \end{aligned}} \tag{10.64}$$

where $u = u(x, y, t)$ and $v = v(x, y, t)$. In this exercise, you will study the bifurcation of traveling wave solutions to Eq. (10.64) through the following tasks.

(a) Apply the transformation

$$u(\eta) = u(x, y, t), \quad v(\eta) = v(x, y, t), \quad \eta = px + qy - ct, \quad pq \neq 0,$$

to rewrite the original model Eq. (10.64) as

$$\begin{aligned} -qcu'' + p^2 v'' + pq(uu'' + (u')^2) &= 0 \\ -cv' + p(uv + u + pqu'')' &= 0, \end{aligned}$$

where ($' = d/d\eta$). Solve the second equation for v' and substitute into the first one, to arrive (after letting $u = v$) to the following version of the original model

$$\alpha u'''' + \beta u'' + \mu(uu'' + (u')^2) = 0, \tag{10.65}$$

where $\alpha = p^4 q$, $\beta = p^3 - c^2 q$, $\mu = 2p^3 + pq$, also

$$u(\eta) = u(\eta + 2\pi), \quad v(\eta) = v(\eta + 2\pi).$$

(b) Apply the Lyapunov-Schmidt reduction to study the bifurcations of Eq. (10.65).

Exercise 10.14 Consider a reaction-diffusion model of the form

$$u_t = Du_{xx} - f(u), \qquad 0 \leq x \leq l, \quad t \geq 0, \tag{10.66}$$

with Dirichlet boundary conditions $u(0, t) = u(l, t)$, which represent the fact that concentrations are fixed.

(a) Let $v(\eta, t) = u(l\eta, t)$, with $0 \leq \eta \leq 1$. Rewrite the original model Eq. (10.66) as

$$v_t = \frac{D}{l} v_{\eta\eta} - f(v), \tag{10.67}$$

with boundary conditions $v(0, t) = v(1, t) = 0$.

(b) Apply the Lyapunov-Schmidt reduction on Eq. (10.67) and study the effect of varying the length l. What kind of bifurcation effects should you expect from the reduced model?

Exercise 10.15 Consider the Brusselator model

$$\begin{aligned} \frac{dX}{dt} &= D_1 X_{\xi\xi} + X^2 Y - (B + 1)X + A \\ \frac{dY}{dt} &= D_2 Y_{\xi\xi} - X^2 Y + BX, \end{aligned} \tag{10.68}$$

where $0 \leq \xi \leq l$, and boundary conditions: $X(0, t) = X(l, t) = 0$, $Y(0, t) = Y(l, t) = B/A$. Observe that $(X, Y) = (A, B/A)$ is an equilibrium point. Apply the Lyapunov-Schmidt reduction to show that the original model reduces to a problem of the form

$$\dot{x} = g(x, \beta) = x(\alpha x + \beta\lambda).$$

Describe the type of bifurcation.

Exercise 10.16 Consider the Kuramoto-Sivashinsky model Eq. (10.36). Let x be a one-dimensional domain. Extend the Galerkin reduction developed in Sect. 10.3 with $N = 4$ modes, i.e., let $k_{\max} = 2$ in the basis the basis:

$$\Phi_{k=1}^{N} = \left\{\sqrt{2}\cos(kx),\ \sqrt{2}\cos(kx),\ k = 1, \ldots, k_{\max} = 2\right\}$$

Exercise 10.17 Burger's model appears in various fields such as acoustics, fluid mechanics, traffic flow and gas dynamics, to mention only a few cases. The model is

$$\frac{\partial u}{\partial t} + u\frac{\partial u}{\partial x} = \frac{1}{P_e}\frac{\partial^2 u}{\partial x^2}. \tag{10.69}$$

Let $\{\Phi_k\}_{k=1}^{N}$ be a generic POD basis for the decomposition of $u(x, t)$. Derive a reduced order model in terms of the POD modes.

Exercise 10.18 Consider the following models

(i)

$$\frac{dx}{dt} = 27 + 3\mu + \mu x + 12x + x^2.$$

(ii)

$$\frac{dx}{dt} = 24 - 3\mu - \mu x + 11x + x^2.$$

Rewrite the models in normal form. Then study the bifurcations and draw bifurcation diagrams for each case.

Exercise 10.19 Compute the normal form for a mathematical model on $\mathbb{R}^2$ in the neighborhood of an equilibrium point, where the linear part of the model is given by

$$L = \begin{bmatrix} 0 & 1 \\ 0 & 0 \end{bmatrix}.$$

Exercise 10.20 Compute the normal form for a mathematical model on $\mathbb{R}^2$ in the neighborhood of an equilibrium point, where the linear part of the model is given by

$$L = \begin{bmatrix} 1 & 1 \\ 0 & 1 \end{bmatrix}.$$

Compare the resulting normal form with that of the previous exercise.

References

1. G. Teschl, *Ordinary Differential Equations and Dynamical Systems*, vol. 40 (American Mathematical Society, 2012)
2. S. Wiggins, *Introduction to Applied Nonlinear Dynamical Systems* (Springer, New York, 1990)
3. J. Carr, *Applications of Centre Manifold* (Springer, Berlin, 1981)

4. M. Golubitsky, D.G. Schaeffer, *Singularities and Groups in Bifurcation Theory Vol. I*, vol. 51 (Springer, New York, 1984)
5. Y. Kutnetsov, *Elements of Applied Bifurcation Theory* (Springer, Berlin, 2004)

Appendix A
MATLAB Programs

This appendix provides the programs cited in the main text.

A.1 Algebraic Programs

Below we present a MatLab function to efficiently create the Vandermode matrices, A_m, from the data:

$$N = [N_1, N_2, ..., N_n]^T.$$

This MATLAB function receives as input the data vector **x** and the degree, m, of the approximation. The function can then be called by an external MATLAB code to compute the desired polynomial fit.

```
1 function A = VandermodeA(x,m)
2 A = [ones(length(x),1)];
3 for i = 1:m
4      A = [A,x.^i];
5 end
6 end
```

Below is a MatLab code for finding the parameters for fitting an allometric model to the metabolism data set.

```
1 %  Allometric Model for Metabolism vs Weight
2 clear all;
3 close all;
4 clc;
5
6 load 'metabolism.data'
7 xdata  = metabolism(:,1);
8 ydata  = metabolism(:,2);
```


A. Palacios, *Mathematical Modeling*, Mathematical Engineering,
https://doi.org/10.1007/978-3-031-04729-9

```
9
10 % Linear Least Squares Fit to Logarithmic Data
11 Y = log(ydata);         % Logarithm of y-data
12 X = log(xdata);         % Logarithm of x-data
13
14 % Find Parametmeters k, mu to Model y = k*x^mu
15 p = polyfit(X,Y,1);  % Linear fit to X and Y
16 mu = p(1)              % Scaling exponent
17 k  = exp(p(2))         % Multiplicative factor
```

Below is a MatLab code for the logarithmic fit to data with a linear least squares.

```
1  %  Allometric Model
2  close all;
3  clear all;
4  clc
5
6  load 'metabolism.data'
7  xdata  = metabolism(:,1);
8  ydata  = metabolism(:,2);
9
10 x = linspace(0,700,50);
11
12 % Power law fit for model y = k*x^a
13 Y = log(ydata);         % Logarithm of y-data
14 X = log(xdata);         % Logarithm of x-data
15
16 p = polyfit(X,Y,1);    % Linear fit to X and Y
17 a = p(1)               % Scaling exponent
18 k = exp(p(2))          % Multiplicative factor
19 y = k*x.^a;
20
21 figure(1);
22 plot(xdata,ydata,'ko','LineWidth',3);
23 hold on
24 plot(x,y,'b-','LineWidth',3);
25 grid on;
26 axis([0 700 0 10000]);
27
28 % Set up fonts and labels for the Graph
29 fontlabs = 'Times New Roman';
30 xlabel('Weight (Kg)','FontSize',16);
31 ylabel('Metabolism(Kcal)','FontSize',16);
32 set(gca,'FontSize',40);
```

A.2 Discrete Model Programs

A.2.1 Population Models for the United States

In Sect. 3.1 a Malthusian growth model, with a constant growth rate, was derived to fit the U.S. census data from 1790 to 2010. Equation (3.1) represents the model in the form of a discrete-time system. Later in Sect. 3.3.2, the Malthusian growth is revised to include a linearly varying rate, which allows accounting for the declining growth rate in U.S. population over the period of the census data.

Below is the MatLab code for finding the best parameter fits for three discrete models: a Malthusian model with constant rate and two Malthusian models with two different linearly varying rates. In the latter cases, a best linear fit is used to estimate, first, two parameters, a and b, then three parameters a, b, and initial condition P_0. The results are compared through Fig. 3.2

```
1  function uspop_model
2  close all;
3  clc;
4
5  xlab = '$t$ (Years after 1790)';
6  ylab = 'Population ($\times 10^6$)';
7  xxpop = 1790:10:2010;
8  xxmod = 1790:10:2050;
9  yycpop = [3.929 5.308 7.240 9.638 12.866 17.069 23.192...
10      31.433 39.818 50.189 62.948 76.212 92.228 106.022...
11      122.775 132.165 150.697 179.323 203.302 226.546...
12      248.710 281.422 308.746];
13 N = length(yycpop);
14
15 % Best linear fit
16 for i=1:N-1
17     grow(i) = yycpop(i+1)/yycpop(i) - 1;
18 end
19 tgrow = 0:10:210;
20 a = polyfit(tgrow,grow,1);
21 tt = linspace(0,250,100);
22 gg = a(1)*tt + a(2);
23
24 figure(1);
25 plot(tgrow,grow,'bo','MarkerSize',8);
26 hold on
27 plot(tt,gg,'k-','LineWidth',2.5);
28 grid
29
30 xlim([0,250]);
31 ylim([0.05 0.4]);
32 fontlabs = 'Times New Roman';
33 xlabel(xlab,'FontSize',16,'FontName',fontlabs,...
34     'interpreter','latex');
35 ylabel(ylab,'FontSize',16,'FontName',fontlabs,...
36     'interpreter','latex');
37 set(gca,'FontSize',36);
38 print -depsc usgrow.eps
```

```
39  hold off
40
41  tmod = 0:10:220;
42  tN = 0:10:250;
43
44  % Best fitting Malthusian growth model
45  [p1m,Jm] = fminsearch(@ssemal,[4,0.2],[],yycpop);
46  mal(1) = p1m(1);
47  for i=2:26
48      mal(i) = (1+p1m(2))*mal(i-1);
49  end
50
51  % Fitting nonautonmous growth model with 2 parameters
52  [p1n,Jn] = fminsearch(@ssenonmal,4,[],a,yycpop);
53  nonmal(1) = p1n(1);
54  for i=2:26
55      nonmal(i) = (1+a(2)+a(1)*10*(i-2))*nonmal(i-1);
56  end
57
58  % Fitting nonautonmous growth model with 3 parameters
59  [p1n3,Jn3] = ...
        fminsearch(@ssenonmal3,[4,-0.001,0.37],[],yycpop);
60  nonmal3(1) = p1n3(1);
61  for i=2:26
62      nonmal3(i) = (1+p1n3(3)+p1n3(2)*10*(i-2))*nonmal3(i-1);
63  end
64
65  figure(2);
66  plot(tmod,yycpop,'bo','MarkerSize',8);
67  hold on
68  plot(tN,mal,'g-','LineWidth',2.5);
69  plot(tN,nonmal,'r-','LineWidth',2.5);
70  plot(tN,nonmal3,'k-','LineWidth',2.5);
71  grid;
72  legend('U.S. Population','Malthusian','Nonautonomous (1)',...
73      'Nonautonomous (2)','location','northwest');
74  xlim([0,250]);
75  ylim([0 400]);
76  fontlabs = 'Times New Roman';
77  xlabel(xlab,'FontSize',16,'FontName',fontlabs,...
78      'interpreter','latex');
79  ylabel(ylab,'FontSize',16,'FontName',fontlabs,...
80      'interpreter','latex');
81  set(gca,'FontSize',36);
```

The following MatLab functions are used to calculate the sum of the square errors (SSE), which is commonly employed as a criterion to find the best fitting parameters.

```
1  function J = ssemal(pm,uscen)
2  %SSE for Malhusian growth
3  N = length(uscen);
4  pop(1) = pm(1);
5  for i=2:N
6      pop(i) = (1+pm(2))*pop(i-1);
7  end
8  err = pop - uscen;
9  J = err*err';
10 end
```

```
1  function J = ssenonmal(pnm,a,uscen)
2  %SSE for nonautonomous Malhusian model from ...
       growth data
3  N = length(uscen);
4  pop(1) = pnm;
5  for i=2:N
6      pop(i) = (1+a(2)+a(1)*10*(i-2))*pop(i-1);
7  end
8  err = pop - uscen;
9  J = err*err';
10 end
```

```
1  function J = ssenonmal3(pnm3,uscen)
2  %SSE for best fit nonautonomous Malhusian model
3  N = length(uscen);
4  pop(1) = pnm3(1); % initial P0
5  for i=2:N  % time simulation of nonautonomous ...
       model
6      pop(i) = ...
           (1+pnm3(3)+pnm3(2)*10*(i-2))*pop(i-1);
7  end
8  err = pop - uscen;
9  J = err*err';    % SSE
10 end
```

Similarly, the *logistic growth* and *Beverton-Holt* autonomous models have three parameters to fit. In MatLab the nonlinear minimizer `fminsearch`is employed to find the least squares best fit of the model to the census data.

```
1  function uspop_auto
2  clear all;
3  close all;
4  clc;
5
6  hold off
```

```
7  xlab = '$t$ (Years after 1790)';
8  ylab = 'Population ($\times 10^6$)';
9
10 xxpop = 1790:10:2010;
11 xxmod = 1790:10:2050;
12 yycpop = [3.929 5.308 7.240 9.638 12.866 17.069 23.192...
13     31.433 39.818 50.189 62.948 76.212 92.228 106.022...
14     122.775 132.165 150.697 179.323 203.302 226.546...
15     248.710 281.422 308.746];
16 N = length(yycpop);
17 tmod = 0:10:220;
18 tN = 0:10:250;
19
20 % 3 parameter best fit to nonautonmous growth model
21 [p1n3,Jn3] = ...
       fminsearch(@ssenonmal3,[4,-0.001,0.37],[],yycpop);
22 nonmal3(1) = p1n3(1);
23 for i=2:26
24     nonmal3(i) = (1+p1n3(3)+p1n3(2)*10*(i-2))*nonmal3(i-1);
25 end
26
27 % 3 parameter best fit to logistic growth model
28 [p1lg,Jl3] = fminsearch(@sselog3,[4,0.23,450],[],yycpop);
29 plg3(1) = p1lg(1);
30 for i=2:26
31     plg3(i) = (1+p1lg(2)*(1-plg3(i-1)/p1lg(3)))*plg3(i-1);
32 end
33
34 % 3 parameter best fit to BH growth model
35 [p1bh,Jbh3] = fminsearch(@ssebh3,[4,1.23,2100],[],yycpop);
36 pbh3(1) = p1bh(1);
37 for i=2:26
38     pbh3(i) = p1bh(2)*pbh3(i-1)/(1+pbh3(i-1)/p1bh(3));
39 end
40
41 figure(1);
42 plot(tmod,yycpop,'ko','MarkerSize',8);
43 hold on
44 plot(tN,nonmal3,'k-','LineWidth',2.5);
45 plot(tN,plg3,'g-','LineWidth',2.5);
46 plot(tN,pbh3,'r-','LineWidth',2.5);
47 grid;
48 legend('U.S. Population','Nonautonomous','Logistic',...
49     'Beverton-Holt','location','northwest');
50
51 xlim([0,250]);
52 ylim([0 400]);
53 fontlabs = 'Times New Roman';
54 xlabel(xlab,'FontSize',16,'FontName',fontlabs,...
55     'interpreter','latex');
56 ylabel(ylab,'FontSize',16,'FontName',fontlabs,...
57     'interpreter','latex');
58 set(gca,'FontSize',36);
```

```
1  function J = sselog3(pl3,uscen)
2  %SSE for best fit logistic model
3  N = length(uscen);
4  pop(1) = pl3(1); % initial P0
5  for i=2:N  % time simulation of logistic model
6      pop(i) = (1+pl3(2)*(1-pop(i-1)/pl3(3)))*pop(i-1);
7  end
8  err = pop - uscen;
9  J = err*err';   % SSE
10 end
```

```
1  function J = ssebh3(pbh3,uscen)
2  %SSE for best fit Beverton-Holt model
3  N = length(uscen);
4  pop(1) = pbh3(1); % initial P0
5  for i=2:N  % time simulation of Beverton-Holt model
6      pop(i) = pbh3(2)*pop(i-1)/(1+pop(i-1)/pbh3(3));
7  end
8  err = pop - uscen;
9  J = err*err';   % SSE
10 end
```

A.2.2 *Bifurcations in the Discrete Logistic Model*

The discrete logistic model Eq. (3.20) undergoes a sequence of transitions that include period-doubling bifurcations and chaos. The MATLAB code used to generate the bifurcation diagram of Fig. 3.11 is shown below. The visualization window of the bifurcations can be changed by adjusting the parameters *lambda min* and *lambda max*.

```
1  clear all;
2  close all;
3
4  Niterates    = 700;
5  Nlambda      = 1000;
6  Ntransients  = 500;
7
8  %   ---------   Initial Conditions   ---------
9  lambda_min = 0.0;
10 lambda_max = 3.9;
11 xmin       = 0.0;
12 xmax       = 1.0;
13
14 for k=1:Nlambda
15    lambda = lambda_min + ...
          (lambda_max-lambda_min)*(k-1)/(Nlambda-1);
16
17    %   ---------   Transients   ---------
```

```
18      x0 = 0.1237;
19      for i=1:Ntransients
20        x1 = lambda*x0*(1.0-x0);
21        x0 = x1;
22      end;
23
24      %  ---------  Iterate  ---------
25      for j=1:Niterates
26        x1     = lambda*x0*(1.0-x0);
27        x0     = x1;
28        t(j,k) = lambda;
29        v(j,k) = x1;
30      end;
31  end;
32
33  plot(t,v,'k.','Markersize',4);
34  xlabel('{r}');
35  ylabel('{x_n}');
36  set(gca,'FontSize',30);
37  grid on;
38  axis([lambda_min lambda_max xmin xmax]);
```

A.2.3 Sensitive Dependence in the Logistic Model

The Lyapunov exponents of a dynamcial system can be used to determine the presence of chaotic behavior. The MATLAB code used to compute the Lyapunov exponents of the logistic growth model of Fig. 3.15 is shown below. Positive exponents are indicative of chaos.

```
1   clear all;
2   close all;
3
4   lyap=zeros(1,1000);
5   j=0;
6   for(r=1:0.001:4)
7       xn1=rand(1);
8       lyp=0;
9       j=j+1;
10      for(i=1:10000)
11          xn=xn1;
12          %logistic map
13          xn1=r*xn*(1-xn);
14         %wait for transient
15         if(i>300)
16             % calculate teh sum of logaritm
17             lyp=lyp+log(abs(r-2*r*xn1));
18         end
19      end
```

```
20        %calculate lyapun
21        lyp=lyp/10000;
22        lyap(j)=lyp;
23    end
24    r=1:0.001:4;
25    plot(r,lyap,'k','LineWidth',2);
26
27    xlabel('{r}','Fontsize',12);
28    ylabel('{h(x)}','Fontsize',12);
29    set(gca,'FontSize',30);
30    grid on;
```

A.3 Continuous Model Programs

In this part of the appendix we include MATLAB Codes for the Chapter on Continuous Models.

A.3.1 Yeast Growth Models

In the analysis of yeast cultures, we studied their growth by finding parameter values for exponential growth. One common method to find such parameters is through the *linear least squares best fit* to approximate the logarithm of data. To do so, we can create, in MATLAB, two vectors, one for the time data, t, and one for the population data, P. Respectively:

```
t = [0 1.5 9 10]; P = [0.37 1.63 6.2 8.87];
```

Then a straight line is fitted through the time data and the logarithm of the population data using the MATLAB command:

```
p = polyfit(t,log(P),1)
```

which gives the slope $r = 0.2690$ and intercept $\ln(P_0) = -0.5034$. The model can then be written as:

$$P(t) = 0.6045e^{0.2690t}.$$

The second method for studying the Malthusian growth model employs a *nonlinear least squares best fit*. The MATLAB codes for minimizing the sum of square errors between the model and the data use the same data vectors from above. Then the program listed below finds the sum of square errors:

```
1 function J = yst_lstm(p, tdata, pdata)
2 % Least Squares fit to Logistic Growth
3 N = length(tdata);
4 yst = p(1)*exp(p(2)*tdata);
5 err = pdata - yst;
6 J = err*err'; % Sum of square errors
7 end
```

This program is executed within MATLAB's nonlinear solver, `fminsearch`, to find the best fitting parameters. The `fminsearch` needs a reasonable guess for P_0 and r, which can be obtained from either our algebraic solution or the linear fit to the logarithm of the population data. The specific MatLab command is:
`[p,J,flag] = fminsearch(@yst_lstm,p0,[],t,P)`
where `p0 = [0.6 0.27]`, and the result is `p = [0.6949 0.2511]`, giving the best continuous Malthusian growth model,

$$P(t) = 0.6949\, e^{0.2511\, t}.$$

The second study of growth of yeast cultures uses the logistic growth model. Once again, time, $tdata$, data, $pdata$ and population data vectors are created. We considered two time and population data sets, depending on the species of yeast being studied. The nonlinear solver employs the MATLAB program `fminsearch`, which also needs a reasonable initial guess for the parameters. An initial guess for the parameters $p_0 = [P_0, r, M]$ would be to take $P_0 = P_d(t_0)$ (the first yeast volume), r equal the value from the Malthusian growth model, and $M = P_d(t_N)$ (the last yeast volume). The sum of square errors code, which is used to find the best parameter fit, is:

```
1 function J = yst_lst(p, tdata, pdata)
2 % Least Squares fit to Logistic Growth
3 N = length(tdata);
4 yst = p(1)*p(2)./(p(1) + (p(2)-p(1))*exp(-p(3)*tdata));
5 err = pdata - yst;
6 J = err*err'; % Sum of square errors
7 end
```

The best fitting parameters are obtained by running the code `fminsearch`:

```
p1 = fminsearch(@yst_lst,p0,[],tdata,pdata)
```

In this way, the best fitting parameters for *S. cerevisiae* are found to be:

$$P_0 = 1.2343, \;\; r = 0.25864, \;\; M = 12.7421,$$

and for *S. kephir* are:

$$P_0 = 0.67807, \;\; r = 0.057442, \;\; M = 5.8802$$

with least *SSE* = *4.9460* and *SSE* = *1.3850*, respectively. The best fitting solutions satisfy:

$$P_{sc}(t) = \frac{12.742}{1 + 9.323\, e^{-0.2586t}} \qquad \text{and} \qquad P_{sk}(t) = \frac{5.880}{1 + 7.672\, e^{-0.05744t}}.$$

A.3.2 Two Species Competition

In Sect. 4.5 we described the main steps for finding the best fitting parameters to the two yeast species competition model. Then in Appendix A.3.1 we found the parameters for the monocultures, which reduces the search for the remaining parameters, a_3 and b_3, and initial conditions, X_0 and Y_0. Fitting the model (4.19) to the data in Table 4.3 uses the sum of square errors (SSE) formula (4.24). This is included in the MatLab function:

```
1  function J = leastcomp2(p,tdata,xdata,ydata)
2  global A1 A2 B1 B2
3  [t,y] = ...
       ode23(@compet,tdata,[p(1),p(2)],[],A1,A2,p(3),B1,B2,p(4));
4  errx = y(:,1)-xdata';
5  erry = y(:,2)-ydata';
6  J = errx'*errx + erry'*erry;
7  end
```

Input of the data in Table 4.3 is required for this function. In addition, the parameters from the monocultures, values for the initial conditions, X_0 and Y_0 (`p(1),p(2)`), and the competition parameters, a_3 and b_3 (`p(3),p(4)`) are also required.

Recall that the competition model (4.19) does not have an algebraic solution. Consequently, a numerical solution is required, and we see the previous program calls MatLab's *Runge-Kutta-Fehlberg ODE solver*, ODE23, which calls the model:

```
1  function dydt = compet(t,y,a1,a2,a3,b1,b2,b3)
2  % Competition Model for Two Species
3  tmp1 = a1*y(1) - a2*y(1)^2 - a3*y(1)*y(2);
4  tmp2 = b1*y(2) - b2*y(2)^2 - b3*y(1)*y(2);
5  dydt = [tmp1; tmp2];
6  end
```

Because the actual data has different population values at the same times, there are some technical complications, which required special modifications to the program above.

The main program is a script that downloads the data, sets up *Global variables* from the *monoculture logistic models*, gives a good initial guess for the parameters, and calls the `fminsearch`routine.

```
1  load yeast
2  global A1 A2 B1 B2;
3  A1  =  0.25864;  A2  =  0.020298;
4  B1  =  0.057442;  B2  =  0.0097687;
5  p  =  [0.4  0.63  0.057  0.0048];
6  p1  =  fminsearch(@leastcomp2,p,[],tdmix,scdmix,skdmix)
```

The MatLab nonlinear solver `fminsearch` is used to find the *least sum of square of errors* by adjusting the parameters in the ODE solver to fit the experimental data. This MatLab code gives the best fitting interspecies competition parameters for the competition model:

$$a_3 = 0.057011 \quad \text{and} \qquad b_3 = 0.0047576$$

and initial conditions:

$$X(0) = X_0 = 0.41095 \qquad \text{and} \qquad Y(0) = Y_0 = 0.62578.$$

The least sum of square errors is **19.312**.

A.3.3 Forced Linear Oscillator

The various modes of oscillations of a linear oscillator, shown in Fig. 4.20, were obtained with the following MATLAB code:

```
1   function forced_lin_oscillator
2
3   clear all;
4   close all;
5   clc;
6
7   %   ----------   Parameters   ----------
8   w0      = 3.0;
9   m       = 1.0;
10  f0      = 1.0;
11
12  w_min   = 0.0;
13  w_max   = 2.0;
14  N       = 99;
15  h       = (w_max - w_min) / N;
16  %w        = [w_min:h:w_max];
17
18  zeta0   = 0.0;
19  zeta1   = 0.1;
20  zeta2   = 0.2;
21  zeta3   = 0.3;
22  zeta4   = 0.5;
23  zeta5   = 1.0;
```

```
24
25  for k = 1:N;
26       w(k) = w_min + (k-1)*w_max/(N-1);
27       A0(k) = 1 / (sqrt((1 - w(k)^2)^2 + (2*zeta0*w(k))^2));
28       A1(k) = 1 / (sqrt((1 - w(k)^2)^2 + (2*zeta1*w(k))^2));
29       A2(k) = 1 / (sqrt((1 - w(k)^2)^2 + (2*zeta2*w(k))^2));
30       A3(k) = 1 / (sqrt((1 - w(k)^2)^2 + (2*zeta3*w(k))^2));
31       A4(k) = 1 / (sqrt((1 - w(k)^2)^2 + (2*zeta4*w(k))^2));
32       A5(k) = 1 / (sqrt((1 - w(k)^2)^2 + (2*zeta5*w(k))^2));
33  end;
34
35
36  figure(1);
37  plot(w,A0,'k--',w,A1,'g',w,A2,'r',w,A3,'c',w,A4,'m',w,A5,...
38       'b','LineWidth',4);
39  leg1=legend('\zeta=0.0','\zeta=0.1','\zeta=0.2',...
40       '\zeta=0.3','\zeta=0.5','\zeta=1.0');
41  xlabel('w/w_0','FontSize',16);
42  ylabel('Amplitude','FontSize',16);
43  set(gca,'FontSize',36);
44  set(leg1,'FontSize',36);
45  grid on;
46  %title('Numerical Solution');
47  axis([0 2 0 6]);
```

A.3.4 Weakly Forced van der Pol Oscillator

Equation (4.52) is a model for a weakly forced van der Pol oscillator. The synchronization of the natural frequency with that of the external periodic force is demonstrated in Fig. 4.29. The MATLAB code that was written to conduct the simulations is shown below:

```
1   function vderpol_averaging
2
3   clear all;
4   close all;
5   clc;
6
7   %  ----------   Parameters   ----------
8   epsilon = 1.0;
9   w0      = 1.0;
10  wd      = 1.1;
11  k       = 1.0;
12
13  %  ----------   Intial Conditions  ----------
14  V0 = [0.26, 0.1];
15  W0 = [2.0, 3.0];
16
17  %  -----------   Integrate ODEs   -----------
18  tmin = 0.0;
19  tmax = 30;
20  xmin = -4.0;
```

```
21  xmax = 4.0;
22  h    = 0.1;
23  tspan = [tmin:h:tmax];
24
25
26  % ----------- Numerical Solution -------------
27  [t,v]=ode45(@vderpol_ODE,tspan,V0,[],epsilon,w0,wd,k);
28
29  x = v(:,1);
30  y = v(:,2);
31
32  figure(1);
33  subplot(2,1,1);
34  plot(t,x,'b',t,y,'r--','LineWidth',2),
35  legend('x','y');
36  xlabel('Time','FontSize',16);
37  ylabel('x,y','FontSize',16);
38  title('Numerical Solution');
39  axis([tmin tmax xmin xmax]);
40
41  subplot(2,1,2)
42  plot(x,y);
43  xlabel('x','FontSize',16);
44  ylabel('y','FontSize',16);
45
46
47  % ----------- Averaging ODE -------------
48  [t,w]=ode45(@transformed_ODE,tspan,V0,[],epsilon,w0,wd,k);
49
50  a1 = w(:,1);
51  a2 = w(:,2);
52
53  figure(2);
54  subplot(2,1,1);
55  plot(t,a1,'b',t,a2,'r--','LineWidth',2),
56  legend('a_1','a_2');
57  xlabel('Time','FontSize',16);
58  ylabel('a_1,a_2','FontSize',16);
59  axis([tmin tmax -1.5 3.0]);
60
61  subplot(2,1,2)
62  plot(a1,a2);
63  xlabel('a_1','FontSize',16);
64  ylabel('a_2','FontSize',16);
65
66
67  % ----------- Comparison -------------
68  figure(3);
69
70  u = a1.*cos(wd*t) + a2.*sin(wd*t);
71
72  plot(t,x,'b',t,u,'k--','LineWidth',2.5),
73  legend('x_{num}','x_{avg}');
74  xlabel('Time','FontSize',16);
75  ylabel('x_{(t,\epsilon)}','FontSize',16);
76  %title('Averaging Solution');
77  axis([tmin tmax xmin xmax]);
78  set(gca,'FontSize',30);
79  grid on;
```

```
80
81
82
83  %  ---------   Right-Hand-Side of van der Pol ODEs  ----------
84  function f = vderpol_ODE(t,v,epsilon,w0,wd,k)
85  x     = v(1);
86  y     = v(2);
87
88  f(1)  = y;
89  f(2)  = -w0^2*x + epsilon*(1.0 - x^2)*y + ...
         epsilon*k*wd*cos(wd*t);
90
91  f=f';
92
93
94  %  ---------   Right-Hand-Side of Transformed ODEs  ----------
95  function g = transformed_ODE(t,v,epsilon,w0,wd,k)
96  a1    = v(1);
97  a2    = v(2);
98  rho   = (a1^2 + a2^2)/4.0;
99
100 g(1)  = epsilon*(1 - rho)*a1 - ((wd^2-w0^2)/(2.0*wd))*a2;
101 g(2)  = epsilon*(1 - rho)*a2 + ((wd^2-w0^2)/(2.0*wd))*a1 + ...
         epsilon*k;
102
103 g=g';
```

The code compares the numerical solution obtained from integrating the original model Eq. (4.52) with a solution obtained through the method of averaging.

A.4 Bifurcation Theory

In this section we include a few MATLAB codes for the Chapter on Bifurcation Theory. The codes are for both discrete and continuous systems.

A.4.1 Sand Dollar Pattern

Figure 5.2 showcases the bifurcations of a discrete model Eq. (5.5) with DN_5-symmetry. The long-term behavior of this model is in the form of a sand-dollar pattern. The MATLAB code for performing the simulations is shown below

```
clear all;
close all;

% ----  Parameters  ----
m           = 5.0;
lambda      = -2.34;
alpha       = 2.0;
beta        = 0.4;
gamma       = 0.1;

Niterates    = 60000;
Ntransients = 100;

%  ---------  Initial Conditions  ---------
x_n = 0.1;
y_n = 0.1;

z_n = x_n + y_n*i;

%  ---------  Transients  ---------
for i=1:Ntransients
    z_np1 = (lambda + alpha*z_n*conj(z_n) + ...
        beta*real(z_n^m))*z_n + ...
             gamma*conj(z_n)^(m-1);

    z_n = z_np1;
end;


%  ---------  Iterates  ---------
for i=1:Niterates
    z_np1 = (lambda + alpha*z_n*conj(z_n) + ...
        beta*real(z_n^m))*z_n + ...
             gamma*conj(z_n)^(m-1);

    x(i) = real(z_n);
    y(i) = imag(z_n);

    z_n = z_np1;
end;

plot(x,y,'.k','Markersize',4);
xlabel('{x_n}');
ylabel('{y_n}');
set(gca,'FontSize',30);
grid on;
axis square;
axis([-1.2 1.2 -1.2 1.2]);
```

A.4.2 Neimark-Sacker Bifurcation

This type of bifurcation scenario is equivalent to that of the Hopf bifurcation but for discrete systems. It occurs when a pair of complex-valued eigenvalues cross the unit circle. The representative example shown in Fig. 5.14 was simulated with the following MATLAB code.

```
1  clear all
2  close all
3  clc
4
5  N = 400;    % Number of iterations
6
7  %  ----------  Model Parameters  ----------
8  a = 40.0;
9  r = 3.0;
10
11 %  ----------  Initial Conditions  ----------
12 x(1)=1.10;
13 y(1)=1.10;
14
15 for k=1:N-1
16
17     x(k+1)=r*x(k)*exp(-y(k));
18     y(k+1)=x(k)*(1 - exp(-a*y(k)));
19
20 end
21
22 %  ----------  Plotting  -----------
23 figure(1);
24 clf
25 plot(x,y,'o','MarkerEdgeColor',...
26            'b','MarkerFaceColor',...
27            'r','markersize',5)
28 xlabel('{x_n}');
29 ylabel('{y_n}');
30 axis([0 2.5 0 2.5]);
31 set(gca,'FontSize',30);
32 grid on;
```

A.5 Hybrid Model Programs

In this part of the appendix we include MATLAB Codes for the Chapter on Hybrid Models. The computer simulations of the solutions of the spring-mass model Eq. (6.5) shown in Fig. 6.2 were obtained through the MATLAB code below.

```
1  % ----------------------------------------
2  %            Two Spring-Mass System
3  % ----------------------------------------
4  function sm_system
5
6  close all;
7  clear all;
8  clc;
9
10 % ------- Parameters -------
11 m = 1.0;
12 c = 0.1;
13 k = 1.0;
14 par = [m;c;k];
15
16 tmin        = 0.0;
17 tmax        = 60.0;
18 h           = 0.01;
19 trans       = 10.0;
20 tspan_trans = [tmin:h:trans];
21 tspan       = [tmin:h:tmax];
22
23 % --------------- Integrate ODE ...
      ---------------
24 options = odeset('RelTol',1e-10);
25 V0 = [0.1,0.3,0.1,0.3]';
26 [t_trans,v_trans] = ...
      ode45(@rhs_ode,tspan_trans,V0,options,par);
27
28 V0    = v_trans(end,:)';
29 [t,v] = ode45(@rhs_ode,tspan,V0,options,par);
30
31 % --------------- Plot Results --------------
32 xmin = -2.0;
33 xmax = 2.0;
34
35 % ------ Time Series -------
36 figure(1);
37 plot(t,v(:,1),'b',t,v(:,3),'k','LineWidth',3);
38 legend('x_1','x_3');
39 ylabel('{x_1, x_3}');
40 set(gca,'FontSize',40);
41 grid on;
42 axis([tmin tmax -0.25 0.25]);
43
44 % ------ Phase-Space -------
45 figure(2);
46 plot(v(:,1),v(:,3),'LineWidth',3);
47 xlabel('{x_1}'); ylabel('{x_3}');
48 set(gca,'FontSize',50);
49 grid on;
50
51 function f = rhs_ode(t,v,par)
```

```
52      x1  = v(1);
53      x2  = v(2);
54      x3  = v(3);
55      x4  = v(4);
56      m   = par(1);
57      c   = par(2);
58      k   = par(3);
59
60      %  --------  Write RHS of ODEs  --------
61      f(1) = x2;
62      f(2) = -(k/m)*x1 - (c/m)*x2 + k*(x3 - x1);
63      f(3) = x4;
64      f(4) = -(k/m)*x3 - (c/m)*x4 + k*(x1 - x3);
65
66      f = f';
```

A.6 Delay Model Programs

This section of the appendix documents MATLAB programs for the Chapter on Delay Models. The code employ the numerical solver `dd23` developed by Shampine and Thompson [1].

A.6.1 Epidemic Model Programs

In Sect. 7.4 we introduced a delay model for the spread of an epidemic. The computer simulations of the solutions, with and without delay, shown in Fig. 7.5 were obtained through the MATLAB code below.

```
1   function epidemic;
2
3   close all;
4   clear all;
5   clc;
6
7   % Parameters
8   B = 2;
9   C = 1;
10
11  history = [0.8];
12  tspan   = [0 80];
13
14  % Solve the ODEs that arise when there is no delay.
15  sol1 = dde23(@ddefun,[],history,tspan,[],B,C);
16
17  % Solve the DDEs that arise when there is a delay of r.
18  sol2 = dde23(@ddefun, [7], history, tspan,[], B,C);
```

```
19
20  figure(1)
21  plot(sol1.x,sol1.y,'b-','Linewidth', 3), hold on;
22  plot(sol2.x,sol2.y,'k-','Linewidth', 3);
23  xlabel('Time t');
24  ylabel('x(t)');
25  legend('No Delay','Delay','Location','NorthEast');
26  set(gca,'FontSize',40);
27  grid on;
28
29  function dxdt = ddefun(t,x,Z,B,C) % equation being solved
30  if isempty(Z)      % ODEs
31     dxdt = B*x(1)*(1 - x(1)) - C*x(1);
32  else
33    xlag = Z(:,1);
34
35    dxdt = B*xlag(1)*(1 - x(1)) - C*x(1);
36  end
```

A.6.2 Lotka-Volterra Model

In Sect. 7.5 we studied the effects of a time delay in a predator-prey model. In particular, we used the model to study the emergence of small amplitude oscillations via Hopf bifurcations driven by delay.

The computer simulations of the solutions, with and without delay, shown in Fig. 7.4 were obtained through the MATLAB code below.

```
1   function lotka_volterra;
2
3   close all;
4   clear all;
5   clc;
6
7   % Parameters
8   a = 1.0;
9   b = 1.0;
10  c = -2.0;
11  d = 1.0;
12  mu1 = 1;
13  mu2 = -1;
14
15  history = [0.5; 0.8];
16  tspan   = [0 500];
17  opts    = ddeset('RelTol',1e-5,'AbsTol',1e-8);
18  % Solve the ODEs that arise when there is no ...
        delay.
19  sol1 = dde23(@ddefun,[],history,tspan,[], ...
        a,b,c,d,mu1,mu2);
20
```

```
21  % Solve the DDEs that arise when there is a ...
        delay of r.
22  sol2 = dde23(@ddefun,[2.2],history, ...
        tspan,opts, a,b,c,d,mu1,mu2);
23
24  figure(1)
25  plot(sol1.x,sol1.y(1,:),'b-','Linewidth', 3), ...
        hold on;
26  plot(sol2.x,sol2.y(1,:),'k-','Linewidth', 3);
27  xlabel('t');
28  ylabel('y(t)');
29  legend('No Delay','Delay','Location','NorthEast');
30  set(gca,'FontSize',40);
31  grid on;
32
33  figure(2)
34  plot(sol1.y(1,:),sol1.y(2,:),'b-','Linewidth', ...
        3), hold on;
35  plot(sol2.y(1,:),sol2.y(2,:),'k-','Linewidth', 3);
36  xlabel('x(t)');
37  ylabel('y(t)');
38  legend('No Delay','Delay','Location','NorthEast');
39  set(gca,'FontSize',40);
40  grid on;
41
42  % equation being solved
43  function v = ddefun(t,x,Z,a,b,c,d,mu1,mu2)
44  v = zeros(2,1);
45  if isempty(Z)      % ODEs
46      v(1) = x(1)*(mu1 - a*x(1) - b*x(2));
47      v(2) = x(2)*(mu2 - c*x(1) - d*x(2));
48  else
49     xlag = Z(:,1);
50
51     v(1) = x(1)*(mu1 - a*x(1) - b*xlag(2));
52     v(2) = x(2)*(mu2 - c*xlag(1) - d*x(2));
53  end
```

A.6.3 Nyquist Plots

In Sect. 7.7 we introduced the Nyquist method as a visual technique to study the stability of delay differential equations.

Below is the *MatLab code* that was written to create the Nyquist plots for the first example shown in Fig. 7.11.

```
function Nyquist_ex1

close all;
clear all;
clc;

tau = 1.0;

% ==============  Example 1  ===========
q1 = 1;
p1 = [1 0];
L1 = tf(q1,p1,'InputDelay',tau);

nyquist(L1);

set(gca,'FontSize',40);
axis([-2 2], [-2 2]);
```

We also applied the Nyquist technique to study the delay-induced oscillations in a Lotka-Volterra model. Below is the *MatLab code* that was written to create the Nyquist plots for the third example shown in Fig. 7.13.

```
function Nyquist_ex3

close all;
clear all;
clc;

tau = 0.5;

q3 = 4/9;
p3 = [1 1 2/9];
L3 = tf(q3,p3,'InputDelay',2*tau);

nyquist(L3);

set(gca,'FontSize',40);
```

A.7 Stochastic Models

This section of the appendix documents MATLAB programs for the Chapter on Stochastic Models.

A.7.1 Stochastic Model of Stock Prices

In Sect. 9.2.2 we introduced a stochastic model for stock prices. Actually, the model describes the fluctuations in the relative change of price dP/P, where $P(t)$ represents the price of a stock at time t. The computer simulations of the deterministic and stochastic solutions shown in Fig. 9.1 were obtained through the Euler-Maruyama code below.

The command `pd=makedist('Normal',0,sqrt(dt))`, creates a normal (Gaussian) distribution with mean 0 and standard deviation $\sqrt{dt}$. The distribution is stored in the variable `pd`. The command `random(pd)`, returns a random number from the previously created probability distribution.

```
function stock_prices

close all;
clear all;
clc;

% Model Parameters
mu    = 2;
sigma = 1.0;

randn('state',100);
tspan = [0 1];          % The bounds of t
N     = 1000;           % Compute 1000 grid points
dt    = (tspan(2) - tspan(1)) / N;
T     = N*dt;

pi0   = 1.0;  % Initial Stochastic Price condition
qi0   = 1.0;  % Initial Deterministic Price condition
pi    = zeros(1,N);       % 1xN Matrix of zeros
qi    = zeros(1,N);
pi(1) = pi0;
qi(1) = qi0;

pd = makedist('Normal',0,sqrt(dt));
dW = random(pd);

for j = 2:N
   dW    = random(pd);
   pi(j) = pi(j-1) + mu*pi(j-1)*dt + sigma*pi(j-1)*dW;
   qi(j) = qi(j-1) + mu*qi(j-1)*dt;
end

figure(1)
hold on;
plot([0:dt:T],[pi0,pi], 'k-', 'Linewidth', 3), hold on;
plot([0:dt:T],[qi0,qi], 'b-', 'Linewidth', 3);
xlabel('t','FontSize',12);
ylabel('P(t)','FontSize',12);
legend('Stochastic','Deterministic');

set(gca,'FontSize',40);
grid on;
```

A.7.2 Ornstein-Uhlenbeck Process

In Sect. 9.3 we introduced the Ornstein-Uhlenbeck process as a method to model stochastic systems subject to colored noise. The formulation employs a Langevin form. The simulation shown in Fig. 9.2 was created through the Euler-Maruyama algorithm. The code is shown below.

```
1  function ou
2
3  close all;
4  clear all;
5  clc
6
7  randn('state',100)
8  tau = 1;
9  xi0 = 1;
10 dt  = 0.01;
11 D   = 0.5;
12 N   = 1000;
13 T   = N*dt;
14
15 pd = makedist('Normal',0,sqrt(dt));
16 dW = random(pd);
17
18 xi     = zeros(1,N);     % preallocate for ...
       efficiency
19 xi(1)  = xi0 - dt*xi0/tau + sqrt(2*D)*dW/tau;
20
21 for j=2:N
22     dW    = random(pd);
23     xi(j) = xi(j-1) - dt*xi(j-1)/tau + ...
           sqrt(2*D)*dW/tau;
24 end
25
26 plot([0:dt:T],[xi0,xi],'k-', 'LineWidth',3);
27 xlabel('t','FontSize',12)
28 ylabel('\eta(t)','FontSize',16,'Rotation',0,...
29 'HorizontalAlignment','right');
30 set(gca,'FontSize',40);
31 grid on;
```

A.7.3 Fokker-Planck Equation

In Sect. 9.5 we introduced the Fokker-Planck equation for the probability density function, $p(x, t)$, which measures the probability that a stochastic system is in state x at time t. An example of a 2D stochastic model with a limit cycle solution was studied. The evolution around the limit cycle shown in Fig. 9.5 was obtained with the following code.

```
1  function Limit_Cycle
2
3  close all;
4  clear all;
5  clc;
6
7  % Model Parameters
8  mu     = 0.1;
9  omega  = 0.1;
10
11 randn('state',100);
12 tspan  = [0 10];          % The bounds of t
13 N      = 1000;          % Compute 1000 grid points
14 dt     = (tspan(2) - tspan(1)) / N;
15 T      = N*dt;
16
17 x1     = zeros(1,N);       % 1xN Matrix of zeros
18 x2     = zeros(1,N);
19 y1     = zeros(1,N);       % 1xN Matrix ot zeros
20 y2     = zeros(1,N);
21 x1(1)  = 1.0;
22 x2(1)  = 0.0;
23 y1(1)  = 1.0;
24 y2(1)  = 0.0;
25
26 pd1  = makedist('Normal',0,0.3*sqrt(dt));
27 pd2  = makedist('Normal',0,0.3*sqrt(dt));
28
29 for j = 2:N
30   x1t = x1(j-1);
31   x2t = x2(j-1);
32   y1t = y1(j-1);
33   y2t = y2(j-1);
34
35   dW_1  = random(pd1);
36   dW_2  = random(pd2);
37   x1(j) = x1t + mu*(1-x1t^2-x2t^2)*x1t - omega*x2t + dW_1;
38   x2(j) = x2t + mu*(1-x1t^2-x2t^2)*x2t + omega*x1t + dW_2;
39
40   y1(j) = y1t + mu*(1-y1t^2-y2t^2)*y1t - omega*y2t;
41   y2(j) = y2t + mu*(1-y1t^2-y2t^2)*y2t + omega*y1t;
42 end
43
44 figure(1)
45 plot(x1,x2,'k*', 'Linewidth', 3), hold on;
46 plot(y1,y2,'b-', 'Linewidth', 3);
47 xlabel('x_1','FontSize',12);
48 ylabel('x_2','FontSize',12);
49 legend('Stochastic','Deterministic');
50
51 set(gca,'FontSize',40);
52 grid on;
```

Appendix B
Computations of Phase Drift

Colored noise is employed to simulate fluctuations due to electronic components. This means that the noise is assumed to be Gaussian, band-limited, having a zero mean, a variance σ^2, and have a specific correlation time, τ_c. The noise is assumed to not drive the dynamics of the system, this corresponds to $\tau_f \ll \tau_c$, where τ_f is the time-constant of each crystal oscillator [2, 3]. These assumptions lead us to rewrite the averaged equations for a network of N crystals with unidirectional and bidirectional coupling in a more general Langevin form:

$$\begin{aligned}
\dot{z}_k &= f(z_k, a, b, R_1) + \sum_{j \to k}^{N} \lambda_{kj} h(z_k, z_j) + z_k(\eta_k^A(t) + \eta_k^P(t)i) \\
\dot{\eta}_k^A &= -\frac{\eta_k}{\tau_c} + \frac{\sqrt{2D}}{\tau_c}\xi_k^A \\
\dot{\eta}_k^P &= -\frac{\eta_k}{\tau_c} + \frac{\sqrt{2D}}{\tau_c}\xi_k^P,
\end{aligned} \tag{B.1}$$

where, for convenience, we have replaced z_{k1} with simply z_k (since $z_{k2} = 0$ at all times), f is a smooth function in z_k which defines the (averaged) internal dynamics of each mode of oscillation in each crystal, and h is the coupling function between those crystals j that are coupled to each crystal k, with coupling strength λ_{kj}, N is the total number of crystals in the network, τ_c and D are correlation time and intensity of noise, respectively.

The functions η_k^A and η_k^P describe the noise applied, respectively, to the amplitude and phase of each kth oscillator, ξ_k^A and ξ_k^P are Gaussian distributed random variables with zero mean, and standard deviation σ. Each colored noise function is characterized by $\langle \eta_i^A(t) \rangle = 0$, $\langle \eta_i^P(t) \rangle = 0$ and $\langle \eta_i^A(t)\eta_j^A(s) \rangle = (D/\tau_c) \times \exp[-|t-s|/\tau_c]$, $\langle \eta_i^P(t)\eta_j^P(s) \rangle = (D/\tau_c) \times \exp[-|t-s|/\tau_c]$, where $D = \sigma^2\tau_c^2/2$ [4]. As $\tau_c \to 0$ the noise becomes white, however in practice all noise is band limited [5].

A. Palacios, *Mathematical Modeling*, Mathematical Engineering,
https://doi.org/10.1007/978-3-031-04729-9

Phase error is calculated by first locating the zeros of the solution. The zeros are approximated using a standard three point quadratic interpolation method. From the location of the zeros, the periods of oscillation are calculated. Let $P = \{p_i\}_{i=1}^{n}$ be the sequence of periods and σ_p be the mean absolute deviation of the periods. Then phase error is defined as $\sigma_p/E(P)$—the mean absolute deviation divided by the expected period length. Under normal conditions the standard deviation is normally used for phase error; however due to the natural length of the periods, the squared error for each period is smaller than machine epsilon leaving those measurements unreliable. The mean absolute deviation does not square the values so that the calculations stay away from machine epsilon, $\mathcal{O}(1 \times 10^{-16})$.

When noise is removed from the equation the phase error is 0. The size of the sequence of periods is dictated by the saved integration time. The integration time used for the data sets considered in this section was 3.5×10^{-5} s. The corresponding period for the 22 MHz solution is approximately 4.5×10^{-08} s. Therefore, the phase error is the sample standard deviation of approximately 778 cycles. Since there is a significant variation between the uncoupled scaling and the scaling for a rotating wave pattern, RW_1, the phase error analysis was expanded to examine the phase reduction along the coupling interval $0 < \lambda < 0.99$. Figure 9.11 illustrates the performance with respect to the scaling exponent, i.e., this figure is a log plot phase error, $Err(N, \lambda) = N^{m(\lambda)}$.

Samples are taken for 100 values of λ. For each value of λ, the mean phase error for 50 repeated simulations is calculated for $N = 3, 5, \ldots, 21$. Then a least squares regression is performed on the log of these values, producing the scaling exponents depicted in Fig. 9.11. This analysis suggests that strong coupling is preferable to weak coupling to produce optimal scaling. From Fig. 9.11, the optimal scaling is found at $\lambda = 0.99$ with $m = -0.8947$. Notice that for $\lambda \in (0, 0.387)$, the coupled system performs poorly compared to the uncoupled standard, having a scaling exponent $m(\lambda) > -0.5$. When the circuit is coupled the inherent noise of each node is amplified by the coupling, and in the case of $0 < \lambda < 0.387$ the coupling is too weak to overcome the amplification in noise.

Appendix C
Proper Orthogonal Decomposition

In Chap. 8, the Proper Orthogonal Decomposition (POD) was introduced as technique for the analysis of spatio-temporal data sets. In this appendix, we discuss additional properties of the POD decomposition and its relation to systems with symmetry.

C.1 Properties

Since the kernel is Hermitian, $r(\mathbf{x}, \mathbf{y}) = r^*(\mathbf{y}, \mathbf{x})$, it admits according to Riesz Theorem [6], a diagonal decomposition of the form

$$r(\mathbf{x}, \mathbf{y}) = \sum_{k=1}^{N} \lambda_k \, \Phi_k(\mathbf{x}) \, \Phi_k^*(\mathbf{y}). \tag{C.1}$$

This fact is particularly useful when finding the POD modes analytically. They can be read off from the diagonal decomposition (C.1).

The temporal coefficients, $a_k(t_i)$, are calculated by projecting the data set on each of the eigenfunctions

$$a_k(t_i) = (u(\mathbf{x}, t_i), \Phi_k(\mathbf{x})) \ , i = 1, \ldots, M. \tag{C.2}$$

It can be shown that both temporal coefficients and eigenfunctions are uncorrelated in time and space, respectively [7, 8].

Proposition C.1 The POD modes, $\{\Phi_k(\mathbf{x})\}$, with corresponding temporal coefficients, $\{a_k(t_i)\}$, satisfy the following orthogonality properties

(i) $\Phi_j^*(\mathbf{x}) \, \Phi_k(\mathbf{x}) = \delta_{jk}$

A. Palacios, *Mathematical Modeling*, Mathematical Engineering,
https://doi.org/10.1007/978-3-031-04729-9

(ii) $\langle a_j(t_i)\, a_k^*(t_i)\rangle = \delta_{jk}\, \lambda_j$

where δ_{jk} represents the Kronecker delta function.

Property **(ii)** is obtained when the terms in the diagonal decomposition (C.1) are compared with the expression

$$r(\mathbf{x}, \mathbf{y}) = \sum \langle a_j(t_i)\, a_k^*(t_i)\rangle\, \Phi_j(\mathbf{x})\, \Phi_k^*(\mathbf{y}).$$

The nonnegative and self-adjoint properties of $r(\mathbf{x}, \mathbf{y})$ imply that all eigenvalues are nonnegative and can be ordered accordingly: $\lambda_1 \geq \lambda_2 \ldots \geq \ldots \geq 0$. Statistically speaking, λ_k represents the variance of the data set in the direction of the corresponding POD mode, $\Phi_k(\mathbf{x})$. In physical terms, if u represents a component of a velocity field, then λ_k measures the amount of kinetic energy captured by the respective POD mode, $\Phi_k(\mathbf{x})$. In this sense, the energy measures the contribution of each mode to the overall dynamics.

Definition C.1 *The total energy captured in a POD decomposition of a numerical or experimental data set is defined as the sum of all eigenvalues*

$$E = \sum_{k=1}^{N} \lambda_k\,. \tag{C.3}$$

The relative energy captured by the kth *mode, E_k, is defined by*

$$E_k = \frac{\lambda_k}{\sum_{j=1}^{N} \lambda_j}\,. \tag{C.4}$$

The cumulative sum of relative energies, $\sum E_k$, approaches one as the number of modes in the reconstruction increases.

Spatiotemporal systems are capable of producing different kinds of behavior including periodic, quasiperiodic and nonperiodic motion in space and time. In some cases, the POD decompositions of qualitatively different states may produce seemingly similar spectra. However, the decomposition can still be used to differentiate between different solutions. One possibility is to apply the POD decomposition to the state of interest and then use the POD energy spectrum to calculate the entropy of the data set. The entropy is a measure of order or disorder and provides an objective way of classifying the complexity in experimental or numerical data.

Definition C.2 *The entropy of a POD decomposed data set, u, can be calculated from its energy spectrum according to*

$$Entropy(u) = -\lim_{N\to\infty} \frac{1}{\log N} \sum_{k=1}^{N} E_k \log E_k \, . \tag{C.5}$$

where $\log N$ *is a normalization factor which allows comparisons between different data sets.*

The entropy, as defined by (C.5), measures the energy distribution among the modes in the POD spectra and varies between 0 and 1, as the number of modes increases. The entropy is low when the energy is concentrated in a few modes. A zero entropy indicates that only one eigenfunction, with maximal energy $E_1 = 1$, is needed to reproduce the dynamics. The entropy approaches 1 when the energy spreads across a large number of modes, indicating complex behavior.

Equation (8.46) states that POD decomposition produces a basis that minimizes the least square truncation error. This property can also be stated in terms of the energy captured by the POD modes.

Proposition C.2 Let $\{a_k(t_i), \Phi_k(\mathbf{x})\}$ be the POD-basis pairs obtained from a scalar field $u(\mathbf{x}, t_i)$, satisfying Eq. *(8.45)*, *(C.1)* and *(C.2)*. Let $\{b_k(t_i), \Psi_k(\mathbf{x})\}$ be any arbitrary orthonormal basis pair satisfying *(8.45)*. The POD-basis is optimal in the sense that the total cumulative energy captured by the sequence $\{a_k(t_i), \Phi_k(\mathbf{x})\}$ is always greater or equal to the total cumulative energy captured by $\{b_k(t_i), \Psi_k(\mathbf{x})\}$, provided that the number of eigenfunctions (respecting their ordering from most to least energetic) employed is the same. Formally

$$\sum_{k=1}^{N} E_k = \sum_{k=1}^{N} \langle a_k(t_i)\, a_k^*(t_i)\rangle = \sum_{k=1}^{N} \lambda_k \geq \sum_{k=1}^{N} \langle b_k(t_i)\, b_k^*(t_i)\rangle \tag{C.6}$$

C.2 Consequences of Symmetry

One motivation for applying POD decomposition is to obtain information about the long-term behavior of the system. Suppose that this behavior is captured by an attractor, denoted by $\mathcal{A}$ (see [9] for a precise definition). Assume also that scalar measurements of the system, $g(\mathbf{x}, t_i)$, $i = 1, \ldots, M$, are provided. In practice, one must first compute the average $\bar{g}(\mathbf{x}) = \frac{1}{M}\sum_{i=1}^{M} g(\mathbf{x}, t_i)$, in order to produce a new set of measurements, $u(\mathbf{x}, t_i) = g(\mathbf{x}, t_i) - \bar{g}(\mathbf{x})$, with zero average. Let Γ denote the group of symmetries of the system of interest. The symmetries of the attractor form a subgroup of Γ defined by

$$\Gamma_{(\mathcal{A})} = \{\gamma \in \Gamma \mid \gamma A = A\} \, . \tag{C.7}$$

The critical observation is that the symmetries of the attractor, $\mathcal{A}$, appear as symmetries of the time-average, $\bar{g}(\mathbf{x})$, independent of the symmetries of the instantaneous scalar field $g(\mathbf{x}, t_i)$ [10]. Unfortunately, the converse is not always true. The symmetries of the time-average do not necessarily reflect the symmetries of the underlying attractor. Furthermore, the POD decomposition satisfies the following symmetry properties.

Proposition C.1 Let, $\{\Phi(\mathbf{x})\}$, be the POD eigenfunctions satisfying the eigenvalue problem $\langle u(\mathbf{x}, t)\, u^*(\mathbf{y}, t)\rangle\, \Phi(\mathbf{y}) = \lambda\Phi(\mathbf{x})$. Then

(i) $\langle \gamma\, u(\mathbf{x}, t)\, \gamma\, u^*(\mathbf{y}, t)\rangle\, \gamma\, \Phi(\mathbf{y}) = \lambda\, (\gamma\, \Phi(\mathbf{x}))$, for all $\gamma \in \Gamma$.

(ii) $\langle \sigma\, u(\mathbf{x}, t)\, \sigma\, u^*(\mathbf{y}, t)\rangle = \langle u(\mathbf{x}, t)\, u^*(\mathbf{y}, t)\rangle$, for all $\sigma \in \Gamma_{(\mathcal{A})}$.

(iii) $\langle\, u(\mathbf{x}, t)\; u^*(\mathbf{y}, t)\rangle\; \sigma\, \Phi(\mathbf{y}) = \lambda\, (\sigma\, \Phi(\mathbf{x}))$, for all $\sigma \in \Gamma_{(\mathcal{A})}$.

Property (**i**) establishes that the eigenfunctions in the POD decomposition of $\gamma\, u(\mathbf{x}, t)$ are those of $u(\mathbf{x}, t)$ under the action of γ. This property explains the observation that the POD decomposition of a periodic data set is not unique. If $\Phi(\mathbf{x})$ is an eigenfunction, so is $\gamma\, \Phi(\mathbf{x})$, for all $\gamma \in \Gamma$. Which one is then chosen? In the case of experimental or computational data, the answer depends on how the data is collected. Performing the decomposition with different initial conditions may produce a rotated version of $\Phi(\mathbf{x})$. Nevertheless, the important point is to realize that they are all symmetrically related. Properties (**ii**) and (**iii**) indicate that the POD kernel and its eigenvectors have at least the same symmetries of the attractor.

References

1. L.F. Shampine, S. Thompson, Solving ddes in matlab. Appl. Numer. Math. **37**, 441–458 (2001)
2. C.W. Gardiner, *Handbook of Stachastic Methods*, 3rd edn. (Springer, Complexity, 2003)
3. M. Lopez S. Wio, R. Deza, *An Introduction to Stochastic Processes and Nonequilibrium Statistical Physics* (World Scientific Publishing, 2012)
4. B. Hajek, *Random Processes for Engineers* (Cambridge University Press, 2015)
5. G. Adomian, *Nonlinear Stochastic Systems Theory and Applications to Physics*. Mathematics and Its Applications, vol. 46 (Springer, Netherlands, 1989)
6. F. Riesz, B. Sz-Nagy, *Functional Analysis* (Dover Publications, 1990)
7. P. Holmes, J. Lumley, G. Berkooz, *Turbulence, Coherenet Structures, Dynamical Systems and Symmetry* (University Press, Cambridege, 1996)
8. L. Sirovich, Turbulence and the dynamics of coherent structures. Part I: Coherent structures. Q. Appl. Math. **5**, 561 (1987)
9. J. Hale, *Ordinary Differential Equations* (Dover Publications, 2009)
10. M. Dellnitz, M. Golubitsky, M. Nicol, Symmetry of attractors and the karhunen-loeve decomposition, in *Trends and Perspectives in Applied Mathematics*, ed. by L. Sirovich, vol. 100 (Springer, 1990), p. 73

Index

A. Palacios, *Mathematical Modeling*, Mathematical Engineering,
https://doi.org/10.1007/978-3-031-04729-9

D

Printed in the United States
by Baker & Taylor Publisher Services